Advances in Flow Modeling for Water Resources and Hydrological Engineering

Advances in Flow Modeling for Water Resources and Hydrological Engineering

Editors

Carmelina Costanzo
Tommaso Caloiero
Roberta Padulano

MDPI • Basel • Beijing • Wuhan • Barcelona • Belgrade • Manchester • Tokyo • Cluj • Tianjin

Editors

Carmelina Costanzo
Enviromental Engeeniring-DIAM
University of Calabria
Rende-Cosenza
Italy

Tommaso Caloiero
CNR-ISAFOM
Rende
Italy

Roberta Padulano
IAFES and REMHI divisions
Fondazione Centro Euro-Mediterraneo sui Cambiamenti Climatici
Lecce
Italy

Editorial Office
MDPI
St. Alban-Anlage 66
4052 Basel, Switzerland

This is a reprint of articles from the Special Issue published online in the open access journal *Hydrology* (ISSN 2306-5338) (available at: www.mdpi.com/journal/hydrology/special_issues/flow_modeling).

For citation purposes, cite each article independently as indicated on the article page online and as indicated below:

LastName, A.A.; LastName, B.B.; LastName, C.C. Article Title. *Journal Name* **Year**, *Volume Number*, Page Range.

ISBN 978-3-0365-6512-5 (Hbk)
ISBN 978-3-0365-6511-8 (PDF)

Contents

About the Editors

Carmelina Costanzo

Carmelina Costanzo was born on September 13th, 1974 in Cosenza, Italy. She is an assistant professor of the University of Calabria, Department of Environmental Engineering (Italy). She hold Master's Degree in Environmental Engineering (specialization in Hydraulic Protection of the Territory) from the University of Calabria in 2001. In 2005 she received a Ph.D. in Hydraulic Engineering from the same university. Her research topics are: implementation of rain-on-grid numerical models; impact of the climate change in urban areas; flood propagation in compound channel; high unsteady flood modelling over erodible beds; generation and propagation of dam break wave; flood-risk communication; and management of water resources in agriculture. She teaches an Environmental Engineering Course: "Hydraulic Works"; "Hydraulic Protection of the Territory", and "Dynamics of Slope and Fluvial Processes". She is the author of many papers in ISI international journals and international and national conference proceedings.

Tommaso Caloiero

Tommaso Caloiero was born on May 6th, 1977 in Catanzaro, Italy. He graduated in Civil Engineer (specialization in Hydraulic) from the University of Calabria, Italy in 2002. In 2005 he received a Second level Master Degree in Mathematical Modelling of Hydrogeological Disaster from the same University. In 2009 he received a Ph.D. in Hydraulic Engineering from the Politecnico of Milan.

Main Research Interest: Hydrology, Climatology, Climate Change, Natural Hazards, Hydrologic and Water Resource Modelling and Simulation, Geographic Information Systems, Civil Engineering, Environmental Engineering, Ecological Engineering, Land-Use Change, and Forest Ecology.

Publications: More than 80 scientific papers published international WOS/SCOPUS academic journals.

Roberta Padulano

She holds a Master's degree in Environmental Engineering (2010) and a Ph.D. in Hydraulic Engineering (2014) from the Federico II University of Naples (IT). At the same University, she was involved as a research fellow for several research activities mainly concerning water demand modelling, sewer systems (both theoretical and laboratory experience) and catchment hydrology. She has been working at CMCC for four years as a post-doc and then junior researcher for both the Regional Models and geo-Hydrological Impacts (REMHI) and the Impacts on Agriculture, Forests and Ecosystem Services (IAFES) divisions. At CMCC, her main research topics concern, among others, the quantification of local impacts of climate changes in terms of extreme rainfall regime and flooding, with particular interest for urban contexts, soil erosion and droughts. She has wide experience in extreme value analysis, probabilistic and AI-based pattern detection, hydraulic structures design, flood modelling, and flood hazard analysis.

Preface to "Advances in Flow Modeling for Water Resources and Hydrological Engineering"

This book presents a print version the Special Issue of the journal *Hydrology* dedicated to "Advances in Flow Modeling for Water Resources and Hydrological Engineering". The overall goal of this Special Issue was to consider the recent advances on models and methods for water resource modelling. In particular, basin-wide water resources planning, watershed management, flood forecasting, droughts, climate changes impacts on flood risk and water resources, reservoir operation and management, river morphology and sediment transport, river water quality, and irrigation were the main issues that the papers published in this Special Issue aimed to discuss. These original objectives were achieved, and in the 21 papers collected in this volume, readers will find a collection of scientific contributions providing a sample of the state-of-the-art and forefront research in these fields. Among the articles published in the Special Issue 1 is a Technical Note and 20 are Research Articles. In total, 79 authors from five different continents (Africa, America, Asia, Europe, and Oceania) contributed to the Special Issue, showing the results of case studies and demonstration sites from the same five continents. The geographic distribution of the case studies is wide enough to attract the interest of an international audience of readers. The articles collected here will hopefully provide different, useful insights into advancements in computer techniques that allow the water scientists to develop complex models at different scales to support water resources planning and management.

Carmelina Costanzo, Tommaso Caloiero, and Roberta Padulano
Editors

hydrology

MDPI

Editorial

Advances in Flow Modeling for Water Resources and Hydrological Engineering

Carmelina Costanzo [1,*], Roberta Padulano [2] and Tommaso Caloiero [3]

1 Department of Environmental Engineering, University of Calabria, 87036 Rende, Italy
2 Regional Models and Geo-Hydrological Impacts Division, Fondazione Centro Euro-Mediterraneo sui Cambiamenti Climatici, 81100 Caserta, Italy
3 National Research Council—Institute for Agricultural and Forest Systems in Mediterranean (CNR-ISAFOM), Via Cavour 4/6, 87036 Rende, Italy
* Correspondence: carmen.costanzo@unical.it

Abstract: Surface and ground waters can be considered the main sources of water supply for agricultural, municipal, and industrial consumers. Over the centuries, the combination of both naturally occurring conditions and humanity's actions has placed increasing pressure on these water resources. As an example, climate change and natural variability in the distribution and occurrence of water are among the natural driving forces that complicate the sustainable development of water resources. Recent advances in computer techniques have allowed scientists to develop complex models at different scales to support water-resource planning and management. The Special Issue "Advances in Flow Modeling for Water Resources and Hydrological Engineering" presents a collection of scientific contributions providing a sample of the state-of-the-art research in this field.

Keywords: water resources modelling; flood forecast; climate-change impacts; drought; river quality; river morphology; watershed hydrology; watershed management; reservoir management

Citation: Costanzo, C.; Padulano, R.; Caloiero, T. Advances in Flow Modeling for Water Resources and Hydrological Engineering. *Hydrology* **2022**, *9*, 228. https://doi.org/10.3390/hydrology9120228

Received: 9 November 2022
Accepted: 23 November 2022
Published: 19 December 2022

1. Introduction

Water resource systems planning and management issues are often very complex. The pressures on water resources are increasing with the expansion of global development, involving ecological and hydrological consequences in river basins and groundwater aquifers and water-quality deterioration. All this leads to the growing need to investigate the effects of different human influences and impacts on the hydrological regime and water quality, such as land-use changes, climatic variability and climate change, and intensified water and land-use practices. Moreover, economic, environmental, and social issues have gained considerable attention in water resources research. In this context, computer-based models can help to choose the most impactful plans, designs and policies. Over the last few years, advances in computer techniques have allowed scientists to develop complex models at different scales to support water resource planning and management.

The Special Issue "Advances in Flow Modeling for Water Resources and Hydrological Engineering" focuses on recent advances in models and methods for water resource modelling. In particular, the following issues have been discussed: basin-wide water resources planning; watershed management; flood forecasting; droughts; climate change impacts on flood risk and water resources; reservoir operation and management; river morphology and sediment transport; river-water quality.

2. Some Data of the Special Issue

From early December 2019 to late September 2022, a total of 36 papers were submitted to this Special Issue. After a rigorous editorial check and peer-review process, involving external and independent experts in the field, 15 papers were rejected and 21 papers were accepted (about 57%). Of the 21 articles published in the Special Issue, 1 is a Technical

Note [1], and 20 are Research Articles [2–21]. Figure 1 compares the geographic distribution of the first authors of the research teams publishing in the Special Issue (Figure 1a), as well as that of the case studies and demonstration sites (Figure 1b). The analysis of this figure allows one an overview of the scientific community working on flow modelling for water resources and hydrological engineering, although it is just a sample and thus not an exhaustive representation. Seventy-nine authors from five different continents (Africa, America, Asia, Europe and Oceania) contributed to the Special Issue, showing the results of case studies and demonstration sites from the same five continents.

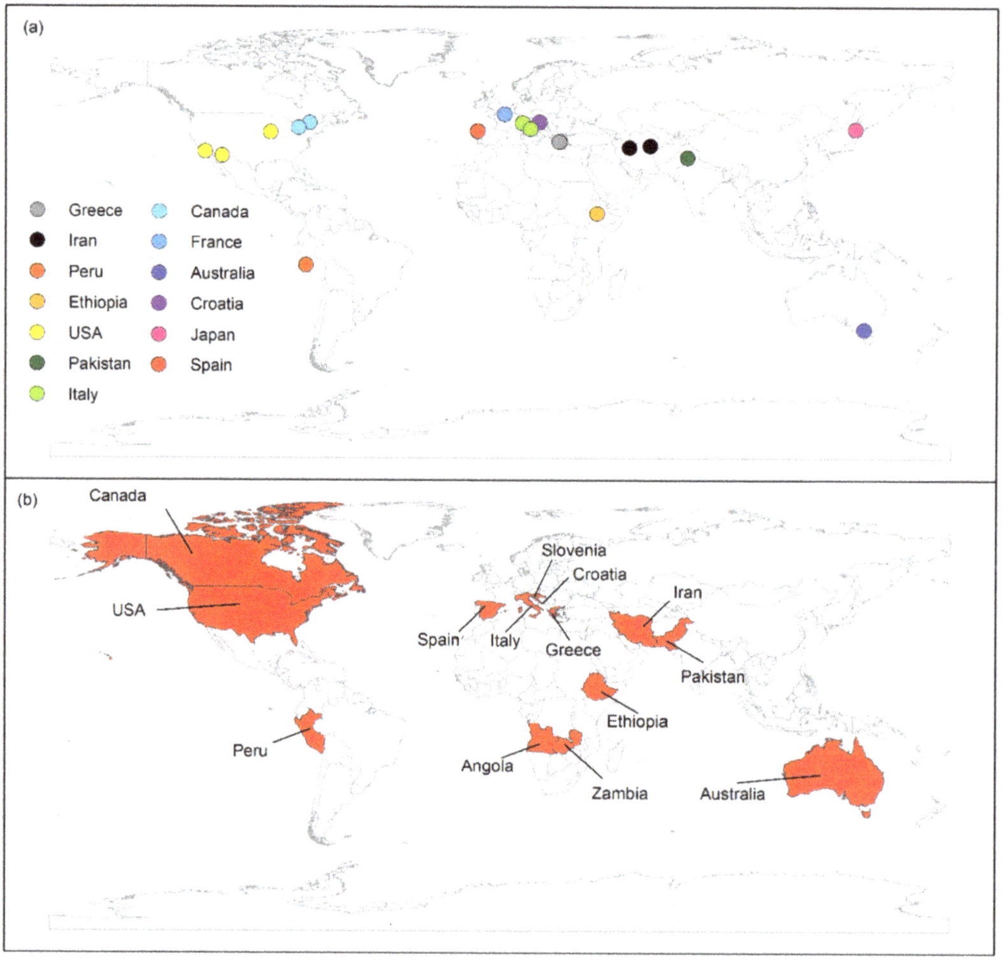

Figure 1. Geographic distribution of (**a**) first authors of research teams publishing in the Special Issue; (**b**) case studies and demonstration sites that are discussed in the papers.

Figure 2 shows the main keywords of the papers in Special Issue, which reflect the scope of scientific content on the subject. The relevant themes are numerous, ranging from hydraulic laws to hydraulic numerical models, to climate change, to hydrological models and forecasting models. All these themes refer to applications to experimental cases or actual rivers or catchment areas. "Modelling" is the predominant keyword, cited in 5 out of 20 articles, among which three referred to "Hydrological Modelling".

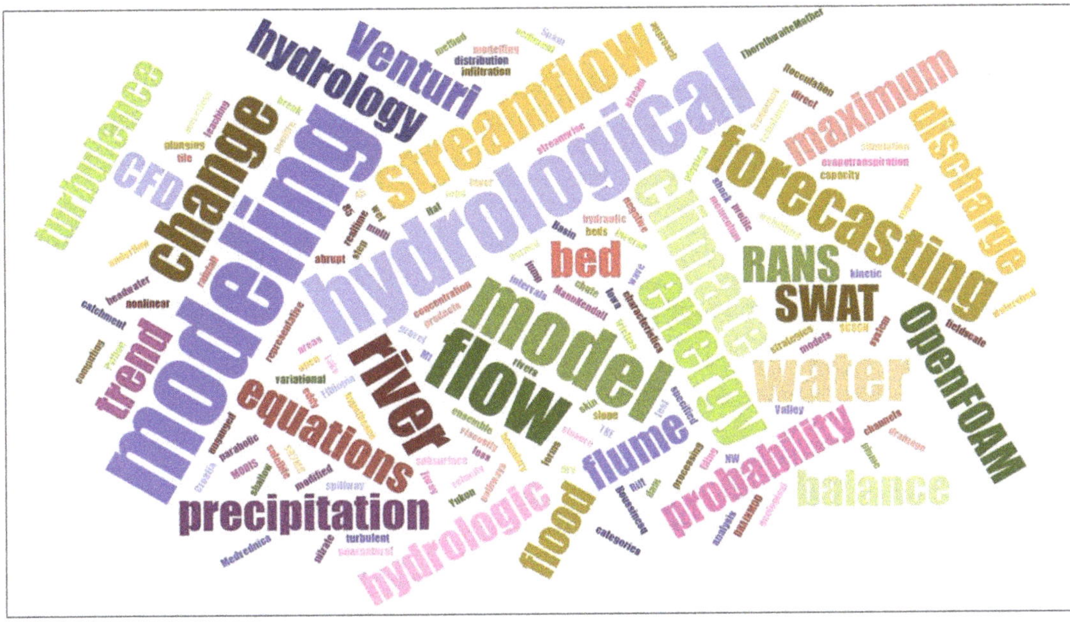

Figure 2. Word cloud of the keywords published in the Special Issue.

3. Overview of the Special Issue Contributions

The paper [1] presents a web app for a rapid, smart and smooth computation of catchment-scale water balance, relying on the Thornthwaite–Mather method, which only requires simple input information such as temperature, precipitation and location of the site. By properly setting, or possibly calibrating, input parameters such as the runoff coefficient, flow values are simulated, and they can be compared to observed values for validation. The app is tested on two different catchments: one in Northern Italy and the other in Slovenia, with different performances due to the underestimation of snowfall and snow melting in the latter.

Paper [2] presents an in-depth trend analysis of the main hydro-climatic variables (rainfall, temperature, low and high flows) recorded in the headwaters of the Mero catchment (NW Spain). The case study is a near-natural regime stream, negligibly influenced or modified by human activities, and it is particularly significant in terms of water resources for the nearby area. The study detected statistically significant trends in mean seasonal stream discharge for autumn and summer. In addition, a significant upward trend in the number of days with low flow was particularly evident in spring and summer. Additionally, a falling trend in high flows was observed in autumn. Such trends often proved to be inconsistent with rainfall trends: however, this behaviour can be explained by also accounting for temperature trends.

In paper [3], the authors address the possible role of climate change in altering groundwater quality in terms of tile drainage and consequent nitrogen yield, using a test case located in the Great Lakes area (Canada). A well-known field-scale model, DRAINMOD, is calibrated, relying on literature parameters and site observations for rainfall and temperature, and then applied to climate projections under a future emission scenario. Tile flow is estimated to increase, especially in winter; however, the projected rise in nitrate loss through tiles was uncorrelated with tile flow. This could be explained by an increase in minimum spring temperature, leading to enhanced nitrification and excessive nitrate loss.

The goal of paper [4] is to assess the potentialities and limitations of CFD models in simulating the performance of the Parshall flume, a variation on the Venturi flume often used to collect laboratory and field discharge data. By using OpenFOAM, which is associated with seven different turbulence models, a selection of geometries for the Parshall flume are numerically set up, and numerical results are compared to the measured data collected in corresponding experimental scenarios. Results show that, although their performances vary, all the tested turbulence models are able to satisfactorily capture the actual flow in the flume. However, additional tests are recommended to further explore the range of simulated discharge values.

In paper [5], the estuarine area of the Yukon River (Alaska) is analysed with particular reference to the surface sediment plumes formed by glacier-melt and rainfall sediment runoff, with the aim of exploring the mechanisms behind plume plunging at the boundary between river water and marine water. Analysis relies on discharge and sediment measurements, as well as on plume observations conducted from a boat. It was found that both the suspended sediment concentration and sediment load of the Yukon River were relatively high in the glacier-melt and rainfall runoffs of July-September. Hence, temporal variations of glacier-melt and rainfall could change the behavior of the sediment plume in the coastal region.

The goal of paper [6] is to set up a flood prediction model based on the concept of "Probability of Success". The model, developed for the Croatian catchment referring to the Gornja Kašina hydrological station, assesses the probability of flooding as the overlap of five statistical categories describing the most relevant factors affecting the rainfall–runoff transformation (climatological, geological and geographical features). Comparison with past flood observations for the test case showed that the model could capture flood events that caused significant damages, although they were not registered as "floods" by the involved stream gauges.

In paper [7], a real-scale dam-break wave was simulated using the 2D finite volume Roe-TVD method. For this purpose, a numerical code was developed to solve the 2D depth average, shallow water equations on unstructured triangular cells considering turbulence terms and a dry bed front. To validate the code, initially, available experimental data were considered. After verifying the model, the real-scale dam break was simulated, and the flow behaviour from encountering the two bridges was analysed along the pathway. The flood wave arrival time to the bridges, the flooded area and the duration of flooding of the bridges were studied.

Paper [8] aims to provide an explanation and a theoretical foundation for the empirical well-known eddy viscosity profiles. The eddy viscosity is defined as a product between a velocity scale and a length scale. From this definition, two analytical eddy viscosity models are proposed. The proposed analytical models are validated through the computation of velocity profiles, obtained from the resolution of the momentum equation, and comparison with experimental data.

In paper [9], the spatial variability of the main water balance components in an intensively agricultural area in the headwaters of Upper East Fork White River in Indiana, USA, was analysed. Extensive data collection was necessary to provide the best possible input for a SWAT model set up for the simulation. To optimise the data outputs, a spatial calibration approach was implemented in four gauging sites. It was confirmed that in areas with intensified agricultural development—an activity that heavily disturbs the land phase of the hydrological cycle—it is critical for hydrological models to incorporate factors such as water use and relevant agricultural management practices.

The authors of paper [10] use a computational fluid dynamics (CFD) approach to simulate flows in Parshall flumes, which are used to measure flowrates in channels. The objective of this research was to study the reliability of numerical simulations of a Parshall flume using various nonlinear turbulence models. The numerical results are compared with the experimental data, which show that choosing the right turbulence model is the key element in accurately simulating Parshall flumes.

Paper [11] aims to develop a robust and rational methodology to assess the change in the hydrological response of a post-fire watershed, especially where the scarcity or absence of hydrometric data do not allow the calibration of a more complex rainfall–runoff model. Thus, this study proposes an integrated approach that combines spatial information on burned areas and levels of fire severity, direct soil infiltration measurements and rainfall–runoff modelling. This approach was applied to a burned forest catchment in Italy to explain the repercussions of fire on the hydrological response of a natural watershed. Flood peak and volume were computed through the application of the Soil Conservation Service-Curve Number method (SCS-CN); the flow propagation was simulated through a lag-time approach based on the time–area curve of the catchment.

Paper [12] presents the physical model study of shock waves at the Mohmand Dam Spillway project in Pakistan. In this study, a hydraulic analysis of shock waves was carried out to investigate its generation mechanism. Different experiments were performed to analyse the rooster tail on a flat spillway chute and to examine the factors affecting the characteristics of the rooster tail. The results revealed that shock wave height is influenced by spillway chute slope, pier shape and flow depth. Moreover, the height of the shock wave can be minimized by installing a semi-elliptical pier on the tail part of the main pier.

Paper [13] evaluates the potential for a newly proposed nonlinear subsurface flux equation to improve the performance of the hydrological Hillslope Link Model. The equation contains parameters that are functionally related to the hillslope steepness and the presence of tile drainage. To assess performance improvements, they compare simulation results to streamflow observations during a 17-year period (2002–2018) at 140 U.S. Geological Survey (USGS) gauging stations. The new equation provides a better representation of hydrograph recession curves, hydrograph timing and total runoff volume. However, the authors found discrepancies in the spatial distribution of hillslope scale parameters.

Paper [14] describes the main stages and processes required to implement and improve an operational hydrologic forecasting system in the Upper Zambezi River Basin and its sub-basins. The process of implementation was complex, and several decisions needed to be made about the input data (precipitation from satellites or climate products), the hydrological models to be included along with their optimal parameter sets and the timescales required for the generation of streamflow forecasts. Once the system was completely operational, additional developments were required to improve its performance and reduce the spread of total hydrologic uncertainty into the final streamflow forecast products.

Paper [15] assesses historical (1983–2005) and future (2026–2100) rainfall, maximum temperature and minimum temperature trends in the Ziway Lake Basin (Ethiopia). Simulated historical and future climate data were obtained from the CMIP5 datasets considering the RCP4.5 and RCP8.5 emission scenarios. The modified Mann–Kendall trend test was applied to estimate the trends of annual rainfall, Tmax and Tmin in historical and future periods. Rainfall experienced no clear trends, while Tmax, and Tmin significantly increased in both RCP 4.5 and 8.5 scenarios, especially in the central part of the basin at the end of the 21st century.

In paper [16], the potential effects of climate change and variability on the maximum precipitation, temperature and hydrological regime in Devil's Creek, Tacna, Peru were analyzed. For this purpose, the outputs of the meteorological variables of fifteen regional climate models were used as inputs for the hydrological model considering the RCP4.5 and RCP8.5 emission scenarios. The results showed an increase in the maximum annual precipitation by more than 30% for both the RCP4.5 and RCP8.5 scenarios for the 2021–2050 period with reference to the 1981–2005 period. Moreover, the maximum flows could increase by 220% and 154% for the RCP4.5 scenarios for the 2021–2050 and 2051–2080 terms, respectively, and 234% and 484% for the RCP8.5 scenarios and for the 2021–2050 and 2051–2080 terms, respectively.

The aim of paper [17] is to develop a modified model that improves the accuracy of the determination of skin friction factors in gravel-bed rivers. With this aim, 100 velocity profile data obtained from eight gravel-bed rivers were utilized to develop an analytical method that considers the momentum thickness of the boundary layer and its deviation in large-scale topographic bedforms in a 1D force-balance model. The proposed model showed high accuracy in the prediction of skin friction factors for energy slopes between 0.001 and 0.1. Additionally, the model was used to modify the classic Einstein–Strickler equation, allowing an improvement of the accuracy of the predicted skin friction factors in nonuniform flow conditions even when velocity profiles and energy slope were unavailable.

In paper [18], the transition from supercritical to subcritical flow around a fully submerged abrupt negative step in a horizontal rectangular open channel was investigated in a laboratory experiment. As a result, five different types of rapidly varying flow were observed by varying the subcritical downstream tailwater depth. Moreover, the numerical results showed that the Boussinesq equations can simulate the basic flow characteristics with acceptable accuracy.

In paper [19], monthly streamflow and satellite-based actual evapotranspiration data (AET) were used to evaluate the Soil and Water Assessment Tool (SWAT) model for the calibration of an experimental sub-basin with mixed land-use characteristics in Athens, Greece. Three calibration scenarios were conducted with streamflow, AET and streamflow–AET data to evaluate the simulated outputs. The sensitivity analysis showed that the most sensitive parameters for streamflow are related to groundwater flow, runoff generation and channel routing, and for actual evapotranspiration, they are all connected to soil properties. This research showed that combining streamflow and MODIS satellite-based AET data in the calibration process can improve model performance regarding streamflow and water balance and contribute to understanding the hydrological processes in a mixed land-use catchment.

Paper [20] tests the suitability of the RFFE approaches within smaller headwater catchments in the Pilbara (Australia) and evaluates them through a comparison of peak discharge values derived from a 2D hydrodynamic direct rainfall model. This paper provides the first comparative study of The RFFE approaches for the Pilbara using updated ARR values to validate their use within smaller catchments in the same region.

Paper [21] deals with how the Colorado River may respond to future climates. Historical and future streamflow projections for the Colorado River basin were evaluated with a perspective of drought and surplus periods.

4. Conclusions

This Special Issue on numerical methods and models for water resource modelling is very interesting and constitutes a point of reference for future developments on the topic.

In particular, the Special Issue illustrates that the use of hydraulic models can assist with hydraulic constructions for planning the negative effects of shock waves on spillway operations [12], to determine the best combination of different turbulence models to design Parshall flumes [4,10], to model the eddy viscosity in surface flows [8], to determine hazard maps downstream of a dam [7] and to improve the accuracy of the determination of skin friction factors in gravel-bed rivers [17] and the performance of the hydrological Hillslope Link Model [13].

Similarly, the presented hydrological models contribute to future development regarding the study of the hydrological response after a fire [11], environmental sustainability ensuring socioeconomic stability and the production of critical crops in agriculture [9] and to determine the peak flow through regionalization techniques [20]. Significant attention is paid to the impacts of climate change on the performance of hydrological models in terms of water quantity and quality, using observed trends [3,21] and future climate scenarios [2], and to the role of extreme flood events [5,6], also by means of simplified [1] or empirical [6] prediction models.

Finally, the use of high-resolution Satellite Precipitation Products [14] and emission scenarios [15,16] as inputs for the hydrological model allowed us to improve the existing hydrologic forecasting system and detect the impacts of climate changes on the hydrological regime.

Acknowledgments: The Guest Editors thank all the authors, *Hydrology*'s editors and the reviewers for their contributions and commitment to this Special Issue. A special thank you goes to *Hydrology*'s Managing Editor, for her dedication to this project and her valuable collaboration in the design and setup of the Special Issue.

Conflicts of Interest: The authors declare no conflict of interest.

References

1. Mammoliti, E.; Fronzi, D.; Mancini, A.; Valigi, D.; Tazioli, A. WaterbalANce, a WebApp for Thornthwaite–Mather Water Balance Computation: Comparison of Applications in Two European Watersheds. *Hydrology* **2021**, *8*, 34. [CrossRef]
2. Rodríguez-Blanco, M.; Taboada-Castro, M.; Taboada-Castro, M. An Assessment of the Recent Evolution of the Streamflow in a Near-Natural System: A Case Study in the Headwaters of the Mero Basin (Galicia, Spain). *Hydrology* **2020**, *7*, 97. [CrossRef]
3. Golmohammadi, G.; Rudra, R.; Parkin, G.; Kulasekara, P.; Macrae, M.; Goel, P. Assessment of Impacts of Climate Change on Tile Discharge and Nitrogen Yield Using the DRAINMOD Model. *Hydrology* **2021**, *8*, 1. [CrossRef]
4. Heyrani, M.; Mohammadian, A.; Nistor, I.; Dursun, O. Numerical Modeling of Venturi Flume. *Hydrology* **2021**, *8*, 27. [CrossRef]
5. Chikita, K.A.; Wada, T.; Kudo, I.; Saitoh, S.-I.; Toratani, M. Effects of River Discharge and Sediment Load on Sediment Plume Behaviors in a Coastal Region: The Yukon River, Alaska and the Bering Sea. *Hydrology* **2021**, *8*, 45. [CrossRef]
6. Sudar, V.; Malvić, T.; Vujnović, T.; Ivšinović, J. Modeling of the Geological Probability Procedure for the Prediction of High Flows in Small Streams, Case Study of Medvednica Mt., Croatia. *Hydrology* **2021**, *8*, 83. [CrossRef]
7. Alamatian, E.; Dadar, S.; Đurin, B. Simulation of Dam Breaks on Dry Bed Using Finite Volume Roe-TVD Method. *Hydrology* **2021**, *8*, 88. [CrossRef]
8. Absi, R. Reinvestigating the Parabolic-Shaped Eddy Viscosity Profile for Free Surface Flows. *Hydrology* **2021**, *8*, 126. [CrossRef]
9. Bariamis, G.; Baltas, E. Hydrological Modeling in Agricultural Intensive Watershed: The Case of Upper East Fork White River, USA. *Hydrology* **2021**, *8*, 137. [CrossRef]
10. Heyrani, M.; Mohammadian, A.; Nistor, I. Numerical Simulation of Flow in Parshall Flume Using Selected Nonlinear Turbulence Models. *Hydrology* **2021**, *8*, 151. [CrossRef]
11. Folador, L.; Cislaghi, A.; Vacchiano, G.; Masseroni, D. Integrating Remote and In-Situ Data to Assess the Hydrological Response of a Post-Fire Watershed. *Hydrology* **2021**, *8*, 169. [CrossRef]
12. Sarwar, M.K.; Tariq, M.A.U.R.; Farooq, R.; Abbasi, H.K.J.; Haq, F.U.; Ahmad, I.; Shah, M.I.; Ng, A.W.M.; Muttil, N. A Hydraulic Analysis of Shock Wave Generation Mechanism on Flat Spillway Chutes through Physical Modeling. *Hydrology* **2021**, *8*, 186. [CrossRef]
13. Velásquez, N.; Mantilla, R.; Krajewski, W.; Fonley, M.; Quintero, F. Improving Hillslope Link Model Performance from Non-Linear Representation of Natural and Artificially Drained Subsurface Flows. *Hydrology* **2021**, *8*, 187. [CrossRef]
14. Valdés-Pineda, R.; Valdés, J.B.; Wi, S.; Serrat-Capdevila, A.; Roy, T. Improving Operational Short- to Medium-Range (SR2MR) Streamflow Forecasts in the Upper Zambezi Basin and Its Sub-Basins Using Variational Ensemble Forecasting. *Hydrology* **2021**, *8*, 188. [CrossRef]
15. Hordofa, A.T.; Leta, O.T.; Alamirew, T.; Chukalla, A.D. Spatiotemporal Trend Analysis of Temperature and Rainfall over Ziway Lake Basin, Ethiopia. *Hydrology* **2021**, *9*, 2. [CrossRef]
16. Pino-Vargas, E.; Chávarri-Velarde, E.; Ingol-Blanco, E.; Mejía, F.; Cruz, A.; Vera, A. Impacts of Climate Change and Variability on Precipitation and Maximum Flows in Devil's Creek, Tacna, Peru. *Hydrology* **2022**, *9*, 10. [CrossRef]
17. Kazem, M.; Afzalimehr, H.; Nazari-Sharabian, M.; Karakouzian, M. Determination of Skin Friction Factor in Gravel Bed Rivers: Considering the Effect of Large-Scale Topographic Forms in Non-Uniform Flows. *Hydrology* **2022**, *9*, 58. [CrossRef]
18. Retsinis, E.; Papanicolaou, P. Supercritical Flow over a Submerged Vertical Negative Step. *Hydrology* **2022**, *9*, 74. [CrossRef]
19. Koltsida, E.; Kallioras, A. Multi-Variable SWAT Model Calibration Using Satellite-Based Evapotranspiration Data and Streamflow. *Hydrology* **2022**, *9*, 112. [CrossRef]
20. Flatley, A.; Rutherfurd, I. Comparison of Regionalisation Techniques for Peak Streamflow Estimation in Small Catchments in the Pilbara, Australia. *Hydrology* **2022**, *9*, 165. [CrossRef]
21. Bedri, R.; Piechota, T. Future Colorado River Basin Drought and Surplus. *Hydrology* **2022**, *9*, 227. [CrossRef]

hydrology

MDPI

Article

Future Colorado River Basin Drought and Surplus

Rama Bedri [1] and Thomas Piechota [1,2,*]

[1] Schmid College of Science and Technology, Chapman University, Orange, CA 92866, USA
[2] Fowler School of Engineering, Chapman University, Orange, CA 92866, USA
* Correspondence: piechota@chapman.edu; Tel.: +1-714-628-2897

Abstract: Historical and future drought and surplus periods in the Colorado River basin are evaluated based on eight climate scenarios. Unimpaired streamflow from 17 stations in the Colorado River are evaluated based on U.S. Geological Survey, Bureau of Reclamation, and Coupled Modeled Intercomparison Projection 5 downscaled data from 1950–2099. Representative Concentration Pathway (RCP) 4.5 and 8.5 emission scenarios are considered for four climate models (HadGEM2-ES, CNRM-CM5, CanESM2, MI-ROC5). Drought (surplus) quantities, magnitudes, severities, and water year flows are compared for the historical and future periods. Results indicate that there is a significant difference between the historical record and future projections. The results are not consistent in terms of increase of drought or surplus; however, the intensity (as measured by magnitude and duration) will likely increase for both RCP 4.5 and 8.5. The CanESM2 and CNRM-CM5 models project wetter scenarios, and HadGEM2 and MI-ROC5 models project drier scenarios. For the critical Lees Ferry station, models indicate a chance of higher drought and surplus length and magnitude on the order of two times the historical period. In addition, basin wide flow at Lees Ferry had a shift in the future mean ensemble of approximately 3–10% for the water year. Future hydrologic changes will heighten the need for appropriate management and infrastructure options available to adapt to these changes.

Keywords: drought; climate change; water; hydrology; streamflow

Citation: Bedri, R.; Piechota, T. Future Colorado River Basin Drought and Surplus. *Hydrology* **2022**, *9*, 227. https://doi.org/10.3390/hydrology9120227

Academic Editor: Luca Brocca

Received: 10 November 2022
Accepted: 11 December 2022
Published: 14 December 2022

1. Introduction

The Colorado River basin is one of the most important basins in the United States and provides critical water resource for seven States (Colorado, Utah, Wyoming, New Mexico, Nevada, Arizona, and California) and Mexico. It serves a population of over 36 million people, supports an economy of $26 billion based on recreation, provides water for 4 million acres of farmland and sustains 30 endemic fish species [1]. It is also a highly managed system with over 50 million acre-ft of storage available between the two largest reservoirs (Lake Mead and Lake Powell). This large amount of storage is critical for sustaining water supplies during droughts, hydropower generation, recreation, and environmental productivity. Much of the southwestern United States has experienced sustained drought over the past 20 years and this has resulted in declines in water supplies in many basins and reservoirs. For instance, the Colorado River basin experienced Tier 1 and 2 shortage for the first time in 2022 and 2023 which trigger mandatory water consumption cuts for southwestern states. In addition, drought continency plans will need to be developed to ensure long term sustainability of water supplies and reservoir [2]. There is a need to understand if this is the "new normal" for the future or are these changes part of drought and surplus cycles that are experienced over years.

Extensive research has been conducted on the impact of climate change on the water conditions of the Colorado River basin. Early studies [3] used specific warming scenarios (e.g., 1, 2, 3 °C warming) along with hydrologic and system models to evaluate changes. Rajagopalan et al. [4] demonstrated how increased demand and changing climate were taxing the reliability of water supply. As Coupled Model Intercomparison Project (CMIP) projections of future climate have become available, studies have focused on impacts in headwater basins [5,6], uncertainties in precipitation projections [7], the role of increasing temperature on reduction of streamflow [8,9], and implications of changes in future flows in the basin [10]. Others such as Woodhouse et al. [11] have evaluated hypothetical droughts from 1–4 °C warming and implications for reliability of Colorado River flows. Finally, the Bureau of Reclamation [12] has studied this extensively using paleo records and future flows into major reservoirs (Lake Mead and Lake Powell). Most studies that have evaluated the future projections, use the entire suite of models from CMIP5 (on the order of 32 models) and usually for Representative Concentration Pathway (RCP) 4.5 and 8.5.

This work fills a gap in research where both drought and surplus periods from the historical records and future projections that are likely in the Colorado River. The contributions from the work presented here include the evaluation of both drought and surplus periods from 2021–2099 based on specific climate model projections that are most likely for this region. It is important to note that paleo records are not considered here as the focus is on evaluating the impact of future streamflow conditions in relation to historical (as measured from gauge records). Finally, observations and conclusions are drawn on changes in extremes (both droughts and surpluses) to stations that contribute to the entire basin and those for specific states (e.g., Arizona).

2. Materials and Methods

2.1. Data

The main data used in this study is yearly streamflow for various locations in the Colorado River basin from past periods and for the future. The selected stations represent a combination of headwater locations that are important for understanding high water generating portions of the basin, a main river station (e.g., Lees Ferry) that is important for water management, and lower basin stations that contribute only to Arizona flows and are important for understanding monsoonal impacts. A total of 17 stations noted below are used in the analysis of droughts and surpluses.

2.1.1. Historical Streamflow

Unimpaired streamflow stations are needed to conduct this analysis and remove any anthropogenic effects. Unimpaired stations were initially identified by Wallis et al. [13] updated by Tootle et al. [14]. For this study, those locations within the Colorado River basin are used from the larger U.S. data set. This results in 16 stations that were updated from the U.S. Geological Survey (USGS) NWISWeb Data retrieval (Available online: https://waterdata.usgs.gov/nwis/, accessed on 1 January 2022) to water year 2021, so the analysis is performed using the time period water year 1951–2021. This time period was selected by Tootle et al. [13] as being ideal for having a large number of streamflow stations and sufficient number of years of data for long-term analyses. In addition, the use of NWIS streamflow data allows for these results to be compared in the future with other analyses that may be conducted in the west on other unimpaired streamflow data. In addition, the Lees Ferry data was obtained for water years 1906–2021 from the Bureau of Reclamation who maintains natural flow that accounts for consumptive uses and losses. (Available online: https://www.usbr.gov/lc/region/g4000/NaturalFlow/index.html, accessed on 1 January 2022). In the analysis performed for all the stations (Section 3.1), the period of 1951–2021 was used for Lees Ferry to allow for appropriate comparisons. For specific analysis of Lees Ferry (Section 3.2), additional analysis was performed using the extended record 1906–2021.

Monthly data for all the stations is used to compute the water year cumulation defined as the months October–September. The 17 stations noted above are summarized and show in Table 1 and Figure 1. These are also locations where future flow data is available (see Section 2.1.2). For these stations, water year streamflow data is calculated in million acre-ft (MAF) which is a common unit used for water management. (Note: 1 MAF = 1233 million cubic meters MCM).

Table 1. List of Stations with characteristics.

Station Name	Location/Name	Elevation (m)	Drainage Area (km^2)
Lees Ferry	Upper (U1)	940	289,561
Piney River	Upper (U2)	2217	219
East River	Upper (U3)	2440	749
Lake Fork	Upper (U4)	2386	878
Slater Fork	Upper (U5)	2012	391
Rock Creek	Upper (U6)	2210	381
Yellowstone River	Upper (U7)	2265	342
White River	Upper (U8)	1951	1678
Fish Creek	Upper (U9)	2338	156
Muddy Creek	Upper (U10)	1983	272
Smiths Fork	Upper (U11)	2044	427
Gila River	Lower (L1)	1419	4828
San Francisco River	Lower (L2)	1047	7156
Gila River at Safford	Lower (L3)	957	20,350
Salt River Chrysotile	Lower (L4)	1023	7379
Salt River Roosevelt	Lower (L5)	664	11,152
Tonto Creek	Lower (L6)	769	1748

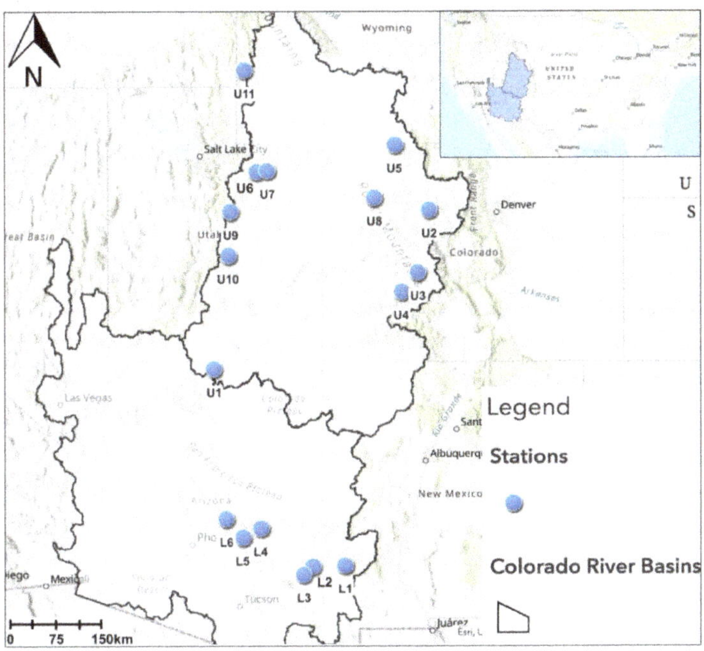

Figure 1. Map of stations used in this study for Upper and Lower Colorado River basin.

Of the 17 stations identified in Table 1, 11 stations are from the Upper basin and six are from the Lower basin. It is noteworthy that 10 of the Upper basin stations contribute flows to Lees Ferry which is a measure of total Upper basin. The inclusion of all the stations allows for a spatial analysis of how flows may change in the future and eventually contribute to the overall flow at Lees Ferry. In addition, the inclusion of the six Lower basin stations allows for further spatial analysis of the entire Colorado River (Upper and Lower).

2.1.2. Future Projections of Streamflow

Future water year streamflow values were obtained from the comprehensive database of downscaled hydrology projections using Bias Correction and Spatial Disaggregation (BCSD) and made available by Bureau of Reclamation [14]. These projections represent downscaled climate projections (e.g., wind, temperature, precipitation) of the Coupled Modeled Intercomparison Project 5 (CMIP) projections to grid scales of $\frac{1}{2}$ degree 12 km using BCSD [15]. The downscaled climate data is then used in the VIC hydrologic models for the western U.S. and values aggregated to subbasin scale [16]. The VIC hydrologic model used to develop the downscaled streamflow projections is physically based for various subbasins of the Colorado River basin and routes to key points that provide streamflow projections. This results in 97 projections of monthly and daily hydrology at specified locations (e.g., streamflow stations) over the contiguous U.S. using BCSD on CMIP5 projections. In the Reclamation archive [14], there are 18 CMIP5 models and Representative Concentration Pathways (RCP) 4.5 and 8.5 scenarios available to the year 2099.

A total of eight projections were used that represent four models (of the 18 noted above) at two emission levels as RCP 4.5 and 8.5. The emission level of RCP 4.5 is an intermediate climate change scenario that minimizes greenhouse gas emissions. RCP 8.5 is the highest level of greenhouse gas emissions resulting from a large population with high energy [17]. The Fourth National Climate Assessment [18] identified RCP 4.5 and 8.5 as the core scenarios representing the appropriate range of future conditions. The four model simulations are those used by Pierce et al. [19] and Lynam and Piechota [20] to evaluate various scenarios of California streamflow and include HadGEM2-ES, CNRM-CM5, CanESM2, and MIROC5. While there is no direct evidence that these models behave well for the Colorado River basin, these models are selected here based on their ability to simulate historical climate spatial and temperature structure at the global, Southwestern U.S. and California scales [19]. In addition, the range of these models will give a representative sample of the scenarios while still allowing for a detailed analysis of what specific futures may look like in terms of water supply. For this study there is no assumption about the model bias (e.g., water, wet, average) and direct reference is made to the models that best represent climate of the region.

To best represent the overall impact of the models for the two climate scenarios (RCP 4.5 and 8.5), ensemble water year means are calculated based on the average of the four models. This is a common practice in climate change studies where multiple models are evaluated and there is a need to establish a best estimate of the "average" of the models, e.g., [21].

2.2. *Drought and Surplus Definitions*

Following a definition used by Lynam and Piechota [20] drought was defined as periods where there was two or more consecutive years of deficit flow (below average flow) with the period ending only when two consecutive years of surplus occurred. The average flow for the historical period is used for all analyses to ensure analyses are evaluating the impact in relation to current hydrologic conditions. Similarly, surplus periods were defined as two or more consecutive years of excess flow (above average flow) with the period ending only when two consecutive years of deficit occurred. Other studies in the west have used drought definitions where in any given year a drought can occur if the index (i.e., PDSI) drops below zero or alternatively using a three-year moving average to determine drought years [12]. The approach used here recognizes the varied nature of drought (and surplus) definitions while using a standard statistical measure appropriate for a basin that has significant storage (i.e., 4 years of storage in the two major reservoirs). For both drought and surplus periods, duration, magnitude (defined as the cumulative of the departure from the average), and the severity (defined as magnitude/duration) were compiled.

2.3. Testing of Differences

The difference in the total number of droughts and surplus periods for each scenario (i.e., historical vs. future periods) was evaluated using the F-test. In this test, the variance in two populations is evaluated to see if there is a significant difference. In this study, the population consist of the number of drought and surplus periods. For all tests, the historical population of drought and surplus periods is evaluated against the future population of drought and surplus periods determined in the analysis. Results are displayed and discussed in Section 3.

The sample populations of drought and surplus for each scenario are also presented as box and whisker plots where the middle of the box represents the median, the top and bottom represent the 75th and 25th percentiles of the population and the top and bottom of the whisker represent the 90th and 10th percentiles of the population (see Section 3.2).

3. Results

3.1. Station Specific Results for the Basin

The station specific results of the entire basin for historical and future droughts (surpluses) are shown in Tables 2–4, and Figures 2 and 3. These results are shown for future flows including drought and surplus periods (duration and magnitude) along with changes in the mean water year flow. For all analysis presented in this Section 3.1, the historical time period of 1951–2021 is used for all stations.

Table 2. Duration (in years) of drought and surplus for Upper (U) and Lower (L) Basin stations based on 1951–2021 for all stations and also the time period 1907–2021 for Lees Ferry. For drought duration, red values are higher than the historical record and blue values are lower. For surplus duration, blue values are higher than the historical record and red values are lower.

Drought Duration		**RCP 4.5**				**RCP 8.5**			
Station	Historical	CanESM2	CNRM-CM5	HadGEM2-ES	MI-ROC5	CanESM2	CNRM-CM5	HadGEM2-ES	MI-ROC5
Lees Ferry (U) 1907–2021	22	10	6	10	17	13	17	32	14
Lees Ferry (U) 1951–2021	11	8	6	10	17	11	8	32	14
Piney River (U)	11	10	11	9	12	7	17	14	19
East River (U)	9	28	13	12	25	16	12	37	21
Lake Fork (U)	17	5	4	7	5	3	4	20	13
Slater Fork (U)	9	3	2	3	4	2	4	2	4
Rock Creek (U)	22	6	5	18	22	5	11	32	8
Yellowstone River (U)	9	30	27	54	50	18	28		
White River (U)	11	15	8	17	24	13	17	54	24
Fish Creek (U)	11	10	11	24	27	12	21	31	27
Muddy Creek U)	15	4	6	14	27	4	4	17	13
Gila River (L)	21			2	7	2		10	4
San Francisco River (L)	27				2	2		3	3
Gila at Safford (L)	28								
Salt River Chrysotile (L)	27		12	50	44	24	12		65
Salt River Roosevelt (L)	27	19	10	30	44	24	12	57	35
Tonto Creek (L)	23	38	12	23	44	24	17	73	45
Smiths Fork (L)	14	5	11	15	22	5	8	12	13

Table 2. *Cont.*

Surplus Duration		RCP 4.5				RCP 8.5			
Station	Historical	CanESM2	CNRM-CM5	HadGEM2-ES	MI-ROC5	CanESM2	CNRM-CM5	HadGEM2-ES	MI-ROC5
Lees Ferry (U) 1907–2021	18	16	20	8	4	17	23	6	8
Lees Ferry (U) 1951–2021	10	16	20	10	5	18	23	6	11
Piney River (U)	7	17	20	10	6	17	15	7	11
East River (U)	10	9	11	6	4	17	15	6	11
Lake Fork (U)	10	44	65	20	13	27	40	8	11
Slater Fork (U)	9	36	20	31	20	38	33	43	30
Rock Creek (U)	20	16	20	6	5	21	27	6	8
Yellowstone River (U)	9	3	3	2	2	8	3		
White River (U)	9	9	20	4	5	15	15	2	3
Fish Creek (U)	9	8	15	4	3	15	4	6	8
Muddy Creek U)	9	19	18	13	5	19	14	6	3
Gila River (L)	8			30	16	25		15	13
San Franciso River (L)	6				69	70		21	37
Gila at Safford (L)	6								
Salt River Chrysotile (L)	11		6	2	2	9	10		2
Salt River Roosevelt (L)	11	7	7	6	2	10	16	2	8
Tonto Creek (L)	8	7	6	2	2	7	12	4	3
Smiths Fork (L)	10	14	12	4	10	32	8	11	7

Table 3. Magnitude (in MAF) of drought and surplus periods for Upper (U) and Lower (L) Basin stations based on 1951–2021 for all stations and also the time period 1907–2021 for Lees Ferry. Blue values are higher than the historical record and red values are lower (Note: 1 MAF = 1233 million cubic m).

Drought Magnitude		RCP 4.5				RCP 8.5			
Station	Historical	CanESM2	CNRM-CM5	HadGEM2-ES	MI-ROC5	CanESM2	CNRM-CM5	HadGEM2-ES	MI-ROC5
Lees Ferry (U) 1907–2021	−52.3	−14.3	−20.0	−35.7	−48.8	−26.9	−30.7	−123.4	−56.3
Lees Ferry (U) 1951–2021	−20.9	−9.2	−14.8	−27.1	−34.1	−12.3	−16.8	−95.8	−44.2
Piney River (U)	−0.08	−0.06	−0.10	−0.09	−0.12	−0.10	−0.12	−0.18	−0.26
East River (U)	−0.33	−0.89	−0.44	−0.77	−1.02	−0.60	−0.63	−2.56	−1.59
Lake Fork (U)	−0.18	−0.05	−0.07	−0.21	−0.16	−0.18	−0.04	−0.71	−0.37
Slater Fork (U)	−0.13	−0.05	−0.05	−0.02	−0.07	−0.03	−0.02	−0.05	−0.07
Rock Creek (U)	−0.87	−0.06	−0.11	−0.38	−0.56	−0.13	−0.14	−0.96	−0.23
Yellowstone River (U)	−0.11	−0.95	−0.70	−2.85	−2.15	−0.78	−1.01		
White River (U)	−0.52	−1.04	−1.06	−2.61	−2.85	−1.41	−1.89	−10.20	−3.73
Fish Creek (U)	−0.08	−0.08	−0.13	−0.25	−0.38	−0.15	−0.19	−0.49	−0.33
Muddy Creek U)	−0.06	−0.02	−0.04	−0.11	−0.22	−0.04	−0.03	−0.22	−0.14
Gila River (L)	−0.69			−0.03	−0.08	−0.06		−0.14	−0.10
San Franciso River (L)	−0.90				−0.05	−0.04		−0.06	−0.07
Gila at Safford (L)	−1.99								
Salt River Chrysotile (L)	−3.11		−1.44	−7.68	−9.54	−2.03	−1.64		−9.70
Salt River Roosevelt (L)	−4.14	−1.83	−1.21	−4.02	−10.77	−1.34	−1.32	−9.64	−5.13
Tonto Creek (L)	−0.78	−0.94	−0.52	−0.89	−2.79	−0.49	−0.42	−3.12	−1.55
Smiths Fork (L)	−0.32	−0.13	−0.24	−0.47	−0.47	−0.14	−0.19	−0.49	−0.27

Table 3. *Cont.*

Surplus Magnitude			RCP 4.5				RCP 8.5			
Station	Historical	CanESM2	CNRM-CM5	HadGEM2-ES	MI-ROC5	CanESM2	CNRM-CM5	HadGEM2-ES	MI-ROC5
Lees Ferry (U) 1907–2021	60.9	54.6	98.2	21.8	27.9	142.9	89.1	19.2	20.6
Lees Ferry (U) 1951–2021	45.0	66.7	115.5	25.3	29.8	157.9	109.0	24.3	29.8
Piney River (U)	0.12	0.21	0.30	0.17	0.12	0.32	0.21	0.12	0.14
East River (U)	0.55	0.85	0.52	0.24	0.36	1.17	0.66	0.22	0.41
Lake Fork (U)	0.36	4.01	6.51	1.25	1.11	2.55	5.12	0.53	1.08
Slater Fork (U)	0.21	1.58	1.02	0.94	0.87	1.56	1.67	1.30	0.82
Rock Creek (U)	0.65	0.48	0.61	0.35	0.25	1.70	1.01	0.15	0.35
Yellowstone River (U)	0.18	0.04	0.04	0.11	0.09	0.35	0.11		
White River (U)	1.04	0.85	2.72	0.46	0.88	2.12	1.68	0.58	0.50
Fish Creek (U)	0.18	0.11	0.13	0.09	0.07	0.55	0.13	0.05	0.03
Muddy Creek U)	0.12	0.40	0.48	0.18	0.10	0.79	0.31	0.06	0.08
Gila River (L)	0.48			2.01	1.79	2.88		0.40	0.98
San Francisco River (L)	0.71				8.33	12.03		1.96	5.13
Gila at Safford (L)	1.86								
Salt River Chrysotile (L)	3.32		1.74	1.62	0.85	1.28	3.71		0.86
Salt River Roosevelt (L)	4.39	2.20	3.02	3.05	2.33	3.30	9.14	0.17	1.44
Tonto Creek (L)	1.21	0.30	0.39	0.72	0.13	0.57	1.51	0.01	0.10
Smiths Fork (L)	0.17	0.80	0.42	0.30	0.20	3.36	0.31	0.42	0.29

Table 4. Table of historical and future annual water year mean flow (MAF) for all eight scenarios and historical period of 1951–2021 for all stations. For Lees Ferry, the historical mean is also show for the time period 1907–2021. Ensemble means are also provided as the average of the four models for each RCP 4.5 and 8.5. Blue values are higher than the historical record and red values are lower. (Note: 1 MAF = 1233 million cubic m).

Station	Historical	RCP 4.5 CanESM2	RCP 4.5 CNRM-CM5	RCP 4.5 HadGEM2-ES	RCP 4.5 MI-ROC5	RCP 4.5 Ensemble	RCP 8.5 CanESM2	RCP 8.5 CNRM-CM5	RCP 8.5 HadGEM2-ES	RCP 8.5 MI-ROC5	RCP 8.5 Ensemble
Lees Ferry (U) 1906–2021	14.7	16.7	16.8	13.6	13.8	15.2	17.7	17.3	12.3	13.6	15.2
Lees Ferry (U) 1951–2021	13.8	16.7	16.8	13.6	13.8	15.2	17.7	17.3	12.3	13.6	15.2
Piney River (U)	0.055	0.061	0.060	0.056	0.056	0.058	0.060	0.061	0.053	0.054	0.057
East River (U)	0.228	0.239	0.230	0.195	0.207	0.218	0.233	0.240	0.174	0.199	0.212
Lake Fork (U)	0.162	0.252	0.259	0.197	0.207	0.229	0.241	0.268	0.155	0.194	0.215
Slater Fork (U)	0.057	0.093	0.093	0.086	0.083	0.089	0.101	0.096	0.085	0.084	0.092
Rock Creek (U)	0.096	0.120	0.116	0.084	0.093	0.103	0.140	0.114	0.076	0.094	0.106
Yellowstone River (U)	0.098	0.074	0.073	0.050	0.059	0.064	0.087	0.072	0.044	0.060	0.066
White River (U)	0.432	0.437	0.457	0.348	0.363	0.401	0.439	0.452	0.288	0.340	0.380
Fish Creek (U)	0.034	0.035	0.033	0.026	0.024	0.030	0.045	0.033	0.024	0.024	0.032
Muddy Creek U)	0.027	0.042	0.041	0.027	0.025	0.034	0.050	0.039	0.024	0.023	0.034
Gila River (L)	0.118	0.183	0.228	0.168	0.167	0.187	0.200	0.244	0.140	0.172	0.189
San Francisco River (L)	0.153	0.301	0.382	0.273	0.266	0.306	0.319	0.407	0.227	0.274	0.307
Gila at Safford (L)	0.338	0.754	0.925	0.688	0.665	0.758	0.808	0.982	0.582	0.686	0.765
Salt River Chrysotile (L)	0.447	0.388	0.488	0.335	0.307	0.379	0.406	0.538	0.280	0.329	0.388
Salt River Roosevelt (L)	0.579	0.596	0.746	0.520	0.456	0.580	0.633	0.834	0.433	0.494	0.599
Tonto Creek (L)	0.110	0.096	0.117	0.085	0.066	0.091	0.108	0.137	0.070	0.074	0.097
Smiths Fork (L)	0.135	0.170	0.146	0.126	0.131	0.143	0.206	0.142	0.132	0.139	0.155

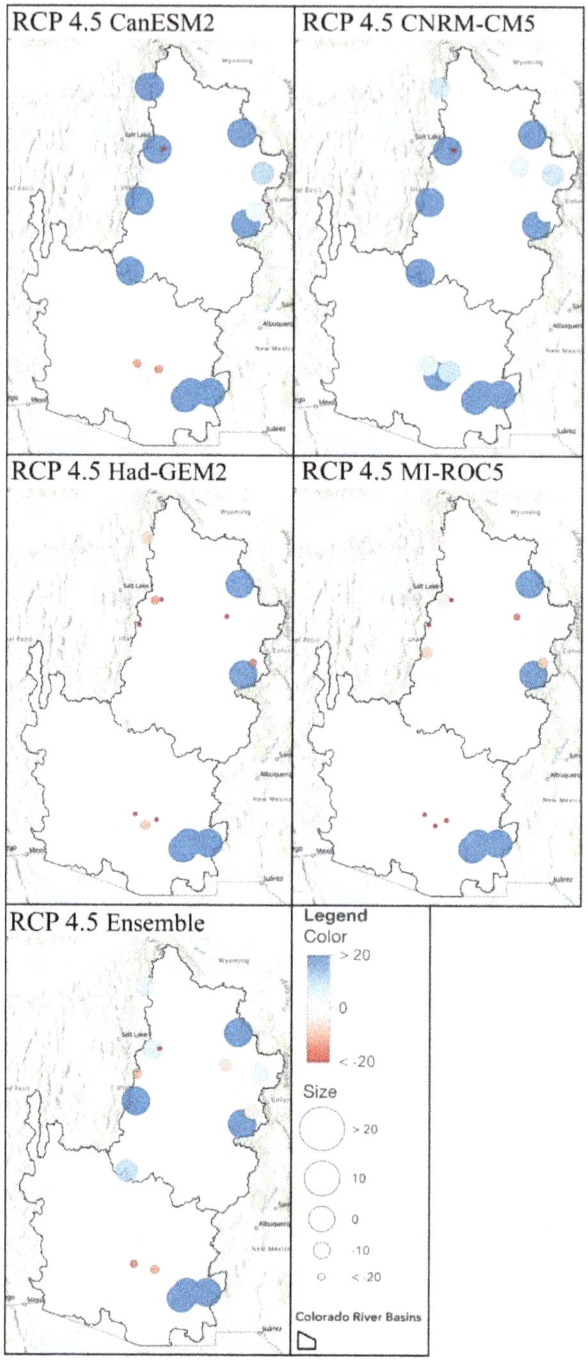

Figure 2. Spatial distribution of percentage change in mean projected water year streamflow for the period 2022–2099 for 4.5 RCP scenario, four models and ensemble.

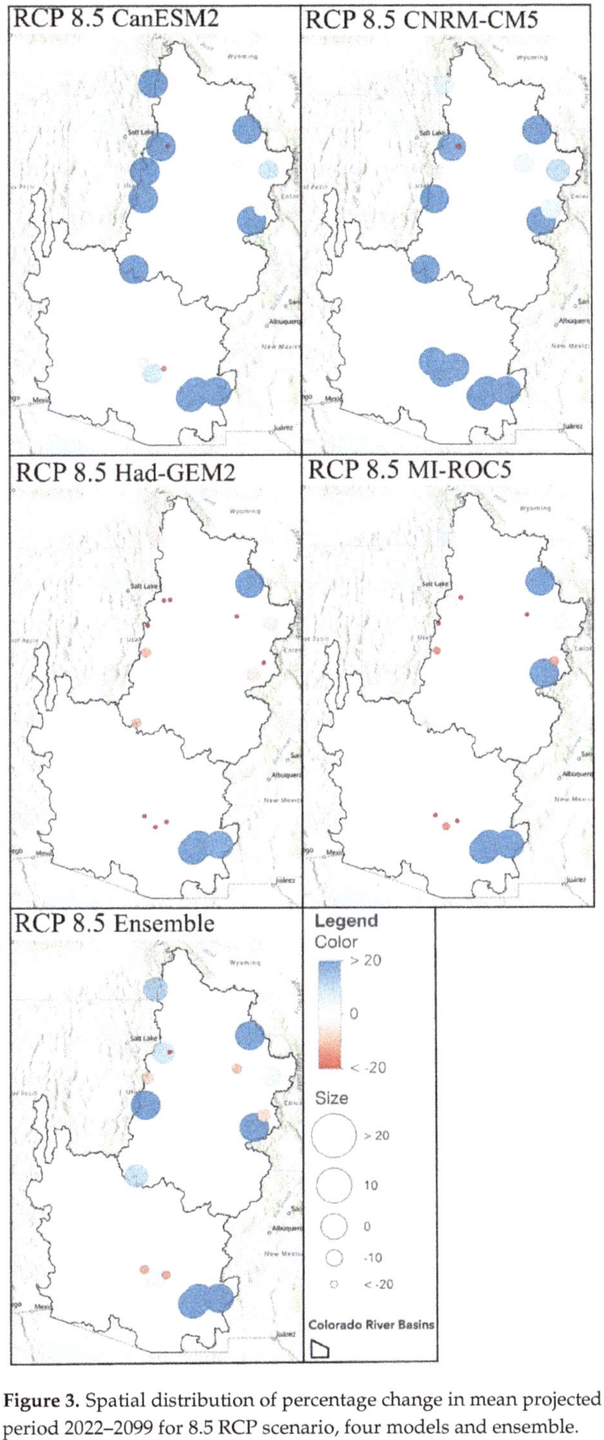

Figure 3. Spatial distribution of percentage change in mean projected water year streamflow for the period 2022–2099 for 8.5 RCP scenario, four models and ensemble.

3.1.1. Drought (Surplus) Duration

For the evaluation of drought duration and magnitude, the changes appear to be larger under the RCP 8.5 scenario as compared to the historical record and RCP 4.5 (Tables 2 and 3). The longest duration droughts occur in the Lower basin (e.g., Salt Creek stations). Historically, the longest drought was 27 years at Salt Creek, and in the future (MI-ROC5 scenario), droughts may be up to 44–65 years long (Table 2). In the Upper basin, Lees Ferry station has two models (HadGEM2-ES and MI-ROC5) where the drought duration is longer (14–32 years compared to 11 years in the historical record. All other models show shorter drought periods than the historical record.

The duration of surplus periods is longer for many of the stations in the Upper basin. For instance, Lake Fork has seven of the eight scenarios with longer duration, and Slater Fork has all the scenarios with longer surplus duration (increasing from 9 years in the historical record to 20–43 years depending on the model scenario). In comparison, Lees Ferry six of the eight scenarios had longer duration surplus periods than the historical record (increasing in duration from 10 years in the historical record to 11–23 years in the future).

3.1.2. Drought (Surplus) Magnitude

The largest changes in drought magnitude (as indicated by negative values in Table 3) appear in four Upper Basin stations (East River, Yellowstone River, White River, Fish Creek) that show a higher magnitude in all scenarios corresponding to the longer drought periods (Table 2). The change in magnitude is very large and, in some cases, shows droughts that are 5–20 times larger than the historical record. It should be noted that these are much smaller headwater basins as compared to Lees Ferry that represents the entire Upper basin. For Lees Ferry, two of the models (HadGEM2-ES and MI-ROC5) had drought magnitudes larger than the historical record (on the order of two to four times larger). For instance, HadGEM2-ES2 and MI-ROC5 models under the RCP 8.5 scenarios had largest drought magnitudes of −95.8 MAF and −44.2 MAF over multiple years, compared to the largest drought in the historical record that was −20.9 MAF.

For the magnitudes of future surplus, the CanESM2 and CNRM-CM5 models' conditions were likely to be wetter. Similar to the results for surplus duration periods, Lake Fork and Slater Fork had all the scenarios with higher surplus magnitudes (increasing from 2–20 times in magnitude depending on scenario). For Lees Ferry, two of the models had higher surplus magnitudes (CanESM2 and CNRM-CM5) on the order of two to three times larger than the historical record.

3.1.3. Changes in Water Year Mean Flow

To understand the implications of changes in the mean water supply for a given year, it is important to understand how the water year (October–September) flows change from the historical record to the future. This is different from the results show in Table 3 which are the average drought (surplus) that is expected over a longer period for the historical record and the future projections. Table 4 presents the historical water year average flow for each station and the change in water year average flow that would occur under each of the future climate scenarios. Of all the scenarios (RCP 4.5 and 8.5 and four models), 60% of the results indicated a future mean water year flow that would be higher than the historical mean. As noted earlier, the CanESM2 and CNRM-CM5 models are wetter than the HadGEM2-ES and MI-ROC5 that tend to have drier conditions as noted in the water year flows. For the RCP 4.5 and 8.5 scenarios using CanESM2 and CNRM-CM5 models, 85% of analyses had higher (or wetter) water year means. Similarly, for the HadGEM2-ES and MI-ROC5 models, 66% had lower (or drier) water year means.

When comparing the RCP 4.5 and 8.5 results in Table 4, there does not appear to be much of a difference. For instance, both RCPs have similar changes in the ensemble water year mean-six (6) stations are lower (drier) and 11 are higher (wetter). Of note, is the approximate 10% (from 13.8 MAF to 15.2 MAF) increase in water year flow at Lees Ferry in Table 4 based on the historical record (1951–2021) and 3% increase (from 14.7 MAF to 15.2 MAF) based on the historical record (1907–2021). It is important to note that this is an ensemble from two models that produce higher averages (on the order of 17 MAF) and two models that produce lower average (on the order of 12–13 MAF). If realized, this would have significant implications for water management in the region.

Figures 2 and 3 present the spatial changes in the basin and the magnitude of the water year flows. There is a slight tendency for higher magnitude of changes in water year flows under the RCP 8.5 scenarios than 4.5 scenarios. For instance, at Lees Ferry, the ensemble for RCP 4.5 and 8.5 is 15.2 MAF. Figures 2 and 3 also show the tendency for wetter conditions in CanESM2 and CNRM-CM5 streamflow projections and generally drier in the Had-GEM2 and MI-ROC5 streamflow projections (for both RCP 4.5 and 8.5). This is consistent with results seen in California climate change analysis [19,20]. However, the spatial results for the Had-GEM2 and MI-ROC5 were not as consistent as the CanESM2 and CNRM-CM5 scenarios. For the Had-GEM2 and MI-ROC5 scenarios (both RCP 4.5 and 8.5) there was a mix of wet and dry signals in the streamflow projections in both the Upper and Lower basin.

3.2. Further Analysis for Lees Ferry

Specific results of historical flows (using the entire record from 1907–2021) and climate projections (RCP 4.5 and 8.5) are presented in Figures 4–6 for Lees Ferry. This is the most critical station in the Colorado River basin and represents the dividing point between the Upper and Lower basin located downstream of Lake Powell (Glen Canyon Dam).

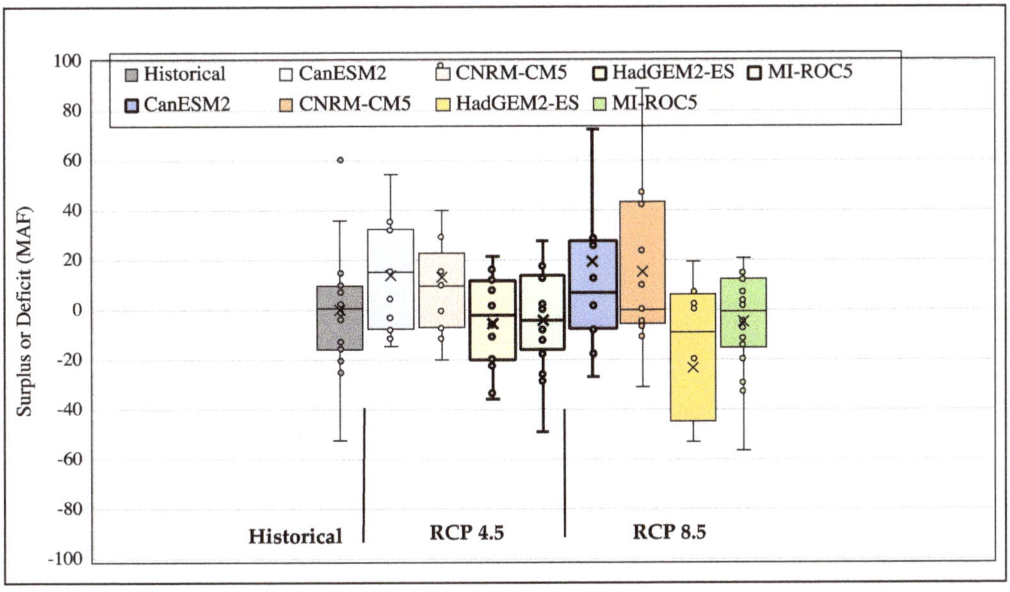

Figure 4. Boxplot of cumulative deficit and surplus (MAF) for all years and all RCP 4.5 and 8.5 scenarios and all models (CanESM2, CnRM-CM5, HadGEM2-ES, MI-ROC5) at Lees Ferry. (Note: 1 MAF = 1233 million cubic m).

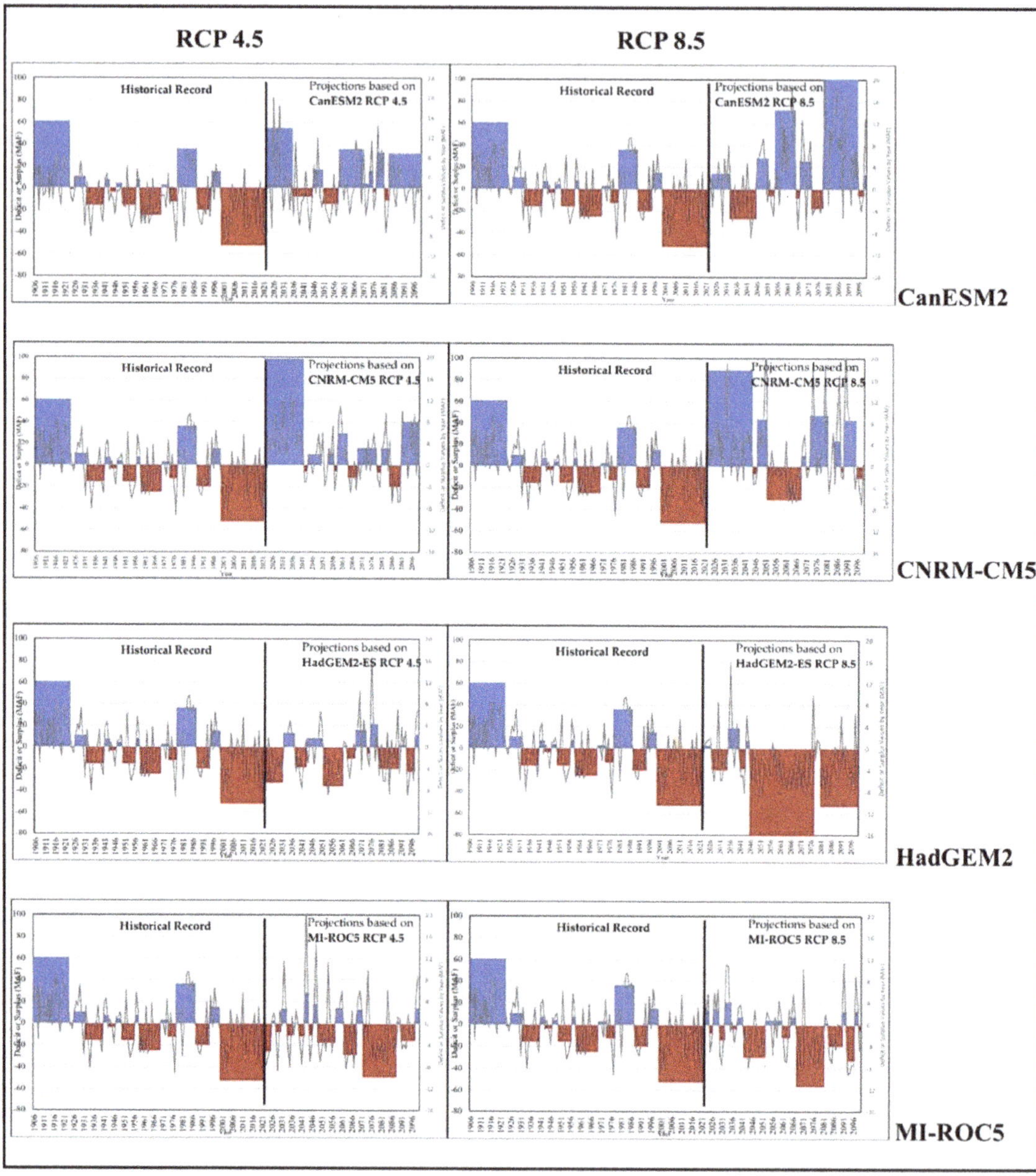

Figure 5. Time series of historical and future projections at Lees Ferry for all scenarios. Blue areas are surplus periods and red areas are deficit periods. (Note: 1 MAF = 1233 million cubic m).

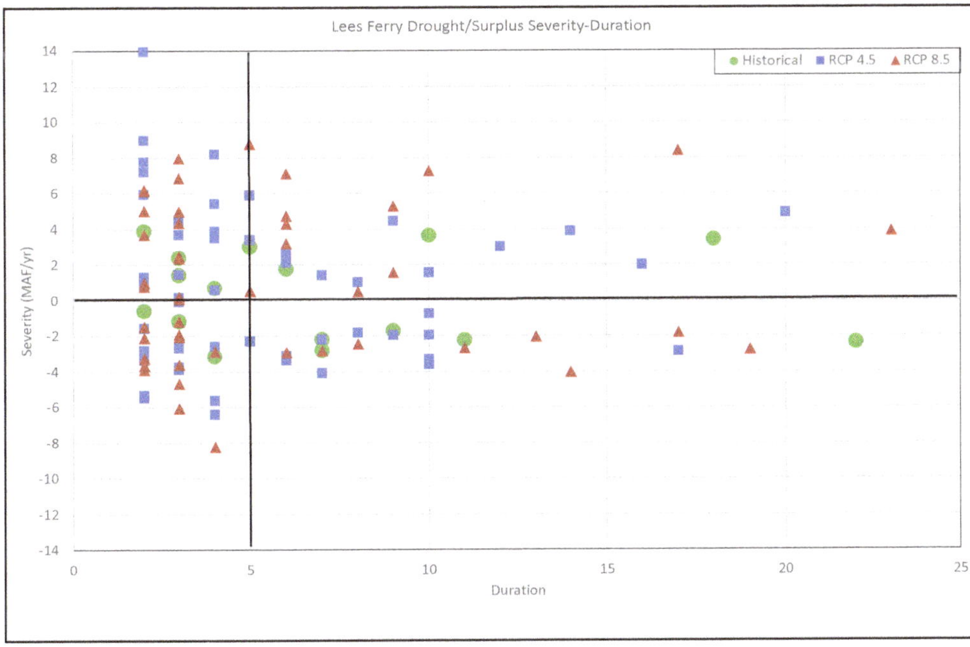

Figure 6. Scatter plot of duration vs. severity for historical and all RCP 4.5 and 8.5 scenarios at Lees Ferry. (Note: 1 MAF = 1233 million cubic m).

The variation in surplus and deficit quantities is presented with boxplots in Figure 4. Each boxplot represents the population of droughts and surplus for the specified scenario. For the CanESM2 and CNRM-CM5 models (RCP 4.5 and 8.5), the overall tendency is more surplus (positive flows) periods. For instance, the 25th, 50th and 75th percentiles change from historical values of −15, 0, 10 MAF, respectively, to values under CNRM-CM5 (RCP 8.5) of −5, 0, 43 MAF, respectively. For the HadGEM2-ES and MI-ROC5 (RCP4.5 and 8.5) scenarios, the overall tendency was more deficit (negative flows) periods. For instance, the 25th, 50th and 75th percentiles change from historical values of −15, 0, 10 MAF, respectively, to values under HadGEM2-ES (RCP 8.5) of −44, 0, 6 MAF, respectively. In Figure 4, it appears there is a large range in values (i.e., larger box) in the RCP 8.5 scenarios as compared to RCP 4.5. This is further evaluated in Figure 6.

Further representation of model specific variations at Lees Ferry is shown in Figure 5 for all models and RCP 4.5 and 8.5. As noted earlier, the CanESM2 and CNRM-CM5 are wetter scenarios, and HadGEM2 and MI-ROC5 are drier. Example scenarios include a range of surplus periods for CanESM2 (RCP 8.5) with durations of 17 to 2 years and magnitudes of 143 to 2 MAF compared to historical surpluses with durations of 18 to 3 years and magnitudes of 61 to 3 MAF. Similar example scenarios include a range of deficit periods for HadGEM2 (RCP 8.5) with durations of 32 to 3 years and magnitudes of −123 to −18 MAF compared to historical surpluses with durations of 22 to 2 years and magnitudes of −52 to −1 MAF. This indicates more extreme surplus and deficit periods in the future when an ensemble of all models is used.

The severity (defined as magnitude divided by duration) of drought and surplus for all scenarios in comparison to the historical data is presented in Figure 6. The green circles represent the historical drought and surplus periods (16 total) ranging from −3.1 to 3.9 MAF/year with durations from 2 to 22 years. Under future conditions, maximum drought and surplus severity increases to levels of −8.2 to 14.0 MAF/year. The number of drought and surplus periods is also noteworthy. In the historical record, there are eight

(8) drought and eight (8) surplus periods. Where in the future record for RCP 4.5 (8.5), there are 26 (22) drought periods and 30 (26) surplus periods. Thus, the ensemble of model scenarios shows a slightly wetter condition for Lees Ferry and more extreme drought and surplus periods. Many of these are under 5 years in duration reflecting the rapidly changing conditions in the future.

4. Discussion

In this study, historical and future streamflow projections for the Colorado River basin were evaluated with a perspective of drought and surplus periods. The results of this study conform with past studies of the region. For instance, Lynam and Piechota [20] and Pierce et al. [19] found that the HadGEM2-ES model had drier conditions (lower streamflow) in California and CNRM-CM5 model to have wetter conditions (higher streamflow). The results for the Colorado River basin in this research have similar signatures in the model output for streamflow projections at Lees Ferry (i.e., surplus flows for CNRM-CM5 and drier flows for HadGEM2-ES). This does highlight the potential of either connections between California basins and Colorado River, and/or similarities in which models before well in the western United States for climate change studies. This also highlights the range of climate scenarios that can be provided from GCMs. In general, Pierce et al. [19] found that all the GCMs had warming in the future, but some of the models were warming at higher levels than others and the cooler/wetter models were less warm. This range in model scenarios could be viewed as contradictory, or it could be viewed as a measure of the uncertainty of potential future projections. This does have implications on water management as the range of future conditions are planned for and assigned levels of risk.

Other research in the Colorado River basin has shown the potential for declines at Lees Ferry under future climate ensemble mean projections below a critical threshold of 13.8 MAF [11]. The results in the study presented here show many scenarios are above this threshold (i.e., wetter conditions) and highlight the importance of understanding the appropriate models for the region. It is noteworthy that some of the previous studies [12] have shown the change in seasonal runoff with higher streamflow (10–20%) occurring during the December–March period and less during the critical April–July period. This may be important in establishing the timing of future streamflow under warmer conditions. In addition, other studies use many models (32) from CMIP5 which adds to the variability shown in results. It is believed that the study presented here is a more focused approached that has specific results for certain model scenarios. This will lead to a better understanding of climate change impacts in the western United States.

Various studies have confirmed the wet signal identified in the research presented here. Hoerling et al. [9] noted an increase in the median precipitation from CMIP5 projections for both RCP 4.5 and 8.5 precipitation over Colorado by the middle of the 21st century. Ayers et al. [21] showed higher precipitation projections in CMIP5 (compared to CMIP3) along with streamflow for the Upper Colorado River basin. Lukas et al. [22] also noted this potential for higher streamflow along with an earlier snowmelt

5. Conclusions

The research presented here provides interesting findings related to how the Colorado River may respond to future climates. While the results are not always consistent in terms of how drought or surplus periods will change in the future, the intensity (as measured by magnitude and duration) will likely increase for both RCP 4.5 and 8.5. There did not appear to be large difference between the results for the RCP 4.5 and RCP 8.5 scenarios. There were more pronounced differences between models where the CanESM2 and CNRM-CM5 models had a wet signal, and HadGEM2 and MI-ROC5 models had a dry signal. The spatial results for the Had-GEM2 and MI-ROC5 were not as consistent (i.e., mix of dry and wet) as the CanESM2 and CNRM-CM5 scenarios which were mostly wet. Most interesting results were for the Lees Ferry station where models indicate a chance of higher drought and surplus length and magnitude on the order of two times the historical period. It was a surprise that the future mean ensemble water flow for the Lees Ferry was approximately 3–10% of the historical and highlights the sensitivity of record length used in the analysis.

The hydroclimatology and future water supply of the Colorado River basin is an active area of research, discussion and management. While there remains a lot of uncertainties with future projections, the future will likely have more extremes (wet and dry). These uncertainties include the large-scale climate models that are downscaled for local climate conditions and the different hydrologic modeling methods (e.g., physically based, distributed, semi-distributed, statistical). Regardless, the results of this study, highlight that the future may have extended dry or wet periods that were not always seen in the historical (or paleo) record and may question the use of paleo records given that future climates may look very different. Having the appropriate management and infrastructure options available will be critical to adapt to these changes.

Author Contributions: Conceptualization, R.B. and T.P.; methodology, R.B. and T.P.; formal analysis, R.B.; writing—original draft preparation, R.B. and T.P.; writing—R.B. and T.P.; All authors have read and agreed to the published version of the manuscript.

Funding: This research received no external funding.

Institutional Review Board Statement: Not applicable.

Informed Consent Statement: Not applicable.

Data Availability Statement: Data for this study was access at the U.S. Geological Survey (USGS) NWISWeb Data retrieval (Available online: https://waterdata.usgs.gov/nwis/, accessed on 1 January 2022) and available at the Bureau of Reclamation who maintains natural flow that accounts for consumptive uses and losses. (Available online: https://www.usbr.gov/lc/region/g4000/NaturalFlow/index.html, accessed on 1 January 2022).

Acknowledgments: We would like to thank Chapman University for providing the opportunity to author this research as part of the Research and Creative Activity Course (491 in Spring 2022). Piechota acknowledges the sabbatical support from Chapman University.

Conflicts of Interest: The authors declare no conflict of interest.

References

1. Western Resources Advocates. Protecting the Colorado River Through Smarter Water Management Throughout the West. 2022. Available online: https://westernresourceadvocates.org/colorado-river-basin/ (accessed on 1 June 2022).
2. Bureau of Reclamation. Interior Department Announces Actions to Protect Colorado River System, Sets 2023 Operating Conditions for Lake Powell and Lake Mead. Available online: https://www.usbr.gov/newsroom/news-release/4294 (accessed on 1 June 2022).
3. Christensen, N.S.; Wood, A.W.; Voisin, N.; Lettenmaier, D.P.; Palmer, R.N. The Effects of Climate Change on the Hydrology and Water Resources of the Colorado River Basin. *Clim. Chang.* **2004**, *62*, 337–363. [CrossRef]
4. Rajagopalan, B.; Nowak, K.; Prairie, J.; Hoerling, M.; Harding, B.; Barsugli, J.; Udall, B. Water Supply Risk on the Colorado River: Can Management Mitigate? *Water Resour. Res.* **2009**, *45*. [CrossRef]
5. Miller, W.P.; Piechota, T.C. Trends in Western U.S. Snowpack and Related Upper Colorado River Basin Streamflow. *J. Am. Water Resour. Assoc.* **2011**, *47*, 1197–1210. [CrossRef]
6. Miller, W.P.; Butler, R.A.; Piechota, T.C.; Prairie, J.; Grantz, K.; DeRosa, G.M. Water Management Decisions using Multiple Hydrologic Models within the San Juan River Basin Under Changing Climate Conditions. *J. Water Resour. Plan. Manag.* **2012**, *138*, 412–420. [CrossRef]
7. Dettinger, M.; Udall, B.; Georgakakos, A. Western Water and Climate Change. *Ecol. Appl.* **2015**, *25*, 2069–2093. [CrossRef] [PubMed]
8. McCabe, G.J.; Wolock, D.M.; Pederson, G.T.; Woodhouse, C.A.; McAfee, S. Evidence that recent warming is reducing upper Colorado River flows. *Earth Interact.* **2017**, *21*, 1–14. [CrossRef]
9. Hoerling, M.; Barsugli, J.; Livneh, B.; Eischeid, J.; Quan, X.; Badger, A. Causes For the Century-Long Decline in Colorado River Flow. *J. Clim.* **2019**, *32*, 8181–8203.
10. Udall, B.; Overpeck, J. The Twenty-First Century Colorado River Hot Drought and Implications for the Future. *Water Resour. Res.* **2017**, *53*, 2404–2418. [CrossRef]
11. Woodhouse, C.A.; Smith, R.M.; McAfee, S.A.; Pederson, G.T.; McCabe, G.J.; Miller, W.P.; Csank, A. Upper Colorado River Basin 20th Century Droughts Under 21st Century Warming: Plausible Scenarios for the Future. *Clim. Serv.* **2021**, *21*, 100206. [CrossRef]

12. Bureau of Reclamation. West-Wide Climate and Hydrology Assessment, Technical Memorandum No. ENV-2021-011. March 2021. Available online: https://www.usbr.gov/climate/secure/docs/2021secure/westwidesecurereport.pdf (accessed on 1 June 2022).
13. Wallis, J.R.; Lettenmaier, D.P.; Wood, E.F. A Daily Hydroclimatological Data Set for the Continental United States. *Water Resour. Res.* **1991**, *27*, 1657–1663. [CrossRef]
14. Tootle, G.A.; Piechota, T.C.; Singh, A.K. Coupled Interdecadal and Interannual Oceanic/Atmospheric Variability and United States Streamflow. *Water Resour. Res.* **2005**, *41*, W12408.
15. Bureau of Reclamation. Downscaled CMIP3 and CMIP5 Hydrology Projections—Release of Hydrology Projections. 2014. Available online: https://gdo-dcp.ucllnl.org/downscaled_cmip_projections/techmemo/BCSD5HydrologyMemo.pdf (accessed on 1 January 2022).
16. Maurer, E.P.; Wood, A.W.; Adam, J.C.; Lettenmaier, D.P. A Long-Term Hydrologically Based Dataset of Land Surface Fluxes and States for the Conterminous United States. *J. Clim.* **2002**, *15*, 3237–3251. [CrossRef]
17. Riahi, K.; Rao, S.; Krey, V.; Cho, C.; Chirkov, V.; Fischer, G.; Rafaj, P. RCP 8.5—A Scenario of Comparatively High Greenhouse Gas Emissions. *Clim. Chang.* **2011**, *109*, 33–57.
18. U.S. Global Change Research Program (USGCRP). *Impacts, Risks, and Adaptation in the United States: Fourth National Climate Assessment, Volume II*; U.S. Global Change Research Program: Washington, DC, USA, 2018.
19. Pierce, D.; Kalansky, J.; Cayan, D. *Climate, Drought, and Sea Level Rise Scenarios for California's Fourth Climate Change Assessment*; Scripps Institution of Oceanography: La Jolla, CA, USA, 2018; pp. 52–56. Available online: https://www.energy.ca.gov/sites/default/files/2019-11/Projections_CCCA4-CEC-2018-006_ADA.pdf (accessed on 1 January 2022).
20. Lynam, L.; Piechota, T. California Drought Outlooks Based on Climate Change Models' Effects on Water Availability. *Water* **2021**, *13*, 3211. [CrossRef]
21. Ayers, J.; Ficklin, D.L.; Stewart, I.T.; Strunka, M. Comparison of CMIP3 and CMIP5 Projected Hydrologic Conditions over the Upper Colorado River Basin. *Int. J. Climatol.* **2016**, *36*, 3807–3818. [CrossRef]
22. Lukas, J.; Barsugli, J.; Doesken, N.; Rangwala, I.; Wolter, K. *Climate Change in Colorado: A Synthesis to Support Water Resources Management and Adaptation*; University of Colorado; Colorado Water Conservation Board; Western Water Assessment, 2014; 114p., Available online: https://wwa.colorado.edu/climate/co2014report/Climate_Change_CO_Report_2014_FINAL.pdf (accessed on 1 June 2022).

hydrology

MDPI

Article

Comparison of Regionalisation Techniques for Peak Streamflow Estimation in Small Catchments in the Pilbara, Australia

Alissa Flatley [1,*] **and Ian Rutherfurd** [1,2]

[1] School of Geography Earth and Atmospheric Sciences, University of Melbourne, Parkville, VIC 3010, Australia
[2] Alluvium Consulting, Level 1, 105-115 Dover Street, Cremorne, VIC 3121, Australia
* Correspondence: alissa.flatley@unimelb.edu.au

Abstract: Arid and semi-arid regions typically lack high-resolution river gauging data causing difficulties in understanding rainfall-runoff patterns. A common predictive method for discharge estimation within ungauged catchments is regional flood frequency estimation (RFFE), deriving peak discharge estimates from similar, gauged catchments and applying them to the catchment of interest. The majority of RFFE equations are developed for larger catchments where flow events may be larger and of greater interest. We test a series of RFFE methods derived for the Pilbara region, applying them to new ungauged small catchments under 10 km^2. Rainfall values are derived from a guideline Australian design rainfall database, Australian Rainfall and Runoff 2019 (ARR2019) which was recently updated with an additional 30 years of rainfall data. RFFE equations are compared to a direct rainfall model to evaluate their performance within small catchments, identifying key limitations and considerations when modelling small headwater catchments.

Keywords: regional flood frequency; ungauged catchments; direct rainfall modelling; headwater catchments; Australia

Citation: Flatley, A.; Rutherfurd, I. Comparison of Regionalisation Techniques for Peak Streamflow Estimation in Small Catchments in the Pilbara, Australia. *Hydrology* **2022**, 9, 165. https://doi.org/10.3390/hydrology9100165

Academic Editors: Carmelina Costanzo, Tommaso Caloiero, Roberta Padulano and Aronne Armanini

Received: 25 August 2022
Accepted: 22 September 2022
Published: 24 September 2022

Publisher's Note: MDPI stays neutral with regard to jurisdictional claims in published maps and institutional affiliations.

1. Introduction

Flood frequency analysis is commonly undertaken to identify and estimate flood quantiles corresponding to a given return period using the available streamflow observations in a catchment [1]. Where streamflow records are absent, or catchments are widely ungauged, prediction of streamflow involves applying regional flood frequency estimation (RFFE) methods, which are data-driven empirical procedures that attempt to compensate for a lack of temporal data at a given location with spatial data, obtained from other locations within a homogenous region [2]. RFFE approaches are frequently used throughout Australia for the prediction of flood events. Accurately predicting the magnitude of flood events is essential for the planning of water resource systems [3] in addition to adhering to the design standards of engineered structures designed to convey a certain flow [4,5]. A lack of long-term rainfall and streamflow data within arid and semi-arid regions is a major issue for run-off modelling [6] and this challenge is highlighted here within the Pilbara region in Western Australia.

Western Australia accounts for 39% of the global supply of iron ore [7] but has a notable lack of rainfall and streamflow data owing to (a) the high spatial and temporal variability of streamflow, (b) the high cost of establishing dense hydrologic gauging sites (c) the likelihood of gauging sites being destroyed by flash flooding events and (d) the disproportionate interest of flow events in larger river channels resulting in a scarcity of data smaller catchments. This data scarcity results in a fewer opportunities for validation of hydrological models used and therefore there are limited opportunities to demonstrate that a given-site specific model is capable of making accurate predictions for periods outside a calibration period [8]. Nevertheless, increasing engineering modifications are carried out in small catchments in the Pilbara, such as culvert installation for roads and railways or mining river diversions—such as those found in our study region. River diversions are expensive engineering structures and their design and performance rely on adequately estimating both frequent and rare flow events [9]. There is also increased recognition of the importance to estimate and subsequently preserve natural hydrologic flow conditions resulting in an interest in estimating frequent flow events (e.g., 1EY, 0.5EY and 0.2EY where EY stands for the average number of exceedances per year) particularly in these semi-arid areas where flow is sparse and poorly understood.

Current practices for the design and prediction of peak discharges for river relocation designs are based on several estimation methods such as regionalisation methods, numerical and hydrologic modelling. The outcome of these methodological provides wide-ranging, typically heuristic results. However, the improvement of two-dimensional hydrodynamic models is changing how peak discharges are being estimated. Direct rainfall modelling (or rain-on-grid) has increasingly become a standard approach for predicting design flood behaviour [10,11]. The application of rainfall directly onto a 2D domain allows for the simulation of catchment runoff by applying it directly to the modelling grid [12,13]. Catchment runoff within direct rainfall models is dependent on the grid or mesh cell area, the rainfall depth, grid cell roughness, rainfall losses (IL/CL) and the slope between neighbouring cells. This distributed rainfall approach directly onto the 2D domain can give considerably more detail, particularly in the upper parts of sub-catchments [14] and has been found to provide a better representation of minor overland flowpaths than conventional modelling methods [12]. In the absence of stream gauges, this paper examines the suitability of direct rainfall modelling to test the suitability of RFFE approaches on small ungauged catchments within the Pilbara. The tested RFFE approaches were previously designed for the semi-arid Pilbara region and have previously been validated using gauged flow data within medium to large catchments [15–18]. Most direct-rainfall models are used as a means of indirect calibration or comparison with traditional hydrological predictive methods and can be used to elucidate discrepancies in other models [19]. Direct rainfall models are useful in the modelling of design flood simulations providing appropriate checks and quality assurance procedures are implemented [13]. There are many sources of uncertainty that can have a strong influence on flood mapping and flow hydrographs, including synthetic rainfall estimation with IDF curves [20], initial soil moisture conditions [21,22] the basin response model, modelling grid sizes [23] and the difference between storm return time and the correspondent flood return time [24] to name a few.

This paper tests the suitability of these RFFE approaches within smaller headwater catchments in the Pilbara and evaluates them through a comparison of peak discharge values derived from a 2D hydrodynamic direct rainfall model. This paper provides the first comparative study of RFFE approaches for the Pilbara using updated ARR (2019) [25] values to validate their use within smaller catchments in the same region. To achieve this, we firstly (a) select existing RFFE techniques previously applied and validated in larger catchments, (b) simulate direct rainfall events within a series of smaller catchments (c) use the resulting discharge values to quasi-validate their application within small headwater catchments to provide a range of likely predicted peak discharges for a range of annual exceedance probabilities.

2. Study Area

The Pilbara region of Western Australia is a large arid to semi-arid region with a wide expanse of ungauged catchments with limited streamflow records within its catchments [26,27]. The region in a transitional location between the Eyrean (central desert) and the southern Torresian (tropical) bioclimatic regions [28]. The Pilbara region is classified as either Arid, desert hot (*Bwh*) and Arid steppe hot (*Bsh*) by the updated Köppen–Geiger climate classification [29]. Temperatures exceed 30 °C for most of the year, and rarely dip below 10 °C. Because of its positioning, tropical depressions and recurrent cyclonic events comprise most the regions' total rainfall [30,31]. The Pilbara has a low total rainfall averaging between 250–300 mm annually [26,32,33] but the majority of flow events are concentrated in short duration floods of high magnitude [34]. The region is characterized by extreme hydroclimatic conditions, in which the rainfall is highly sporadic [31], driven by infrequent tropical cyclones (Figure 1) and thunderstorms occurring mainly within the summer months between January to March [26,35]. Winter rainfall is typically from low pressure trough systems [26]. For smaller streams, such as those within headwater channels, flow events typically last around 5 days of the year.

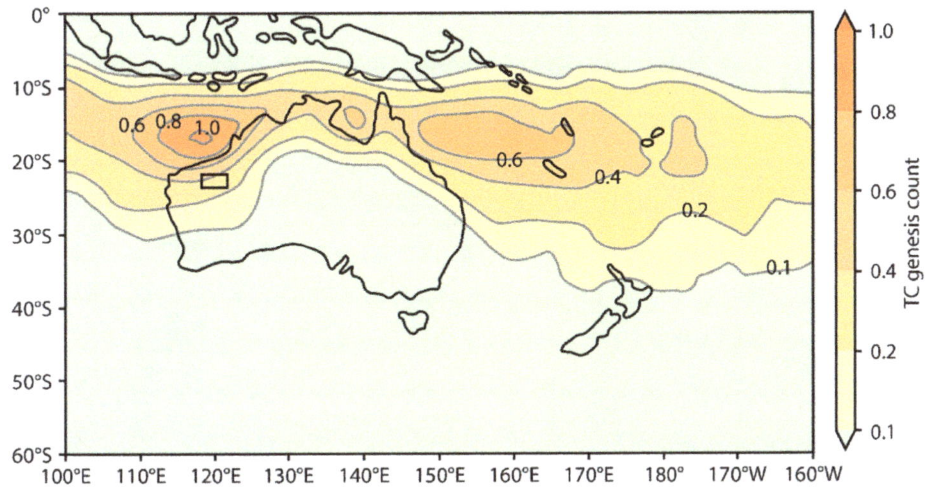

Figure 1. Observed Southern Hemisphere Tropical Cyclone genesis counts (all seasons) indicating the 122 average annual number of tropical cyclones through Australian region in El Niño, La Niña, and neutral 123 years (modified from Australian Government Bureau of Meteorology, 2019 [36]).

Rainfall is very localised, causing issues for the correlation of rainfall and runoff. In addition, the Pilbara has very high evaporative losses, the annual potential evaporation is also 10 times higher than the annual rainfall [37]. Runoff is also highly variable, and only 2–13% of mean annual rainfall becomes runoff in the Pilbara [38]. Higher percentages of rainfall almost certainly run off in small headwater catchments, with a lot of runoff infiltrating into streambeds and therefore failing to reach downstream gauging stations [38]. Most gauging sites are located within larger catchments, however most of these initial stream gauges were not operational until 1967 [33,39].

The Pilbara region is also rich in iron ore and open-cut mines resulting in a wide array of engineering structures built to service mine sites, in addition to many watercourse modifications such as culverts, drains and river diversion channels constructed both within large and smaller channels. River diversion channels for mining in the Pilbara region are designed conservatively to convey rare flow events driven by large cyclonic events or infrequent 100 and 1000-year ARI floods (or the 1 to 0.1AEP (%)) [9]. Many river diversions are constructed within smaller catchments that lack gauged rainfall or streamflow data resulting in a poor understanding of the peak flood discharges and more frequent events experienced within these catchments.

3. Materials and Methods

3.1. General Approach

This research tests a series of RFFE methods to calculate predicted peak discharge (Q_{peak}) previously used within the Pilbara and applies them specifically to smaller sized catchments using updated ARR2019 IFD rainfall values. The selected RFFE methods were chosen based on their satisfactory performance when applied in larger catchments in the Pilbara. The methods tested include a QRT and PRT method from Taylor et al.,

2011 [15], QRT and PRT methods from Rahman et al., 2012 [2], a Regional Flood Frequency Procedure (RFFP2000) from Flavell., 2012 [17], a IFM method from Davies and Yip, 2014 [18] and the ARR 2016 RFFE (Table 1). Further details on the RFFE approaches and their prior performance is provided in Table S1 of the Supplementary Material. Next, the RFFE methods are quasi-validated against a TUFLOW direct rainfall model, acting as the observation of rainfall within the catchments lacking gauging infrastructure. The results yield a range of peak discharge estimates for each AEP. The direct rainfall model (and associated sensitivity analysis) is used to provide validation for the most appropriate RFFE for small catchments. Figure 2 shows the full sequence of steps. Similar approaches have been undertaken to assess the performance of ARR 2016 RFFE using RORB modelling [40] or to incorporate it into a direct rainfall model of complex urban catchments [41]. However, few studies have used direct rainfall modelling approaches to benchmark RFFE methods within small semi-arid headwater channels.

Table 1. Regional Flood Frequency Estimation methods applied to headwater catchments.

Method	Equation
ARR (Australian Rainfall and Runoff Regional Flood Frequency Estimation Model) RFFE Model	$Qx = Q_{10} \times GFx$ with Q_{10} as: $\log_{10} = b_0 + b_1 \log_{10}(\text{area}) + b_2 \log_{10}(I_{6,50})$ Where b_0, b_1 and b_2 are regression coefficients, estimated using OLS regression, area is the catchment area in km^2 and $I_{6,50}$ is the design rainfall intensity at catchment centroid for a 6 h duration and 50% AEP. The values of b_0, b_1 and b_2 and the regional Growth Factors (GFx) are embedded into the RFFE Model 2015.
Index Flood Method (IFM) (Davies and Yip, 2014) [18]	For a small catchment area: $Q_5 = 7.32 \times 10^{-8} A^{0.651} I_{1hr,\,2\,yrs}{}^{5.251}$ Frequency Factors: 2ARI = 0.31, 5ARI = 1.0 10 ARI = 1.70, 20ARI = 2.58, 50ARI = 4.15, 100ARI = 5.82
Parameter Regression Technique (PRT) (Taylor et al., 2011) [15]	$M = -11.411 + 0.527 \times \ln(\text{area}) + 7.765 \times \ln(I_{12hr,2})$ $S = C_1{:}g = C_2$ where C_1 and C_2 are regional average $M = 2.54 + 0.52[\ln(\text{area}) - 4.71] + 8.08[\ln(I_{12,2}) - 1.47]$
Fixed Region Parameter Regression Technique (PRT) (Rahman et al., 2012a) [2]	$M = 2.54 + 0.52[\ln(\text{area}) - 4.71] + 8.08[\ln(I_{12hr,2}) - 1.47]$ stdev = 1.45 + 0.10(zarea) + 0.07(zforest) (4.8.17) skew = −0.49 − 0.08(zarea) − 0.64(zsden) (4.8.18)
Quartile Regression Technique QRT (Taylor et al., 2011) [15]	$\ln(Q_2) = -11.366 + 0.521 \times \ln(\text{area}) + 7.858 \times \ln(I_{12hr,2})$ $\ln(Q_5) = -15.913 + 0.486 \times \ln(\text{area}) + 5.336 \times \ln(I_{1hr,2})$ $\ln(Q_{10}) = -14.285 + 0.465 \times \ln(\text{area}) + 5.055 \times \ln(I_{1hr,2})$ $\ln(Q_{20}) = -12.949 + 0.445 \times \ln(\text{area}) + 4.824 \times \ln(I_{1hr,2})$ $\ln(Q_{50}) = -4.914 + 0.431 \times \ln(\text{area}) + 5.705 \times \ln(I_{12hr,2})$ $\ln(Q_{100}) = -4.072 + 0.413 \times \ln(\text{area}) + 5.412 \times \ln(I_{12hr,2})$
Quartile Regression Technique QRT (Rahman et al., 2012) [2]	$\ln(Q_2) = 2.66 + 0.51(\ln(\text{area}) - 4.71) + 8.08\,[\ln(I_{12,2}) - 1.47]$ $\ln(Q_5) = 3.90 + 0.48[\ln(\text{area}) - 4.71] + 7.20\,[\ln({}_{I12,2}) - 1.47]$ $\ln(Q_{10}) = 4.51 + 0.45[\ln(\text{area}) - 4.71] + 6.74\,[\ln(I_{12,2}) - 1.47]$ $\ln(Q_{20}) = 5.01 + 0.44[\ln(\text{area}) - 4.71] + 6.19\,[\ln(I_{12,2}) - 1.47]$ $\ln(Q_{50}) = 5.59 + 0.41[\ln(\text{area}) - 4.71] + 5.66\,[\ln(I_{12,2}) - 1.47]$ $\ln(Q_{100}) = 5.87 + 0.39[\ln(\text{area}) - 4.71] + 5.34\,[\ln(I_{12,1}) - 1.47]$
RFFP (Flavell, 2012) [17]	$Q_2 = 1.72 \times 10^{-64}(AS_e{}^{0.5})^{0.8}\, LAT^{-12.17}\, LONG^{38.77}\, (L2/A)^{-1.05}$ $Q_5 = 7.47 \times 10^{-46}(AS_e{}^{0.5})^{0.81}\, LAT^{-14.62}\, LONG^{31.40}\, (L2/A)^{-0.68}$ $Q_{10} = 2.36 \times 10^{-34}(AS_e{}^{0.5})^{0.81}\, LAT^{-15.24}\, LONG^{26.28}\, (L2/A)^{-0.39}$ With the largest value from two Q_{20} equations being adopted for the Q_{20} value: $Q_{20} = 1.98 \times 10 - 23(AS_e{}^{0.5})^{0.79}\, LAT^{-15.08} LONG^{20.91}$ $Q_{20} = Q_{10} = (13.21A^{0.61})/(8.74A^{0.60})$ $Q_{50} = Q_{20} \times$ frequency factor (Q_{50}/Q_{20}) $Q_{100} = Q_{20} \times$ frequency factor (Q_{100}/Q_{20}) A = catchment area (km^2), S_e = equivalent uniform slope (m/km) and L = mainstream length (km)

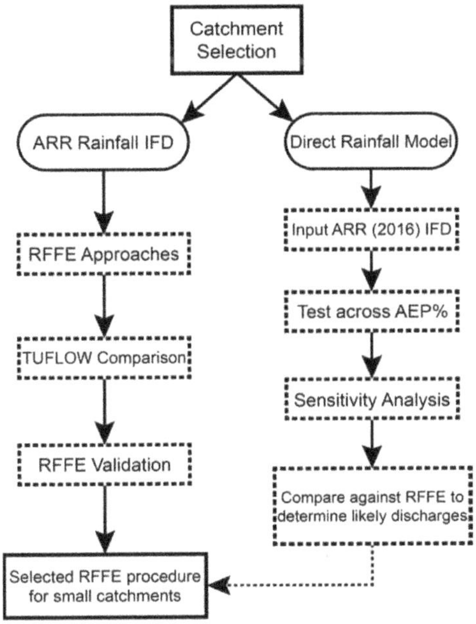

Figure 2. Workflow of steps to determine Regional Flood Frequency Estimation (RFFE) selection and for small headwater channels.

3.2. Regionalisation Approaches

Regionalisation refers to the process of transferring hydrological information from one catchment to another. RFFE approaches assume a statistical relationship between observable catchment properties and flood discharge characteristics, allowing the construction of flood hydrographs by applying relationships developed for gauged catchments with similar properties [42]. Regionalization approaches are commonly based either on spatial proximity or similar catchment attributes [43]. Within regionalization methods, model parameters are used as an instrument to transfer hydrological information from gauged to ungauged basins [44]. In general, a regional model can be stated in a simplified form defined by Wagener and Wheater (2006) as:

$$\hat{\theta}_L = H_R(\theta_R|\phi) + v_R \tag{1}$$

where $\hat{\theta}_L$ is the estimated hydrological variable of interest at the ungauged site (it can be an estimated model parameter, probability or cumulative distribution function parameter, or hydrological response such as streamflow or flow events), H_R is a functional relation for $\hat{\theta}_L$ using a set of catchment attributes—physiographic or meteorological attributes ϕ, θ_R is a set of regional hydrological variables of interests and v_R is an error term [45]. Regionalisation approaches may be satisfactory if the catchments are similar in some sense, but error prone if they are not [46]. Razavi and Coulibaly (2013) provide a review of methodology for streamflow prediction in ungauged basin using regionalization methods, concluding that most model-dependent methods in arid to warm-temperate climates (e.g., Australia) indicate that physical similarity and spatial proximity appear to be the best approach to estimating streamflow. However, the most regionalisation methods are highly site-specific, and therefore, a comparative study between suitable approaches is suggested before selecting the regionalisation method for a given site or region.

3.3. Regional Flood Frequency Estimation (RFFE)

Regional flood frequency estimation (RFFE) is widely used to estimate flood quartiles in ungauged catchments. RFFE approaches provide an alternative method to flood frequency analysis (FFA) where a lack of temporal data is substituted with spatial data to make more accurate flood estimates at ungauged sites [15]. Common RFFE techniques include Probabilistic Rational Method (PRM), Quantile Regression Techniques (QRT) the Index Flood Method (IFM) and a Parameter Regression Technique (PRT). Regression based RFFE methods are more commonly applied to recent studies within Australia. The following section details selected regression based RFFE methods, which are developed from a longer record of data and are considered to give a more reliable estimation of design flows.

3.3.1. Quartile Regression Technique

The quartile regression technique (QRT) is used frequently within ungauged catchments. The method estimates flood quartiles through a multiple regression between recorded streamflow data and a set of climatic and catchment characteristics within a region [47]. The QRT regression technique is expressed as:

$$Q_T = aB^bC^CD^d \tag{2}$$

where B, C and D are catchment and climatic characteristics variables (predictors); a, b, c, d are the regression coefficients and Q_T is the flood magnitude with T-year ARI (flood quantile) [48].

3.3.2. Parameter Regression Technique

The parameter regression technique (PRT) is similar to the QRT. However, instead of quartiles, the first three moments of the log-Pearson type 3 (LP3) distribution are taken as dependent variables in regression analysis against catchment characteristics [15]. Let Q be the annual maximum flood series at a site and X = ln(Q), then the mean (M), standard deviation (S) and skew (g) of the X series are taken as dependent variables:

$$\ln Q_T = M + K_T S \tag{3}$$

where Q_T is a flood quantile of T years ARI and K_T is the standardized LP3 frequency factor (which is a function of skew) and can be obtained from ARR or can be approximated [15].

3.3.3. Index Flood Method

The index flood method (IFM) assumes that the exceedance probability distribution of annual peak discharge is identical, except for a site-specific scaling factor called the index flood (average likely flood) [49]. The IFM method is expressed as:

$$Q_T = q_T \mu_i \tag{4}$$

where Q_T is the flood quantile, μ_i is the function basin area, slope and q_T is a regional growth factor (a dimensionless frequency distribution quantity common to all sites within each homogeneous region).

3.4. Australian Rainfall and Runoff (ARR) RFFE

The Australian Rainfall and Runoff (ARR) guidelines offer predicted estimates of rainfall intensity, frequency, and duration (IFD) values for Australia. Additionally, ARR also has a regional flood frequency estimation model which is widely used and recommended for design flood estimation [11,50]. The ARR guidelines were updated in 2016 and again in 2019. Five predictor values were adopted for the RFFE technique [51,52]. These predictor values are: catchment area (in km^2); design rainfall intensity at catchment centroid (in mm/h) for the 6 h duration and 50% AEP ($^{50\%}$ I_{6h}); design rainfall intensity at catchment centroid (in mm/h) for the 6 h duration and 2% AEP ($^{2\%}$ I_{6h}); ratio of design rainfall intensities of AEPs of 2% and 50% for duration of 6 h ($^{2\%}$ I_{6h}/$^{50\%}$ I_{6h}); and catchment shape factor (S_f), which is defined as the shortest distance between catchment outlet and centroid divided by the square root of catchment area. The RFFE technique used in ARR is adapted for different regions throughout Australia (Figure 3).

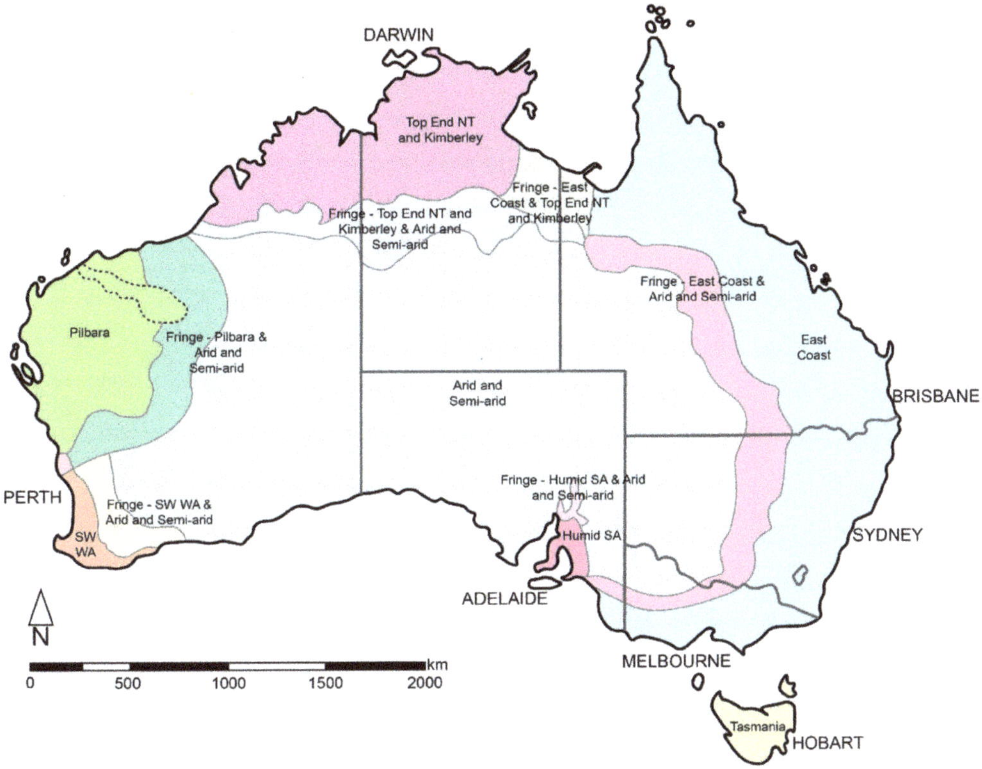

Figure 3. Adopted regions for RFFE technique in Australia. Modified from Rahman et al., (2019) [52]. The Fortescue Catchment, Pilbara is outlined.

Each region is determined based on a Region of Interest *(ROI)* approach based on geographical proximity of gauging stations, with fringe zones between regions defined by the 500 mm and 400 mm isohyet to delineate between humid and arid/semi-arid regions [53]. The Pilbara region was characterized as an alternative sub-region distinct from the other arid and semi-arid regions of Australia [51]. This was due to (a) concentrations of stream gauging stations in three parts of Western Australia which are separated by long distances (e.g., Kimberley region, Pilbara region and South-West WA) and (b) notable differences in region hydrologic conditions [2]. The approach used for the Pilbara region was modified to an IFM as suggested by Farquharson et al., (1992) [54]. This recommended approach is an IFM with Q_{10} as an index variable and a dimensionless growth factor (*GF*) for X% AEP (GF_x):

$$Q_x = Q_{10} \times GF_x \quad (5)$$

A prediction equation was developed for Q_{10} as a function of catchment characteristics, and regional growth factors were developed based on the estimated at-site flood quartile. A Bayesian parameter estimation procedure with LP3 distribution was used to estimate flood quantiles for each gauged site for AEPs of 50%, 20%, 10%, 5%, 2% and 1%. Rahman et al.,

(2019) provide further information surrounding this process [52]. The adopted predictive equation for the index variable Q_{10} has the form:

$$log_{10}(Q_{10}) = b_0 + b_1 log_{10}(area) + b_2 log_{10}(I_{6,\ 50}) \tag{6}$$

where b_0, b_1 and b_2 are regression coefficients, estimated using OLS regression, area is the catchment area (in km^2) and $I_{6,50}$ is the design rainfall intensity at catchment centroid for a 6 h duration and 50% AEP. The values of b_0, b_1 and b_2 and the regional growth factors (GFx) are embedded into the RFFE Model 2015.

3.5. Direct Rainfall Modelling

Direct Rainfall Modelling (also known as rain-on-grid) was undertaken in TUFLOW HPC, a 2D fixed-grid, adaptive time-step, hydrodynamic solver that uses an explicit finite volume solution [55]. TUFLOW HPC reduces the run time of models. The direct-rainfall approach applied the rainfall hyetograph (mm versus time) uniformly to active cells within the defined grid of the catchment of interest. Each hyetograph value represents the rainfall that fell per increment. The double precision version of TUFLOW Classic was used in initial model set-up to minimize initial model errors (such as deficient or erroneous data) before running greater numbers of simulations in the GPU for faster run-times after model establishment. TUFLOW Classic uses a fixed time step and will highlight any initial errors with the model runs. Instabilities in the model highlight bad data or poor model set up [56]. TUFLOW HPC can hide poor model set up through its adaptive time-stepping to ensure the model remains stable. Model simulation parameters (grid size, time-step) were established to optimise the accuracy, run-time, and stability of the model. To effectively resolve flow events within the channel it is recommended to provide at least 5 grid/mesh elements laterally across the river channel [57].

The time-step for the model runs was maintained at 1/5 of the selected model grid size in meters (a time-step appropriate for TUFLOW classic) [55] with a cell wet/dry depth of 0.0002 m to account for the high proportion of shallow flow with a direct rainfall model. Due to the small catchment size, rainfall values are small and the reported IL/CL values for larger Pilbara river channels are scaled to these larger catchment hydrological inputs. Simply, the observed losses in larger catchments are larger than the grid-averaged rainfall inputs in small headwater channels. Therefore, IL/CL values were not applied during the final modelled scenarios. Hall (2015) provides a description of the advantages and disadvantages of direct rainfall modelling through model construction, calibration validation and sensitivity analysis [13].

3.5.1. Inputs

Catchment DEMs

Headwater channels (first-to-third order) are the areas from which water originates within a channel network and are closely coupled to hillslope processes [58]. This study used a high-resolution dataset around the periphery of the Yandi iron ore mine in the Pilbara, WA (Figure 4). Catchments were selected based on the high-resolution dataset covering the full extent of the catchment. Surface catchment DEMs were derived from a wide area semi-global matching (SGM) survey [59] (horizontal accuracy = 0.5 m and vertical accuracy = 0.25 m) undertaken in the Upper Fortescue catchment. Additionally, the catchments were screened for minimal catchment disturbance, such as mining, agriculture of the hydraulic alteration of the waterway from bridges and culverts. Catchments with engineering infrastructure, such as railways, culverts and main roads were modelled until the upstream contact with these features. Ten catchments (ranging from 0.96 km^2 to 9.23 km^2) matched these requirements and were selected to test the RFFE approaches (Table 2).

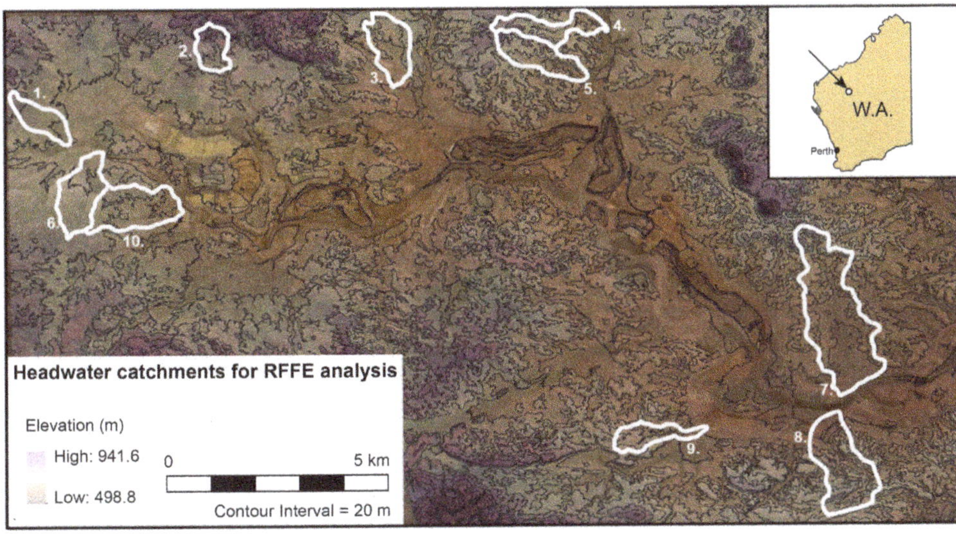

Figure 4. Map of headwater catchments used to test RFFE methods. Numbered catchments border the Yandi Mine in Eastern Pilbara, WA.

Table 2. Catchment description for RFFE and direct rainfall analysis.

Catchment	Area (km^2)	Latitude	Longitude	S_e (m/km)	L (km)
1	1.05	−22.711924	118.955892	18.54	1.13
2	0.96	−22.694436	119.001585	40.57	0.71
3	1.68	−22.695323	119.045882	16.65	1.10
4	1.48	−22.690506	119.087058	22.97	1.64
5	1.71	−22.697491	119.084533	37.61	1.55
6	1.95	−22.732084	118.965195	12.49	1.37
7	5.99	−22.765762	119.159322	22.48	2.06
8	3.23	−22.804315	119.161401	22.72	1.30
9	1.1	−22.795682	119.109156	26.71	1.17
10	2.42	−22.735894	118.980309	16.85	1.21

Rainfall

The Bureau of Meteorology, Australia (BOM) 2016 Design Rainfall Data System [60] was used to provide rainfall intensities for RFFE analysis and as a modelling input for direct rainfall modelling. The BOM 2016 IFD values provide a rainfall intensity (mm/h) or depth (mm) based on the latitude, longitude, and catchment size within a set location (Figure 5). BOM2016 IFD values replaced the older 1987 IFD and interim 2013 IFD values providing 30 additional years of hydrological data and adjustments to the approach. Additionally, the direct rainfall was applied to the 2D model over the entire catchment as time-series data as mm versus hours. The time-series data has a histogram stair-step shape) therefore rainfall was applied as a stepped approach holding the rainfall constant during the allocated time interval (BMT, 2018) meaning each rainfall value is the amount of rain that fell in mm between the previous time and current time. Rainfall was applied to every active cell within the digitised catchment. Ten varied areal hyetograph patterns were used for each catchment (Figure 6). Podger et al., (2018) provide a detailed description of the creation of these hyetograph patterns [61].

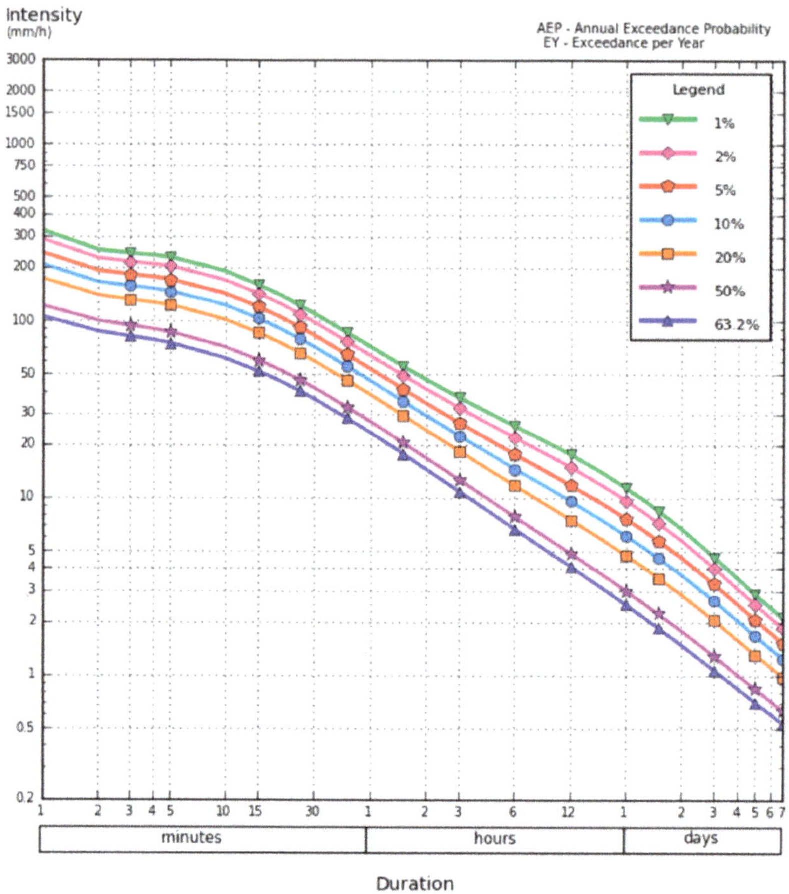

Figure 5. Single location IFD design rainfall plot (Catchment 1). Legend shows Annual Exceedance Probability (AEP) expressed as a percentage. Output from BOM Design Rainfall Data System [60].

3.5.2. Roughness

Manning's n values were assigned to account for runoff conditions. The roughness values assigned within a direct rainfall model can affect the timing of runoff. Additionally, constant roughness values may underestimate the effective roughness and ignores the role of spatially varied roughness within the catchment. Catchment floodplains were assigned a Manning's value of $n = 0.02$ and the channel was assigned a default Manning's value of 0.035. Adjustments of the roughness values were made in the sensitivity analysis of the TUFLOW models.

3.5.3. Output

Discharge

For each catchment a cross section was delineated near the catchment outlet, (Figure 7). At this cross section, a plot output was created in the form of a time-series hydrograph. This time-series provides the flood flow through the catchment during each model scenario and is used to identify peak discharge (Q_{peak}) values for each rainfall event. Backwater development has been reported during flood modelling of channel confluences within the larger Marillana Catchment [62]. Therefore, cross sections were delineated slightly upstream from the catchment outlet to reduce backwater flow events impacting peak discharge values during flood modelling.

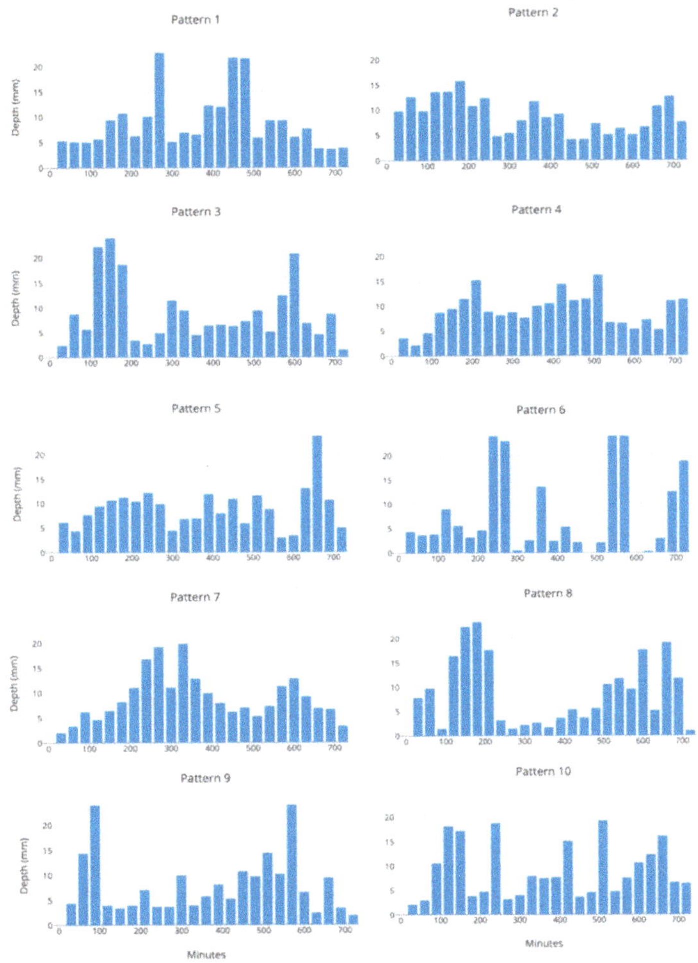

Figure 6. Design rainfall patterns for 5% AEP 12 h storm.

Figure 7. Catchment with a digitized white line showing a cross section (Plot output) for the TUFLOW model.

In addition to using the plot output for each simulation, maps of water depth (d) and velocity (V) were constructed at each of the model time-steps. The output from this was used to make an independent estimate of discharge using a velocity-area method:

$$Q = AV \quad (7)$$

where Q is the discharge expressed in cubic meters per second ($m^3\ s^{-1}$ or cumecs), A is the stream cross sectional area (m^2) and V is the mean velocity of stream flow ($m\ s^{-1}$). Measurements of stream water depth (stage) are typically measured at sites within the Pilbara, however continuous flow measurement of river discharge is expensive and logistically unwise given the annual frequency of flow events within the region.

3.6. Sensitivity Analysis

Sensitivity analysis or calibration methods are critical steps in rainfall-runoff and developing useful models of complicated hydrologic systems [63,64]. Models can be calibrated to observed data to demonstrate that the model can produce an observed flow time series with an acceptable level of accuracy [64]. Alternatively, a model may be available that has been previously calibrated for a catchment as part of another study. Sensitive model parameters should be recognised and appropriately evaluated to ensure they are constrained within acceptable ranges. Prediction in ungauged basins is challenging to validate owing to the data limitations within the area. When models are not able to be calibrated to measurements sensitivity testing should at least be carried out to assess the sensitivity of the model to variations in the main model parameters [11]. Model evaluation may not be limited to how accurately model predictions match historical observations, but how well the model represents the hydrological system. In this paper, we use a direct rainfall model to quasi-validate and select the most appropriate RFFE procedure for headwater, small size catchments. All RFFE procedures used in this investigation have been previously calibrated on larger catchments in the Pilbara.

There is no standard method for estimating uncertainty in streamflow in ungauged basins using regionalisation techniques [44]. Uncertainty is estimated here in the sensitivity analysis of the direct rainfall modelling approach by adjusting model parameters on three catchments. (Figure 8). Sensitivity analysis was carried out to determine the optimum model running conditions and explore the suitability of direct rainfall modelling within these small headwater catchments. Catchments were selected on the basis of channel gradient to encompass a range of catchment types within the analysis. Catchment 5 was the steepest, with greater expected areas of supercritical flow and areas of complex terrain which would challenge model performance [56]. Other catchments included catchment 1; a small, shallow, unconfined channel with multiple flow paths and 10; a larger catchment with a predominantly unconfined single-thread channel representative of many of the headwater channels in the vicinity.

Existing direct rainfall models have been found to be most sensitive to Manning's roughness and rainfall [13] and therefore sensitivity analysis was carried out to address the parameters that have larger uncertainties within the model. Sensitivity analysis was carried out to address: rainfall hyetograph shape, Manning's roughness (adjusted to +/− 20% or the upper and lower bounds for characteristic minor natural streams [65] and grid sizing, assessed with a 1 m, 2 m, 5 m and 10 m spacing.

3.7. Procedure for Evaluating RFFE Approaches

To evaluate the RFFE approaches, we use the result of the direct rainfall model to act as a measured series in this "quasi-validation". This quasi-validation uses joint plots of the RFFE output and simulated rain-on-grid discharge to compare the output of predictive methods. The ARR RFFE approach provides confidence limits of 5% Lower confidence limit (LCL) and a 95% Upper confidence limit (UCL) which are used as the absolute cut off for RFFE values. The evaluation procedure was as follows: (1) Was the RFFE output within the 5–95% confidence limit range for the ARR approach? (2) Did the approach demonstrate appropriate hydrological scaling across space and flood return period? (i.e., did the results increase with an increase in rainfall magnitude or catchment area) and lastly, (3) Did RFFE results display good agreement with findings from the rain-on-grid model? If not, did any approaches align with other tested RFFE methods.

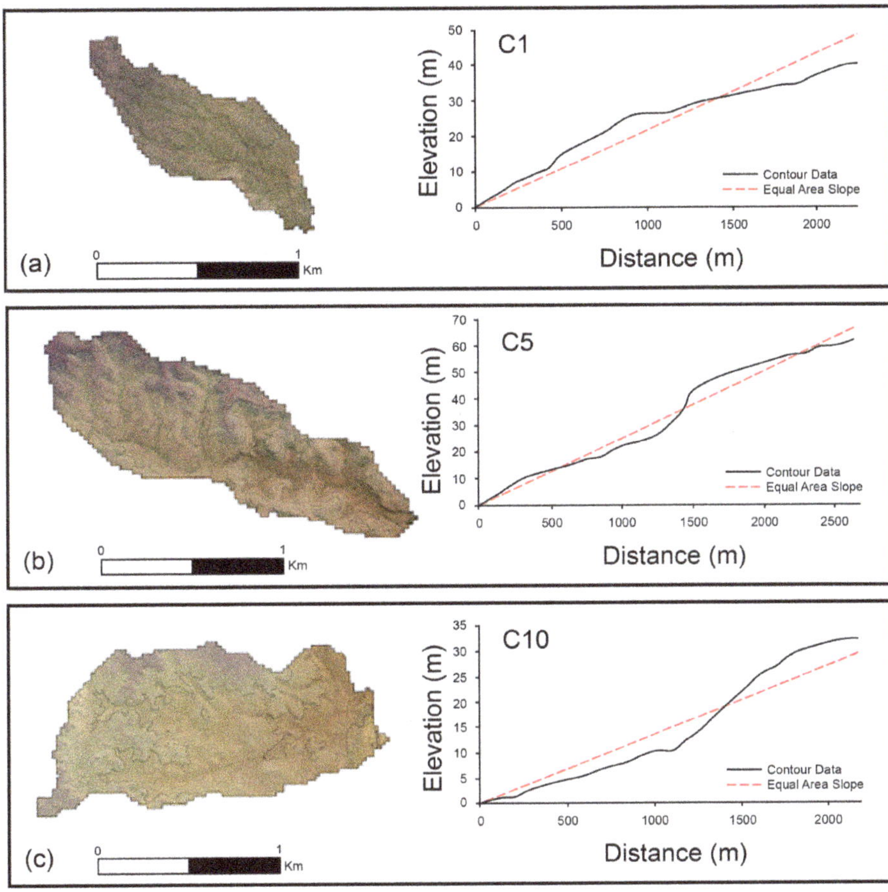

Figure 8. Catchments used in sensitivity analysis. (**a**) Catchment 1—A small shallow unconfined catchment with multiple flow paths, (**b**) Catchment 5—A steep, confined catchment and (**c**) Catchment 10—An unconfined single thread channel within a larger catchment. Graphs show the contour data of the longitudinal profile of the river channel. The red dashed line shows the equal area slope.

4. Results

4.1. Estimates of Peak Discharge for RFFE Methods

Rainfall values for RFFE approaches were obtained from BOM Design Rainfall Data System (2016) using 2016 Intensity, Frequency and Duration (IFD) values. The rainfall output showed a relatively homogenous depth for catchments ranging between 0.96 km^2 and 2 km^2. Larger catchments (sized 5.99 and 3.23 km^2) were predicted to have higher rainfall depths for the 12 h storm across the range of AEP (Table 3). A wide range of Q_{PP} values were obtained using the RFFE equations (Figure 9). The Flavell RFFP2000 procedure produced higher Q_{PP} values for Q_2 events. Between Q_5 and Q_{100}, the ARR values were highest, with greatest agreement with the IFM and Flavell RFFP2000. The PRT and QRT approach from Taylor et al., (2011) show nearly identical values with no increase in peak discharge values for an increase in catchment size. The fixed region PRT (Rahman et al., 2012) provided the lowest estimates of predicted peak discharge. Both QRT and PRT approaches produced low discharge values across Q_{10}–Q_{100} flow events.

Table 3. ARR rainfall depths (mm) for a 12 h storm per AEP (%).

Catchment	Area (km²)	50AEP	20AEP	10AEP	5AEP	2AEP	1AEP
1	1.05	2.35	6.75	11.0	16.0	23.3	29.0
2	0.96	2.37	6.78	11.0	16.1	23.4	29.1
3	1.68	3.34	9.59	15.6	22.7	33.1	41.2
4	1.48	3.32	9.52	15.5	26.6	32.8	40.9
5	1.71	3.59	10.3	16.8	24.4	35.5	44.3
6	1.95	3.22	9.24	15.1	21.9	31.9	39.7
7	5.99	7.21	20.7	33.7	49.0	71.3	88.9
8	3.23	4.70	13.5	22.0	32.0	46.5	58.0
9	1.1	2.59	7.44	12.1	17.6	25.6	32.0
10	2.42	3.72	10.7	17.4	25.3	36.8	45.9

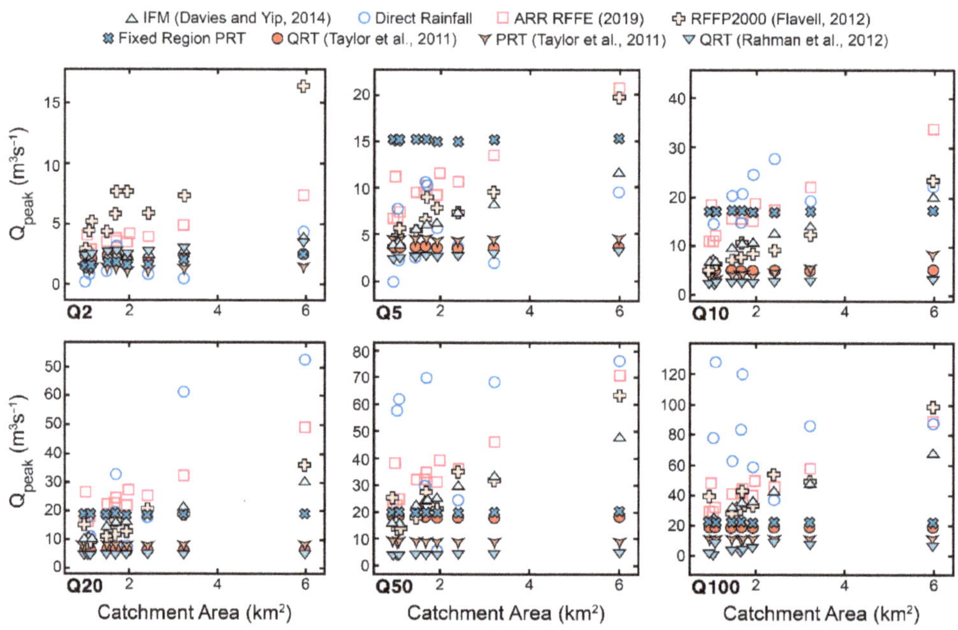

Figure 9. Comparison of small flood predicted peak discharges using RFFE approaches [2,15,17,18] for Q_2, Q_5, Q_{10}, Q_{20}, Q_{50} and Q_{100} return intervals.

4.2. Direct Rainfall Catchment Modelling

Table 4 shows the output values from direct rainfall modelling from the 10 catchments. Figure 10 show the results of direct rainfall modelling. These values have also been plotted next to the RFFE approaches with similar output (ARR RFFE (2019), IFM (Davies and Yip, 2014) and the Flavell RFFP (Flavell, (2012)) selected from deductive reasoning, omitting PRT and QRT approaches that yielded uncharacteristically small flood magnitudes across all annual exceedance probability.

Table 4. Modelled peak discharges using the direct rainfall modelling approach for a 12 h rainfall event across average return intervals. Peak discharge values are in $m^3\ s^{-1}$.

Catchment	2ARI	5ARI	10ARI	20ARI	50ARI	100ARI
1	0.51	7.75	16.56	17.58	58.10	77.79
3	2.24	10.63	12.45	19.39	30.37	83.30
4	0.80	2.68	3.50	18.41	21.22	62.69
5	2.88	10.11	27.40	32.56	70.25	120.11
6	1.38	5.75	7.32	7.50	6.32	58.53
7	4.16	9.60	52.0	72.66	76.73	87.60
8	0.20	1.95	52.29	61.26	68.82	85.86
9	2.20	2.37	9.78	10.97	62.24	128.20
10	0.64	3.72	10.7	17.4	25.3	36.80

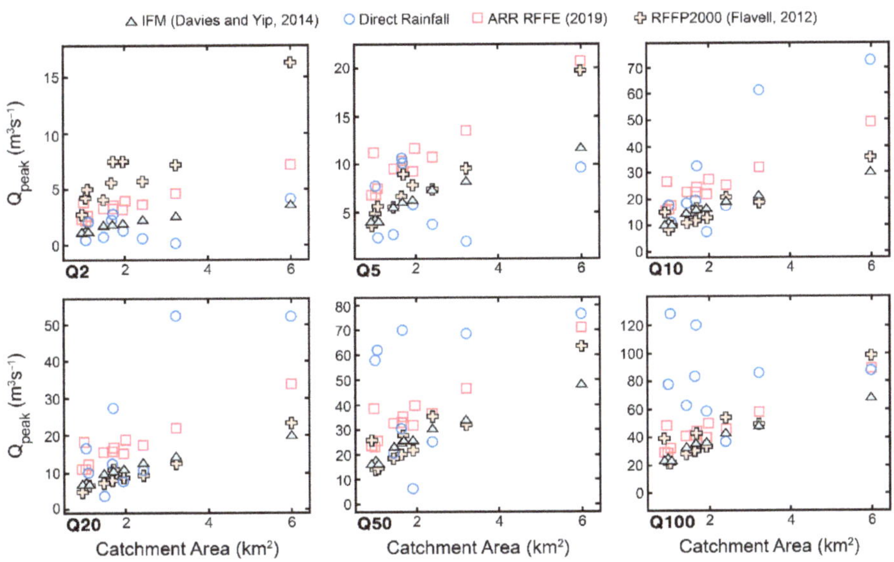

Figure 10. IFM [18], ARR RFFE and RFFP2000 [17] approaches compared with results from direct rainfall modelling from TUFLOW for the modelled headwater catchments.

4.3. *TUFLOW Sensitivity Analysis*

The results of the sensitivity analysis are found in Figure 11. The grid size of the model caused large variability in peak discharge values. Changes in peak velocity values were sensitive to this change in grid size, resulting in higher velocities (as one would expect) for a higher resolution DEM (e.g., 2 m) compared to the 10 m grid size. This increase in velocity is attributed to the more detailed topographic representation of channel constrictions and channel bed heterogeneity in the higher resolution grid size (Figure 12). C5 was the steepest and most topographically varied catchment, with steep confining areas. Within this catchment there was greater variability in the peak discharge from adjustments to Manning's n value, hyetograph shape, (and most prominently) grid size; where Q_{pp} values were doubled between a grid size increment increase of 5 m to 10 m. Final model scenarios used a grid size of 2 m, rainfall pattern 6, which resulted in the highest peak discharge in sensitivity analysis and a channel-wide Manning's roughness value of 0.035. The range of output values for the sensitivity analysis are found in Table 5.

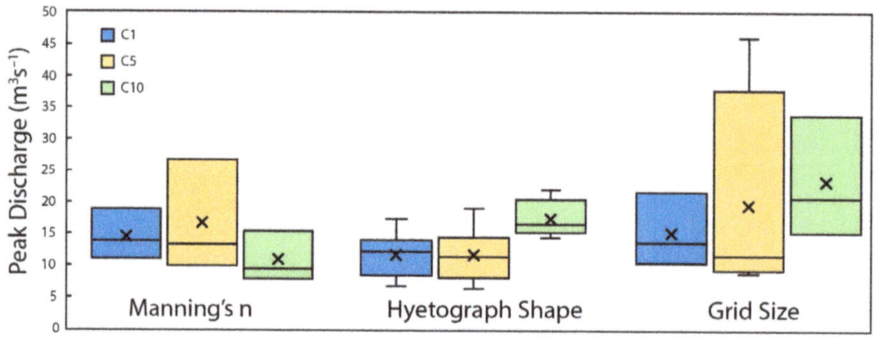

Figure 11. Sensitivity analysis box plot to show results for Catchments 1, 5 and 10 for a Q_{20} flood where Manning's n, hyetograph shape and grid size were adjusted.

Table 5. Results of sensitivity analysis catchments 1, 5 and 10 for the 20ARI.

Catchment	Grid Size	Hyetograph	Manning's n	Q_{peak} (m^3 s^{-1})
1	5	7	0.028	18.76
1	5	7	0.035	13.71
1	5	7	0.042	11.02
1	5	1	0.035	17.22
1	5	2	0.035	7.38
1	5	3	0.035	14.24
1	5	4	0.035	8.67
1	5	5	0.035	10.70
1	5	6	0.035	6.68
1	5	7	0.035	13.71
1	5	8	0.035	9.98
1	5	9	0.035	13.58
1	5	10	0.035	13.66
1	2	7	0.035	10.54
1	5	7	0.035	13.71
1	10	7	0.035	21.61
5	5	7	0.028	26.62
5	5	7	0.035	13.30
5	5	7	0.042	9.86
5	5	1	0.035	7.87
5	5	2	0.035	10.75
5	5	3	0.035	17.36
5	5	4	0.035	18.90
5	5	5	0.035	13.01
5	5	6	0.035	11.60
5	5	7	0.035	13.30
5	5	8	0.035	7.92
5	5	9	0.035	9.25
5	5	10	0.035	6.29
5	1	7	0.035	8.93
5	2	7	0.035	10.10
5	5	7	0.035	13.30
5	10	7	0.035	46.21
10	5	7	0.028	9.48
10	5	7	0.035	15.45
10	5	7	0.042	7.80
10	5	1	0.035	14.28
10	5	2	0.035	14.66
10	5	3	0.035	16.96

Table 5. *Cont.*

Catchment	Grid Size	Hyetograph	Manning's n	Q_{peak} ($m^3\ s^{-1}$)
10	5	4	0.035	16.00
10	5	5	0.035	17.55
10	5	6	0.035	20.34
10	5	7	0.035	15.45
10	5	8	0.035	20.22
10	5	9	0.035	15.27
10	5	10	0.035	21.91
10	2	7	0.035	20.82
10	5	7	0.035	15.45
10	10	7	0.035	33.98

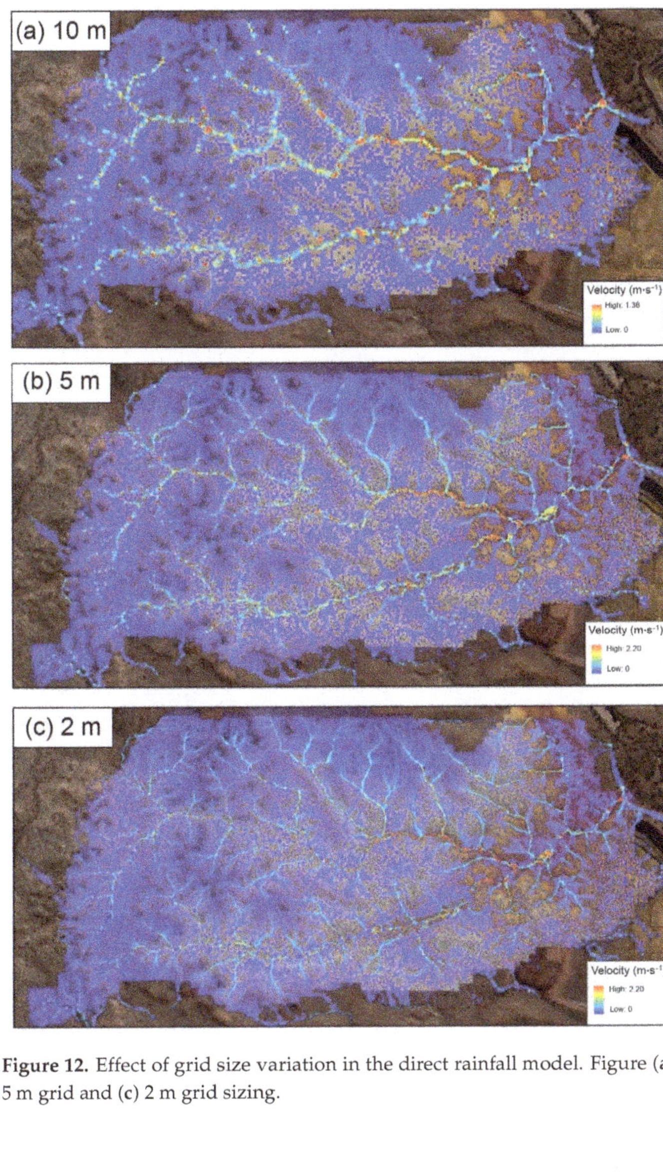

Figure 12. Effect of grid size variation in the direct rainfall model. Figure (**a**) shows 10 m grid, (**b**) 5 m grid and (**c**) 2 m grid sizing.

5. Discussion

5.1. Direct Rainfall Model Performance

5.1.1. Sensitivity Analysis

Sensitivity analysis of the direct rainfall model has indicated that the steeper, confined C5 is more sensitive to alterations in grid size, Manning's n and hyetograph shape. Further research and integration of IL/CL values would improve the model's representation of headwater channel processes. A grid size of 2 m was selected to avoid excessive computation time but to integrate detailed topographic forms such as smaller flowpaths and to capture the heterogeneity of the natural environment. Higher within a catchment, flowpaths become smaller and may be poorly represented by the model if they exist on a sub-grid scale, affecting the timing of runoff routing within the catchment [57] (ARR, 2012). This is likely a key contributor to the increased Q_{PP} values for coarser resolution model grid sizes (5–10 m). Mesh (grid-size) resolution has been shown to have a high impact on a models output flow volume in other studies conducting direct rainfall in small catchments [23]. Additionally, the slope between neighbouring cells has a key influence on the behaviour of catchment runoff, and therefore peak discharge values. Care should be taken to adequately understand the potential variability in peak discharge estimation using direct rainfall models. Additionally, explicit consideration of local variation in rainfall patterns in peak flood discharge modelling efforts is advised.

5.1.2. Applicability of Direct Rainfall Modelling in Headwater Catchments

Direct rainfall models are subject to higher levels of mass error when using the direct rainfall approach particularly where the model has areas of steep, complex flow or the model is located at a high elevation above sea level and experience relatively small inflows [56]. The catchments modelled here have many of these challenges. They are small catchments with small inflows, and experience sudden changes in slope conditions. Drainage pathways are not clearly defined and for some catchments, the hydrological boundaries between channel and floodplain are difficult to model. Flow is routed along the floodplain or within a claypan environment culminating in diffusive and complex drainage patterns. For these catchments, there were unexpected peak discharge values which may not fully represent what would occur in these catchments outside of where flow is rapidly dispersed away from channel setting (e.g., catchment 10).

The catchment outlet of several small headwater channels was prone to backwater effects. Such backwater effects have been reported for other channels in the Pilbara region from constrictions within the channel [34]. Backwaters can decrease velocities, raise the water surface elevation (WSE) and extend upstream [34]. Peak discharge values obtained from the catchment outlet were based on a cross section beyond the extent of backwater effects from outlet constrictions or tributary junctions. Care was taken to inspect the mapped output to interpret the water surface elevation at catchment cross sections.

Despite these warnings, the use of direct rainfall modelling is useful within small catchments with well-defined drainage pathways to quasi-calibrate selected RFFE approaches in the complete absence of rainfall and streamflow data. In doing so, an approximation of likely flood flows within these ungauged settings can be made. However, direct rainfall modelling with these catchment characteristics should not replace a hydrologic modellers perspicacity in dealing with flow estimation within dryland headwater environments. Even when calibration is properly done, models tend to have greater predictive strengths over shorter timeframes than longer timeframes as the system over short time scales is more similar to the one which it was calibrated for. Therefore, greater trust can be placed on the more frequent flow predictions than longterm flow dynamics within these headwater catchments.

5.2. RFFE Evaluation

The evaluation procedure was as follows: (1) Was the RFFE output within the 5–95% confidence limit range for the ARR approach. (2) Did the approach demonstrate appropriate hydrological scaling across space and flood return period (i.e., did the results increase with an increase in rainfall magnitude or catchment area) and lastly, (3) Did RFFE results display good agreement with findings from the rain-on-grid model and if not, did any approaches align with other tested RFFE methods. Many of the RFFE approaches for the Pilbara region are created using far larger catchment sizes. For example, the Flavell RFFP2000 approach is derived from an average catchment size of 5570 km^2 and this research tests the application of this RFFE in catchments with an average size of 2.16 km^2. Despite this magnitude variation in catchment size, these RFFE approaches are used in these smaller catchments for design discharge calculations. However, the direct modelling highlights that catchment representativeness remains an issue. This issue is exacerbated by sparse gauging coverage and continues to be a high priority for future research [57].

The RFFP2000 (Flavell, 2012), IFM (Davies and Yip, 2014) and ARR RFFE (2019) methods produce higher estimates of peak discharge. These three methods diverge from the other approaches in estimating higher peak discharge estimates and this deviation is more pronounced in less frequent flood flows (Q_{20}, Q_{50} and Q_{100}). These three methods are also more sensitive to changes in catchment area with larger catchments (above 5 km^2) predicted to have peak discharges double that of smaller catchments (around 2 km^2).

Both PRT methods estimate the lowest peak discharge values across all ARI, with little increase in discharge across less frequent flood flow events. These estimation techniques are less sensitive to catchment size and may overlook flood flow magnitudes within smaller catchments as they were created with a dataset comprised of larger catchments. Additionally, parameter regression techniques give emphasis to the mean and therefore frequent flows [66]. These methods provide a lower estimate of peak discharges, with emphasis on frequent flow events and are likely not suited to smaller headwater catchments such as those in this study.

The QRT methods show slightly higher predicted peak discharges but still do not provide convincing flood discharges at higher return intervals. PRT and QRT methods should be used with caution on smaller headwater channels. The IFM (Davies and Yip, 2014) method uses greater weighted frequency factors which results in the larger range of peak flood values across return intervals. This greater range of frequency factors for the IFM method compared to the PRT methods is likely to better represent higher magnitude flow events within these headwater catchments.

The ARR RFFE (2019) [25] approach is industry standard for larger catchments and is useful for small headwater catchments even if it is likely to provide a conservative estimate. The ARR RFFE model is noted to have large uncertainty with mean relative errors of 50–60% [57]. Alternative methods developed by Flavell (RFFP2000 and RFFP2006) and by Davies and Yip (2014) have been suggested as viable locally developed replacements to ARR2015 methods for the Pilbara region [57]. The Flavell RFFE approach is relatively complicated in comparison to the IFM and ARR (RFFE) 2019 approaches. The Flavell RFFP approach requires some catchment analysis to provide equation values (such as slope, latitude, longitude, length of catchment) but it is likely these values would be easy to populate for any catchment of interest. The Davies and Yip (2014) IFM approach uses a wide spatial distribution and catchment range in developing design equations making it applicable throughout the Indian Ocean drainage division [18]. The RFFP2000, IFM and ARR RFFE (2019) are therefore suggested for the estimate of peak discharge values within small ungauged headwater channels.

6. Conclusions

The role of this research was to identify peak discharge values from a direct rainfall model, and to compare these findings with RFFE approaches conducted within the region. Within the TUFLOW direct rainfall model simulations, the model was predominantly sensitive to grid size and Manning's n, with the rainfall pattern having a smaller impact on Q_{peak}. It is worth noting however that even the variation in the shape of the hyetograph produced up to 50% variation in the final Q_{peak} value within the sensitivity analysis. Therefore, awareness of the limitations of direct rainfall modelling within these steeper catchments is necessary when using this approach or selecting an RFFE approach. Despite this caution, the Flavell RFFP2000 and Davies and Yip, (2014) IFM appear the most reasonable estimates of peak discharge within the channels. These approaches provide lower estimates than the ARR2019 RFFE models, which are suggested to provide higher flow rates and are commonly used as part of a conservative approach to waterway design within the Pilbara. Whilst the direct rainfall modelling approach has many barriers when applied to headwater channels, this approach is useful to provide a frame of reference when selecting an RFFE approach for ungauged headwater catchments. It is common for a RFFE approach to be applied without first considering the weaknesses and limitations of the derived equation, or its suitability within small headwater catchments. Here, we show the potential for direct rainfall modelling and its applicability in an environment with limited opportunities to validate peak flood discharge estimation methods. The lack of gauged real-world measurements within these catchments continues to present the main obstacle in improving our understanding of flow conditions and rainfall values within the Pilbara region.

Supplementary Materials: The following supporting information can be downloaded at: https://www.mdpi.com/article/10.3390/hydrology9100165/s1, Table S1: Supplementary material highlighting key method details for the RFFE estimation methods applied to headwater catchments in the Pilbara, WA.

Author Contributions: Conceptualization, A.F. and I.R.; methodology, A.F. and I.R.; software, A.F., writing—original draft preparation, A.F.; writing—review and editing, A.F. and I.R. All authors have read and agreed to the published version of the manuscript.

Funding: This research was funded the University of Melbourne Research Scholarship with additional funding from BHP, Australia (Project 041639).

Data Availability Statement: Data available on request from the authors.

Acknowledgments: The authors would like to thank the 3RG research group for their discussions surrounding this work. We would also like to acknowledge the support of Iain Rea of BHP for this research.

Conflicts of Interest: The authors declare no conflict of interest.

References

1. Hailegeorgis, T.T.; Alfredsen, K. Regional flood frequency analysis and prediction in ungauged basins including estimation of major uncertainties for mid-Norway. *J. Hydrol. Reg. Stud.* **2017**, *9*, 104–126. [CrossRef]
2. Rahman, A.; Haddad, K.; Zaman, M.; Ishak, E.; Kuczera, G.; Weinmann, E. Australian Rainfall and Runoff Revision Project 5: Regional Flood Methods: Stage 2 Report. P5/S2/015. 2012. Available online: http://www.arr.org.au/wp-content/uploads/2013/Projects/ARR_Project5_Stage2_Report_Final_.pdf (accessed on 24 August 2022).
3. Feldman, A.D. *Flood Hydrograph and Peak Flow Frequency Analysis*; US Army Corps of Engineers: Davis, CA, USA, 1979.
4. Canterford, R.P.; Pescod, N.R.; Pearce, N.H.; Turner, L.H.; Atkinson, R.J. Frequency analysis of Australian rainfall data as used for flood analysis and design. In *Regional Flood Frequency Analysis: Proceedings of the International Symposium on Flood Frequency and Risk Analyses, Baton Rouge, LA, USA, 14–17 May 1987*; Springer: Dordrecht, The Netherlands, 1987; pp. 293–302.
5. Ladson, A.R.; Weinmann, E. Hydrology—An Australian Introduction. *Australas. J. Water Resour.* **2008**, *12*, 71–72. [CrossRef]
6. Pilgrim, D.H.; Chapman, T.G.; Doran, D.G. Problems of rainfall-runoff modelling in arid and semi-arid regions. *Hydrol. Sci. J.* **1988**, *33*, 379400. [CrossRef]
7. Chamber of Minerals and Energy of Western Australia (CME). Iron Ore. 2022. Available online: https://www.cmewa.com.au/about/wa-resources/iron-ore/ (accessed on 24 August 2022).
8. Refsgaard, J.C.; Knudsen, J. Operational Validation and Intercomparison of Different Types of Hydrological Models. *Water Resour. Res.* **1996**, *32*, 2189–2202. [CrossRef]
9. Flatley, A.; Markham, A. Establishing effective mine closure criteria for river diversion channels. *J. Environ. Manag.* **2021**, *287*, 112287. [CrossRef]
10. Huxley, C.; Syme, B. TUFLOW GPU-best practice advice for hydrologic and hydraulic model simulations. In Proceedings of the 37th Hydrology & Water Resources Symposium, Queenstown, New Zealand, 28 November–2 December 2016; pp. 195–203.
11. Ball, J.E.; Babister, M.K.; Nathan, R.; Weinmann, P.E.; Weeks, W.; Retallick, M.; Testoni, I. *Australian Rainfall and Runoff—A Guide to Flood Estimation*; Institution of Engineers: Barton, Australia, 2016.

12. Caddis, B.M.; Jempson, M.A.; Ball, J.E.; Syme, W.J. Incorporating hydrology into 2D hydraulic models–the direct rainfall approach. In Proceedings of the 9th National Conference on Hydraulics in Water Engineering, Darwin, Australia, 23–26 September 2008.
13. Hall, J. Direct rainfall flood modelling: The good, the bad and the ugly. *Aust. J. Water Resour.* **2015**, *19*, 74–85. [CrossRef]
14. Taylor, H.; Kerr, T. Designing for Mining: Challenges of Hydrological Design in the Pilbara. *Hydrol. Water Resour. Symp.* **2014**, *2014*, 4452.
15. Taylor, M.; Haddad, K.; Zaman, M.; Rahman, A. Regional flood modelling in Western Australia: Application of regression-based methods using ordinary least squares. In Proceedings of the 19th International Congress on Modelling and Simulation, Perth, Australia, 12–16 December 2011; pp. 3803–3810.
16. Rahman, A.; Zaman, M.; Haddad, K.; Kuczera, G.; Weinmann, E.; Weeks, W.; Rajaratnam, L.; Kemp, D. Development of a new regional flood frequency analysis method for semi-arid and arid regions of Australia. In Proceedings of the 34th Hydrology and Water Resources Symposium Sydney, Sydney, Australia, 19–22 November 2012; pp. 1433–1440.
17. Flavell, D. Design flood estimation in Western Australia. *Australas. J. Water Resour.* **2012**, *16*, 1–20. [CrossRef]
18. Davies, J.R.; Yip, E. Pilbara Regional Flood Frequency Analysis. In Proceedings of the Hydrology and Water Resources Symposium, Perth, Australia, 24–27 February 2014; pp. 182–189.
19. Oreskes, N.; Shrader-Frechette, K.; Belitz, K. Verification, validation, and confirmation of numerical models in the earth sciences. *Science* **1994**, *263*, 641–646. [CrossRef] [PubMed]
20. Annis, A.; Nardi, F.; Volpi, E.; Fiori, A. Quantifying the relative impact of hydrological and hydraulic modelling parameterizations on uncertainty of inundation maps. *Hydrol. Sci. J.* **2020**, *65*, 507–523. [CrossRef]
21. Ahmadisharaf, E.; Kalyanapu, A.J.; Bates, P.D. A probabilistic framework for floodplain mapping using hydrological modeling and unsteady hydraulic modeling. *Hydrol. Sci. J.* **2018**, *63*, 1759–1775. [CrossRef]
22. Grillakis, M.G.; Koutroulis, A.G.; Komma, J.; Tsanis, I.K.; Wagner, W.; Blöschl, G. Initial soil moisture effects on flash flood generation—A comparison between basins of contrasting hydro-climatic conditions. *J. Hydrol.* **2016**, *541*, 206–217. [CrossRef]
23. David, A.; Schmalz, B. Flood hazard analysis in small catchments: Comparison of hydrological and hydrodynamic approaches by the use of direct rainfall. *J. Flood Risk Manag.* **2020**, *13*, e12639. [CrossRef]
24. Viglione, A.; Blösch, G. On the role of storm duration in the mapping of rainfall to flood return periods. *Hydrol. Earth Syst. Sci.* **2009**, *13*, 205–216. [CrossRef]
25. Ball, J.; Babister, M.; Nathan, R.; Weeks, W.; Weinmann, E.; Retallick, M.; Testoni, I. (Eds.) *Australian Rainfall and Runoff: A Guide to Flood Estimation*; Geoscience Australia: Canberra, Australia, 2019.
26. Ruprecht, J. Arid Zone Hydrology: Pilbara region of Western Australia. In Proceedings of the 23rd Hydrology and Water Resources Symposium, Hobart, Australia, 21–24 May 1996; pp. 301–305.
27. Broit, A.; Boytar, G. Development of a methodology for catchments exhibiting sheet flow characteristics in the Pilbara region. *Hydrol. Water Resour. Symp.* **2014**, *2014*, 953–961.
28. BHP Billiton Iron Ore Pty Ltd. BHP Billiton Iron Ore Pilbara Expansion: Strategic Proposal, 650 Environmental Scoping Document. November 2013. Available online: https://www.bhp.com/-/media/bhp/regulatory-information-media/iron-ore/western-australia-iron-ore/0000/referral-and-environmental-scoping-document/160316_ironore_waio_pilbarastrategicassessment_state_environmentalscopingdocument.pdf (accessed on 24 August 2022).
29. Peel, M.C.; Finlayson, B.L.; McMahon, T.A. Updated world map of the Köppen Geiger climate classification. *Hydrol. Earth Syst. Sci.* **2007**, *11*, 1633–1644. [CrossRef]
30. Sudmeyer, R. Climate in the Pilbara. Bulletin 4873, Department of Agriculture and Food, Western Australia, Perth. Available online: https://library.dpird.wa.gov.au/bulletins/220/ (accessed on 24 August 2022).
31. Charles, S.P.; Fu, G.; Silberstein, R.P.; Mpelasoka, F.; McFarlane, D.; Hodgson, G.; Teng, J.; Gabrovsek, C.; Ali, R.; Barron, O.; et al. *Hydroclimate of the Pilbara: Past, Present, and Future. A Report to the Government of Western Australia and Industry Partners from the CSIRO Pilbara Water Resource Assessment*; CSIRO Land and Water: Clayton, Australia, 2015; pp. 1–140.
32. Ruprecht, J.; Ivansecu, S. *Surface Hydrology of the Pilbara Region: Summary Report. Surface Water Hydrology Report*; Series no. 32; Water and Rivers Commission: Perth, Australia, 2000.
33. Wark, B.; Thomas, L. Does your rating curve hold water: The consequence of rating 860 curve errors. In Proceedings of the ANCOLD, Canberra, Australia, 21–22 October 2014; pp. 1–11.
34. Harvey, M.; Pearcey, M.; Price, K.; Devkota, B. Geomorphic, hydraulic and sediment transport modelling for mine related channel realignment—Case Study: Caves Creek, Pilbara, Western Australia. In *Hydrology and Water Resources Symposium*; Engineers Australia: Queensland, Australia, 2014; pp. 259–266.
35. MWH. *Ecohydrological Conceptualisation of the Fortescue Marsh Region*; Project No. 83501069; MWH: Melbourne, Australia, 2015; pp. 1–178.
36. Bureau of Meterology (Australia). Tropical Cyclone Climatology Maps. Product Code: IDCJCM0011. 2019. Available online: http://www.bom.gov.au/climate/maps/averages/tropical-cyclones/ (accessed on 24 August 2022).
37. Haque, M.M.; Rahman, A.; Haddad, K.; Kuczera, G.; Weeks, W. Development of a regional flood frequency estimation model for Pilbara, Australia. In Proceedings of the 21st International Congress on Modelling and Simulation, Gold Coast, Australia, 29 November–4 December 2015; pp. 2172–2178.

38. CSIRO; McFarlane, D.M. *Pilbara Water Resource Assessment: Upper Fortescue Region;* A Report to the Government of Western Australia and Industry Partners from the CSIRO Pilbara Water Resources Assessment; CSIRO Land and Water Flagship: Clayton, Australia, 2015.
39. Doherty, R. Calibration of HEC-RAS models for rating curve development in semi-arid regions of Western Australia. In Proceedings of the AHA 2010 Conference, Perth, Australia, 5–9 July 2010; pp. 1–25.
40. Kemp, D.; Hewa, G. *An Investigation into the Efficacy of Australian Rainfall and Runoff 2016 Procedures in the Mount Lofty Ranges, South Australia;* Australian Flow Management Group, University of South Australia: Adelaide, Australia, 2019.
41. Coombes, P.J.; Colegate, M.; Buchanan, S. Use of direct rain as an investigation process and design of basins using ARR2016 methods. In Proceedings of the Stormwater 2018, Sydney, Australia, 8–12 October 2018.
42. Schumann, G.; Bates, P.D.; Apel, H.; Aronica, G.T. *Global flood hazard mapping, modeling, and forecasting: Challenges and perspectives In Global Flood Hazard: Applications in Modeling, Mapping, and Forecasting;* John Wiley & Sons, Inc.: Hoboken, NJ, USA, 2018; pp. 239–244.
43. Merz, R.; Bloschl, G.; Parajka, J. Regionalisation methods in rainfall-runoff modelling using large catchment samples. In *Large Sample Basin Experiments for Hydrological Model Parameterization: Results of the Model Parameter Experiment—MOPEX;* IAHS Publication: Wallingford, UK, 2006; Volume 307, pp. 117–125.
44. Razavi, T.; Coulibaly, P. Improving streamflow estimation in ungauged basins using a multi-modelling approach. *Hydrol. Sci. J.* **2016**, *61*, 2668–2679. [CrossRef]
45. Wagener, T.; Wheater, H.S. Parameter estimation and regionalization for continuous rainfall-runoff models including uncertainty. *J. Hydrol.* **2006**, *320*, 132–154. [CrossRef]
46. Blöschl, G.; Sivapalan, M. Scale issues in hydrological modelling: A review. *Hydrol. Process.* **1995**, *9*, 251–290. [CrossRef]
47. Pandey, G.R.; Nguyen, V.T.T. A comparative study of regression-based methods in regional flood frequency analysis. *J. Hydrol.* **1999**, *225*, 92–101. [CrossRef]
48. Rijal, N.; Rahman, A. Design flood estimation in ungauged catchments: Quantile regression technique and Probabilistic Rational Method compared. In Proceedings of the Modsim05: International Congress on Modelling and Simulation: Advances and Applications for Management and Decision Making, Melbourne, Australia, 12–15 December 2005.
49. Mishra, B.K.; Takara, K.; Yamashiki, Y.; Tachikawa, Y. An assessment of predictive accuracy for two regional flood frequency estimation methods. *Annu. J. Hydraul. Eng. JSCE* **2010**, *54*, 7–12.
50. Austroads. Guide to Bridge Technology Part 8: Hydraulic Design of Waterway Structures. 2018, pp. 1–157. Available online: https://austroads.com.au/publications/bridges/agbt08-18/media/AGBT08-18_Guide_to_Bridge_Technology_Part_8_Hydraulic_Design_of_Waterway_Structures.pdf (accessed on 24 August 2022).
51. Rahman, A.; Haddad, K.; Kuczera, G. Features of regional flood frequency estimation (RFFE) model in Australian Rainfall and Runoff. In Proceedings of the 21st International Congress on Modelling and 792 Simulation, Gold Coast, Australia, 29 November–4 December 2015; pp. 2207–2213.
52. Rahman, A.; Haddad, K.; Kuczera, G.; Weinmann, E. Chapter 3. Regional Flood Methods. In *ARR Australian Rainfall and Runoff;* Geoscience Australia: Canberra, Australia, 2019.
53. Rahman, A.; Haddad, K.; Haque, M.M.; Kuczera, G.; Weinmann, E. *Australian Rainfall and Runoff, Project 5: Regional Flood Methods: Stage 3 Report;* Geoscience Australia: Canberra, Australia, 2015; ISBN 978-0-85825-796 869-3.
54. Farquharson, F.A.K.; Meigh, J.R.; Sutcliffe, J.V. Regional Flood Frequency Analysis in arid and semi-arid areas. *J. Hydrol.* **1992**, *138*, 487–501. [CrossRef]
55. BMT. TUFLOW Classic/HPC User Manual. Build 2018-03-AD. 2018, pp. 1–443. Available online: https://www.tuflow.com/Download/TUFLOW/Releases/2018-03/TUFLOW%20Manual.2018-65703.pdf (accessed on 20 June 2021).
56. Syme. TUFLOW HPC. TUFLOW UK Conference, Bristol. 2018, pp. 1–63. Available online: https://tuflow.com/media/4964/201804-tuflow-hpc-overview-tuflow-conference-uk.pdf (accessed on 24 August 2022).
57. Engineers Australia. *Australian Rainfall and Runoff Revision Project 15: Two Dimensional Modelling in Urban and Rural Floodplains. Stage 1 and 2 Draft Report;* Geoscience Australia: Canberra, Australia, 2012. Available online: https://arr.ga.gov.au/__data/assets/pdf_file/0019/40573/ARR_Project15_TwoDimensional_Modelling_DraftReport.pdf (accessed on 24 August 2022).
58. Gomi, T.; Sidle, R.C.; Richardson, J.S. Undersftanding processes and downstream linkages of headwater systems: Headwaters differ from downstream reaches by their close coupling to hillslope processes, more temporal and spatial variation, and their need for different means of protection from land use. *BioScience* **2002**, *52*, 905–916.
59. Hirschmuller, H. Stereo processing by semiglobal matching and mutual information. *IEEE Trans. Pattern Anal. Mach. Intell.* **2007**, *30*, 328–341. [CrossRef]
60. Bureau of Meteorology (Australia). Design Rainfall System 2016. 2016. Available online: http://www.bom.gov.au/water/designRainfalls/revised-ifd/ (accessed on 10 July 2022).
61. Podger, S.; Babister, M.; Ward, M. Determination of pre-burst rainfalls for Australian rainfall and runoff. In Proceedings of the Hydrology and Water Resources Symposium, Melbourne, Australia, 3–6 December 2018; pp. 649–660.
62. RioTinto. *Resource Development: Marillana Creek Regional Flow Balance and Catchment Hydrology;* Appendix A9; Rio Tinto: London, UK, 2010; pp. 1–46.
63. Foglia, L.; Hill, M.C.; Mehl, S.W.; Burlando, P. Sensitivity analysis, calibration, and testing of a distributed hydrological model using error-based weighting and one objective function. *Water Resour. Res.* **2009**, *45*. [CrossRef]

64. Vaze, J.; Jordan, P.; Beecham, R.; Frost, A.; Summerell, G. *Guidelines for Rainfall-Runoff Modelling: Towards Best Practice Model Application*; eWater Cooperative Research Centre: Canberra, Australia, 2011; ISBN 978-1-921543-51-7.
65. Department of Primary Industries and Regional Development. Typical Values for Manning's Coefficient (n) for Bare Soil Waterways. 2018. Available online: https://www.agric.wa.gov.au/water-management/mannings-roughness-coefficient (accessed on 24 August 2022).
66. Rogers, A.D.; Davies, J.R. ARR 2015 Unpacked—Implications for stormwater design in WA. In Proceedings of the IPWEA State Conference, Freemantle, Australia, 9–11 March 2016; pp. 1–12.

hydrology MDPI

Article

Multi-Variable SWAT Model Calibration Using Satellite-Based Evapotranspiration Data and Streamflow

Evgenia Koltsida and Andreas Kallioras *

Laboratory of Engineering Geology and Hydrogeology, School of Mining and Metallurgical Engineering, National Technical University of Athens, Heroon Polytechniou Str. 9, 15780 Athens, Greece; ekoltsida@metal.ntua.gr

* Correspondence: kallioras@metal.ntua.gr; Tel.: +30-210-772-2098

Abstract: In this study, monthly streamflow and satellite-based actual evapotranspiration data (AET) were used to evaluate the Soil and Water Assessment Tool (SWAT) model for the calibration of an experimental sub-basin with mixed land-use characteristics in Athens, Greece. Three calibration scenarios were performed using streamflow (i.e., single variable), AET (i.e., single variable), and streamflow–AET data together (i.e., multi-variable) to provide insights into how different calibration scenarios affect the hydrological processes of a catchment with complex land use characteristics. The actual evapotranspiration data were obtained from the Moderate Resolution Imaging Spectroradiometer (MODIS). The calibration was achieved with the use of the SUFI-2 algorithm in the SWAT-CUP program. The results suggested that the single variable calibrations showed moderately better performance than the multi-variable calibration. However, the multi-variable calibration scenario displayed acceptable outcomes for both streamflow and actual evapotranspiration and indicated reasonably good streamflow estimations (NSE = 0.70; R^2 = 0.86; PBIAS = 6.1%). The model under-predicted AET in all calibration scenarios during the dry season compared to MODIS satellite-based AET. Overall, this study demonstrated that satellite-based AET data, together with streamflow data, can enhance model performance and be a good choice for watersheds lacking sufficient spatial data and observations.

Keywords: SWAT; streamflow; MODIS; evapotranspiration; hydrological modeling; multi-variable calibration

Citation: Koltsida, E.; Kallioras, A. Multi-Variable SWAT Model Calibration Using Satellite-Based Evapotranspiration Data and Streamflow. *Hydrology* **2022**, *9*, 112. https://doi.org/10.3390/hydrology9070112

Academic Editors: Tommaso Caloiero, Carmelina Costanzo and Roberta Padulano

Received: 26 April 2022
Accepted: 17 June 2022
Published: 21 June 2022

1. Introduction

Hydrological models have been extensively utilized to estimate the consequences of climate variability, land management practices, and policy directions at various temporal and spatial scales [1]. Model development requires a good comprehension of the watershed characteristics to achieve accurate model simulation [2,3]. Nonetheless, most basins are ungauged or inadequately gauged [4]. The absence of adequate observations affects the calibration process and further model improvement [5].

Hydrological model calibration is typically achieved with flow data at the outlet of the basin by choosing the most suitable values for input parameters and comparing simulated outcomes with observed data [6]. However, calibration focused on one variable only may aggregate all watershed processes together and intensify the occurrence of the equifinality problem (i.e., multiple parameter sets can reproduce a similar output) [7–9]. Using multiple variables (e.g., streamflow, evapotranspiration, soil moisture) in the calibration process attempts to overcome equifinality across multiple parameter sets [10–12].

In addition, unknown procedures to the modeler, such as unidentified discharges, agricultural activities, and dumping of construction materials, interfere with the natural behavior of the system and increase the uncertainty in streamflow calibration [13,14]. Therefore, incorporating remote sensing data in model calibration can increase model accuracy, capture the spatial and temporal heterogeneity of hydrological processes, and be a promising alternative for catchments lacking sufficient observations [13,15,16]. Satellite-based actual evapotranspiration (AET) data can be used to constrain hydrological parameters associated with the water balance [17,18]. For instance, a study in southern India [13] used MODIS satellite-based AET data to calibrate the SWAT model and suggested that satellite-based AET data with a monthly temporal resolution can be used to reduce equifinality obtained from traditional streamflow calibration. In addition, a study in Myanmar [16] calibrated the SWAT model using (Global Land Evaporation Amsterdam Model (GLEAM) satellite-based AET data and measured streamflow. This study suggested that constraining the model with actual evapotranspiration and streamflow data (i.e., multi-variable calibration) can produce good results for both variables. Finally, a study in a catchment in Michigan [18] calibrated the SWAT model using observed streamflow data and remotely sensed based AET data from the Simplified Surface Energy Balance (SSEBop) model and the Atmosphere-Land Exchange Inverse (ALEXI) model. The results of this study suggested that incorporating satellite-based AET data in the hydrological model calibration can maintain a satisfactory performance for streamflow while improving evapotranspiration estimations.

The complexity of spatially distributed model applications in mixed land-use watersheds (i.e., blended combinations of land use) has been explored in past studies [19–23]. Nevertheless, the use of remote sensing data for hydrological model calibration in mixed land-use watersheds has not yet been thoroughly analysed. Urban and peri-urban environments are characterized by high variability in land use, soil types, management practices, and diverse hydrological processes, which increase issues of model uncertainties and make the calibration process challenging [20,24]. The SWAT model is a physically-based model that incorporates the spatial distribution of land use, topography, and soil and allows different hydrological processes in a watershed to interconnect [6]. This makes the model able to estimate how the hydrological components are affected by land management methods in catchments with complex land uses and heterogeneity in soil formations. This study used streamflow and satellite-based actual evapotranspiration (AET) data to calibrate the SWAT model of an urban/peri-urban catchment characterized by a typical Mediterranean climate. Three calibration scenarios were developed using (i) streamflow data, (ii) AET data, and (iii) both streamflow and AET data. This study aims to (i) investigate which parameters are more sensitive in the single variable and multi-variable scenarios; (ii) assess the model performance of the different scenarios; (iii) examine the outcomes of major hydrological components between the single-variable and multi-variable scenarios, and (iv) evaluate the suitability of remote sensing data for streamflow simulation. This study is the first attempt to simulate the hydrological components of an experimental sub-basin with complex land use characteristics and will provide insights into how different calibration scenarios affect the hydrological processes for sustainable water resource management. The major innovation of the proposed methodology is that it has been developed for a typical Mediterranean peri-urban area and can be easily applied to catchments with similar hydrological and geomorphological characteristics.

2. Materials and Methods

2.1. Description of the Study Area

The study site is a sub-basin (140 km^2) of the Kifissos River basin (380 km^2), Athens, Greece (Figure 1). The Kifissos River's route is almost 22 km, of which at least 14 km are within an urban area [25]. The elevation varies from 1399 m in the northern part to 94 m in the southern part. The study area has as mild Mediterranean climate [26]. The mean annual temperature is 16.4 °C, and the mean annual rainfall is 643 mm [27]. The mean annual actual evapotranspiration is 483 mm. The annual evapotranspiration ranges from 551 mm (upstream) to 395 mm (downstream) (Figure 2a).

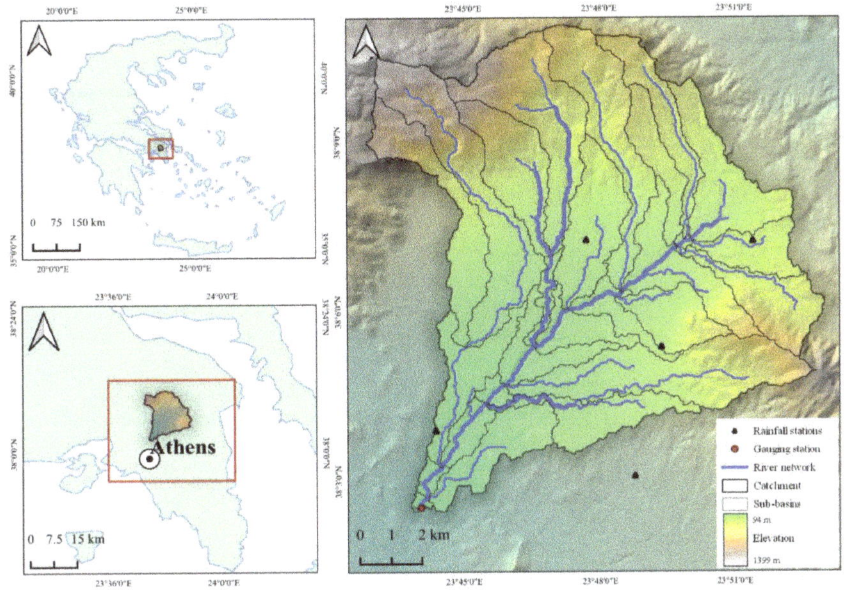

Figure 1. Location of Kifissos River sub-basin and monitoring stations (right), Athens metropolitan area (lower left), and zoom out display of Greece (upper left).

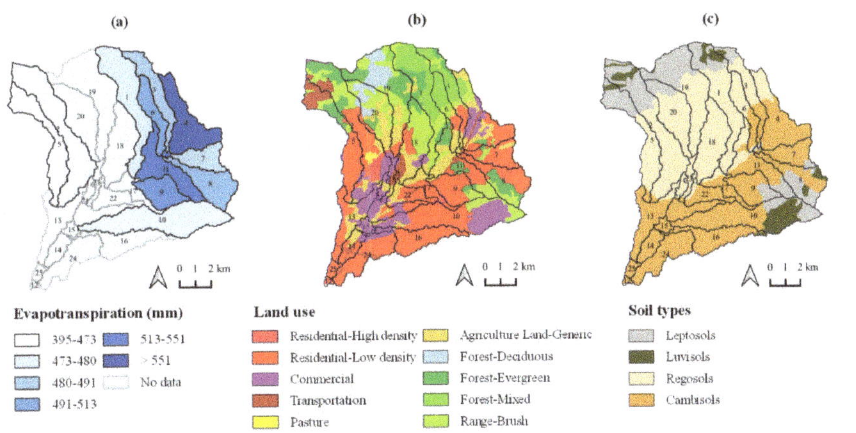

Figure 2. Kifissos River sub-basin (**a**) MODIS average annual evapotranspiration, (**b**) land use, and (**c**) soil types. The study area includes 25 sub-basins of which the sub-basin numbers (1–11) indicate the sub-basins used for actual evapotranspiration calibration.

The sub-basin is an urban/peri-urban area. The dominant land cover types are residential areas (34.1%), shrubland (15.9%), and agriculture (12.4%) (Figure 2b) [28]. Table A1 displays the land use categories of the study area at catchment level and Table A2 displays the land use categories of the study area at sub-basin level. The major soils are Cambisols, Regosols, Leptosols, and Luvisols [29]. These formations are generally high in clay and sand contents with good soil permeability (Figure 2c).

2.2. Data Sources

The input data include a digital elevation model (DEM) at 30 m resolution from the website of the US Geological Survey [30], a land use map from the 100 m 2018 Corine

Land Cover map [28], a soil map from the Food and Agriculture Organization Digital Soil Map of the World (30 arcseconds resolution) [31], and meteorological data from the National Observatory of Athens [27]. Daily rainfall data were obtained from 2015 to 2019. The daily measured streamflow data at the basin outlet (Monastiri gauging station) were available from 2018 to 2019 and were retrieved from Open Hydrosystem Information Network [32]. The actual evapotranspiration (AET) data were collected from the Moderate Resolution Imaging Spectroradiometer Global Evaporation [33] with a pixel resolution of 500 × 500 m [34].

2.3. The SWAT Hydrological Model

The SWAT (Soil and Water Assessment Tool) program is an open-source, physically based, continuous-time river basin model developed to estimate the influence of management practices on discharge, sediments, and agriculture in large complex basins [6,35]. The model runs on a daily time step, and its main variables are hydrology, weather, soil, land use, sediments, nutrients, bacteria, and pathogens.

In SWAT, the basin is divided into sub-basins, then into hydrologic response units (HRUs) with unique land use, soil, and slope characteristics [36]. The water balance is computed separately for each hydrologic response unit [37]. The water balance equation is estimated using the following (Equation (1)):

$$SW_t = SW_o + \sum_{i=1}^{t}\left(R_{day} - Q_{surf} - E_a - W_{seep} - Q_{gw}\right), \tag{1}$$

where SW_t is the soil water content (mm), SW_o is the soil water content on day i in the previous period (mm), t is the time step (days), R_{day} indicates the amount of precipitation on day i (mm), Q_{surf} represents the surface streamflow on day i (mm), E_a indicates the AET on day i (mm), W_{seep} is the percolation and bypass flow on day i (mm), and Q_{gw} represents the return flow on day i (mm).

2.4. Model Setup

The QGIS interface of the SWAT model was utilized for model configuration [38]. The watershed was delineated into 25 sub-basins and 386 hydrological response units (HRUs). A 10% threshold was used for land use, soil, and slope to limit the influence of minor soil and land use types for each sub-basin. The Corine Land Cover land use classes [28] were converted to the SWAT land use classes [6]. The model was simulated from 2015 to 2019 and run on a daily time step. Two years (1 January 2015–31 December 2016) were set as a warm-up period. The potential evapotranspiration was calculated using the Penman–Monteith method, the surface runoff was estimated using the curve number method [39], and the channel routing was computed using the variable storage coefficient method [40].

2.5. Model Calibration and Sensitivity Analysis

The model was calibrated using the Sequential Uncertainty Fitting Algorithm (SUFI-2) in the SWAT-Calibration and Uncertainty Program (SWAT-CUP) [41]. In SUFI-2, the calibrating parameters are set according to literature and sensitivity analysis, and then the parameters sets are generated using Latin hypercube sampling (LHS) [14]. The significance of each parameter is defined with a *t*-test. Parameters with large t-stat and small p-value (p-value < 0.03) are identified as sensitive parameters [42].

Three calibration scenarios were conducted using monthly time series of streamflow from one station at the outlet of the study site and the MODIS satellite-based actual evapotranspiration (AET) from eleven sub-basins based on data availability [33]. The daily streamflow was converted to monthly for comparison reasons. The monthly average values were calculated by a resample function, and the missing values were estimated by an interpolation function using the linear method (Python pandas library). The actual evapotranspiration data from MODIS were in a geotiff format (raster). In addition, to compare the MODIS satellite-based AET (pixel values) to the SWAT simulated AET, an area-weighted averaging approach in QGIS (zonal statistics) was performed to create the aggregated monthly values for each sub-basin [23,43].

The scenarios include: (i) streamflow calibration (i.e., single variable), (ii) AET calibration (i.e., single variable), and (iii) both streamflow and AET calibration (i.e., multi-variable). Based on data availability, streamflow was calibrated from 2018 to 2019, and evapotranspiration was calibrated from 2017 to 2019. All the available data were used for model calibration to represent the wet and dry conditions properly (2017: 487 mm, 2018: 675 mm, 2019: 765 mm). The calibration process used 20 parameters linked to streamflow and evapotranspiration (Table 1), and their sensitivities were estimated. The original value ranges of the parameters and their sensitivities for each calibration scenario are displayed in Table 2. In the single variable calibrations, the two variables (i.e., streamflow and actual evapotranspiration) were calibrated separately, and the performance of the second variable was evaluated. In the multi-variable calibration the two variables were calibrated together using a multi-variable objective function and assigning equal weights to each variable [16,44]. The Nash–Sutcliffe model efficiency (NS) was used as an objective function, and 900 simulations per iteration were performed and up to three iterations.

Table 1. Calibrated parameters. The method "r" (relative) indicates multiplying the current parameter value by a given value, the method "v" (replace) indicates replacing the current parameter value, and the method "a" (absolute) indicates adding a given value to the current parameter [14].

Category	Parameter	Description
Surface runoff	r_CN2.mgt	Curve number
	v_SURLAG.bsn	Surface runoff lag coefficient
Groundwater/Baseflow	v_ALPHA_BF.gw	Baseflow alpha factor
	a_GW_DELAY.gw	Groundwater delay
	v_RCHRG_DP.gw	Deep aquifer percolation fraction
	v_REVAPMN.gw	Threshold depth of water in the shallow aquifer for "revap" to occur
	v_GW_REVAP.gw	Groundwater "revap" coefficient
	v_GWQMN.gw	Threshold depth of water in the shallow aquifer required for return flow to occur
Lateral flow	r_LAT_TTIME.hru	Lateral flow travel time
	r_HRU_SLP.hru	Average slope steepness
Channel	r_OV_N.hru	Manning's coefficient for overland flow
	r_SLSUBBSN.hru	Average slope length
	v_CH_N2.rte	Manning's coefficient for the main channel
	v_CH_K2.rte	Hydraulic conductivity of the main channel alluvium
Soil	v_ESCO.bsn	Soil evaporation compensation coefficient
	v_EPCO.hru	Plant uptake compensation coefficient
	v_CANMX.hru	Maximum canopy storage
	r_SOL_K.sol	Saturated hydraulic conductivity
	r_SOL_BD.sol	Moist bulk density of the soil layer
	r_SOL_AWC.sol	Soil available water storage capacity

The model performance of each scenario was further analyzed using the coefficient of determination $\left(R^2\right)$ [45], Nash–Sutcliffe efficiency (NSE) [46], and percent bias (PBIAS) [47], as shown in Equations (2)–(4).

$$R^2 = \frac{\left[\sum_{i=1}^{n}\left(Q_{obs}(i)-\overline{Q}_{obs}\right)\left(Q_{sim}(i)-\overline{Q}_{sim}\right)\right]^2}{\sum_{i=1}^{n}\left(Q_{obs}(i)-\overline{Q}_{obs}\right)^2\sum_{i=1}^{n}\left(Q_{sim}(i)-\overline{Q}_{sim}\right)^2}, \quad (2)$$

$$NSE = 1-\left[\frac{\sum_{i=1}^{n}\left(Q_{obs}(i)-Q_{sim}(i)\right)^2}{\sum_{i=1}^{n}\left(Q_{obs}(i)-\overline{Q}_{obs}\right)^2}\right], \quad (3)$$

$$\text{PBIAS} = \left[\frac{\sum_{i=1}^{n} (Q_{obs}(i) - Q_{sim}(i)) * 100}{\sum_{i=1}^{n} Q_{obs}(i)} \right], \tag{4}$$

where Q_{obs} is the measured streamflow, Q_{sim} is the simulated streamflow on the day i, $\overline{Q}_{obs}$ is the mean of measured streamflow, and $\overline{Q}_{sim}$ is the mean of simulated streamflow. R^2 varies from 0 to 1, where 0 indicates no correlation and 1 means perfect correlation and less error variance. NSE can vary from $-\infty$ to 1, where values ≤ 0 show that the model is unreliable and values closer to 1 indicate a perfect fit between simulated and measured data. The best PBIAS value is 0. Positive values show that the model results are underestimated, and negative values show that the model results are overestimated. Model performance can be assessed as "satisfactory" for a monthly time step if $R^2 > 0.60$, NSE > 0.50, and PBIAS $\leq \pm 15\%$ for watershed-scale models [48].

Table 2. SWAT calibrated parameters and their sensitivities for each calibration scenario. Numbers in bold indicate the parameters with the highest sensitivity (p-value < 0.03).

Parameters	Initial Ranges		Flow Calibration		AET Calibration		Flow and AET Calibration	
	Min	Max	*t*-Test	*p*-Value	*t*-Test	*p*-Value	*t*-Test	*p*-Value
CN2	−0.10	0.10	−0.85	0.40	1.25	0.21	−0.53	0.59
SURLAG	0.00	10.00	0.38	0.70	0.64	0.53	1.23	0.22
ALPHA_BF	0.00	1.00	−0.07	0.95	0.47	0.64	−0.46	0.65
GW_DELAY	−30.00	90.00	9.89	0.00	−0.82	0.41	9.70	0.00
RCHRG_DP	0.00	0.50	2.78	0.01	−1.24	0.21	2.61	0.01
REVAPMN	800.00	1900.00	0.53	0.60	0.05	0.96	−0.35	0.73
GW_REVAP	0.02	0.20	1.17	0.24	−1.47	0.14	−1.30	0.19
GWQMN	0.00	500.00	0.18	0.86	0.69	0.49	−0.34	0.73
LAT_TTIME	0.00	180.00	18.98	0.00	−0.02	0.99	22.12	0.00
HRU_SLP	−0.50	3.00	6.09	0.00	−7.58	0.00	−8.84	0.00
OV_N	−0.50	3.00	−0.59	0.56	0.37	0.71	0.66	0.51
SLSUBBSN	−0.20	0.20	−2.23	0.03	2.15	0.03	0.17	0.86
CH_N2	0.01	0.30	0.35	0.72	1.56	0.12	1.21	0.23
CH_K2	0.00	127.00	−1.52	0.13	−0.47	0.64	1.60	0.11
ESCO	0.50	0.95	1.70	0.09	29.33	0.00	4.98	0.00
EPCO	0.50	0.95	−0.74	0.46	−5.50	0.00	−0.64	0.52
SOL_K	−0.80	0.80	8.43	0.00	−8.24	0.00	−7.85	0.00
SOL_BD	−0.30	0.30	10.46	0.00	−7.70	0.00	−8.45	0.00
SOL_AWC	−0.05	0.05	−0.34	0.73	0.35	0.72	2.52	0.01

3. Results

3.1. Sensitivity Analysis

The most sensitive parameters for each calibration scenario are shown in Table 2. Sensitive parameters are identified by p-value less than 0.03. Figure A1 shows the relative changes of the five most sensitive parameters for each calibration scenario versus objective function. Most variations of NSE values were found to be in the calibration with streamflow rather than single evapotranspiration and multi-variable calibrations.

In the streamflow calibration, the parameters with the highest sensitivity were lateral flow travel time (LAT_TTIME), moist bulk density of the soil layer (SOL_BD), groundwater delay time (GW_DELAY), saturated hydraulic conductivity (SOL_K), and average slope steepness (HRU_SLP). These parameters were connected to groundwater flow, runoff generation, and channel routing. In the AET calibration, the parameters with the highest sensitivity were soil evaporation compensation coefficient (ESCO), saturated hydraulic conductivity (SOL_K), moist bulk density of the soil layer (SOL_BD), average slope steepness (HRU_SLP), and plant uptake compensation coefficient (EPCO), which are mostly related to soil properties. In the multi-variable calibration, the most sensitive parameters were lateral flow travel time (LAT_TTIME), groundwater delay time (GW_DELAY), average slope steepness (HRU_SLP), moist bulk density of the soil layer (SOL_BD), saturated hydraulic conductivity (SOL_K), and soil evaporation compensation coefficient (ESCO). These parameters referred to both groundwater flow and soil properties.

3.2. Model Performance Evaluation

The model performance was assessed with criteria recommended by Moriasi et al. [48]. In all three calibration scenarios, the deviations from the observed values start to increase during the dry season and decline during the wet season. Table 3 displays the model performance for all three scenarios for all the sub-basins. Figure 3 presents the measured and simulated hydrographs at the outlet of the catchment (Monastiri gauging station) for all three scenarios. Finally, Figure 4 shows the measured and simulated AET for the entire study area at a catchment scale.

Table 3. Model evaluation statistics for each calibration scenario; (a) Streamflow calibration, (b) AET calibration, and (c) Multi-variable calibration. The location of the sub-basins 1–11 is indicated in Figure 2a–c.

Variable	Station/Sub-Basin	NSE			R^2			PBIAS (%)		
		(a)	(b)	(c)	(a)	(b)	(c)	(a)	(b)	(c)
Streamflow	Monastiri station	0.71	0.38	0.70	0.84	0.84	0.86	5.60	8.29	6.10
	Sub-basin 1	0.27	0.49	0.18	0.58	0.75	0.57	11.68	11.60	16.70
	Sub-basin 2	0.30	0.33	0.28	0.72	0.76	0.76	10.96	12.70	13.80
	Sub-basin 3	0.11	0.36	−0.10	0.59	0.69	0.55	15.81	15.00	21.70
	Sub-basin 4	0.37	0.34	0.42	0.73	0.75	0.75	8.48	12.50	13.60
	Sub-basin 5	0.09	0.28	0.15	0.80	0.78	0.81	3.07	6.10	5.30
Evapotranspiration	Sub-basin 6	0.22	0.35	0.26	0.64	0.74	0.68	9.00	10.70	14.50
	Sub-basin 7	−0.17	−0.14	0.19	0.87	0.80	0.86	−12.72	−5.30	−8.40
	Sub-basin 8	0.57	0.56	0.69	0.82	0.82	0.83	2.46	6.40	5.40
	Sub-basin 9	0.42	0.51	0.56	0.83	0.82	0.79	2.69	7.00	7.00
	Sub-basin 10	0.44	0.58	0.52	0.80	0.87	0.81	1.19	6.00	4.00
	Sub-basin 11	0.34	0.54	0.63	0.84	0.82	0.83	−0.53	7.70	2.90

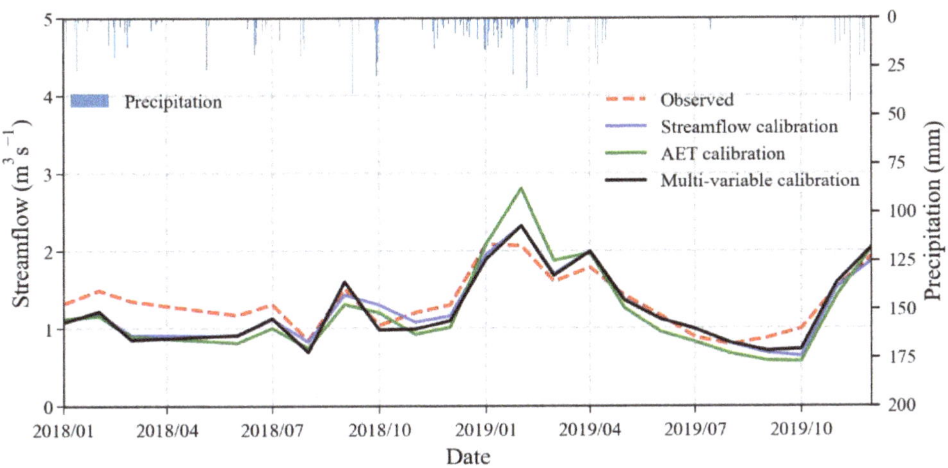

Figure 3. Simulated versus measured monthly streamflow for each calibration scenario at the outlet of the sub-basin (Monastiri gauging station).

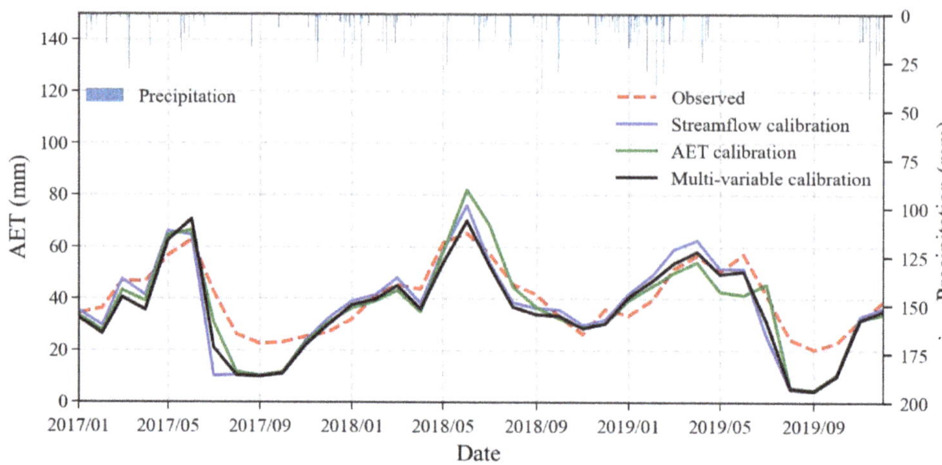

Figure 4. Simulated versus measured MODIS evapotranspiration for each calibration scenario at catchment scale.

3.2.1. Streamflow Calibration

The streamflow calibration showed good results in general for streamflow (NSE = 0.71; R^2 = 0.84; PBIAS = 5.6%) (Table 3). The results regarding evapotranspiration were unsatisfactory for NSE (NSE within −0.17 to 0.57), except for sub-basin 8 (NSE = 0.57), and satisfactory for R^2 and PBIAS (R^2 > 0.58; PBIAS within −12.7% to 15.8%). Figure 3 designates a satisfying match between measured and simulated streamflow except for low flows during the dry season. However, the temporal dynamics of the hydrograph were generated correctly. Figure 4 indicates differences between observed and simulated AET. Streamflow calibration underestimated evapotranspiration in the wet season and overestimated evapotranspiration at the beginning of the dry season.

3.2.2. Actual Evapotranspiration Calibration

The AET calibration presented unsatisfactory performance for NSE for sub-basins 2–7 and satisfactory performance for sub-basins 1, 8, 9, 10, and 11 (NSE within −0.14 to 0.58) (Table 3). In respect of R^2 and PBIAS, the results were satisfactory for all the sub-basins (R^2 > 0.69; PBIAS within −5.3% to 15%). The performance for streamflow was unsatisfactory (NSE = 0.38; R^2 = 0.84; PBIAS = 8.3%). In particular, the observed and simulated AET values did not match well. Nevertheless, they showed a well-matched seasonal variation of evapotranspiration (Figure 4). AET calibration underestimated low flows and overestimated high flows for the simulation period (Figure 3). Performance statistics were generally better for streamflow than evapotranspiration in single-variable calibration scenarios.

3.2.3. Multi-Variable Calibration

The multi-variable scenario, using streamflow and MODIS satellite-based AET, showed satisfactory performance for streamflow (NSE = 0.70; R^2 = 0.86; PBIAS = 6.1%) and unsatisfactory (sub-basins 1–7) to satisfactory (sub-basins 8–11) performance for evapotranspiration (NSE within −0.10 to 0.69; R^2 > 0.55; PBIAS within −8.4% to 21.7%) (Table 3). Simulated and observed streamflow values are much better than those for AET calibration and similar to streamflow calibration (Figure 3). Results for AET are related to those of AET calibration, showing underestimation of the simulated values (Figure 4). Compared to single variable calibration scenarios, multi-variable calibration displayed similar NSE values obtained from single-variable calibration, and R^2 showed good performances (>0.75) for both variables (Table 3).

3.3. Major Water Balance Components

The major water balance components (i.e., actual evapotranspiration, water yield, and precipitation) are displayed in Figure 5. In general, average annual precipitation (643 mm) was slightly greater than combined water yield (WYLD) and actual evapotranspiration (AET) values. Actual evapotranspiration contributed a large amount of water loss from the watershed, about 60% in all scenarios. The total water yield of the multi-variable calibration is higher than the other two modeling scenarios. In particular, the total water yield was estimated to be 171 mm for flow calibration, 151 mm for AET calibration, and 186 mm for multi-variable calibration.

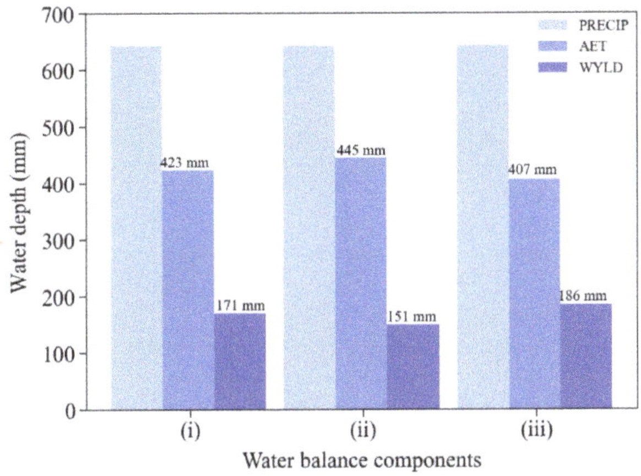

Figure 5. Water balance components; (i) Streamflow calibration, (ii) AET calibration, and (iii) Multi-variable calibration. PRECIP: Precipitation (mm), AET: Actual evapotranspiration (mm), WYLD: Water yield (mm).

4. Discussion

In this study, the SWAT hydrological model was used to interpret the behaviour of an urban/sub-urban environment and analyse its underlying mechanisms. The study area is a typical Mediterranean catchment prone to natural hazards such as floods, forest fires and their combined impact. Therefore, the mechanisms governing surface runoff and the interactions between the hydrological components should be analysed in depth for these vulnerable areas. The main objective was to investigate which parameters are more sensitive in a mixed land-use basin and to propose a multi-variable calibration procedure using both streamflow and satellite-based AET data for SWAT modelling.

The sensitivity analysis results showed that the parameters with the highest sensitivity for streamflow are connected to groundwater flow, runoff generation, and channel routing, and for actual evapotranspiration, they are linked to soil properties, respectively (Table 2, Figure A1). The differences in the sensitivity of the parameters are due to different data used in the calibration process. Similar outcomes were obtained by Sirisena et al. [16] and Moriasi et al. [49]. Sirisena et al. [16] concluded that the most sensitive parameters for evapotranspiration were connected to soil properties. Moriasi et al. [49] pointed out that the high sensitivity of the soil parameters to AET indicated a connection between actual evaporation and soil water.

Both variables present a slightly better performance in the single calibrations with streamflow and evapotranspiration than in the multi-variable calibration (Table 3). Nonetheless, the multi-variable calibration produced satisfactory results for both streamflow and AET and showed reasonably good streamflow estimations (NSE = 0.70). AET showed the best performance for forests (e.g., sub-basins 1, 8, 9, 10, 11). This indicates that the MODIS data probably show better performance at simulating forests and semi-natural areas than in sub-urban areas, including more complex management systems. In addition, evapotranspiration algorithms are characterized by resolution issues, misclassification of land use, and data generation uncertainties [50]. Therefore, these algorithms may not correctly capture the land use changes (especially in sub-urban areas) and the available soil moisture on the ground.

In most sub-basins, MODIS satellite-based AET and SWAT simulated AET show that seasonal patterns match well, although the SWAT model under-predicted AET compared with the MODIS satellite-based AET during the dry season (Figure 4, Table 3). These results are consistent with those of other studies [51,52]. A study in Morocco [51] and a study in Iran [52] suggested that the multi-variable calibration can produce good results for both variables. Nevertheless, the single variable calibrations showed better performance. The differences for all three scenarios intensify during the dry season and decline during the wet season (Figures 3 and 4). The underestimation of AET and the low baseflow, especially during the dry season (Figure 3), could suggest unknown water contributions in the study area. These deviations are probably also connected to soil replenishment and the crops' high water demand during the dry season. Furthermore, it is worth mentioning that MODIS satellite-based AET could include errors and underestimations or overestimations of the "true" AET, altering the model's water balance [53]. For instance, the higher MODIS satellite-based AET values in the AET calibration scenario led to lower water yield values than the streamflow calibration scenario (Figure 5). Satellite-based evapotranspiration datasets use sensor-derived parameters (e.g., surface heat flux, latent heat flux) that may have several uncertainties. Model misrepresentations, errors in the inputs, and spatial and temporal scaling decrease the efficiency of the algorithms [54].

For both single variable calibrations, the simulated second variable (i.e., AET for streamflow calibration and streamflow for AET calibration) is not well represented. The unsatisfactory performance of the second variable in the single variable calibrations indicates the poor representation of the catchment's water balance. Many studies support that incorporating satellite data in the hydrological model calibration improved the estimation of water balance components regardless of the model performance improvement. A study in China [55] calibrated the SWAT model with GLEAM AET data and streamflow. Although streamflow only calibration produced reliable results, this approach grouped the hydrological process. Furthermore, a study by Immerzeel and Droogers [13] pointed out that incorporating AET data in hydrological model calibration reduces equifinality obtained from traditional streamflow only calibration. The water balance is best reproduced when both streamflow and AET are used in the calibration process [52]. Several studies [49,56] reported that satellite AET data could be used to constrain hydrological parameters that are highly sensitive to evapotranspiration.

In general, the calibration process is challenging because of the uncertainties that exist due to model simplification, processes that are not accounted by the model, and processes that are unknown to the modeler [14]. In this study, the main sources of uncertainty are connected to inaccuracies (i) in the quality of input data (climate, soil, and land cover resolution), (ii) in the model set up (aggregation and interpolation methods), (iii) in the choice of objective data and parameterization, (iv) observed data, and (v) processes unknown to the modeler which interfere with the natural system [42]. Observational errors in the precipitation data, MODIS actual evapotranspiration, and discharge, as well as the effects of elevation and topography, increase bias and generate variability. For example, errors in streamflow measurements can vary from 6% to 19% under different combinations of channel types and measurement techniques [57]. In addition, poor resolution and data generation uncertainties in evapotranspiration algorithms can induce biases between AET simulated values and satellite-based AET [17,50,58]. For instance, a study in Ethiopia [50] estimated the major water balance components of the Upper Blue Nile basin using the JGrass-NewAge hydrological system and remote sensing data (GLEAM and MODIS AET). This study concluded that the satellite-based data introduce bias in estimating the water budget. Another study by [58] conducted a multi-objective validation for West Africa river basins using remote sensing data. The results showed that MODIS satellite-based AET underestimated SWAT-simulated AET in arid areas. Finally, Dile et al. [17] evaluated AET outputs derived from AVHRR and MOD16 AET datasets using outputs from a SWAT model for Ethiopia. This study suggested that datasets did not agree well with the precipitation in regions with a bimodal precipitation pattern. Therefore, careful consideration should be given to analyzing data from satellite-based products. Further information is necessary to estimate the uncertainty in model outputs and improve the calibration results at HRU level.

5. Conclusions

This study used monthly streamflow and MODIS satellite-based AET data to calibrate the SWAT model. Three calibration scenarios were conducted with streamflow, AET, and streamflow–AET data to evaluate the simulated outputs. The sensitivity analysis showed that the most sensitive parameters for streamflow are related to groundwater flow, runoff generation, and channel routing, and for actual evapotranspiration, they are all connected to soil properties. The model performance results indicated that the single variable calibrations showed satisfactory performance only for the first variable that was simulated. The multi-variable calibration showed satisfactory performance for both streamflow and AET. The SWAT model generally under-predicted AET in all scenarios compared to MODIS satellite-based AET.

This research showed that combining streamflow and MODIS satellite-based AET data in the calibration process can improve model performance regarding streamflow and water balance and contribute to understanding the hydrological processes in a mixed land-use catchment. Furthermore, the use of satellite data in model calibration, as presented in this study, can be utilized in catchments lacking measured data or in catchments with similar hydrological and geomorphological characteristics. Future work should incorporate discharge, soil moisture, and HRU level AET data in a combined objective function at a high temporal resolution.

Author Contributions: Conceptualization, E.K. and A.K.; methodology, E.K. and A.K.; software, E.K.; validation, E.K.; formal analysis, E.K.; investigation, E.K.; resources, E.K. and A.K.; data curation, E.K.; writing—original draft preparation, E.K.; writing—review and editing, E.K. and A.K.; visualization, E.K.; supervision, A.K.; project administration, A.K.; funding acquisition, A.K. All authors have read and agreed to the published version of the manuscript.

Funding: The research work was supported by the Hellenic Foundation for Research and Innovation (HFRI) under the HFRI PhD Fellowship grant (Fellowship Number: 1586).

Institutional Review Board Statement: Not applicable.

Informed Consent Statement: Not applicable.

Data Availability Statement: The SWAT model is available at the website http://swat.tamu.edu/ (accessed on 10 December 2020) (USDA Agricultural Research Service). The DEM data were obtained from the website https://earthexplorer.usgs.gov/ (accessed on 22 December 2020) (Shuttle Radar Topography Mission, SRTM). The land use data were downloaded from the website https://land.copernicus.eu/ (accessed on 22 December 2020) (Corine Land Cover, CLC 2018). The soil data were downloaded from the website http://www.fao.org/ (accessed on 22 December 2020) (Food and Agriculture Organization, FAO). The meteorological data were downloaded from the website https://www.meteo.gr/ (accessed on 11 January 2021) (National Observatory of Athens, NOA). The streamflow data were obtained from the website https://openhi.net/ (accessed on 20 January 2021) (Open Hydrosystem Information Network). The AET data were downloaded from the website https://modis.gsfc.nasa.gov/ (accessed on 14 March 2021) (Moderate Resolution Imaging Spectroradiometer).

Acknowledgments: The authors are grateful to Nancy Sammons, Information Technology Specialist of the SWAT team for her kind help with the SWAT software. Authors gratefully acknowledge the financial support from the Hellenic Foundation for Research and Innovation (HFRI).

Conflicts of Interest: The authors declare no conflict of interest. The funders had no role in the design of the study; in the collection, analyses, or interpretation of data; in the writing of the manuscript, or in the decision to publish the results.

Appendix A

The following tables display the land use categories of the study area at catchment (Table A1) and sub-basin level (Table A2).

Table A1. Land use categories of the study area at catchment level.

Corine Classification	SWAT Code	SWAT Classification	(%) Catchment
Industrial or commercial units	UCOM	Commercial	11.43
Discontinuous urban fabric	URLD	Residential-Low Density	34.11
Road and rail networks and associated land	UTRN	Transportation	4.07
Continuous urban fabric	URHD	Residential-High Density	1.54
Pastures	PAST	Pasture	0.31
Land principally occupied by agriculture, with significant areas of natural vegetation	AGRL	Agricultural Land-Generic	12.39
Broad-leaved forest	FRSD	Forest-Deciduous	3.11
Coniferous forest	FRSE	Forest-Evergreen	9.59
Mixed forest	FRST	Forest-Mixed	7.51
Sclerophyllous vegetation	RNGB	Range-Brush	15.94

Table A2. Land use categories of the study area at sub-basin level. Artificial surfaces (i.e., urban fabric, industrial, commercial and transport units), agricultural areas (i.e., arable land, pastures and heterogeneous agricultural areas) and forests and semi natural areas (i.e., forests, scrub and herbaceous vegetation associations).

Sub-Basins	Artificial Surfaces (%)	Agricultural Areas (%)	Forests and Semi Natural Areas (%)
Sub-basin 1	1.88	5.02	93.10
Sub-basin 2	53.36	9.95	36.69
Sub-basin 3	21.83	9.28	68.89
Sub-basin 4	56.11	25.05	18.84
Sub-basin 5	76.91	0.74	22.35
Sub-basin 6	18.31	23.25	58.45
Sub-basin 7	76.43	8.34	15.23
Sub-basin 8	56.77	0.29	42.94
Sub-basin 9	75.28	3.06	21.65
Sub-basin 10	80.65	2.72	16.63
Sub-basin 11	28.65	27.38	43.97
Sub-basin 12	100.01	0.00	0.00
Sub-basin 13	74.16	25.83	0.00
Sub-basin 14	73.29	26.71	0.00
Sub-basin 15	81.90	18.10	0.00
Sub-basin 16	99.33	0.67	0.00
Sub-basin 17	64.20	35.76	0.04
Sub-basin 18	29.88	19.13	51.00
Sub-basin 19	7.98	13.54	78.48
Sub-basin 20	9.29	28.47	62.23
Sub-basin 21	80.42	19.58	0.00
Sub-basin 22	73.64	26.36	0.00
Sub-basin 23	83.87	16.13	0.00
Sub-basin 24	100.01	0.00	0.00
Sub-basin 25	100.00	0.00	0.00

The following figure shows the relative changes of the five most sensitive parameters for each calibration scenario versus objective function (Figure A1).

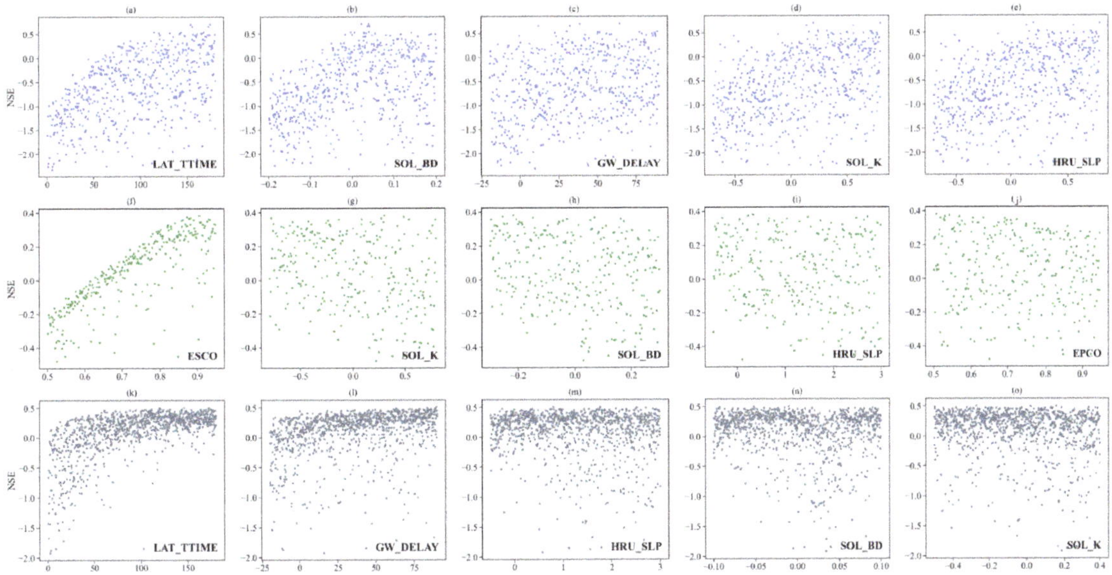

Figure A1. Calibrated values (x-axis) versus objective function NSE value range (y-axis); blue dots symbolize streamflow calibration, green dots symbolize AET calibration, and grey dots symbolize multi-variable calibration. The parameters (**a–e**) are the most sensitive parameters for streamflow calibration, the parameters (**f–j**) are the most sensitive parameters for AET calibration, and the parameters (**k–o**) are the most sensitive parameters for multi-variable calibration.

References

1. Arnold, J.G.; Youssef, M.A.; Yen, H.; White, M.J.; Sheshukov, A.Y.; Sadeghi, A.M.; Moriasi, D.N.; Steiner, J.L.; Amatya, D.M.; Skaggs, R.W.; et al. Hydrological Processes and Model Representation: Impact of Soft Data on Calibration. *Trans. ASABE* **2015**, *58*, 1637–1660. [CrossRef]
2. Daggupati, P.; Yen, H.; White, M.J.; Srinivasan, R.; Arnold, J.G.; Keitzer, C.S.; Sowa, S.P. Impact of model development, calibration and validation decisions on hydrological simulations in West Lake Erie Basin. *Hydrol. Processes* **2015**, *29*, 5307–5320. [CrossRef]
3. Gassman, P.W.; Sadeghi, A.M.; Srinivasan, R. Applications of the SWAT Model Special Section: Overview and Insights. *J. Environ. Qual.* **2014**, *43*, 1–8. [CrossRef] [PubMed]
4. Hrachowitz, M.; Savenije, H.H.G.; Blöschl, G.; McDonnell, J.J.; Sivapalan, M.; Pomeroy, J.W.; Arheimer, B.; Blume, T.; Clark, M.P.; Ehret, U.; et al. A decade of Predictions in Ungauged Basins (PUB)—A review. *Hydrol. Sci. J.* **2013**, *58*, 1198–1255. [CrossRef]
5. Van Emmerik, T.; Mulder, G.; Eilander, D.; Piet, M.; Savenije, H. Predicting the ungauged basin: Model validation and realism assessment. *Front. Earth Sci.* **2015**, *3*, 62. [CrossRef]
6. Arnold, J.G.; Moriasi, D.N.; Gassman, P.W.; Abbaspour, K.C.; White, M.J.; Srinivasan, R.; Santhi, C.; Harmel, R.D.; Van Griensven, A.; Van Liew, M.W.; et al. SWAT: Model Use, Calibration, and Validation. *Trans. ASABE* **2012**, *55*, 1491–1508. [CrossRef]
7. Beven, K.; Binley, A. GLUE: 20 years on. *Hydrol. Processes* **2014**, *28*, 5897–5918. [CrossRef]
8. Kelleher, C.; McGlynn, B.; Wagener, T. Characterizing and reducing equifinality by constraining a distributed catchment model with regional signatures, local observations, and process understanding. *Hydrol. Earth Syst. Sci.* **2017**, *21*, 3325–3352. [CrossRef]
9. Liu, Z.; Yin, J.; Dahlke, H.E. Enhancing Soil and Water Assessment Tool Snow Prediction Reliability with Remote-Sensing-Based Snow Water Equivalent Reconstruction Product for Upland Watersheds in a Multi-Objective Calibration Process. *Water* **2020**, *12*, 3190. [CrossRef]
10. Her, Y.; Seong, C. Responses of hydrological model equifinality, uncertainty, and performance to multi-objective parameter calibration. *J. Hydroinformatics* **2018**, *20*, 864–885. [CrossRef]
11. Rajib, A.; Merwade, V.; Yu, Z. Rationale and Efficacy of Assimilating Remotely Sensed Potential Evapotranspiration for Reduced Uncertainty of Hydrologic Models. *Water Resour. Res.* **2018**, *54*, 4615–4637. [CrossRef]
12. Shah, S.; Duan, Z.; Song, X.; Li, R.; Mao, H.; Liu, J.; Ma, T.; Wang, M. Evaluating the added value of multi-variable calibration of SWAT with remotely sensed evapotranspiration data for improving hydrological modeling. *J. Hydrol.* **2021**, *603*, 127046. [CrossRef]
13. Immerzeel, W.W.; Droogers, P. Calibration of a distributed hydrological model based on satellite evapotranspiration. *J. Hydrol.* **2008**, *349*, 411–424. [CrossRef]

14. Abbaspour, K.C.; Yang, J.; Maximov, I.; Siber, R.; Bogner, K.; Mieleitner, J.; Zobrist, J.; Srinivasan, R. Modelling hydrology and water quality in the pre-alpine/alpine Thur watershed using SWAT. *J. Hydrol.* **2007**, *333*, 413–430. [CrossRef]
15. Parajuli, P.B.; Jayakody, P.; Ouyang, Y. Evaluation of Using Remote Sensing Evapotranspiration Data in SWAT. *Water Resour. Manag.* **2018**, *32*, 985–996. [CrossRef]
16. Sirisena, T.A.J.G.; Maskey, S.; Ranasinghe, R. Hydrological Model Calibration with Streamflow and Remote Sensing Based Evapotranspiration Data in a Data Poor Basin. *Remote Sens.* **2020**, *12*, 3768. [CrossRef]
17. Dile, Y.T.; Ayana, E.K.; Worqlul, A.W.; Xie, H.; Srinivasan, R.; Lefore, N.; You, L.; Clarke, N. Evaluating satellite-based evapotranspiration estimates for hydrological applications in data-scarce regions: A case in Ethiopia. *Sci. Total Environ.* **2020**, *743*, 140702. [CrossRef]
18. Herman, M.R.; Nejadhashemi, A.P.; Abouali, M.; Hernandez-Suarez, J.S.; Daneshvar, F.; Zhang, Z.; Anderson, M.C.; Sadeghi, A.M.; Hain, C.R.; Sharifi, A. Evaluating the role of evapotranspiration remote sensing data in improving hydrological modeling predictability. *J. Hydrol.* **2018**, *556*, 39–49. [CrossRef]
19. Aboelnour, M.; Gitau, M.W.; Engel, B.A. Hydrologic response in an urban watershed as affected by climate and land-use change. *Water* **2019**, *11*, 1603. [CrossRef]
20. Becker, R.; Koppa, A.; Schulz, S.; Usman, M.; aus der Beek, T.; Schüth, C. Spatially distributed model calibration of a highly managed hydrological system using remote sensing-derived ET data. *J. Hydrol.* **2019**, *577*, 123944. [CrossRef]
21. Busico, G.; Colombani, N.; Fronzi, D.; Pellegrini, M.; Tazioli, A.; Mastrocicco, M. Evaluating SWAT model performance, considering different soils data input, to quantify actual and future runoff susceptibility in a highly urbanized basin. *J. Environ. Manag.* **2020**, *266*, 110625. [CrossRef] [PubMed]
22. Jha, M.K.; Afreen, S. Flooding urban landscapes: Analysis using combined hydrodynamic and hydrologic modeling approaches. *Water* **2020**, *12*, 1986. [CrossRef]
23. van Tol, J.; van Zijl, G.; Julich, S. Importance of detailed soil information for hydrological modelling in an urbanized environment. *Hydrology* **2020**, *7*, 34. [CrossRef]
24. Boithias, L.; Sauvage, S.; Lenica, A.; Roux, H.; Abbaspour, K.; Larnier, K.; Dartus, D.; Sánchez-Pérez, J. Simulating Flash Floods at Hourly Time-Step Using the SWAT Model. *Water* **2017**, *9*, 929. [CrossRef]
25. Mamassis, N.; Koukouvinos, A.; Baki, S. *Final Report, Development of a Geographical Information System and an Internet Application for the Supervision of Kephisos Protected Areas*; Department of Water Resources and Environmental Engineering, National Technical University of Athens: Athens, Greece, 2008; Available online: http://www.itia.ntua.gr/el/project_reports/151/ (accessed on 15 January 2021).
26. Zerefos, C.; Repapis, C.; Giannakopoulos, C.; Kapsomenakis, J.; Papanikolaou, D.; Papanikolaou, M.; Poulos, S.; Vrekoussis, M.; Philandras, C.; Tselioudis, G. The climate of the Eastern Mediterranean and Greece: Past, present and future. In *The Environmental, Economic and Social Impacts of Climate Change in Greece*; Bank of Greece: Athens, Greece, 2011; pp. 50–58.
27. Lagouvardos, K.; Kotroni, V.; Bezes, A.; Koletsis, I.; Kopania, T.; Lykoudis, S.; Mazarakis, N.; Papagiannaki, K.; Vougioukas, S. The automatic weather stations NOANN network of the National Observatory of Athens: Operation and database. *Geosci. Data J.* **2017**, *4*, 4–16. [CrossRef]
28. Corine Land Cover, (CLC) Land Use Data. Available online: https://land.copernicus.eu/ (accessed on 15 December 2020).
29. Food and Agriculture Organization, (FAO). HWSD Soil Data. Available online: www.fao.org (accessed on 10 December 2020).
30. U.S. Geological Survey, (USGS). Shuttle Radar Topography Mission (SRTM), DEM Data. Available online: https://earthexplorer.usgs.gov/ (accessed on 5 December 2020).
31. FAO; IIASA; ISRIC; ISSCAS. *Harmonized World Soil Database Version 1.2*; Food & Agriculture Organization of the UN: Rome, Italy; International Institute for Applied Systems Analysis: Laxenburg, Austria, 2012; ISBN 9781626239777.
32. Open Hydrosystem Information Network, (OpenHi.net). Observed Streamflow Data. Available online: https://openhi.net/ (accessed on 20 December 2020).
33. Moderate Resolution Imaging Spectroradiometer, (MODIS). Evapotranspiration Data. Available online: https://modis.gsfc.nasa.gov/ (accessed on 2 March 2021).
34. Running, S.; Mu, Q.; Zhao, M.; Moreno, A. *User 's Guide NASA Earth Observing System MODIS Land Algorithm (For Collection 6)*; National Aeronautics and Space Administration (NASA): Washington, DC, USA, 2019; pp. 1–38.
35. Arnold, J.G.; Srinivasan, R.; Muttiah, R.S.; Williams, J.R. Large Area Hydrologic Modeling and Assessment Part I: Model Development. *J. Am. Water Resour. Assoc.* **1998**, *34*, 73–89. [CrossRef]
36. Neitsch, S.L.; Arnold, J.G.; Kiniry, J.R.; Williams, J.R. *Soil & Water Assessment Tool Theoretical Documentation Version 2009*; Texas Water Resources Institute: Texas, TX, USA, 2011; ISBN Technical Report No. 406.
37. Gassman, P.W.; Reyes, M.R.; Green, C.H.; Arnold, J.G. The Soil and Water Assessment Tool: Historical Development, Applications, and Future Research Directions. *Trans. ASABE* **2007**, *50*, 1211–1250. [CrossRef]
38. Dile, Y.T.; Daggupati, P.; George, C.; Srinivasan, R.; Arnold, J. Introducing a new open source GIS user interface for the SWAT model. *Environ. Model. Softw.* **2016**, *85*, 129–138. [CrossRef]
39. Soil Conservation Service, S. *National Engineering Handbook, Section 4, Hydrology*; Department of Agriculture: Washington, DC, USA, 1972.
40. Jeong, J.; Williams, J.R.; Merkel, W.H.; Arnold, J.G.; Wang, X.; Rossi, C.G. Improvement of the variable storage coefficient method with water surface gradient as a variable. *Trans. ASABE* **2014**, *57*, 791–801. [CrossRef]
41. Abbaspour, K.C.; Johnson, C.A.; van Genuchten, M.T. Estimating Uncertain Flow and Transport Parameters Using a Sequential Uncertainty Fitting Procedure. *Vadose Zone J.* **2004**, *3*, 1340–1352. [CrossRef]

42. Abbaspour, K.C.; Rouholahnejad, E.; Vaghefi, S.; Srinivasan, R.; Yang, H.; Kløve, B. A continental-scale hydrology and water quality model for Europe: Calibration and uncertainty of a high-resolution large-scale SWAT model. *J. Hydrol.* **2015**, *524*, 733–752. [CrossRef]
43. Odusanya, A.E.; Mehdi, B.; Schürz, C.; Oke, A.O.; Awokola, O.S.; Awomeso, J.A.; Adejuwon, J.O.; Schulz, K. Multi-site calibration and validation of SWAT with satellite-based evapotranspiration in a data-sparse catchment in southwestern Nigeria. *Hydrol. Earth Syst. Sci.* **2019**, *23*, 1113–1144. [CrossRef]
44. Franco, A.C.L.; Bonumá, N.B. Multi-variable SWAT model calibration with remotely sensed evapotranspiration and observed flow. *RBRH* **2017**, *22*, 1–14. [CrossRef]
45. Moriasi, D.N.; Arnold, J.G.; Van Liew, M.W.; Bingner, R.L.; Harmel, R.D.; Veith, T.L. Model Evaluation Guidelines for Systematic Quantification of Accuracy in Watershed Simulations. *Trans. ASABE* **2007**, *50*, 885–900. [CrossRef]
46. Nash, J.E.; Sutcliffe, J.V. River flow forecasting through conceptual models part I—A discussion of principles. *J. Hydrol.* **1970**, *10*, 282–290. [CrossRef]
47. Gupta, H.V.; Sorooshian, S.; Yapo, P.O. Status of Automatic Calibration for Hydrologic Models: Comparison with Multilevel Expert Calibration. *J. Hydrol. Eng.* **1999**, *4*, 135–143. [CrossRef]
48. Moriasi, D.N.; Gitau, M.W.; Pai, N.; Daggupati, P. Hydrologic and water quality models: Performance measures and evaluation criteria. *Trans. ASABE* **2015**, *58*, 1763–1785. [CrossRef]
49. Tobin, K.; Bennett, M. Improving Alpine Summertime Streamflow Simulations by the Incorporation of Evapotranspiration Data. *Water* **2019**, *11*, 112. [CrossRef]
50. Abera, W.; Formetta, G.; Brocca, L.; Rigon, R. Modeling the water budget of the Upper Blue Nile basin using the JGrass-NewAge model system and satellite data. *Hydrol. Earth Syst. Sci.* **2017**, *21*, 3145–3165. [CrossRef]
51. López López, P.; Sutanudjaja, E.H.; Schellekens, J.; Sterk, G.; Bierkens, M.F.P. Calibration of a large-scale hydrological model using satellite-based soil moisture and evapotranspiration products. *Hydrol. Earth Syst. Sci.* **2017**, *21*, 3125–3144. [CrossRef]
52. Rientjes, T.H.M.; Muthuwatta, L.P.; Bos, M.G.; Booij, M.J.; Bhatti, H.A. Multi-variable calibration of a semi-distributed hydrological model using streamflow data and satellite-based evapotranspiration. *J. Hydrol.* **2013**, *505*, 276–290. [CrossRef]
53. Vervoort, W.R.; Miechels, S.F.; van Ogtrop, F.F.; Guillaume, J.H.A. Remotely sensed evapotranspiration to calibrate a lumped conceptual model: Pitfalls and opportunities. *J. Hydrol.* **2014**, *519*, 3223–3236. [CrossRef]
54. Ferguson, C.R.; Sheffield, J.; Wood, E.F.; Gao, H. Quantifying uncertainty in a remote sensing-based estimate of evapotranspiration over continental USA. *Int. J. Remote Sens.* **2010**, *31*, 3821–3865. [CrossRef]
55. Jin, X.; Jin, Y. Calibration of a Distributed Hydrological Model in a Data-Scarce Basin Based on GLEAM Datasets. *Water* **2020**, *12*, 897. [CrossRef]
56. Tobin, K.J.; Bennett, M.E. Constraining SWAT Calibration with Remotely Sensed Evapotranspiration Data. *J. Am. Water Resour. Assoc.* **2017**, *53*, 593–604. [CrossRef]
57. Harmel, R.D.; Cooper, R.J.; Slade, R.M.; Haney, R.L.; Arnold, J.G. Cumulative uncertainty in measured streamflow and water quality data for small watersheds. *Trans. ASABE* **2006**, *49*, 689–701. [CrossRef]
58. Poméon, T.; Diekkrüger, B.; Springer, A.; Kusche, J.; Eicker, A. Multi-Objective Validation of SWAT for Sparsely-Gauged West African River Basins—A Remote Sensing Approach. *Water* **2018**, *10*, 451. [CrossRef]

hydrology

MDPI

Article

Supercritical Flow over a Submerged Vertical Negative Step

Eugene Retsinis * and Panos Papanicolaou

School of Civil Engineering, National Technical University of Athens, 9 Iroon Polytexneiou St., 15780 Athens, Greece; panospap@mail.ntua.gr
* Correspondence: retsinis@central.ntua.gr; Tel.: +30-6944-916-187

Abstract: The transition from supercritical to subcritical flow around a fully submerged abrupt negative step in a horizontal rectangular open channel has been investigated. In a laboratory experiment the one-dimensional energy and the momentum conservation equations were studied by means of depth and pressure measurements by piezometers installed along the bottom and the step face. Froude number varied in the range 1.9 to 5.8 while the step height to critical depth ratio was in the range 1.34 to 2.56. The results are presented in dimensionless form using mainly a characteristic length scale that is the sum of critical depth and step height and the Froude number of the supercritical flow upstream. Five different types of rapidly varying flow are observed when the subcritical downstream tailwater depth varied. The supercritical water jet at the top of the step either strikes the bottom downstream of the step when the maximum pressure head is greater, or moves to the surface of the flow when it is lower than tailwater depth, and the separation of the two flow regimes occurs when the tailwater depth to the characteristic length scale is around 1.05. The normalized energy loss and a closure parameter for the momentum equation are presented in dimensionless diagrams for practical use by the design engineer. Finally, the one-dimensional equations of motion including Boussinesq terms are solved numerically and the results found are congruent to the experimental findings.

Keywords: abrupt negative step; hydraulic jump; energy loss; momentum closure; Boussinesq equations; specified intervals

Citation: Retsinis, E.; Papanicolaou, P. Supercritical Flow over a Submerged Vertical Negative Step. *Hydrology* **2022**, *9*, 74. https://doi.org/10.3390/hydrology9050074

Academic Editor: Monzur A. Imteaz

Received: 28 March 2022
Accepted: 26 April 2022
Published: 28 April 2022

1. Introduction

Stilling basins are designed to dissipate the kinetic energy of the flow by means of hydraulic jumps formed in them, for which a thorough investigation of the physical and numerical study is reported in [1]. In several cases an abrupt forward facing (negative) step is introduced to stabilize the jump, so that it is not swept away from the basin. The transition from supercritical to subcritical flow at an abrupt negative step affects the design and construction of a stilling basin. The design usually includes determination of the step height, the required tailwater water elevation, an adequate basin length and all necessary structures in it such as blocks, end sill, etc. Submerged negative steps can also be met in river training works, in canals conveying water, as well as in natural streams. Engineers and scientists have been studying the flow of submerged steps in the laboratory for a long time, measuring mainly the flow depths and pressure distribution at the bottom and the step face, and trying to correlate the flow parameters to energy and momentum conservation, as well as to the various flow profiles that appear there.

In laboratory experiments the flow is usually controlled by a sluice gate upstream, and a sharp crested overflow downstream of the step. Five different rapidly varying flow profiles have been observed around a step with supercritical flow upstream and subcritical downstream [2–4], namely minimum B-jump, B-jump, wave-train, wave-jump and A-jump (Figure 1). Minimum B-jump (Figure 1i) is the hydraulic jump at the toe of the step, B-jump (Figure 1ii) is the submerged jump downstream of the step, wave-train (Figure 1iii) is the surface jet-type flow without formation of a hydraulic jump, wave-jump (Figure 1iv) is the flow of an ascending jet forming a standing wave downstream of the step, before it dives and results in a submerged hydraulic jump, and A-jump (Figure 1v) is the flow where the hydraulic jump is formed upstream of the step. These profiles appear with the above sequence if one increases the downstream tailwater depth continuously.

(i)
(ii)
(iii)
(iv)
(v)

Figure 1. Transition from supercritical to subcritical flow around a vertical step, (**i**) minimum B-jump, (**ii**) B-jump, (**iii**) wave-train, (**iv**) wave-jump and (**v**) A-jump.

Several researchers in the past have studied the transition from supercritical to subcritical flow over a fully submerged negative (abrupt or rounded) step by experiments. Several authors [2–11] have investigated the influence of the pressure distribution at the face of the step of different jump types in order to obtain closure of the one-dimensional momentum equation. Velocity and shear stress distribution was measured [12] at the channel bottom using a Preston tube. In some cases the flow type alternated between B-jump, wave-train and wave-jump every once in a while [4,13,14], thus resulting in varying velocity and pressure fields. Standard deviation of pressure fluctuations at the channel bottom axis was measured by [15], while published experimental results of jumps at an abrupt drop with subcritical and supercritical flow conditions upstream have also been reported [16]. The presence of roughness elements inside a channel with an abrupt drop reduced the pressure forces on the bottom axis of the channel [17,18], if compared to the developed pressure either from the classical hydraulic jump or inside a smooth channel with drop. Also, for upstream Froude number greater than eight, the relative height of the step to upstream depth d/y_1 does not seem to affect the energy loss [5], while the energy dissipation of the jump [6] is higher in the case of a channel with negative step, if compared to the energy loss in a channel with a positive step. The wave-jump type flow was found to dissipate the energy more efficiently than the classical hydraulic jump [7,19], while inclined channels under the presence of negative or positive step with sharp crested B-jumps resulted in higher energy dissipation than A-jumps [20]. The different flow profiles developed in sloping rectangular open channels with an abrupt drop have also been investigated [21], while the highest energy loss was observed in B-jump, if compared to the minimum B-jump and the A-jump [22]. The energy loss was reported [23] in dimensionless form, while jumps formed at a negative step are more stable, energy dissipative efficient and more compact [24], if compared to the jumps appearing in positive steps.

Regarding numerical modeling of the flow in the presence of an abrupt step, the B-jump was studied [25] in an open channel 0.4 m wide with a 9.7 cm abrupt drop high for upstream Froude numbers 1.21 and 1.4. The Reynolds-Averaged Navier Stokes (RANS) equations were solved using the ANSYS-FLUENT commercial software with the finite volume method to compute the free surface profile and the time-averaged velocity field. The B-jump, wave-jump and A-jump in an open channel 2 m long and 0.4 m wide with abrupt drops 3.20 and 6.52 cm high, for Froude number in the range 2.8–3.9, have been studied [26] using Smoothed Particle Hydrodynamics (SPH) to discretize the Navier Stokes equations along with k-ε turbulence model. The results regarded the computation of the instantaneous velocity and vorticity fields.

Systematic pressure measurements have not been reported to date in the area of an abrupt drop in an open channel hence, pressure distribution cannot be linked to the linear characteristics of the flow. Energy losses and momentum equation balance have not been reported conclusively around a step with supercritical flow upstream, which is a result due to lack of use of the appropriate dimensionless representation of them. The aim of the present work is to investigate the rapidly varying flow in a horizontal rectangular open channel with the presence of a fully submerged abrupt negative step, with supercritical flow upstream and subcritical downstream. To obtain it we performed systematic measurements of the linear flow characteristics as well as the pressure at the bottom and face of the step, for supercritical flow upstream with Froude number up to about 6. From these measurements the conditions under which different flow patterns appear, as well as other aspects of the flow, such as the closure of the one-dimensional momentum and energy loss equations will be investigated, and the results will be presented in dimensionless form. The Saint Venant equations will be solved in one dimension including Boussinesq terms that have been usually omitted in the past, to evaluate some of the experimental findings.

2. Theory

The flow under investigation (Figure 2) is that around a vertical forward facing (negative) step where the flow upstream is supercritical and downstream subcritical. The parameters involved are the flow rate per unit width q, the step height d, the upstream and downstream depths y_1 and y_2 and mean velocities V_1 and V_2, respectively. Two parameters that are important in the description of the flow are the critical depth $y_c = (q^2/g)^{1/3}$ and the Froude number of the supercritical flow upstream of the step $Fr_1 = V_1/(gy_1)^{1/2}$, g being the gravitational acceleration.

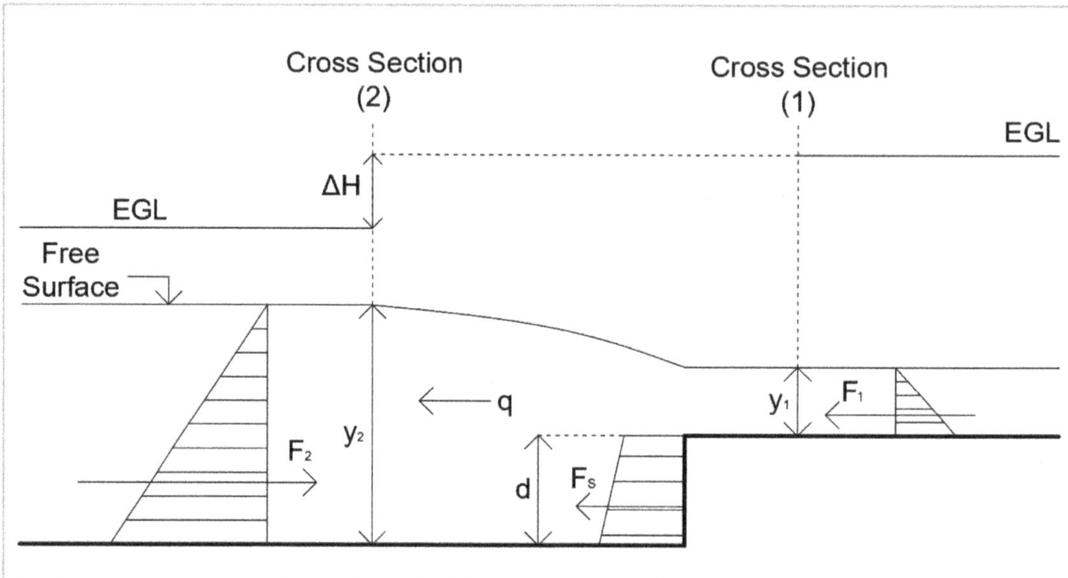

Figure 2. Definition sketch for the momentum and the energy equations.

Let us study the flow theoretically using the one-dimensional momentum and energy conservation equations [27]. We apply the one-dimensional momentum equation along the main flow direction per unit width, in the control volume between upstream Cross Section 1 with supercritical flow and downstream Cross Section 2 with subcritical flow, and depths y_1 and y_2, respectively, to get

$$\frac{1}{2}gy_1^2 + kgd(y_1 + d/2) - \frac{1}{2}gy_2^2 = q(V_2 - V_1) \tag{1}$$

In Equation (1) we assume that the velocity is uniform at Cross Sections 1 and 2, the shear stresses at the wall and the bottom of the channel can be neglected, and the pressure distribution at the face of the step is hydrostatic hence, $F_1 = (gy_1^2)/2$, $F_2 = (gy_2^2)/2$ and $F_s = kgd(y_1 + d/2)$. To make Equation (1) valid, the second term on the left that corresponds to force on step face $\rho gd(y_1 + d/2)$ from the hydrostatic pressure distribution with pressure ρgy_1 at the top and $\rho g(y_1 + d)$ at the bottom must be multiplied by a correction factor k, defined to be the ratio of the real pressure that is due to the curvature of the streamlines near the step, to the anticipated hydrostatic one at the middle of the step $\rho g(y_1 + d/2)$ [3].

If the flow rate q and depths y_1 and y_2 are known, solving Equation (1) for k we get

$$k = \frac{q(V_2 - V_1) - \frac{1}{2}g(y_1^2 - y_2^2)}{gd\left(y_1 + \frac{d}{2}\right)} \tag{2}$$

Equation (2) holds if the flow depth at the step is y_1, in other words for all types of flow but A-jump. For the latter case the term in the denominator must be replaced by $gd(y_2 - d/2)$.

The energy loss ΔH at the step is computed from the one-dimensional energy equation assuming known depths and flow rate, and uniform velocity at Cross Sections 1 and 2.

$$\Delta H = H_1 - H_2 = d + y_1 + V_1^2/2g - y_2 - V_2^2/2g \tag{3}$$

where H_1 and H_2, are the energy heads at Cross Sections 1 and 2, respectively, considering uniform velocity distributions.

3. Experiments

3.1. Setup and Procedure

Experiments were carried out at the Laboratory of Applied Hydraulics of the School of Civil Engineering at the National Technical University of Athens, Greece. The open channel used is 10.50 m long with rectangular cross section 0.255 m wide × 0.50 m deep is shown in Figure 3, and was equipped with a sluice gate upstream and a thin crested weir at downstream end. The section of the channel where measurements were taken has been modified to accommodate the experiments. The steel, nontransparent bottom has been replaced with Lucite with a row of piezometers attached to it, and the vertical side glass walls were replaced with new ones with improved optical properties.

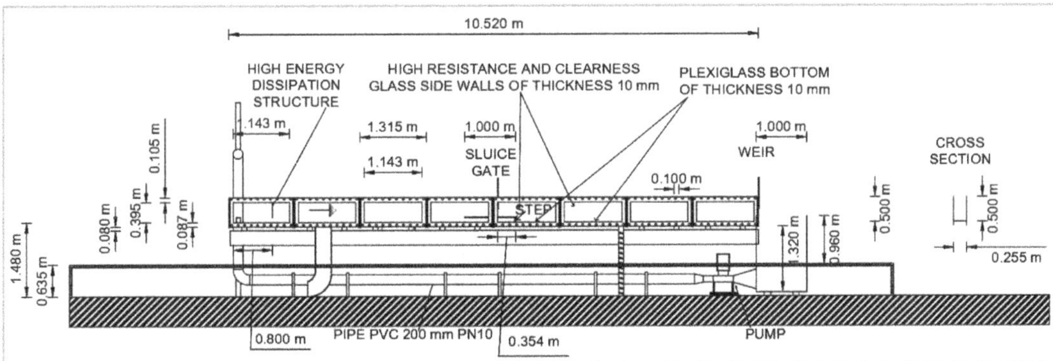

Figure 3. Side view of the open channel.

The water supply was obtained via a recirculation system that consists of a 3 kW pump with variable speed motor and maximum discharge capacity of 40 L/s at 5 m head, which is connected to a 2.65 m^3 water tank at the downstream end of the channel. Water was pumped to the upstream end of the channel through a PVC pipe of nominal diameter 0.2 m. The flowrate was measured with an ultrasonic flow meter of 2–5% accuracy, attached in the horizontal PVC pipe that flowed full about 10 pipe diameters downstream of the pump. A screen system at the entrance of the water in the channel was used to dissipate the kinetic energy of the inflow and reduce waves from agitation, upstream of the sluice gate.

A downstream facing vertical step 10.3 cm high and 1 m long made of Lucite was placed 4.85 m upstream of the channel end. The vertical sluice gate was positioned 0.35 m upstream of the step face, in order to control the flow from upstream. Ten piezometers were placed in the middle of the step and three along the middle of the step face. Twenty-one piezometers were placed along the middle of the Lucite bottom downstream of the step. The piezometers were de-aerated during the course of the experiment, once the step was submerged in the flow.

The flow was controlled with the vertical sluice gate and supercritical flow was obtained on the step surface at the desired inflow Froude number upstream from the drop. Downstream, the flow was controlled with a vertical sharp crested weir at the end of the channel, and the depth was adjusted to the desired level. Once the supercritical flow conditions were set upstream of the drop, the rapidly varying flow around it was adjusted by the tailwater depth. The depth of flow was measured using a point gauge and from the piezometers far downstream of the step, where the flow was parallel to the bottom.

The experimental procedure was as follows. The pump was set to the desired flow rate adjusted by the variable speed motor and measured with the ultrasonic flow meter, then the channel was filled with water above the step, keeping the end weir at the appropriate elevation. The piezometers were de-aerated and the supercritical flow upstream of the drop was adjusted to the desired inflow Froude number with the sluice gate opening. Then, the flow conditions downstream of the submerged step were adjusted with the end weir to the desired type of flow. Increasing the weir height gradually, the flow was set from minimum B-jump to B-jump, to wave-train, to (standing) wave-jump, to A-jump. The flow depths were measured with a point gauge with accuracy ± 0.0001 m. The supercritical flow depth, y_1, was measured at a distance 3 y_1 upstream of the drop, while the tailwater subcritical depth y_2 was measured at a distance 2.5 m downstream of the drop, (Figure 2). In both cases the flow depths y_1 and y_2 were measured at Cross Sections 1 and 2, where the flow was almost uniform. Three hundred (291) different jump and pressure measurements for various upstream and downstream conditions have been performed. The range of the main flow parameters is listed in the following Table 1.

Table 1. Range of initial parameters of supercritical flow upstream of the step.

Q (L/s)	q (L/s/m)	y_1 (cm)	y_2 (cm)	Inflow Fr_1	Inflow Re
6.46–17.50	25.32–68-67	1.4–3.6	25.8–26.8	1.88–5.82	23,000–63,000

3.2. Results

In order to normalize the measured lengths, i.e., the flow depths y_1 and y_2, we must define a length scale which should include information regarding the flow parameters. These parameters must involve information regarding the fluid and the geometry of the flow. A characteristic length scale has been defined to be the sum $y_c + d$, which involves information regarding the flow rate q in terms of the critical depth, and the potential energy of the flow in terms of the step height. This length scale is also greater than $y_1 + d$ and approaches asymptotically the maximum (supercritical) water surface elevation upstream of the drop.

In an effort to distinguish the flow regimes of the different jump types the normalized depth $Fr_1(y_c + d)/y_2$ is plotted in Figure 4 as a function of the inflow Froude number Fr_1 upstream. The triangles correspond to wave-train where the supercritical jet flow remains at the surface downstream of the drop and subsequently diffuses into the tailwater due to turbulent shear stress at the interface between the supercritical and subcritical flow regimes. Downstream of the step, there exists a long recirculation regime up to the point of reattachment. The wave-train separates the following two flow regimes: The first where the supercritical jet flow impinges at the bottom near the toe of the step followed by a (minimum B or B type) hydraulic jump that may be submerged; the second where the supercritical jet flow remains at the surface downstream of the drop (wave-jump or A-jump). An indicative 'line' separating the two regimes may be regarded that for which $Fr_1(y_c + d)/y_2 = 0.71\ Fr_1 + 0.43$.

The normalized depth $y_2/[Fr_1(y_c + d)]$ is plotted in Figure 5 against $y_2/[Fr_1(y_1 + d)]$ for all different types of the flow, and subcritical tailwater depth. It is evident that all data collapse on a single curve for all flow types, with subcritical flow downstream of the step. The second order polynomial regression line (with correlation coefficient $R^2 = 0.98$) shown in the graph that relates the two mononyms, can be of use in the design of stilling basins, because it relates the incoming flow of known discharge q with depth y_1 (Froude number Fr_1), to the tailwater depth y_2 and the step height d. For example, if we consider a 5 m wide orthogonal channel conveying discharge of 30 m^3/s, with known (defined) upstream and downstream depths $y_1 = 0.60$ m and $y_2 = 4.00$ m, the Froude number and critical depth are $Fr_1 = 4.12$ and $y_c = 1.54$ m respectively. Then, from the implicit function with respect to step height, by trial and error one gets d = 2.58 m, corresponding to $y_2/[(y_c + d)Fr_1] = 0.24$ and $y_2/[(y_1 + d)Fr_1] = 0.31$.

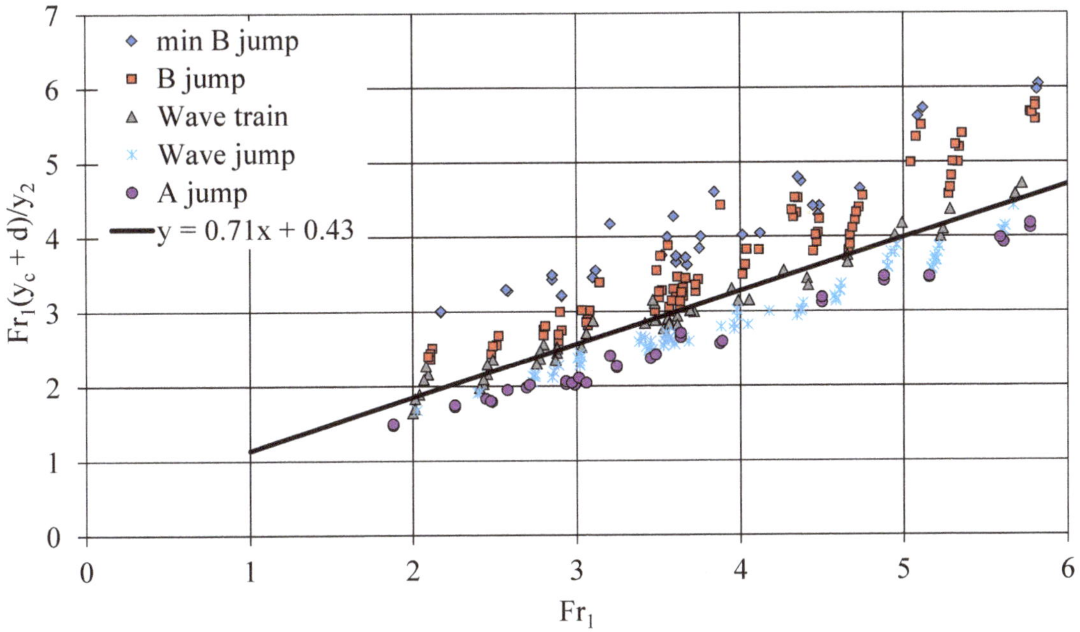

Figure 4. Distinction of the different flow regimes.

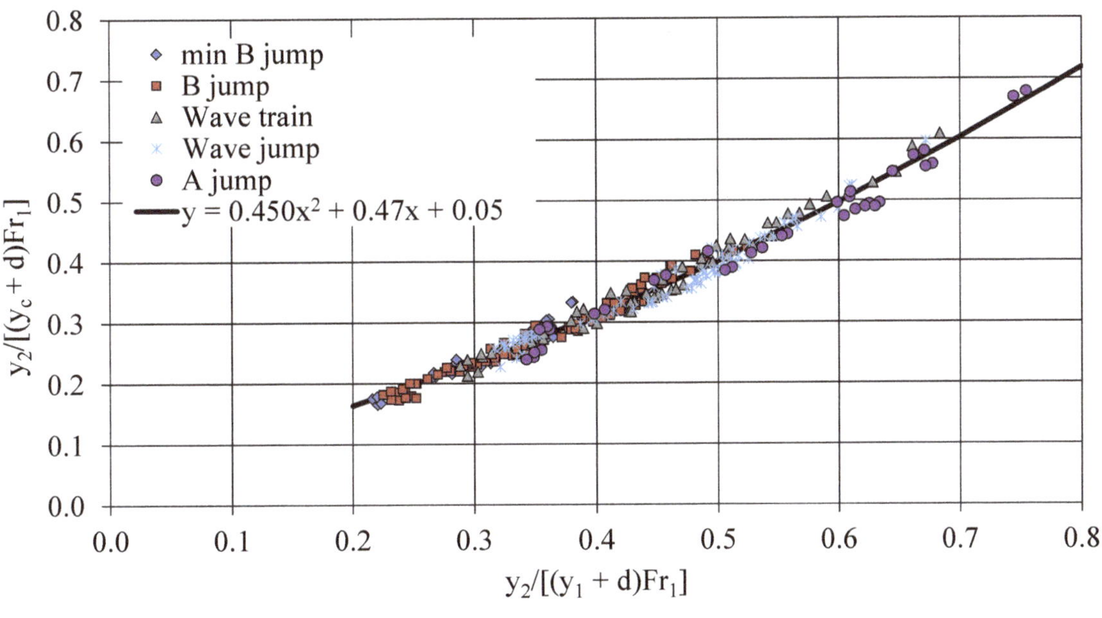

Figure 5. Relationship between y_1, Fr_1, d, y_c and y_2.

The energy loss calculated from the one-dimensional energy Equation (3) where uniform velocity distribution is assumed, is normalized by y_c and plotted versus the normalized length $Fr_1(y_c + d)/y_2$ for all different types of flow in Figure 6. All the data collapse on a second order polynomial:

$$y = -0.047x^2 + 1.38x - 2.10 \quad (4)$$

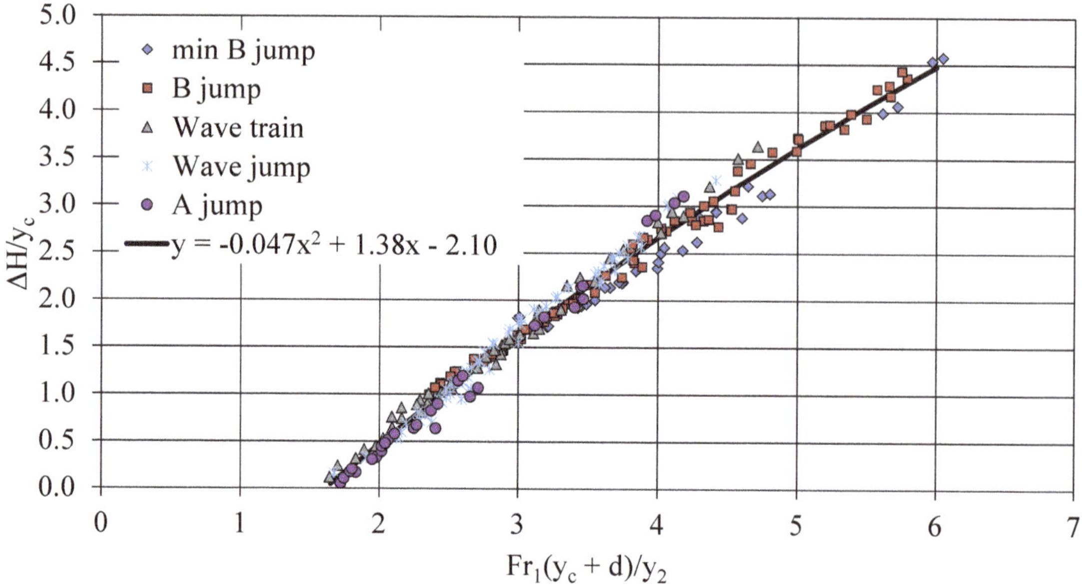

Figure 6. Normalized energy loss at the step $\Delta H/y_c$ versus $Fr_1(y_c + d)/y_2$.

This figure can be used along with Figure 5 to estimate energy loss for the design of a stilling basin that has been selected for construction.

The pressure force at the face of the step is corrected using a pressure correction coefficient k computed from Equation (2). Assuming uniform streamwise velocity and hydrostatic pressure distribution, k is computed using depth y_2 at the step face for the A-jump type of flow and y_1 for all other types of flow considered. Neglecting the momentum loss due to friction, the computed k is plotted against dimensionless length $(y_1 + d)/y_2$ in Figure 7. It is evident that only a few data regarding minimum B-jump are found in the regime $(y_1 + d)/y_2 > 1$, while the majority of the measurements are in the regime $(y_1 + d)/y_2 < 1$. The pressure correction coefficient k takes a value around 0.5 when the flow type is minimum B-jump and 1 for A-jump. If we substitute k = 1/2 in Equation (1) or k = 1 in modified (1) we end up.

$$\begin{gathered}\tfrac{1}{2}gy_1^2 + \tfrac{1}{2}gd\left(y_1 + \tfrac{d}{2}\right) - \tfrac{1}{2}gy_2^2 = q(V_2 - V_1)\\ \tfrac{1}{2}gd(y_1 + d)^2 - \tfrac{1}{2}gd\left(y_1 + \tfrac{d}{2}\right) - \tfrac{1}{2}gy_2^2 = q(V_2 - V_1)\end{gathered} \quad (5)$$

for the minimum B-jump and

$$\begin{gathered}\tfrac{1}{2}gy_1^2 + gd\left(y_2 - \tfrac{d}{2}\right) - \tfrac{1}{2}gy_2^2 = q(V_2 - V_1)\\ \tfrac{1}{2}gy_1^2 - \tfrac{1}{2}g(y_2 - d)^2 = q(V_2 - V_1)\end{gathered} \quad (6)$$

for the A-jump. The first equation shows that if we consider hydrostatic pressure distribution on the step face, the momentum from the left must be reduced by $gd(y_1 + d/2)/2$ for closure, that is equal to one half of the pressure force on the step face, if the water elevation is at $y_1 + d$. The second shows that the pressure distribution on the step face is hydrostatic, since the pressure force of the step face on the control volume is counter balanced by the force of the trapezoidal pressure distribution between elevations 0 and d from bottom (step face).

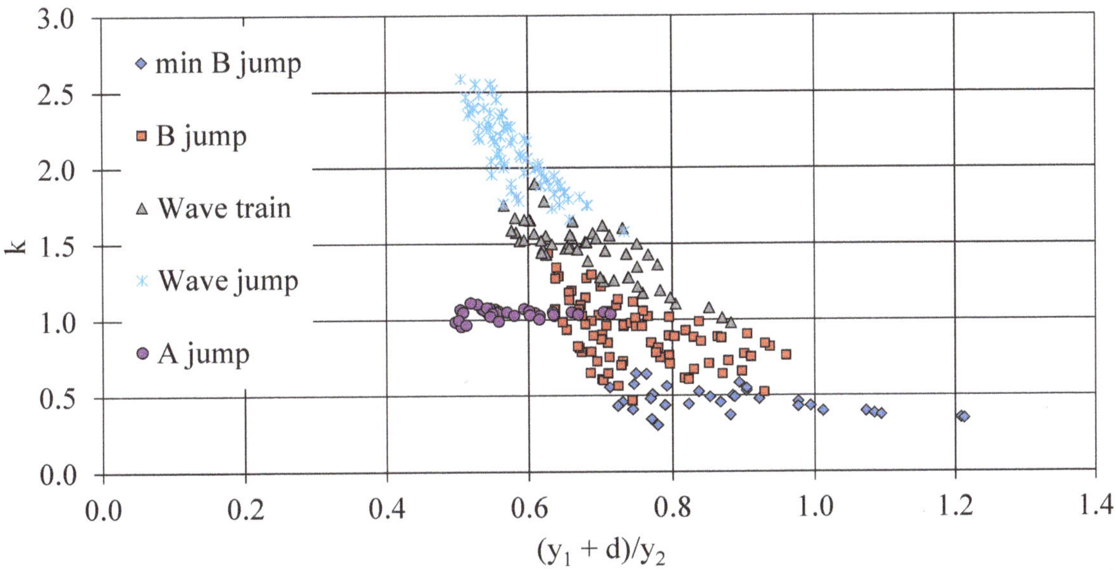

Figure 7. Correction coefficient k to balance the momentum equation versus $(y_1 + d)/y_2$.

When the hydraulic jump is submerged (B-jump) $0.5 < k < 1.5$ for $0.6 < (y_1 + d)/y_2 < 1$. If the flow does not impinge at the bottom, in other words if surface flow is observed downstream of the step face, then $1 < k < 2$ for $0.55 < (y_1 + d)/y_2 < 0.9$ (wave-train) and $k > 1.5$ for $0.5 < (y_1 + d)/y_2 < 0.7$ (wave-jump).

We present the pressure head measurements along the middle of the face of the step measured with three piezometers located 2.0, 4.5 and 7.0 cm above the channel bottom in Figure 8. The lines plotted are linear fits to measured pressure head, where the correlation coefficients were found to be greater than 0.99. The red line is the one corresponding to zero pressure at the top of the step. When the flow is supercritical or the minimum B-jump appears downstream, part of the step face appears to have negative pressure. When the flow type is B-jump, i.e., when the hydraulic jump is submerged, the pressure may be positive all over the face of the step, or negative in part of it, when the tailwater depth is low. For the case of the wave-train the pressure at the face of the step is marginally positive, depending upon the transient characteristic of the flow [4,13], while it is positive all over for the wave-jump and the A-jump.

Assuming linear pressure distributions at the face of the step we can compute the maximum pressure at the bottom. If we normalize the measured pressures with maximum pressure at the step and the vertical distance from the bottom versus the maximum pressure head we end up with a dimensionless graph p/p_{max} against $y/(p/\rho g)_{max}$ shown in Figure 9. One may observe that when the maximum pressure head is equal to the step height, $p/p_{max} \sim 0$, and may occur for three types of flow, supercritical, minimum B-jump and B-jump. For all other types of flow the pressure is positive over the face of the step and the maximum pressure head is greater than d.

We can normalize the horizontal distance x from the toe of the step using the theoretical distance where a free water jet atop the step with velocity V_1 would reach the bottom downstream, namely $2^{1/2}$ Fr_1d. In Figure 10 we have plotted the pressure head $p/\rho g$ at along the bottom downstream from the step normalized by y_2, versus the dimensionless distance $x/(2^{1/2}$ $Fr_1d)$ where the supercritical free water jet would theoretically impinge at the bottom, for all different types of jumps. It is evident that when the water jet strikes the bottom (B-jump and minimum B-jump), there is a peak pressure at some distance from the step that is due to stagnation, as a result of the vertical velocity component. In particular, for the case of the minimum B-jump the peak pressure occurred around $x/(2^{1/2}\ Fr_1 d) = 0.3$. Furthermore, the pressure upstream of the maximum is less than the hydrostatic one for the previous three types of the flow, which is a result of the curved streamlines of the water jet. When the flow runs on the surface (wave-train, wave-jump, A-jump) the peak disappears, the pressure is lower than hydrostatic up to about $x/(2^{1/2}\ Fr_1 d) = 0.5$ and it is attributed to the 'suction' developed from the fast water jet at the top of the step (Bernoulli).

Next, the normalized maximum pressure head $p_{max}/\rho g$ measured at the bottom of the channel downstream of the step by tailwater depth y_2, is plotted versus $y_2/(y_c + d)$ for all types of jumps in Figure 11. Two groups of points appear in this figure showing a sharp discontinuity, the group on the left from pressure measurements of the minimum B-jump and B-jump types of flow where $(p_{max}/\rho g)/y_2 > 1$, while the group on the right from pressure measurements of the wave-train, the wave-jump and the A-jump types of flow where $(p_{max}/\rho g)/y_2 < 1$. This sharp discontinuity occurs at $y_2/(y_c + d)$~1.07 where the flow type from B-jump converts to wave-train, i.e., when the flow becomes a surface jet with a very long recirculation regime downstream of the step. Apparently, the pressure distribution is not hydrostatic in both regimes, on the left because the piezometer measures the dynamic pressure that is due to the vertical velocity component besides the hydrostatic pressure mentioned earlier, while on the right where the fast surface flow 'sucks' the underlying slow recirculating fluid (Bernoulli).

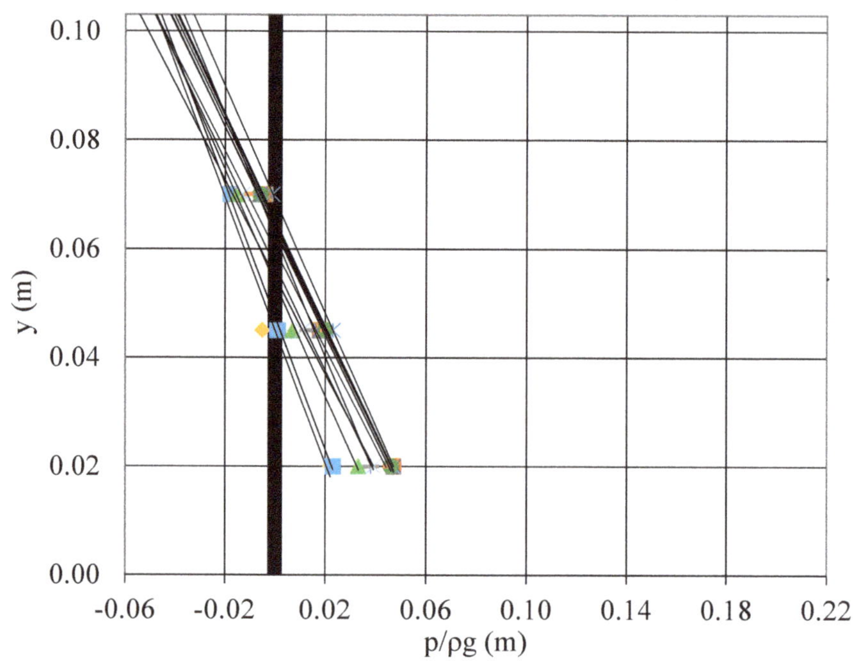

(i) supercritical flow

Figure 8. *Cont.*

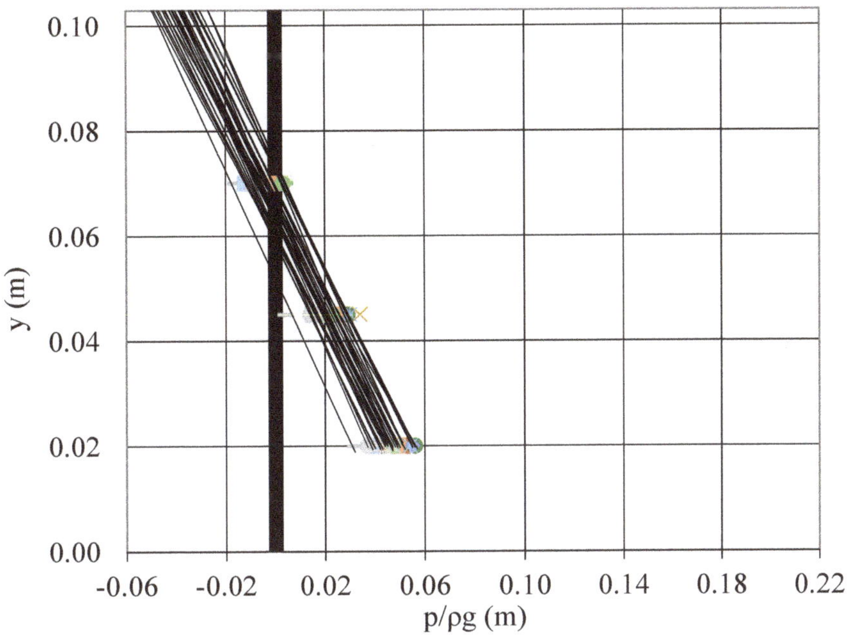

(**ii**) min B-jump

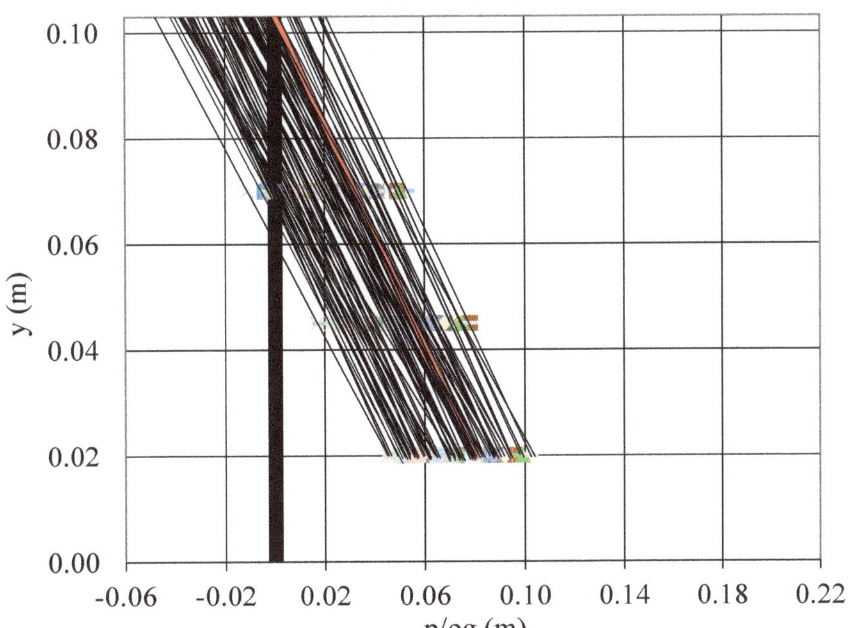

(**iii**) B-jump

Figure 8. *Cont.*

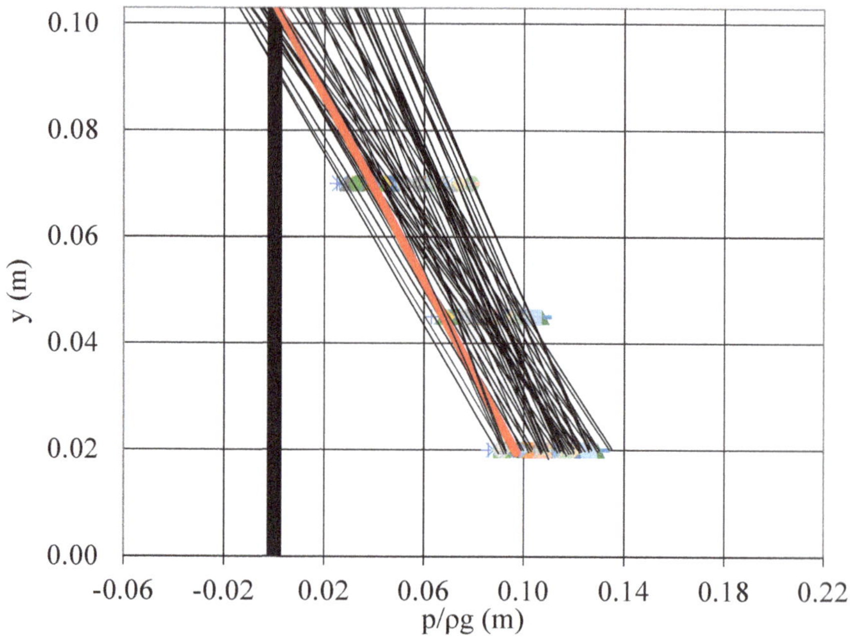

(**iv**) Wave-train

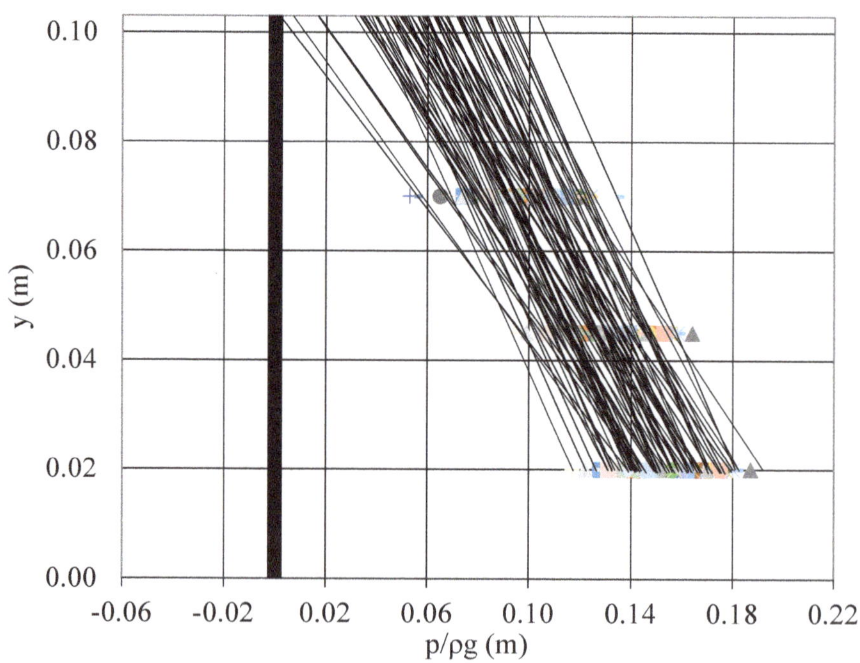

(**v**) Wave-jump

Figure 8. *Cont.*

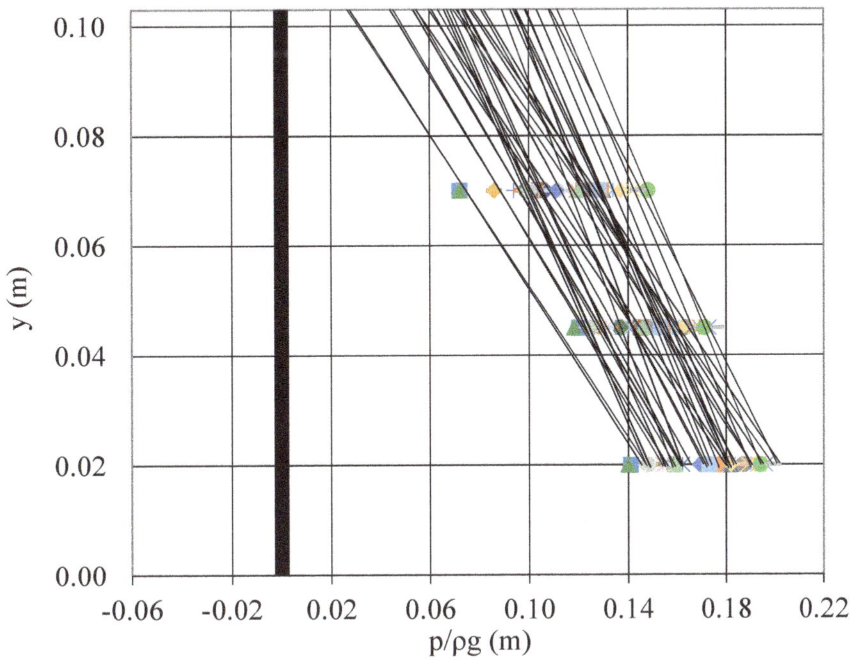

(**vi**) A-jump

Figure 8. Distribution of the measured pressure head at the face of step with three piezometers, extended linearly to the top (**i**) supercritical flow, (**ii**) min B-jump, (**iii**) B-jump, (**iv**) wave-train, (**v**) wave-jump, and (**vi**) A-jump.

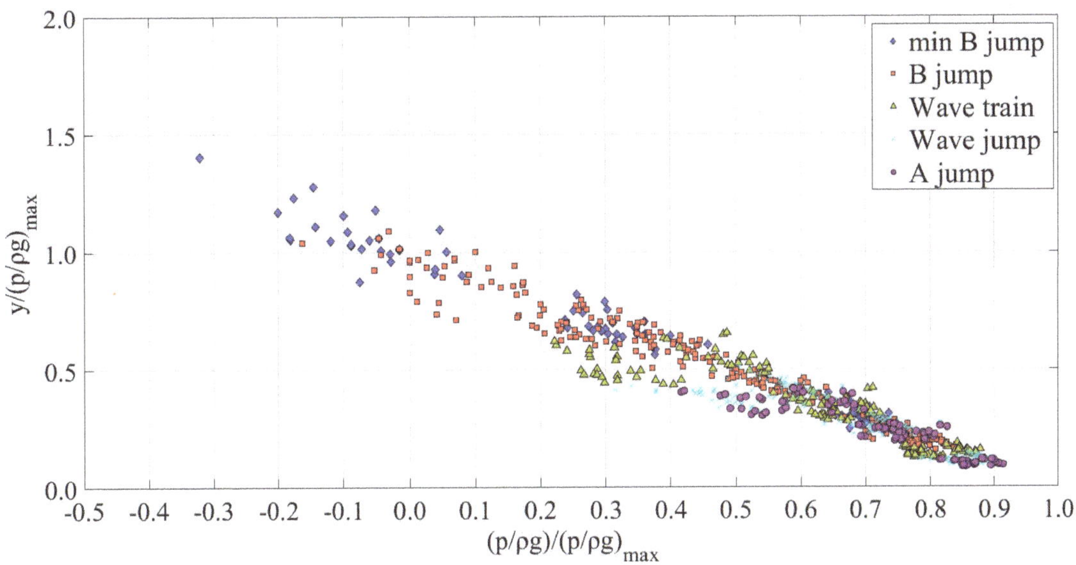

Figure 9. Dimensionless pressure head $(p/\rho g)/(p/\rho g)_{max}$ at the face of the step versus dimensionless vertical distance from bottom $y/(p/\rho g)_{max}$ for all types of flow.

The pressure distribution on the face of the step has been found to be linear along the three piezometers installed 2.0, 4.5 and 7.0 cm above the channel bottom, i.e., the lowest is 2 cm above the bottom while the highest 3.3 cm below the top of the step. If we assume that the pressure distribution is linear over the face of the step and extrapolate the measured linear pressure distribution to the bottom and the top of it, then we can compute the pressure force that the step exerts on the flow. Hence, it is evident to compare the pressure force on the step face estimated from the one-dimensional momentum equation, with the force computed from the hypothetical linear pressure distribution that is measured by the three piezometers and is extrapolated to the top and bottom of the step. Hence, the force estimated from the momentum equation is computed as the difference between pressure force and inertial force at Cross Sections 1 and 2 in Figure 2, assuming uniform velocity and hydrostatic pressure distribution, while the measured force is computed by integrating the assumed 'linear' pressure distribution over the step height. Both forces are normalized by the minimum force obtained for critical depth. The normalized estimated force is plotted versus the measured one in Figure 12, and found to be bigger as expected. Apparently, the differences are resulting from (1) negligence of the friction force from the walls and bottom of the channel and (2) from the hypothesis of linear pressure distribution at the face of the step.

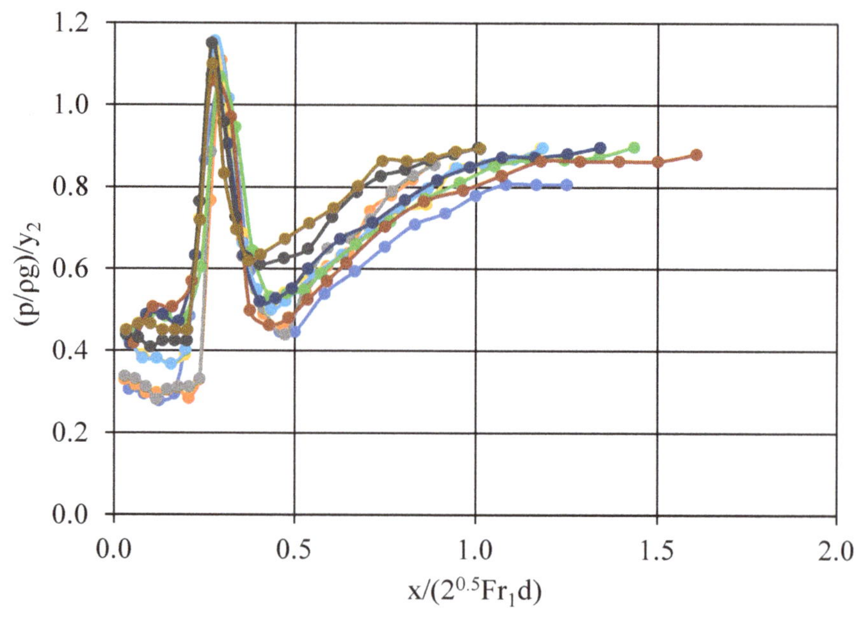

(i) min B-jump

Figure 10. *Cont.*

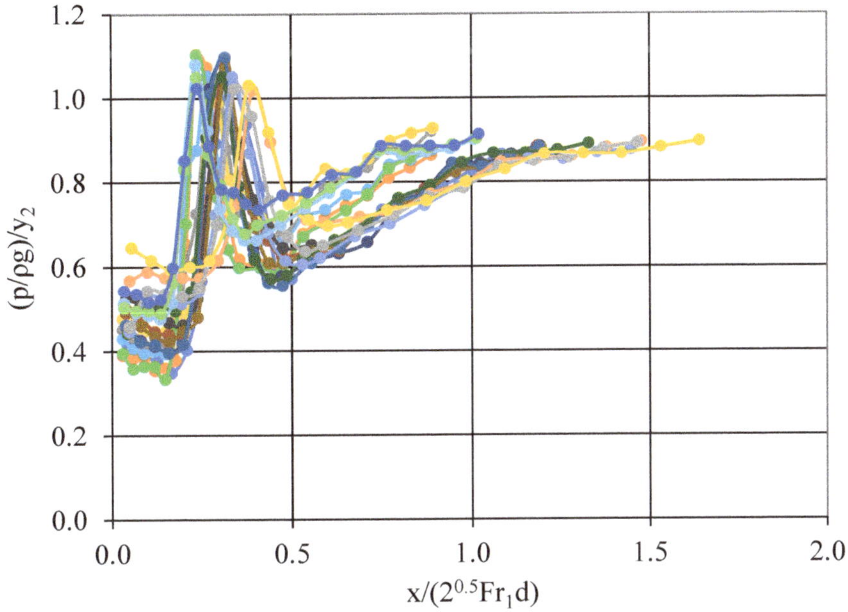

(ii) B-jump

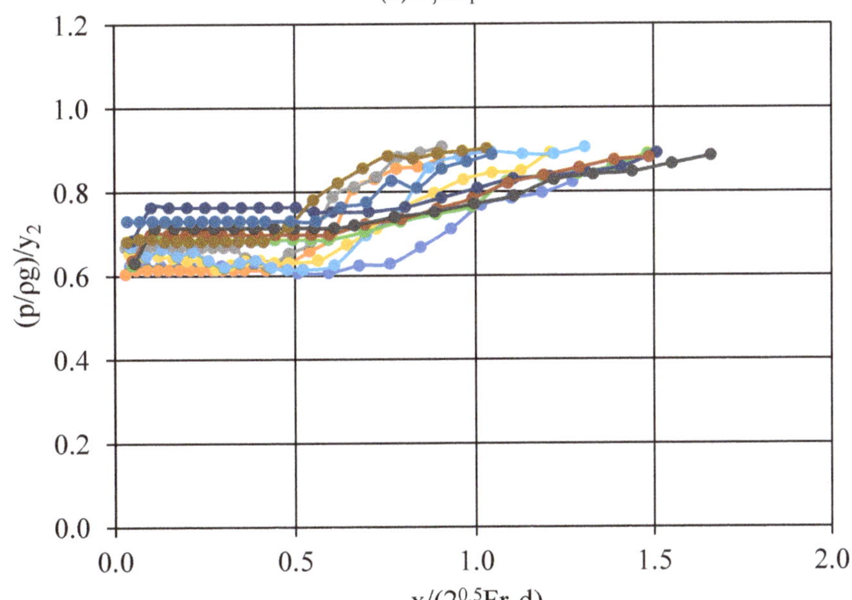

(iii) Wave-train

Figure 10. *Cont.*

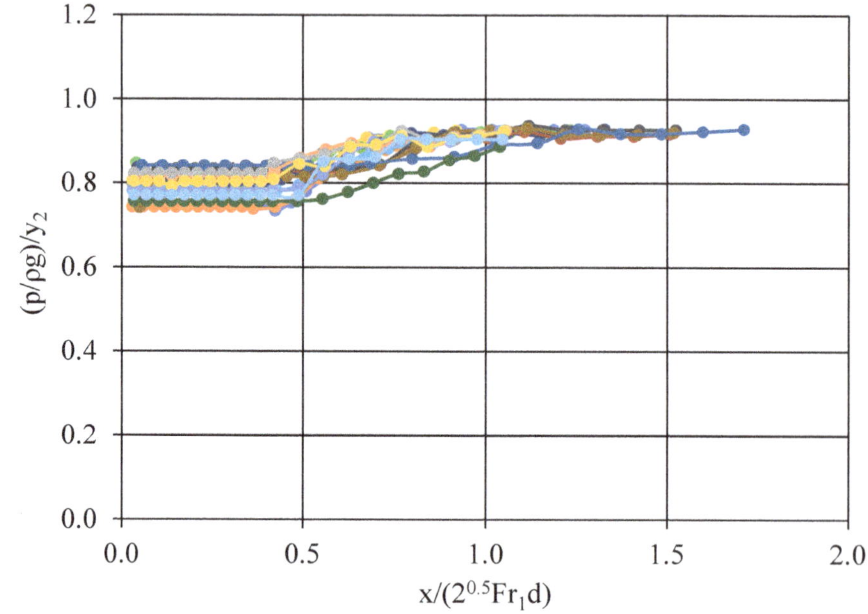

(**iv**) Wave-jump

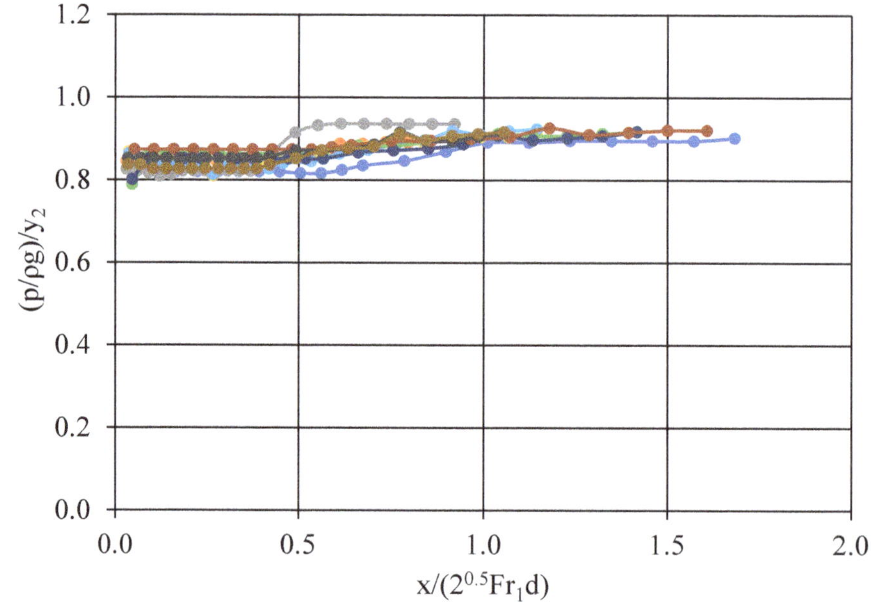

(**v**) A-jump

Figure 10. Distribution of dimensionless measured pressure head along the bottom of the channel versus the dimensionless horizontal distance downstream of the drop (**i**) min B-jump, (**ii**) B-jump, (**iii**) wave-train, (**iv**) wave-jump, and (**v**) A-jump.

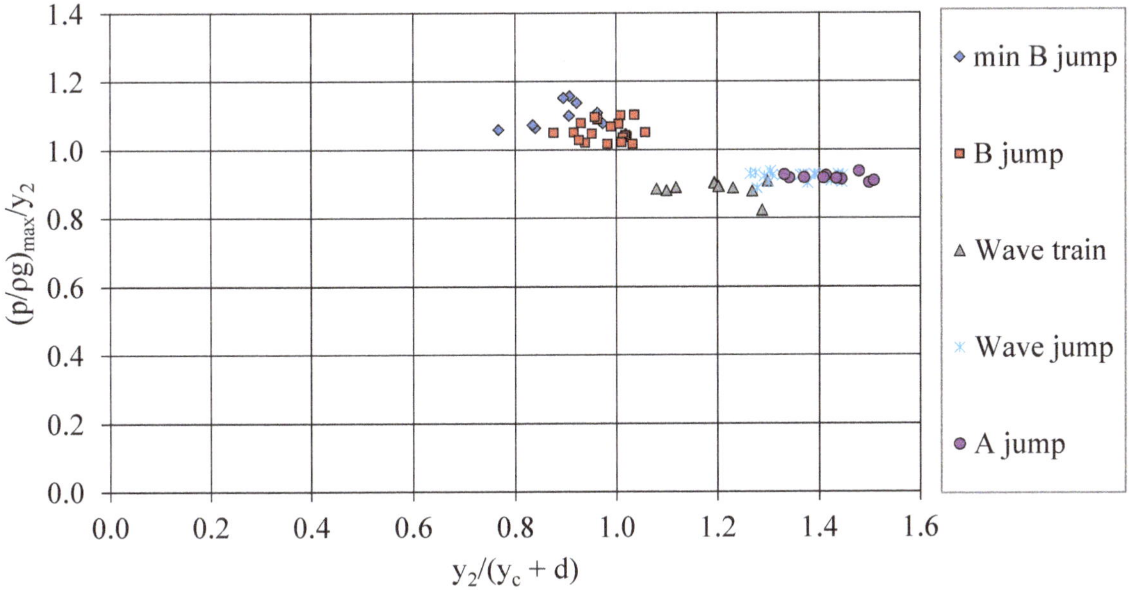

Figure 11. Dimensionless maximum pressure head at the bottom of the channel downstream of the step, versus $y_2/(y_c + d)$ for all types of jumps.

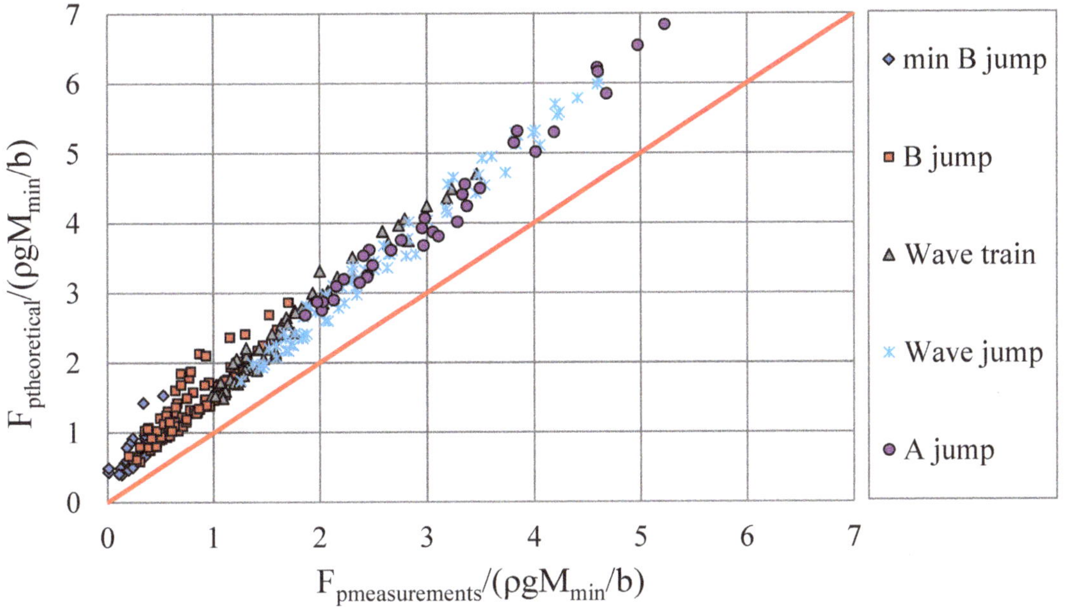

Figure 12. Comparison between the theoretical and the experimental results for the non-dimensionless pressure force at the face of the step for all types of jumps.

4. Numerical Modeling

4.1. Governing Equations

Boussinesq equations have been used to model the unsteady one-dimensional rapidly varied open channel flow [27] and simulate the minimum B jump and A jump. These include additional terms if compared to Saint Venant equations from the non-hydrostatic pressure distribution resulting from the curved streamlines. The channel is prismatic with rectangular cross section and rigid bottom and sides with no lateral flow. The assumptions made are the following: (1) the vertical velocity is zero at the channel bottom and maximum at free surface, (2) the streamwise velocity is uniformly distributed over the depth, (3) the lateral velocity is zero, (4) the fluid is incompressible, (5) the bottom slope is small and (6) the formula for energy friction slope of steady flow is used for the unsteady flow. The one-dimensional Boussinesq equations for mass and momentum conservation in vector form are

$$\frac{\partial G}{\partial t} + \frac{\partial F}{\partial x} = S, \tag{7}$$

where

$$G = \begin{bmatrix} y \\ uy \end{bmatrix}, \; F = \begin{bmatrix} uy \\ u^2y + \left(\frac{1}{2}\right)gy^2 - \left(\frac{1}{3}\right)y^3E \end{bmatrix}, \; S = \begin{bmatrix} 0 \\ gy(S_o - S_f) \end{bmatrix}, \tag{8}$$

$$E = \frac{\partial^2 u}{\partial x \partial t} + u\frac{\partial^2 u}{\partial x^2} - \left(\frac{\partial u}{\partial x}\right)^2 \tag{9}$$

In equations above, x is the longitudinal distance along the channel measured from the sluice gate, t is the time, y = y(x,t) and u = u(x,t) are the unknown depth and average over the cross section velocity in the main flow direction, t S_f the energy grade slope, S_o the longitudinal bottom slope, g the gravitational acceleration and E = E(x,t) the Boussinesq term. The energy slope computed from Manning formula in SI units is $S_f = n_f^2u^2/R^{4/3}$, where n_f is the Manning friction coefficient, u the mean over the wetted cross-section velocity and R the hydraulic radius.

The system of equations can only be solved numerically. The Dissipative Two-Four [28] and the MacCormack [29] finite difference schemes were applied for the discretization of the mass and momentum conservation equations, with the appropriate initial and boundary conditions. The first scheme is fourth order accurate in space and second order accurate in time, while the second is second order accurate both in space and time, both allowing for the proper simulation of the Boussinesq terms as well as the free surface of hydraulic jump. Iterations continue until the depth difference between two successive iterations is less than a fixed value, then the minimum B jump or the A jump form as part of the steady state solution. Implementation of the numerical schemes used for the solution of Equation (7) is presented in the Appendix A.

The input data for the developed algorithm include the geometry of the channel, the flow depths y_{up} and y_{do} as well as the flow rate, Q. The algorithm has been programmed in house using the Matlab® computational environment.

4.2. Results

Four experiments have been selected, two regarding formation of the minimum B-jump and two of the A-jump shown in Table 2, where the flow rate Q, upstream Froude number Fr and depth y_{up} at the vena contracta, and the depth y_{do} upstream of the weir for each jump have been considered. These experiments were modeled using the same flow conditions (upstream and tailwater depths and the flow rate), and the numerical results are compared to the experiments in the following paragraphs.

Table 2. Measured parameters for modeling of the hydraulic jumps.

Test Case/Experiment	Q (L/s)	Fr	y_{up} (m)	y_{do} (m)	Type of Jump
1	8.11	3.59	0.0200	0.1259	minimum B
2	9.88	4.37	0.0200	0.1442	minimum B
3	6.70	3.20	0.0190	0.1922	A jump
4	9.41	4.50	0.0190	0.2234	A jump

The spatial step was 0.025 m resulting in 201 nodes for all cases. The time interval in each iteration was variable for stability purposes, subject to Courant–Friedrichs–Lewy condition. Artificial viscosity was added to the numerical schemes to reduce oscillations in the region of the jump. The dissipation parameter has been set to 0.012 after trial and error, and was applied to all cases. The computed flow depth difference in all nodes between two successive iterations did not exceed 10^{-4} m for convergence.

In Figures 13 and 14 the numerical results are compared to experiments regarding test case 1 (min B-jump) and test case 3 (A-jump) respectively. In the same figures we have plotted the Boussinesq terms computed, using the two different schemes for comparison. From these figures it is evident that Boussinesq term is significant in the region of the jump owing to the non-parallel streamlines where it takes the highest values, while is vanishes everywhere else. The location of the jump is in acceptable agreement with measurements, especially in the case of the A-jump (Figure 14 top), while the jump shape cannot be predicted numerically, as expected. The measured pressure at the bottom of the channel is also plotted upstream and downstream of the step for comparison. It is asymptotically congruent to the computed depth in the uniform (parallel) flow sections upstream and downstream of the hydraulic jump in both cases, while the hydrostatic pressure distribution is confirmed outside the region of the jump, from the almost zero value of the Boussinesq term, as expected.

The computed mean velocity over the cross section along the channel is plotted for test cases 1 and 3 in Figure 15, and is not different for both computational schemes. The MacCormack scheme overestimates slightly the velocity at the upstream end of the channel, while the two numerical schemes produce almost identical results downstream. The 'computational pseudo-time' evolution of the hydraulic jump until steady state is shown in Figure 16 for test cases 2 (min B-jump) and 4 (A-jump), using the Dissipative Two-Four and the MacCormack scheme respectively. In these figures the free surface profile is also shown at time t = 0, resulting from the initial condition, while it can be noted that the jump moves upstream until it is stabilized in its final location. Similar numerical results have been produced for the other test cases.

The required iterations for the algorithm to reach steady state and the maximum percentage error regarding mass conservation are shown in Table 3 for all test cases and both numerical schemes. The depth of flow and mean velocity were used to compute the flow rate. It is evident that the mass conservation error from the MacCormack scheme is smaller if compared to that of the Dissipative Two-Four scheme, except for case 4, while the algorithm reaches steady state solution faster if the MacCormack scheme is used, requiring smaller number of iterations except for test cases 2 and 4.

Table 3. Mass balance error and number of iterations for convergence.

Test Case/Experiment	Numerical Scheme	Maximum Mass Conservation Error (%)	Iterations
1	Dissipative Two-Four	4.01	6117
	MacCormack	3.59	5779
2	Dissipative Two-Four	4.07	4866
	MacCormack	3.64	5067
3	Dissipative Two-Four	2.45	5474
	MacCormack	2.13	4934
4	Dissipative Two-Four	3.92	5965
	MacCormack	4.09	6355

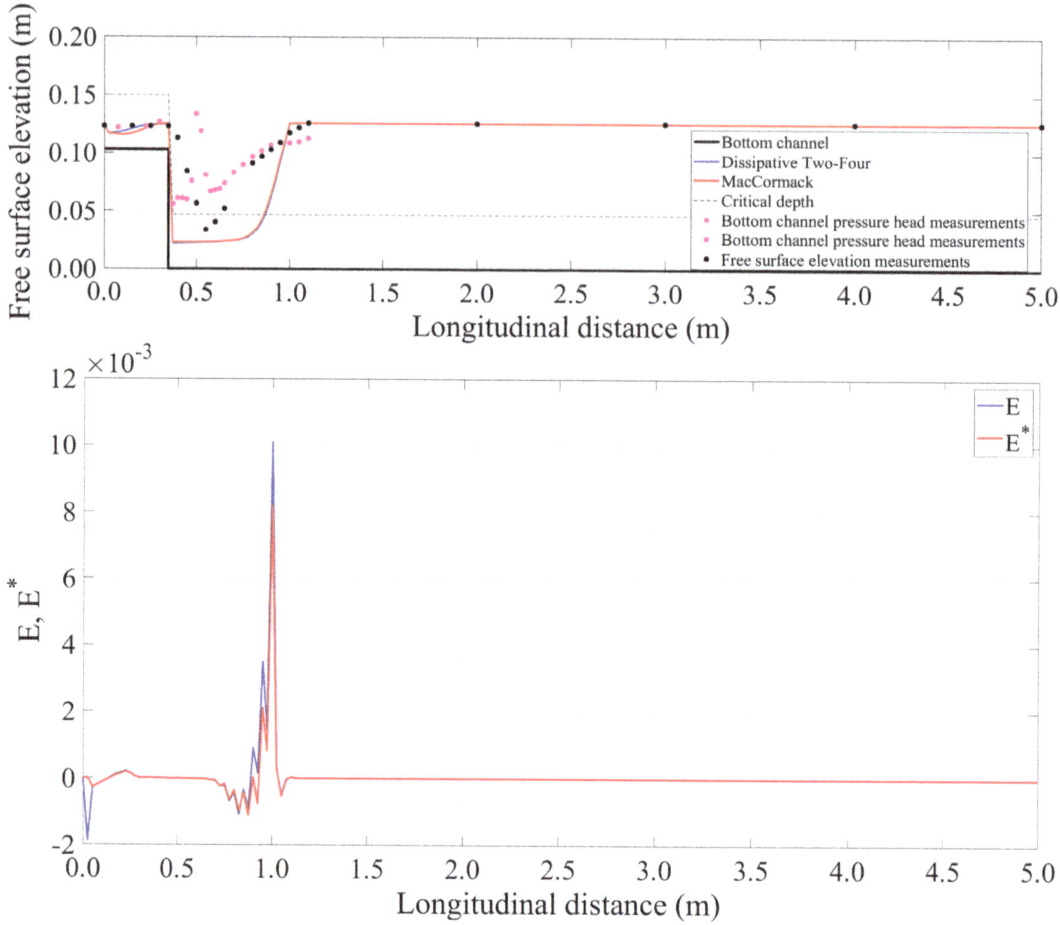

Figure 13. Test case 1: (**Top**) Comparison between the computed free surface profile and the experiment (point gauge and pressure head measurements). (**Bottom**) Numerical results for the Boussinesq term along the channel for the Dissipative Two-Four scheme, prediction and the correction step.

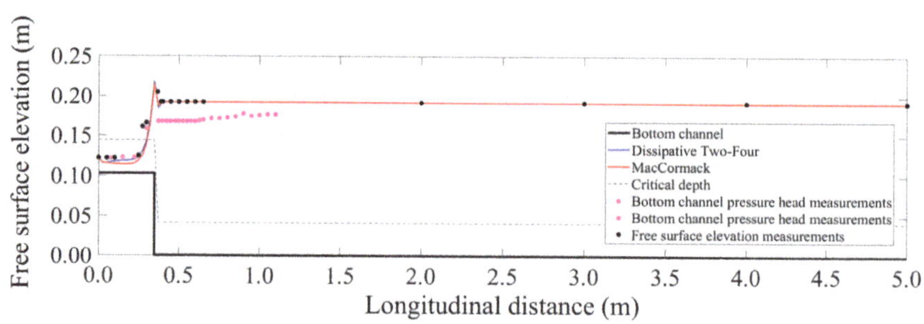

Figure 14. *Cont.*

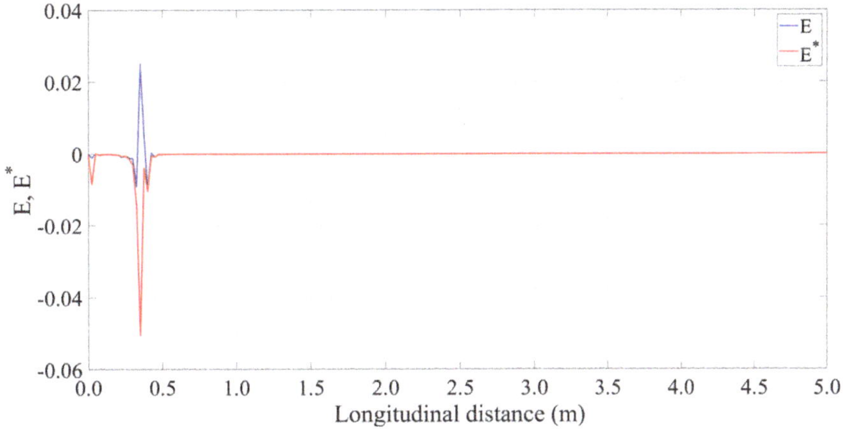

Figure 14. Test case 3: (**Top**) Comparison between the computed free surface profile and the experiment (point gauge and pressure head measurements). (**Bottom**) Numerical results for the Boussinesq term along the channel length for the MacCormack scheme, prediction and the correction step.

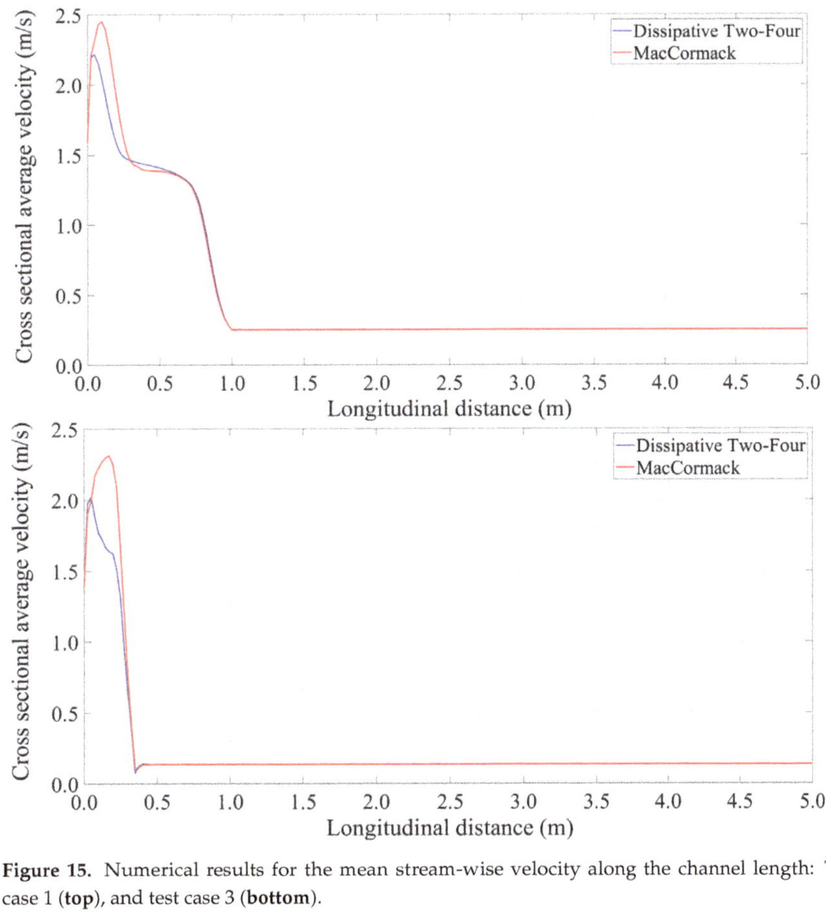

Figure 15. Numerical results for the mean stream-wise velocity along the channel length: Test case 1 (**top**), and test case 3 (**bottom**).

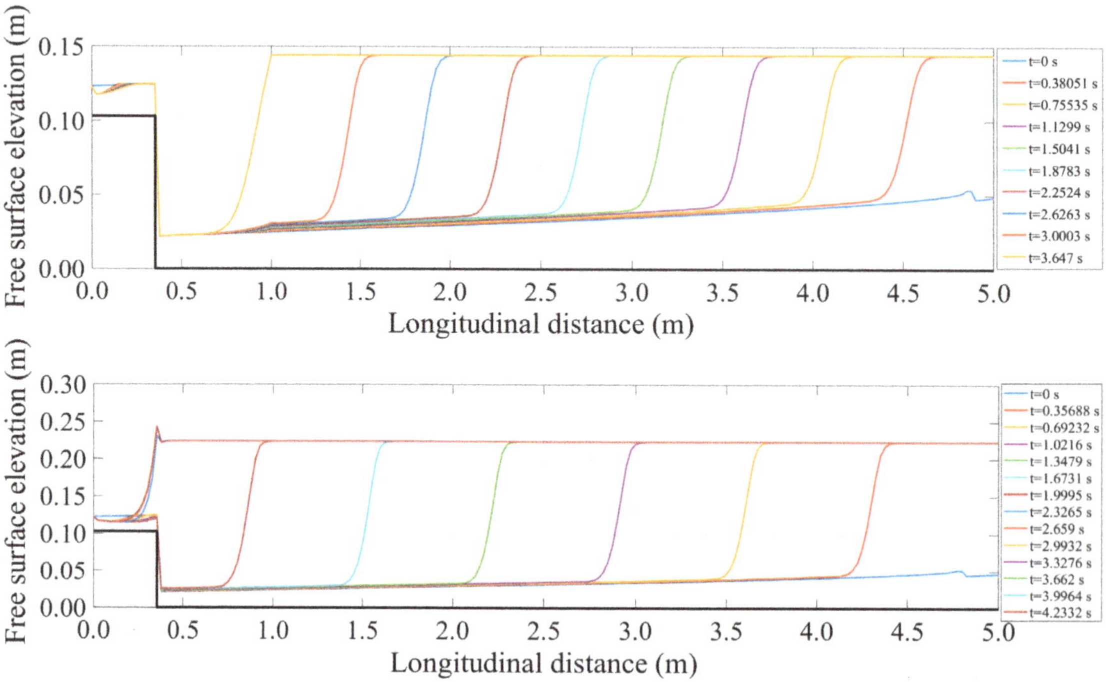

Figure 16. Numerical results for the temporal evolution of the jump: Test case 2 (**top**) for the Dissipative Two-Four scheme, and test case 4 (**bottom**) for the MacCormack scheme.

5. Discussion

The presence of a fully submerged step in an orthogonal channel with supercritical flow upstream and subcritical downstream results in five major types of rapidly varying flow. They are, in the order of appearance for increasing tailwater depth: (i) minimum B-jump, (ii) B-jump, (iii) wave-train, (iv) wave-jump and (v) A-jump. For the first two types of jumps the supercritical water jet impinges at the bottom, while for the other types the water jet moves at the surface.

A characteristic length scale used for the description of the flow and to normalize the measured lengths, was found to be the sum of the step height and critical depth $d + y_c$, regarding the potential energy height and the minimum energy (critical) depth of the flow. To apply the momentum equation in the flow direction, the assumption of the hydrostatic pressure distribution at the face of the step had to be reevaluated, and closure was obtained with the use of a pressure correction coefficient k. The one-dimensional momentum equation has led to an equation for the theoretical calculation of k as a function of measured flow depths y_1 and y_2, the step height d and the discharge per unit width q (Equation (2)). The coefficient k for the case of the downward moving water jet (minimum B-jump and B-jump) was found in general less than one, while for the case of a surface water jet (wave-train, wave-jump and A-jump) was found greater than one (Figure 7). Note that in some measurements regarding B-jump k was greater than the unity, possibly due to the alternating flow characteristics between B-jump and wave-train, something that has been also observed earlier [13]. A comparison of k computed from experiments from earlier investigations is shown in Figure 17, where k has been plotted regardless of the type of flow. From this graph it is evident that k takes the highest values if $(y_1 + d)/y_2 < 0.5$, and the lowest ones if $(y_1 + d)/y_2 > 1$. There are experiments [4,6,12,22] where k computed from the measurements takes negative values, a result that is considered to be ambiguous, since the momentum upstream of the step must always exceed that of tailwater. One issue could be related to the different accuracy of the measurements presented in those studies. Another issue could be related to the size of the step and the type of the flow. In fact, from the analysis of the experiments, it turned out that the types of flow that gave negative values of k are the minimum B-jump and B-jump for small height steps. Those flows that impinge at the bottom if combined with small step height may produce negative pressure profile (see for example Figure 8ii where part of the pressure is negative in a step that is high if compared to y_c). The data of [4] show several points around k = 0 a result that is attributed to small step size, if compared to the critical depth of the flow. The data show a very large scatter for $(y_1 + d)/y_2 < 0.5$, where k in some cases is much larger than 2. The high values of k mean that the pressure at the face of the step is greater than hydrostatic (Figure 8v). A closer look at the experiments showed that all the data that gave k > 2 correspond to wave-jump. Hence, the recirculation roller under the jump where the flow expands in depth increases the pressure abruptly on the step face, resulting in large values of k and ejection of the water jet upwards. Moreover, the discrepancy of the data may be attributed to the different size of the channel width and flow parameters used in different experiments as shown in Table 4.

Apparently, the flow downstream of the step is three-dimensional (3D) based on the ratio of the channel width to the tailwater depth $b/y_2 < 5$. Note that the flow is not affected by the side walls in a wide channel with $b/y_2 > 5$ [27]. It is evident that in a wider channel the flow is rather two-dimensional with no secondary flow, if compared to a narrower one where the tailwater depth is around the same size of channel width, with strong secondary flow and three-dimensional flow characteristics. Moreover, discrepancies may occur from the size of the step height used, if compared to a characteristic length, such as the critical depth y_c. If $d/y_c < 1$, the curvature of streamlines at the drop is small, resulting in a minimal effect of the step in the momentum equation. The effect of the geometrical and hydraulic parameters is also shown in Figure 18 where the present data of Figure 5 are plotted in Figure 18 along with data from earlier investigations. There is a significant spread of the data that is due to the geometric characteristics and flow parameters used.

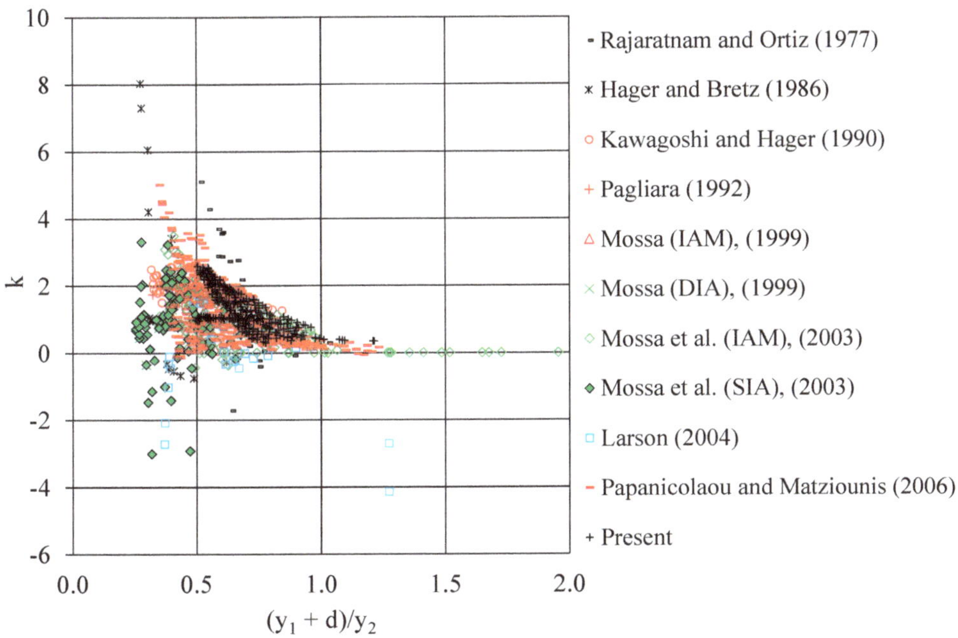

Figure 17. Comparison of k versus $(y_1 + d)/y_2$, from the present and earlier investigations [4,6,7,9,12,13,19,22,23].

Table 4. Main geometric and hydraulic parameters of earlier and the present experiment.

Researchers	b (m)	d (cm)	q (l/s/m)	Fr_1	d/y_c	Flow Dimensionality
Rajaratnam & Ortiz 1977 [12]	0.410	3.60–7.60	35.79–145.24	2.97–10.55	0.40–1.43	3D
Hager & Bretz 1986 [6]	0.500	7.60	60.00–400.00	3.93–5.71	0.36–1.06	3D
Kawagoshi & Hager 1990 [19]	0.500	5.00–7.70	5.98–179.56	1.99–13.68	0.37–5.00	3D
Pagliara 1992 [7,9]	0.500	3.72–8.45	9.80–138.00	1.85–6.90	0.45–2.78	3D
Mossa 1999 [13]	0.300 0.400	5.30–10.00 3.20–6.52	23.33–62.00 33.93–80.30	3.19–8.87 2.77–9.92	0.72–1.68 0.41–1.06	3D
Mossa et al., 2003 [4]	0.300 0.400	5.30–16.00 3.20–6.52	21.47–65.11 33.68–80.37	1.56–10.24 1.78–10.33	0.72–3.58 0.39–1.07	3D
Larson 2004 [22]	0.610	9.72–30.48	95.63–386.22	4.10–6.41	0.39–2.22	3D
Papanicolaou & Matziounis 2006 [23]	0.100	2.50–10.00	29.90–71.50	1.73–4.91	0.31–2.17	3D
Present	0.255	10.30	25.32–67.08	1.88–5.82	1.34–2.56	3D

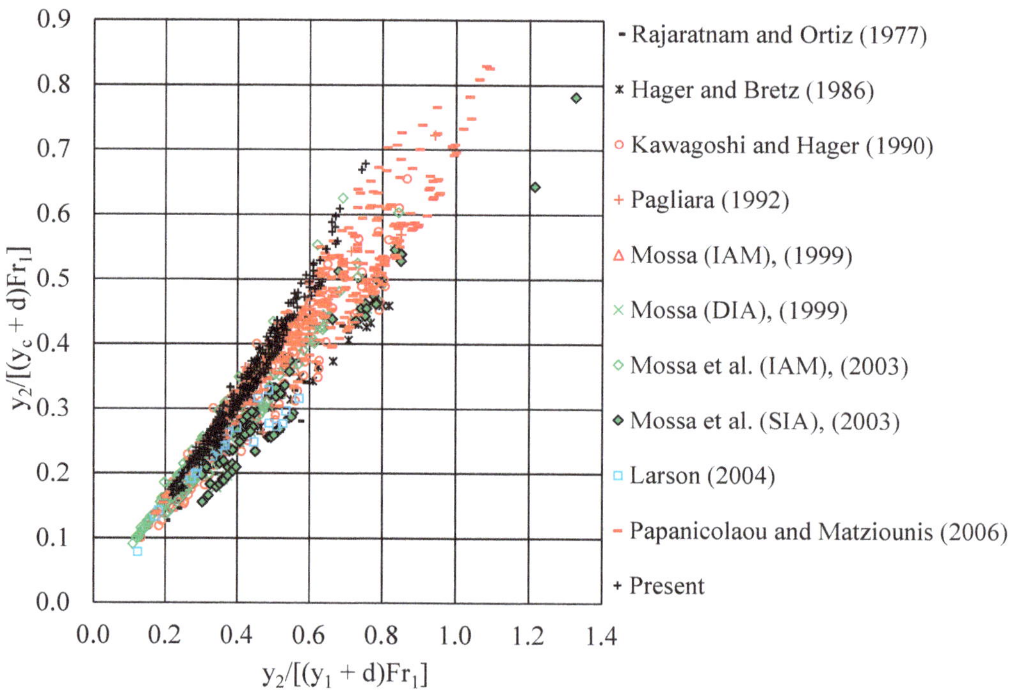

Figure 18. Comparison of $y_2/[(y_c + d)Fr_1]$ versus $y_2/[(y_1 + d)Fr_1]$, from the present and earlier investigations [4,6,7,9,12,13,19,22,23].

The normalized energy loss computed from the one-dimensional energy equation with critical depth $\Delta H/y_c$ is a function of the dimensionless parameter $Fr_1(y_c + d)/y_2$ as shown in Figure 6. A quadratic equation can give an estimate of $\Delta H/y_c$ when the flowrate, step height, Froude number and tailwater depth are given. In Figure 19 we have plotted the present data along with data from earlier experiments for comparison. Data [4] show the same trend as the present ones and those in [22], while those by other authors have shown higher energy loss even in the regime $2 < Fr_1(y_c + d)/y_2 < 6$ of the present data. From Table 4 one may note that the data with higher energy loss correspond to a two-dimensional type of flow where the open channel used was quite wide. The hydraulic jump is an energy loss mechanism where dissipation occurs due to vigorous mixing (mixing energy) at Kolmogorov scale where the viscosity is dominant, converting the high kinetic energy of the flow into heat. In wider channels the side walls do not affect the flow, thus enhancing mixing, while in narrower ones the side wall shear suppresses turbulence and subsequently the energy losses. This is evident in the experiments by Mossa et al. [4], where the normalized loss in the 0.40 m wide SIA channel is higher than that in the 0.30 m wide IAM channel for the same values of $Fr_1(y_c + d)/y_2$ in the horizontal axis. Moreover, the experiments by Larson [22] in a 0.61 m wide channel show higher energy losses than those in measurements in narrow channels [23].

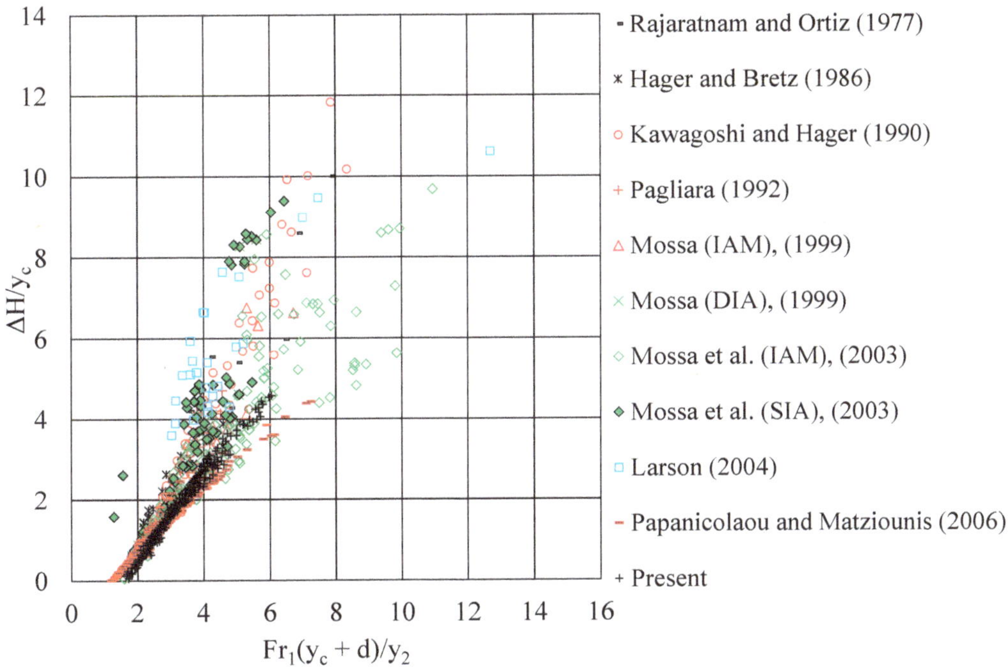

Figure 19. Dimensionless energy loss $\Delta H/y_c$ as a function of $Fr_1(y_c + d)/y_2$ for all types of jumps [4,6,7,9,12,13,19,22,23].

Measurements of the pressure at the step face and downstream of the toe have been performed by piezometers installed along the middle of the channel cross section. The pressure at the face of the step measured at three points was found to vary linearly with distance from bottom. The pressure was extrapolated linearly to the top and bottom of the step, and part of it was found to be negative near the top for the cases of supercritical flow, minimum B-jump and B-jump, while it was positive on the face of the step for the wave-train, wave jump and A-jump. The normalized pressure by the maximum pressure at the bottom was zero at the face of the step at dimensionless height $y/(p/\rho g)_{max}$~1 from bottom, only for minimum B-jump and B-jump.

The pressure head at the bottom downstream of the step has shown a maximum that greater than tailwater depth, and occurred at normalized distance $x/(2^{1/2}Fr_1 d)$~0.25 from the step for the minimum B-jump and B-jump, while the pressure head was lower everywhere else up to $x/(2^{1/2}Fr_1 d)$~2. For the types of flow wave-train, wave-jump and A-jump the pressure was lower than hydrostatic up to about $x/(2^{1/2}Fr_1 d)$~1.50, meaning that the water jet creates a surface flow that sucks the fluid from recirculation zone below it. The normalized maximum pressure head $(p/\rho g)_{max}/y_2$ at the bottom of the channel downstream of the step was found greater than 1 for $y_2/(y_c + d) < 1.07$ for min B-jump and B-jump, and lower than 1 for $y_2/(y_c + d) > 1.07$ for all other types of flow. Hence, one may observe that the water jet flows at the surface once the tailwater depth for $y_2 > 1.07(y_c + d)$. Finally, the normalized force at the face of the step computed from the pressure distribution was found lower than the estimated one from closure of the momentum equation, a result that is expected since in theoretical analysis friction force has been neglected, and pressure distribution at the step is assumed to be linear.

Regarding the numerical modeling of the free surface and the location of the minimum B jump as well as the A jump, the Boussinesq equations were discretized with two finite difference schemes, the Dissipative Two-Four scheme and the MacCormack scheme. Apparently, the RANS equations combined with a turbulence closure model or SPH can capture the turbulent structure of a steady hydraulic jump but the computational cost is high, since at high Reynolds numbers the computational time required for a decent computer is very long. Detailed discussion on this may be found in [30,31], while in [31] it is stated that RANS equations combined with a turbulence closure model can model the mean flow variables with accuracy over 90%, including also air concentrations. Moreover, the high fidelity Eulerian methods such as the LES or DNS or the Langragian method SPH can capture the turbulent structure of a steady hydraulic jump but the computational cost is high. For practical civil engineering applications shallow water modeling is much simpler to use while the one-dimensional shallow water equations can capture the basic characteristics of a hydraulic jump with acceptable accuracy. Computations of the flow depth and the average cross-sectional velocity at the downstream boundary node and at the step were done using the method of characteristics and an iterative convergence algorithm for the flow depth between two successive iterations. Experiment and numerical results regarding the free surface elevation were in agreement, thus validating the numerical algorithm. In the four test cases examined the MacCormack scheme has shown smaller error in mass conservation.

6. Conclusions

From the analysis of experimental data and the appropriate non dimensional representation of the findings in a submerged abrupt drop in an orthogonal channel with supercritical flow upstream where $Fr < 6$ and $1 < d/y_c < 3$, the following conclusions are drawn:

- Five different types of subcritical flow are observed downstream of a submerged vertical step in an orthogonal channel, minimum B-jump and B-jump where the supercritical jet impinges at the bottom if dimensionless depth $y_2/(y_c + d) < 1.07$, and wave-train, wave jump and A-jump when the water jet moves at the surface if $y_2/(y_c + d) > 1.07$. These can be distinguished if the normalized depth $Fr_1(y_c + d)/y_2$ is plotted against Fr_1.
- For all five different types of flow, there is an equation relating the upstream and downstream depths y_1 and y_2, the critical depth and Froude number (Figure 5), from which one may compute the step height that fulfills these data.
- The energy loss in dimensionless form $\Delta H/y_c$ for each type of flow can be estimated using Figure 6, where it is plotted versus the normalized length $Fr_1(y_c + d)/y_2$.
- Regarding the closure of momentum equation, for the limiting case of minimum B-jump the pressure correction coefficient $k = 1/2$ is equivalent to pressure force upstream from a linear pressure distribution extended to depth $y_1 + d$ but reduced by $gd(y_1 + d/2)/2$ for closure; for the limiting case of A-jump the pressure correction coefficient $k = 1$ is equivalent to pressure force downstream from hydrostatic pressure distribution to depth $(y_2 - d)$ from free the surface for closure.
- The pressure distribution measured at the face of the step was linear. If extended to the top of the step, there was a regime of negative pressure for the minimum B-jump and B-jump types of flow. The pressure head at the bottom downstream of the step showed a maximum that exceeded the tailwater depth for the minimum B-jump and B-jump types of flow, while around the toe was less than y_2 for all other types of flow.

- The present experiment can be useful to the hydraulic engineer in the design of stilling basins with abrupt negative steps and other structures relevant to the dissipation of kinetic energy of water.
- The numerical results showed that Boussinesq equations can simulate the basic flow characteristics of the minimum B-jump and the A-jump with acceptable accuracy.

Author Contributions: E.R.: formal analysis and investigation, hardware design, implementation of experiments, code implementation, software, validation, writing—original draft preparation. P.P.: conceptualization, methodology, formal analysis and investigation, supervision, validation, editing—original draft; final approval. All authors have read and agreed to the published version of the manuscript.

Funding: This research received no external funding.

Institutional Review Board Statement: Not applicable.

Informed Consent Statement: Not applicable.

Data Availability Statement: The available datasets analyzed in this study as well as the codes are available upon request from the corresponding author.

Acknowledgments: The financial support by Eugenides Foundation, 387 Sygrou Ave., 17564 Athens, Greece, for the reconstruction of the test section of the channel and construction of the water recirculation system is gratefully acknowledged.

Conflicts of Interest: The authors declare no conflict of interest.

Appendix A

On the Numerical Solution of Boussinesq Equations

The physical and computational grid for the solution of the governing equations is shown in Figure A1. The jump forms inside a horizontal open channel with rectangular cross section of width b and flow rate Q, combining the use of an upstream sluice gate and a downstream sharp crested weir, with the distance between them to be 5.20 m. The longitudinal distance x is measured from the origin set at vena contracta downstream of the sluice gate. The distance between vena contracta and the weir is L = 5.00 m (region where the computational solution is sought) and is discretized by a number of n nodes including the boundary nodes, creating a uniform grid with interval $\Delta x = L/(n-1)$. The depth at vena contracta is y_{up} and the depth upstream of the weir where the flow is almost uniform is y_{do}. The index i denotes the computational grid location, and the upstream and downstream boundaries correspond to nodes i = 1 and i = n with depths y_{up} and y_{do} respectively. The abrupt drop is located at node i = m and it is treated numerically as a boundary.

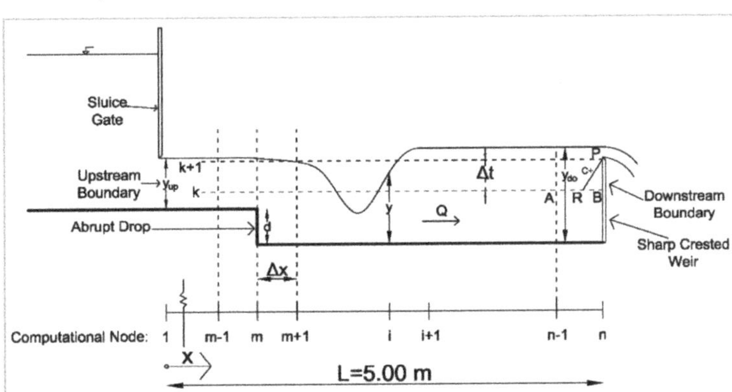

Figure A1. Physical and computational domain of the rapidly varied flow in the region of the drop.

The Boussinesq equations can be solved numerically using two different schemes, the Dissipative Two-Four and the MacCormack scheme. The Dissipative Two-Four scheme consists of a predictor and a corrector step, again for the spatial derivatives of Equation (7), in the predictor step forward finite differences are used and in the corrector step backward finite differences. This scheme includes the spatial nodes i + 2, i + 1 and i in the predictor step (Equation (A1)) and the nodes i, i − 1 and i − 2 in the corrector step (Equation (A2)) respectively, as:

$$G_i^* = G_i^k + \frac{\lambda}{6}\left(F_{i+2}^k - 8F_{i+1}^k + 7F_i^k\right) + \Delta t S_i^k, \tag{A1}$$

$$G_i^{**} = \frac{1}{2}\left(G_i^k + G_i^*\right) + \frac{\lambda}{12}\left(-7F_i^* + 8F_{i-1}^* - F_{i-2}^*\right) + \frac{1}{2}\Delta t S_i^*, \tag{A2}$$

The vector G_i^{k+1} at the next iteration level k + 1 and grid point i is given by:

$$G_i^{k+1} = \frac{1}{2}\left(G_i^k + G_i^{**}\right), \tag{A3}$$

where $\lambda = \Delta t/\Delta x$, Δt being the time step and superscripts k and k + 1 refer to two successive iterations. All variables with asterisk (*) refer to those calculated at the predictor step while all variables with double asterisk (**) refer to those calculated at the corrector step.

The MacCormack scheme is a two-step algorithm scheme. For the spatial derivatives of Equation (7), forward finite differences are used including nodes i + 1 and i in the predictor step (Equation (A4)), while in the corrector step backward finite differences are used including the nodes i and i − 1 (Equation (A5)) as:

$$G_i^* = G_i^k - \lambda\left(F_{i+1}^k - F_i^k\right) + \Delta t S_i^k, \tag{A4}$$

and

$$G_i^{**} = G_i^* - \lambda\left(F_i^* - F_{i-1}^*\right) + \Delta t S_i^*. \tag{A5}$$

The flow variables at the next iteration level k + 1 and grid point i are given by Equation (A3).

Denoting i to be the spatial node and k the iteration number, the second order derivative in the Boussinesq term $\partial^2 u/\partial x^2$ is approximated by a three point central finite difference in both steps, predictor and corrector. Forward finite difference is used in the predictor step (Equation (A6)) and backward finite difference in the corrector step (Equation (A7)) for the first order derivative $\partial u/\partial x$ as:

$$E_i = u_i^k \frac{u_{i+1}^k - 2u_i^k + u_{i-1}^k}{\Delta x^2} - \left(\frac{u_{i+1}^k - u_i^k}{\Delta x}\right)^2, \tag{A6}$$

$$E_i^* = u_i^* \frac{u_{i+1}^* - 2u_i^* + u_{i-1}^*}{\Delta x^2} - \left(\frac{u_i^* - u_{i-1}^*}{\Delta x}\right)^2. \tag{A7}$$

The mixed partial derivative $\partial^2 u/\partial x \partial t$ has been ignored since it is zero at steady state.

The appropriate initial and boundary conditions must be set in order to have a well posed problem [27]. At time t = 0 two characteristic curves enter at the computational domain, so the flow depth and the streamwise velocity must be specified at each grid point. An auxiliary condition has to be specified for each characteristic curve entering the boundaries of the computational domain, which includes the steady, gradually varied supercritical flow, in the entire length of the channel. From the steady gradually varied flow equation:

$$\frac{dy}{dx} = \frac{S_o - S_f}{1 - Fr^2} \tag{A8}$$

the flow depth and velocity are computed at each grid point numerically using the Kutta-Merson method with known initial depth upstream, so the calculations can proceed downstream at each grid point. The initial depth y_{up} is that at the exit of the sluice gate, while $Fr = u/\sqrt{gy}$ is the Froude number of the flow.

The flow conditions at each boundary are fixed. At node i = 1 (Figure A1) the flow is supercritical with depth y_{up} while at node i = n the flow is subcritical with depth y_{do}. During the iteration process the flow depths at nodes i = 1, i = n, are constant and known from the experimental measurements taken downstream of the sluice gate and upstream of the weir, respectively. The velocity at i = 1, u_{up}, is also constant and known, $u_{up} = Q/by_{up}$, while at i = n it has to be computed.

The velocity at the downstream boundary node i = n will be estimated with the method of specified intervals and the positive characteristic equation discretized by finite differences. In Figure A1 points A and B correspond to the nodes n − 1 and n respectively, at time level k, while the positive characteristic passing through the point P with the unknown velocity at the downstream boundary at the time level k + 1 is indicated. The point R is the intersection of the positive characteristic passing through the point P with the grid line of the time level k. With the method of specified intervals the velocity, the celerity and the flow depth at point R respectively, are calculated [32] from Equations (A9)–(A11) as

$$u_R = \frac{u_B + \lambda(c_B u_A - c_A u_B)}{1 + \lambda(u_B - u_A + c_B - c_A)}, \tag{A9}$$

$$c_R = \frac{c_B + \lambda u_R(c_B - c_A)}{1 + \lambda(c_B - c_A)}, \tag{A10}$$

$$y_R = c_R^2/g, \tag{A11}$$

where $c = \sqrt{gy}$ is the celerity of the propagating wave inside a rectangular open channel of small amplitude in shallow water and $u_A = u_{n-1}^k$, $u_B = u_n^k$, $c_A = \sqrt{gy_{n-1}^k}$ and $c_B = \sqrt{gy_n^k}$. The energy line slope at point R is estimated as:

$$S_{f_R} = n_f^2 u_R^2 / R_R^{4/3}, \tag{A12}$$

where $R_R = by_R/(b + 2y_R)$. Then the velocity at point P i.e., the variable u_n^{k+1} at iteration k + 1, can be computed from the following relationship:

$$u_P = u_n^{k+1} = u_R + 2c_R - 2\sqrt{gy_n^{k+1}} - g\Delta t(S_{f_R} - S_o), \tag{A13}$$

The unknown flow variables at the step were computed using the method of specified intervals. In the case of the minimum B jump the flow is supercritical at the step. The depth and the velocity at node, i = m will be calculated using the positive and the negative characteristic curves. In Figure A2a, points A and B correspond to nodes m − 1 and m respectively, at time level k, while the positive and negative characteristics passing through point P with the unknown flow depth and velocity at node m at the time level k + 1 are indicated. The point R is the intersection of the positive characteristic through point P with the grid line of time level k. The method of specified intervals computes the velocity, the celerity, the flow depth and the energy line slope at point R with Equations (A9)–(A12). The point S is the intersection of the negative characteristic passing through point P with the grid line of time level k. The velocity, the celerity, the flow depth and the energy line slope at point S through the negative characteristic curve are computed [32] from Equations (A14)–(A17) as

$$u_S = \frac{u_B - \lambda(c_B u_A - c_A u_B)}{1 + \lambda(u_B - u_A - c_B + c_A)}, \tag{A14}$$

$$c_S = \frac{c_B - \lambda u_S(c_B - c_A)}{1 - \lambda(c_B - c_A)}, \tag{A15}$$

$$y_S = c_S^2/g, \tag{A16}$$

$$S_{f_S} = n_f^2 u_S^2 / R_S^{4/3}, \tag{A17}$$

In Equations (A9), (A10), (A14) and (A15), it is $u_A = u_{m-1}^k$, $u_B = u_m^k$, $c_A = \sqrt{gy_{m-1}^k}$, $c_B = \sqrt{gy_m^k}$ and $R_S = by_S/(b + 2y_S)$. Then, the flow depth and the velocity at point P i.e., the variables y_m^{k+1}, u_m^{k+1}, at iteration k + 1 can be computed from the following equations:

$$y_P = y_m^{k+1} = \left\{0.25\left[u_R - u_S + 2(c_R + c_S) - g\Delta t\left(S_{f_R} - S_{f_S}\right)\right]\right\}^2/g, \tag{A18}$$

$$u_P = u_m^{k+1} = 0.5\left[u_R + u_S + 2(c_R - c_S) - g\Delta t\left(S_{f_R} + S_{f_S}\right)\right], \tag{A19}$$

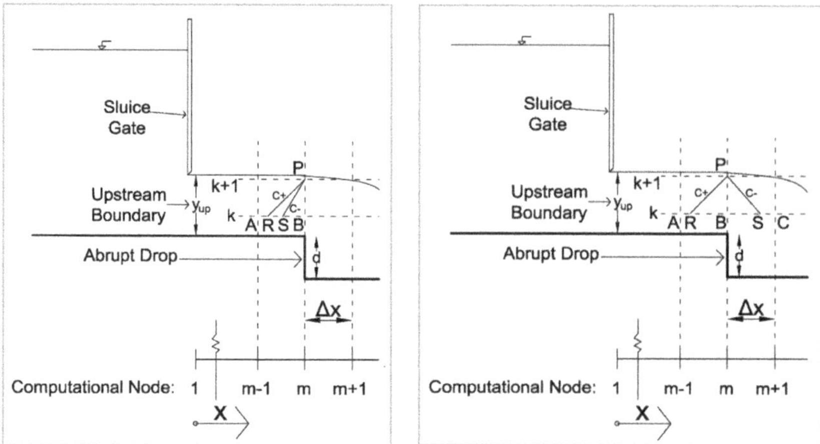

(a) **(b)**

Figure A2. Characteristic curves at the drop: (**a**) Supercritical flow; (**b**) Subcritical flow.

In the case of the A-jump the flow is subcritical at the step. The difference in comparison with the minimum B-jump is the direction of the negative characteristic curve shown in Figure A2b. In Figure A2b, points A, B and C correspond to nodes m − 1, m and m + 1 respectively at time level k, while the positive and the negative characteristics passing through point P with unknown flow depth and velocity at node m at the time level k + 1 are indicated. The point R is the intersection of the positive characteristic passing through point P with the grid line of time level k. The velocity, the celerity the flow depth as well as the energy line slope at point R are calculated from Equations (A9)–(A12) with $u_A = u_{m-1}^k$, $u_B = u_m^k$. Point S is the intersection of the negative characteristic passing through point P with the grid line of time level k. The velocity, the celerity, the flow depth as well as the energy line slope at point S through the negative characteristic curve are computed [32] from Equations (A20)–(A23) as

$$u_S = \frac{u_B + \lambda(c_B u_C - c_C u_B)}{1 + \lambda(u_C - u_B - c_B + c_C)}, \tag{A20}$$

$$c_S = \frac{c_B + \lambda u_S(c_B - c_C)}{1 + \lambda(c_B - c_C)}, \tag{A21}$$

$$y_S = c_S^2/g, \tag{A22}$$

$$S_{f_S} = n_f^2 u_S^2 / R_S^{4/3}, \tag{A23}$$

where $u_B = u_m^k$, $u_C = u_{m+1}^k$, $c_B = \sqrt{gy_m^k}$, $c_C = \sqrt{gy_{m+1}^k}$ and $R_S = by_S/(b + 2y_S)$. Then the flow depth and velocity at point P i.e., the variables y_m^{k+1}, u_m^{k+1} at iteration k + 1 can be computed from Equations (A18) and (A19), respectively.

The time step Δt, was variable in each iteration satisfying the Courant-Friedrichs-Lewy condition for all spatial nodes for stability reasons, calculated from the following relationship:

$$\Delta t = \frac{c_n \Delta x}{\max\left(\left|u_i^k\right| + \sqrt{gy_i^k}\right)}, \tag{A24}$$

where c_n is the Courant number which must be less than or equal to 0.65 [28] and Δx is the constant spatial step as shown in Figure A1.

High oscillations occur in the region of the jump; therefore, in order to filter them out, artificial viscosity had to be added to the numerical schemes. According to Chaudhry [27] we implement the following. First the parameter ξ_i at computational node i and at iteration k + 1 is calculated as:

$$\xi_i^{k+1} = \frac{\left|y_{i+1}^{k+1} - 2y_i^{k+1} + y_{i-1}^{k+1}\right|}{\left|y_{i+1}^{k+1}\right| + 2\left|y_i^{k+1}\right| + \left|y_{i-1}^{k+1}\right|}, \text{ for the interior nodes} \tag{A25}$$

$$\xi_i^{k+1} = \frac{\left|y_{i+1}^{k+1} - y_i^{k+1}\right|}{\left|y_{i+1}^{k+1}\right| + \left|y_i^{k+1}\right|}, \text{ for the upstream end node} \tag{A26}$$

$$\xi_i^{k+1} = \frac{\left|y_i^{k+1} - y_{i-1}^{k+1}\right|}{\left|y_i^{k+1}\right| + \left|y_{i-1}^{k+1}\right|}, \text{ for the downstream end node} \tag{A27}$$

Then at the center of the segment between node i and node i + 1 it is:

$$\xi_{i+(1/2)}^{k+1} = k_{art}\frac{\Delta x}{\Delta t}\max\left(\xi_i^{k+1}, \xi_{i+1}^{k+1}\right), \tag{A28}$$

Similarly between node i − 1 and node i

$$\xi_{i-(1/2)}^{k+1} = k_{art}\frac{\Delta x}{\Delta t}\max\left(\xi_{i-1}^{k+1}, \xi_i^{k+1}\right), \tag{A29}$$

where k_{art} is the coefficient adjusting the amount of dissipation. Finally the flow depth and the velocity are modified to the new ones according to the following equation:

$$f_{new_i}^{k+1} = f_{old_i}^{k+1} + \xi_{i+(1/2)}^{k+1}\left(f_{old_{i+1}}^{k+1} - f_{old_i}^{k+1}\right) - \xi_{i-(1/2)}^{k+1}\left(f_{old_i}^{k+1} - f_{old_{i-1}}^{k+1}\right), \tag{A30}$$

where f is either the flow depth or the velocity.

The developed algorithm for each numerical scheme consists of the following steps:

1. Compute the flow depth and velocity at all computational spatial nodes at initial time (t = 0) according to the initial condition of the problem. At first iteration:
2. Set up the depths y_{up} and y_{do} at the upstream and downstream boundary nodes respectively, known from the experimental measurements.
3. Compute the vector G_i^* in the predictor step, the vector G_i^{**} in the corrector step and the vector G_i for all internal computational spatial nodes except for the node where the drop is placed.
4. Compute the vector G_i at the location of the drop and the velocity at the downstream boundary node with the specified intervals method.
5. Compute ξ_i and $\xi_{i\pm(1/2)}$ and modify the flow depth and velocity according to Equation (A30).

6. Repeat steps 2–5, with the computed depth and velocity of the present iteration to be the starting values for the next iteration. The algorithm iterates until the change of the depth between two successive iterations in all computational spatial nodes is less than a fixed convergence value. Then the minimum B-jump or the A-jump form as part of the steady state solution.

References

1. Gualtieri, C.; Chanson, H. Physical and Numerical Modelling of Air-Water Flows: An Introductory Review. *Environ. Model. Softw.* **2021**, *143*, 105109. [CrossRef]
2. Moore, W.L.; Morgan, C.W. The Hydraulic Jump at an Abrupt Drop. *J. Hydraul. Div. Proc. Am. Soc. Civ. Eng.* **1957**, *83*, 1–21. [CrossRef]
3. Ohtsu, I.; Yasuda, Y. Transition from Supercritical to Subcritical Flow at an Abrupt Drop. *J. Hydraul. Res.* **1991**, *29*, 309–328. [CrossRef]
4. Mossa, M.; Petrillo, A.; Chanson, H. Tailwater Level Effects on Flow Conditions at an Abrupt Drop. *J. Hydraul. Res.* **2003**, *41*, 39–51. [CrossRef]
5. Hager, W.H. B-Jumps at Abrupt Channel Drops. *J. Hydraul. Eng.* **1985**, *11*, 861–866. [CrossRef]
6. Hager, W.H.; Bretz, N.V. Hydraulic Jumps at Positive and Negative Steps. *J. Hydraul. Res.* **1986**, *24*, 237–253. [CrossRef]
7. Pagliara, S. Wave Type Flow at Abrupt Drop: Flow Geometry and Energy Loss, Entropy and Energy Dissipation in Water Resources. In *Water Science and Technology Library*; Kluwer Academic Publishers: Drive Norwell, MA, USA, 1992; Volume 9, pp. 469–479.
8. Ohtsu, I.; Yasuda, Y. Discussion of Hydraulic Jumps at Positive and Negative Steps by Hager and Bretz. *J. Hydraul. Res.* **1987**, *25*, 407–413. [CrossRef]
9. Pagliara, S. Discussion of Transition from Supercritical to Subcritical Flow at an Abrupt Drop by Ohtsu and Yasuda. *J. Hydraul. Res.* **1992**, *30*, 428–432. [CrossRef]
10. Negm, M.A. Analysis of Pressure Distribution Coefficient on Steps Under Hydraulic Jump Conditions in Sloping Stilling Basins. *Trans. Ecol. Environ.* **1998**, *19*, 1–10.
11. Giudice, D.G.; Gisonni, C.; Rasulo, G. Vortex Drop Shaft for Supercritical Flow. In Proceedings of the 16th IAHR-APD Congress and 3rd Symposium of IAHR-ISHS, Nanjing, China, 20–23 October 2008.
12. Rajaratnam, N.; Ortiz, N. Hydraulic Jumps and Waves at Abrupt Drops. *J. Hydraul. Div. Proc. Am. Soc. Civ. Eng.* **1977**, *103*, 381–394. [CrossRef]
13. Mossa, M. On the Oscillating Characteristics of Hydraulic Jumps. *J. Hydraul. Res.* **1999**, *37*, 541–558. [CrossRef]
14. Sunik, S.M. Tailwater Level Effect on Flow Conditions at an Abrupt Drop. In Proceedings of the Nasional Aplikasi Teknologi Prasarana, Wilayah, India, 2 January 2009.
15. Armenio, V.; Toscano, P.; Fiorotto, V. On the Effects of a Negative Step in Pressure Fluctuations at the Bottom of a Hydraulic Jump. *J. Hydraul. Res.* **2000**, *38*, 359–368. [CrossRef]
16. Matzziounis, P.; Papanicolaou, P. Subcritical and Supercritical Flow Conditions at a Submerged Forward Facing Step. In Proceedings of the 1st International Conference on Experiments/Process/System Modelling/Simulation/Optimization, Athens, Greece, 6–9 July 2005.
17. Esfahani, M.; Bajestan, M. Dynamic Force Measurement of Roughened Bed B-jump at an Abrupt Drop. *Arch. Sci.* **2012**, *65*, 47–54.
18. Riazi, R.; Bejestan, M. Analysis Location of Pressure Fluctuation in Hydraulic Jump Over Roughened Bed with Negative Step. *Bull. Environ. Pharmacol. Life Sci.* **2014**, *3*, 103–110.
19. Kawagoshi, N.; Hager, W.H. Wave Type Flow at Abrupt Drops. *J. Hydraul. Res.* **1990**, *28*, 235–252. [CrossRef]
20. Quraishi, A.; Al-Brahim, A.M. Hydraulic Jump in Sloping Channel with Positive or Negative Step. *J. Hydraul. Res.* **1992**, *30*, 769–782. [CrossRef]
21. Ohtsu, I.; Yasuda, Y. Discussion of Hydraulic Jump in Sloping Channel with Positive or Negative Step by Quraishi and Al-Brahim. *J. Hydraul. Res.* **1993**, *31*, 712–716. [CrossRef]
22. Larson, E. Energy Dissipation in Culverts by Forcing a Hydraulic Jump at the Outlet. Master's Thesis, Department of Civil and Environmental Engineering, Washington State University, Pullman, WA, USA, 2004.
23. Papanicolaou, P.; Matziounis, P. Supercritical Flow Conditions Around a Submerged Forward Facing Step. In Proceedings of the 10th National Congress in Management of Water Resources and Protection of Environment, Hellenic Hydrotechnical Association, Xanthi, Greece, 13–16 December 2006.
24. Bakhti, S.; Hazzab, A. Comparative Analysis of the Positive and Negative Steps in a Forced Hydraulic Jump. *Jordan J. Civ. Eng.* **2010**, *4*, 197–204.
25. Simsek, O.; Soydan, N.; Gumus, V.; Akoz, M.; Kirkgoz, M. Numerical Modeling of B-Type Hydraulic Jump at an Abrupt Drop. *Tek. Dergi* **2015**, *24*, 7215–7240.
26. Padova, D.; Mossa, M.; Sibilla, S. SPH Modelling of Hydraulic Jump Oscillations at an Abrupt Drop. *Water* **2017**, *9*, 790. [CrossRef]
27. Chaudhry, H.M. *Open-Channel Flow*, 2nd ed.; Springer: New York, NY, USA, 2008; pp. 1–528.
28. Gottlieb, D.; Turkel, E. Dissipative Two-Four Methods for Time-Dependent Problems. *Math. Comput.* **1976**, *30*, 703–723. [CrossRef]

29. MacCormack, R.W. The Effect of Viscosity in Hypervelocity Impact Cratering. In Proceedings of the AIAA Hypervelocity Impact Conference, Cincinnati, OH, USA, 30 April–2 May 1969. [CrossRef]
30. Valero, D.; Viti, N.; Gualtieri, C. Numerical Simulation of Hydraulic Jumps. Part 1: Experimental Data for Modelling Performance Assessment. *Water* **2019**, *11*, 36. [CrossRef]
31. Viti, N.; Valero, D.; Gualtieri, C. Numerical Simulation of Hydraulic Jumps. Part 2: Recent Results and Future Outlook. *Water* **2019**, *11*, 28. [CrossRef]
32. Chanson, H. *Environmental Hydraulics of Open Channel Flows*, 1st ed.; Elsevier: Oxford, UK, 2004; pp. 1–485. [CrossRef]

hydrology MDPI

Article

Determination of Skin Friction Factor in Gravel Bed Rivers: Considering the Effect of Large-Scale Topographic Forms in Non-Uniform Flows

Masoud Kazem [1], Hossein Afzalimehr [1], Mohammad Nazari-Sharabian [2] and Moses Karakouzian [2,*]

[1] Faculty of Civil Engineering, Iran University of Science and Technology, Tehran 13114-16846, Iran; masoud_kazem@cmps2.iust.ac.ir (M.K.); hafzali@iust.ac.ir (H.A.)
[2] Department of Civil and Environmental Engineering and Construction, University of Nevada Las Vegas, Las Vegas, NV 89154, USA; mohammad.nazari-sharabian@alumni.unlv.edu
* Correspondence: mkar@unlv.nevada.edu; Tel.: +1-702-895-0959

Abstract: Determination of skin friction factor has been a controversial topic, particularly in gravel-bed rivers where total flow resistance is influenced by the existence of small-scale skin roughness and large-scale topographic forms. The accuracy of existing models predicting skin friction factors in conditions where small-scale skin roughness and large-scale topographic forms exist is very low. The objective of this study is to develop a modified model that improves the accuracy of the determination of skin friction factors in gravel-bed rivers. To this end, 100 velocity profile data obtained from eight gravel-bed rivers were utilized to develop an analytical method that considers the momentum thickness of the boundary layer and its deviation in large-scale topographic bedforms in a 1D force-balance model. The results show that the accuracy of the skin friction factors is enhanced when (1) the model is in the form of an exponential function of energy slope, and (2) the deviation of momentum thickness is considered in the model. The proposed model results in high accuracy of the predicted skin friction factors for energy slopes between 0.001 and 0.1, which exist in most gravel-bed rivers with different morphologies. Additionally, this study model was used to modify the classic Einstein–Strickler equation. The modified equation resulted in improved accuracy of the predicted skin friction factors in non-uniform flow conditions even when velocity profiles and energy slope were not available.

Keywords: gravel bed rivers; skin friction; flow resistance; bed forms; energy slope; boundary layer characteristics

Citation: Kazem, M.; Afzalimehr, H.; Nazari-Sharabian, M.; Karakouzian, M. Determination of Skin Friction Factor in Gravel Bed Rivers: Considering the Effect of Large-Scale Topographic Forms in Non-Uniform Flows. *Hydrology* **2022**, *9*, 58. https://doi.org/10.3390/hydrology9040058

Academic Editor: Ezio Todini

Received: 5 March 2022
Accepted: 30 March 2022
Published: 31 March 2022

1. Introduction

Friction factor and its sources have always been a major research topic [1–4]. During the past decades, the decomposition of friction factor has been a prevalent approach in sedimentology. Usually, the total friction factor is assumed to be the sum of the skin friction factor and bedform friction factor [5,6]. Briefly, large-scale impediments, such as bedforms or submerged vegetation patches, are responsible for the formation of large-scale flow structures. These large-scale flow structures transfer large amounts of momentum in the upper and the middle region of flow [7,8]. On the other hand, skin friction is the key factor in the momentum transfer procedures in the near-bed region of rivers. To be brief, not only does the skin friction factor play a key role in the flow resistance, but also it is the main index of shear stress which governs sedimentation and erosion processes. While the determination of skin friction factor has been focused on in the field and experimental studies, the results can also be used in numerical modeling of sedimentation. Such models can be used to predict the changes in bed geometry under different flow scenarios [9].

In the main, velocity profile has been widely used to predict bed friction factor in natural waterways. However, in the presence of a highly rough bed, using of velocity profile may lead to some unexpected results. This condition arises from the multi-layer structure of the boundary layer above the coarse sediments. Indeed, the lower layer of flow (or roughness layer), which is in contact with the coarse sediments, may have different characteristics in closely similar bed materials of river beds with different geometries of bedforms [10]. For small values of h/D, where h is the water depth, and D is a geometry parameter representative of the bed roughness (generally a grain diameter), the streamwise velocity profile is not self-similar inside the roughness layer [11,12]. This characteristic can only be detected by highly sensitive and expensive velocimeter tools, such as acoustic Doppler velocity profiler (ADVP). On the other hand, in a wide range of engineering works, the accuracy and sensitivity of velocimetry are limited by common tools, such as rotary current meters with poor spatial resolution, particularly in the near-bed zone. Consequently, there is a fundamental need for the development of new approaches that be applicable in the absence of sensitive and expensive velocimeter tools.

2. Materials and Methods

Amongst the well-known friction and flow resistance indices, the Darcy–Weisbach friction factor is one of the most applied ones in hydraulic calculations. The total friction factor (f) can be written as the following [13,14]:

$$f = 8\left(\frac{u_*}{u_m}\right)^2 \tag{1}$$

where u_m is the average streamwise velocity in a particular section of flow, and u_* is the shear velocity which can be calculated via different methods. Application of the boundary-layer parameters based on the ASCE Task Force recommendation has led to a series of new approaches, such as the boundary-layer characteristics method (BLCM) [15]. Considering the BLCM, the shear velocity can be calculated as:

$$u_* = \frac{(\delta^* - \theta)u_{max}}{c\delta^*} \tag{2}$$

where C is a constant with a value of 4.4 [16], δ^* is the boundary layer displacement thickness, θ is the momentum thickness of the boundary layer, and, u_{max} is the maximum velocity of a particular velocity profile. δ^* and θ can be calculated as [16]:

$$\delta^* = \int_0^h \left(1 - \frac{u}{u_{max}}\right) dy \tag{3}$$

$$\theta = \int_0^h \frac{u}{u_{max}}\left(1 - \frac{u}{u_{max}}\right) dy \tag{4}$$

Gravel bed rivers have recently been investigated in a distinct topic in which the friction factor can be divided into two parts [13,17–20]:

$$f = f' + f'' \tag{5}$$

where f' is the skin friction factor (emerged from grains) and f'' is the form friction factor (emerged from large-scale topographic forms of the river bed). For plane beds, bed shear stress is the representative stress of the skin friction factor. Figure 1 shows a typical velocity distribution in which bed shear stress can be defined via momentum thickness of the boundary layer:

$$\tau_{bed} = \frac{1}{b}\frac{dD}{dx} = \frac{d}{dx}\left(\rho U^2 \theta\right) \tag{6}$$

where D is the drag force emerged by the bed surface and b is the width of the channel. On the other hand, for a wide channel, the shear stress can be calculated based on the Darcy–Weisbach equation:

$$\tau_{bed} = \gamma h s_f = \gamma h \left(\frac{1}{8} f' \frac{\overline{U}^2}{hg} \right) = \gamma f' \frac{\overline{U}^2}{8g} \tag{7}$$

where γ is the specific gravity of water, h is the flow depth, g is the gravity acceleration, and $\overline{u}$ is the weighted average velocity. However, in the above-mentioned equations, the roughness of the bed surface is not considered directly. For a relatively plane bed and uniform flow, Einstein suggested a modified form of Stickler's Equation to relate the bed roughness (n_b) to the median grain size (d_{50}) [21].

$$n_b = \frac{{d_{50}}^{1/6}}{24} \tag{8}$$

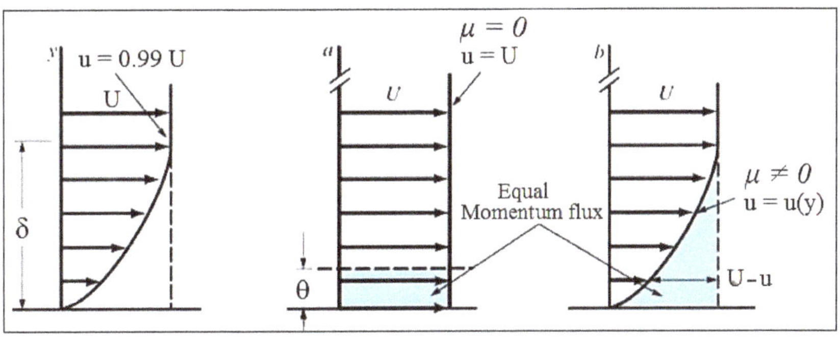

Figure 1. Definition of momentum thickness in a fully developed boundary layer.

By considering Einstein's Equation, the Darcy–Weisbach friction factor can be calculated as:

$$f' = \frac{g}{72} \left(\frac{d_{50}}{d} \right)^{1/3} \tag{9}$$

where d is the flow depth in a wide channel. In the same approach to relative roughness, Keulegan (1938) [22] developed a relatively simple but practical equation:

$$f' = \left[2.03 log \left(\frac{12.2h}{k_s} \right) \right]^{-2} \tag{10}$$

In this equation, h is the depth of flow and k_s is Nikuradse equivalent roughness size. Determination of k_s is a controversial issue and there is a wide range of estimations from $1.23d_{35}$ to $3d_{90}$ or $6.6d_{50}$ [23–25]. In another approach, the Shields parameter can be applied in the Keulegan Equation via a series of calculations. Critical shields parameter (τ^*_{cr}) can be defined as:

$$\tau^*_{cr} = \frac{\tau}{\gamma(SG - 1))d_{50}} = \frac{hS_f}{(SG - 1)d_{50}} \tag{11}$$

where γ is the specific gravity of water, and SG is the ratio of sediments' specific gravity to water specific gravity. This equation is valid for both uniform and non-uniform flow if S_f presents the true gradient of the energy loss. On the other hand, τ^*_{cr} can be written in association with S_f as the following [14,26]:

$$\tau^*_{cr} = 0.15 S_f^{0.25} \tag{12}$$

Assuming $k_s = d_{50}$ in Equation (5), and by comparing Equations (6) and (7), f' can be calculated via logarithm rules. The final equation can be written as [14]:

$$f' = \left[0.9742 - 1.5225log\left(S_f\right)\right]^{-2} \tag{13}$$

While Einstein's Equation (Equations (12) and (13)) is based on field and empirical studies on uniform flow and relatively plane bed, derivation of f' in Equation (13) is not restricted to those conditions. This fundamental difference will lead to considerably different results when the two approaches are applied to a unique case.

3. Motivation and Objective

Generally, skin friction factors calculated via classic approaches (e.g., [21]) are developed based on uniform flow conditions where plane bed morphology is assumed and where no emergent or submerged vegetation and obstacles exist. Accordingly, existing models for predicting the skin friction factors for riverbeds where large-scale topographic forms and vegetation exist yield inaccurate results. Determination of friction factor in non-uniform flow is a relatively complicated task that is very far from the classic approaches and needs advanced algorithms and methodologies [4,15,19].

The objective of this study is to develop a modified model that improves the accuracy of the determination of skin friction factors in gravel-bed rivers. Accordingly, prevalent equations have been modified to increase their accuracy in the presence of large bedforms and non-uniform flow.

4. Methodology and Technical Approach

Since many classic equations of roughness and friction factor were developed based on the uniform flow condition, considering the fundamental assumptions in the application of the uniform flow is essential. According to (1/7)th power velocity profile law of Prandtl, the velocity distribution of a uniform flow can be described as [27,28]:

$$\frac{u}{U} = \left(\frac{y}{\delta}\right)^{1/7} \tag{14}$$

where δ is the boundary layer which is defined as the region adjacent to a surface over which the velocity changes from zero to the free-stream velocity (0.99 U) [29]. The momentum thickness is also calculable as:

$$\theta = \frac{7}{72}\delta = \frac{7}{72}y\left(\frac{U}{u}\right)^7 \tag{15}$$

By considering the definition of δ, the momentum thickness can be defined as a linear function of flow depth:

$$\theta = \frac{7}{72}y\left(\frac{U}{0.99U}\right)^7 \cong 0.1043y \tag{16}$$

where y is the vertical elevation in which u = 0.99 U and is typically equal to the total depth of flow. Inserting Equation (16) into Equation (6) results in a differential form of bed shear stress for uniform flow:

$$\tau_{Uniform} = \frac{d}{dx}\left(0.1043\rho U^2 y\right) \tag{17}$$

Assuming a condition in which skin friction factor is calculated via a classic approach (such as Einstein–Strickler, which relies on uniform flow) and also via a boundary layer characteristic approach such as in Equation (13), the ratio of skin friction factors can be written as:

$$\frac{f'_{bed}}{f'_{Uniform}} = \frac{\tau_{bed}}{\tau_{Uniform}} \tag{18}$$

where f'_{bed} is the calculated value of skin friction factor via momentum thickness and $f'_{Uniform}$ is the skin friction factor calculated via Einstein's Equation. By inserting Equations (6) and (17) into Equation (18), the ratio of skin friction factors can be obtained as:

$$\frac{f'_{bed}}{f'_{Uniform}} = \frac{\frac{d}{dx}(\rho U^2\theta)}{\frac{d}{dx}(0.1043\rho U^2 y)} \cong \frac{9.587}{y_{0.99}}\theta \tag{19}$$

Subsequently, the skin friction in a non-uniform flow can be estimated as:

$$f' = \frac{9.587}{y}\theta f'_{Uniform} \tag{20}$$

Nonetheless, there is a fundamental assumption in this equation: all the momentum thickness of the boundary layer is produced by skin friction. However, despite uniform and even quasi-uniform flow, it is a controversial assumption for non-uniform flow. Figure 2 shows a conceptual scheme of momentum thickness above a large-scale topographic form. Owing to Equation (5), the momentum transition is affected by skin friction and form friction in a complex process. Consequently, a portion of the momentum thickness, which is not restricted to one as the upper limit, must be considered in Equation (20). This phenomenon can be included in Equation (20) in the form of:

$$f' = \frac{9.587}{y}(\varphi\theta) f'_{Uniform} \tag{21}$$

and

$$\frac{\varphi\theta}{d} = 0.1043\,\frac{f'}{f'_{Uniform}} \tag{22}$$

where φ is an indicator representing the portion of momentum thickness which is shaped by the skin friction and d is the flow depth in relatively shallow rivers. Equation (22) and the value of $\frac{\varphi\theta}{d}$ are investigated via data gathered in 100 velocimetry stations of eight gravel-bed rivers, including four reaches in Iran and four in Italy (Table 1). Total and skin friction factors were respectively calculated via BLCM (Equations (1)–(4)) and the Keulegan approach (With a wide range of K_s values). The equivalent skin friction of uniform flow was also calculated via the Einstein–Strickler Equation (Equations (8) and (9)) by applying the value of d_{50}.

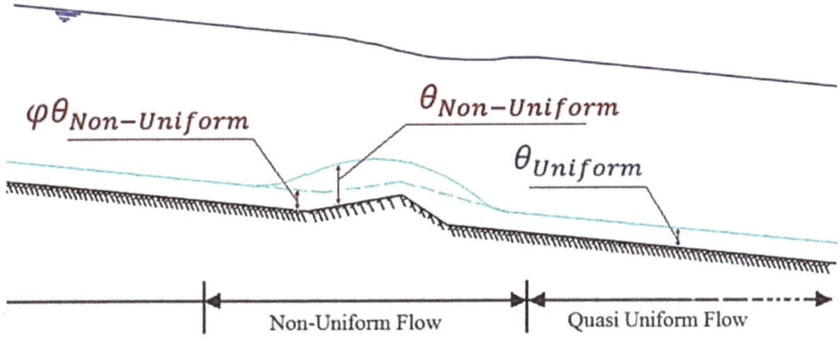

Figure 2. Schematic illustration of momentum thickness near a topographic bed form.

Table 1. The rivers and profiles used to derive Equation (24).

River	Location	No. of Profiles	Avg. Depth (cm)	Avg. U_{mean} (cm/s)	Avg. d_{50} (mm)	Avg. Fr	Avg. f	Average f' (Keulegan)	Avg. S_f
Melodari	Italy	14	13.6	64.9	20	0.554	1.204	0.101	0.038
Cerasia	Italy	15	19.5	56	47	0.416	0.78	0.072	0.016
Valanidi	Italy	8	15.8	61.4	35	0.502	1.089	0.086	0.026
Gallico	Italy	13	24.5	72.2	52	0.470	1.182	0.088	0.028
Zayanderud	Iran	5	73	77	10	0.291	0.092	0.044	0.001
Kaj	Iran	8	27	63.9	10	0.396	0.153	0.042	0.004
Gamasyab	Iran	24	30	84.2	19	0.287	0.107	0.043	0.003
Marbor	Iran	13	22	92.6	17	0.433	0.116	0.050	0.006

Consequently, considering $\frac{\varphi\theta}{d}$ as a nonlinear multivariable function of a series of non-dimensional parameters, this parameter can be defined as:

$$\frac{\varphi\theta}{d} = m * Fr^{c1} * \left(\frac{d_{50}}{d}\right)^{c2} * S_f^{c3} * Re_{grain}^{c4} \tag{23}$$

Subsequently, the modified momentum thickness (θ_{skin}) which is formed by the skin friction would be:

$$\theta_{skin} = \varphi\theta = \left(m * Fr^{c1} * \left(\frac{d_{50}}{d}\right)^{c2} * S_f^{c3} * Re_{grain}^{c4}\right) * d \tag{24}$$

Using the conjugate gradient approach to minimize the root-mean-square deviation (RMSD) for the 100 velocity profile, "m", "$c1$", "$c2$", "$c3$", and "$c4$" were respectively calculated for different K_s values. The results are presented in Table 2. RMSD should always be a non-negative value including 0, which represents the perfect fit, and generally, the lower value of RMSD is considered the better result.

Table 2. Calculated parameters to determine modified momentum thickness emerged by skin friction for different equivalent roughness in the Keulegan Equation.

	$\theta_{skin} = \varphi\theta = \left(m * Fr^{c1} * \left(\frac{d_{50}}{d}\right)^{c2} * S_f^{c3} * Re_{grain}^{c4}\right) * d$						
k_s	m	$c1$	$c2$	$c3$	$c4$	Pearson Correlation Coefficient	Equivalent Equation for Skin Friction Factor
D_{50}	0.013	0	−0.8	0	0	0.89	$0.017 * \left(\frac{d_{50}}{d}\right)^{-0.47}$
1.5 D_{50}	0.014	0	−0.85	0	0	0.93	$0.018 * \left(\frac{d_{50}}{d}\right)^{-0.52}$
2 D_{50}	0.015	0	−0.9	0	0	0.94	$0.019 * \left(\frac{d_{50}}{d}\right)^{-0.57}$
Lamb–Shields Method (Equation (13))	0.225	0	−0.33	0.33	0	0.80	$0.3 * S_f^{0.33}$

While none of the classic variants of the Keulegan Equation contain flow parameters, the Lamb–Shields-based form of the Keulegan Equation (Equation (13)) was considered as the basis for comparison (Figure 3). Normalized root-mean-square deviation (NRMSD) was calculated for this comparison as the following:

$$NRMSD = \frac{1}{\overline{\theta_{skin}}\ (Keulegan - Shields)}\sqrt{\frac{\sum_1^N (\theta_{skin}\ (Keulegan - Shields) - \theta_{skin})^2}{N}} \tag{25}$$

In order to describe the application of the developed equation in a practical situation, a field case study was established to evaluate the skin friction factor where bed topography was captured with high resolution.

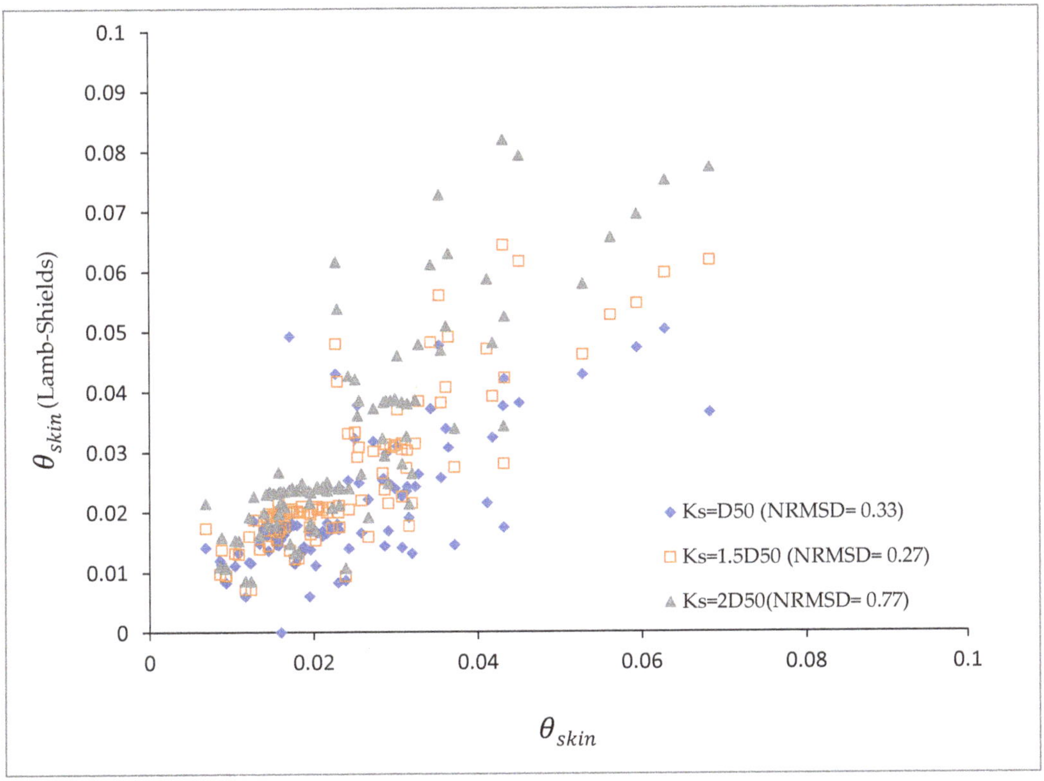

Figure 3. Comparison between values of modified momentum thickness of skin friction and their correlation with the Keulegan–Shields approach.

5. Field Study and Data Collection

The Marbor River is located in the central Zagros Mountains in the Dena region of the Isfahan province in Iran. A relatively straight reach of a local branch was selected for this study. The bed materials consist of gravel size grains with sparse round cobbles and negligible sand-size grains in the banks. The reach width varies from 3 to 4.5 m and the length of the study area was 15.2 m. Grain size distribution and the flow velocity were captured in five sections through 13 stations. The grain size distribution was calculated by using the Wolman method in all stations [30]. Figure 4 shows the general grain size distribution of the reach. Water surface elevation and topographic map of the bed were produced via accurate surveying. The topography of the bed, river plan, and the location of the sections are shown in Figure 5. The topography and velocimetry results represent a completely non-uniform flow condition in the selected reach. Figure 6 shows the profiles of the riverbed and water surface and also velocity distributions in the centroid of the reach. Bed slope has been calculated via accurate surveying and the 2D gravity projections on global coordinates have been included in the water fluxes [31,32]. All in all, data from 13 stations were used for the evaluation of skin friction factor by different methods. The displacement and momentum thicknesses of the boundary layer were calculated for each station through the 2D velocity profile (Table 3).

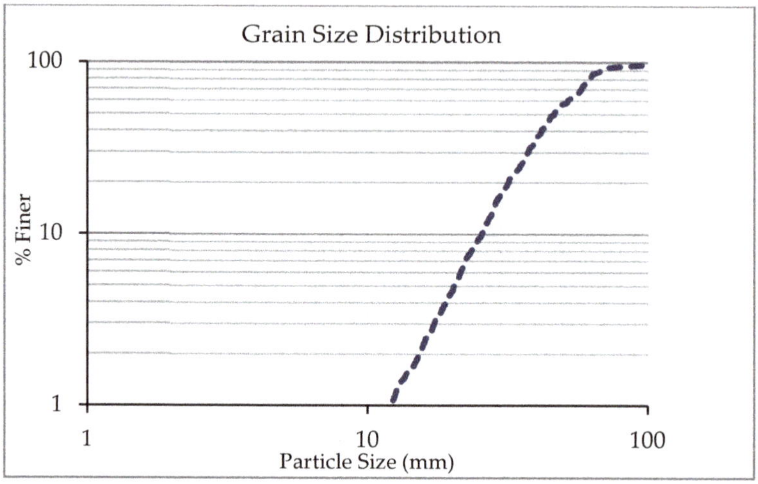

Figure 4. Grain size distribution in the selected reach (Wolman method).

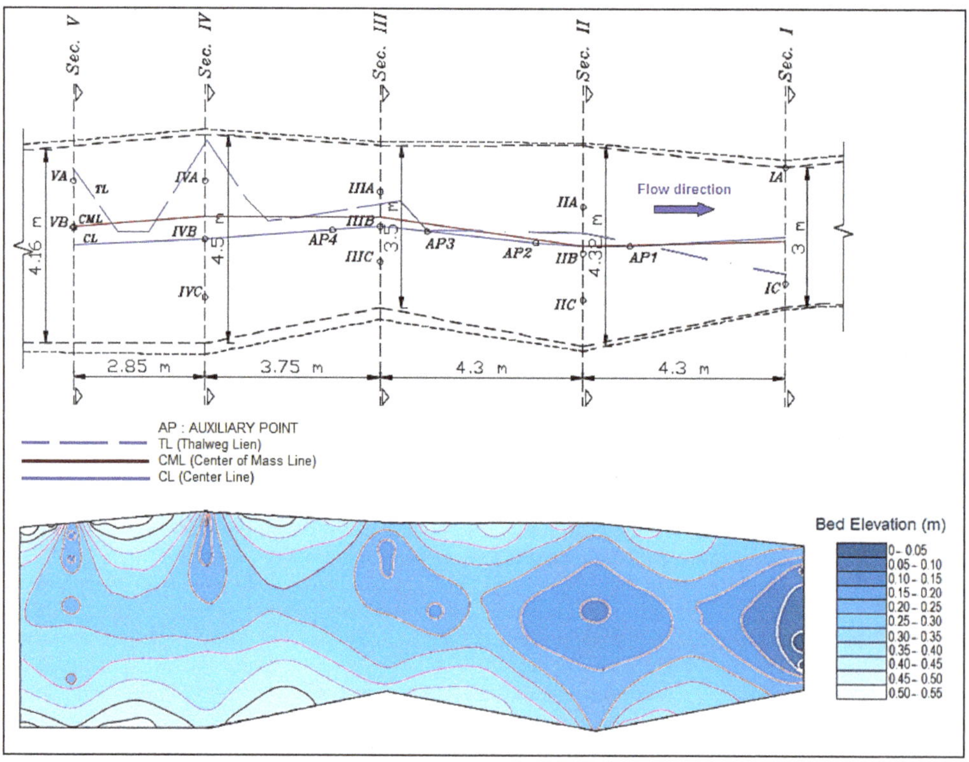

Figure 5. Topography of the bed, river plan, and the location of the sections in the selected reach (The lowest point has been set as the baseline of local elevation code)—Marbor River.

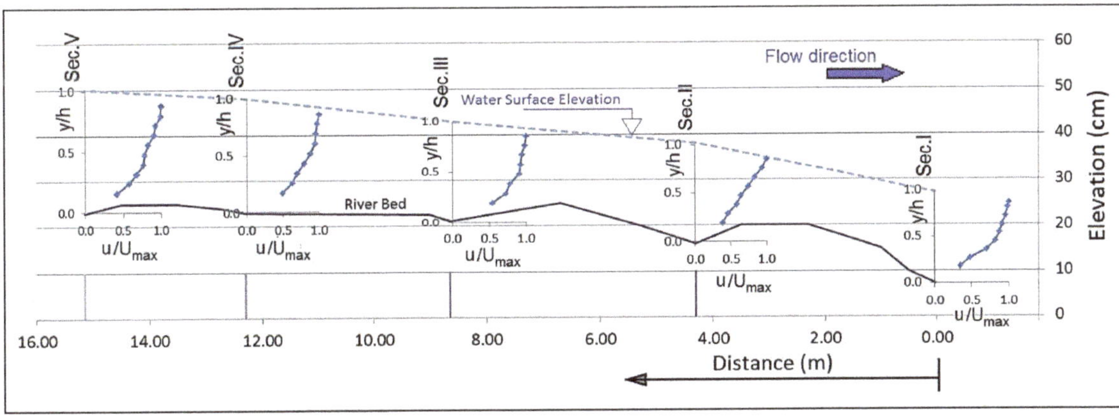

Figure 6. Water surface elevation (light blue dashed line), riverbed elevation (violet line), and velocity profiles (dark blue lines) along the center of mass' line (centroid line)—selected reach of Marbor River.

Table 3. Measured and calculated flow parameters in the Marbor River.

Station	IA	IC	IIA	IIB	IIC	IIIA	IIIB	IIIC	IVA	IVB	IVC	VA	VB
d_{50} (mm)	20	22	21	17	23	17	13	16	20	16	19	19	16
d (cm)	21	23	21	21	17	23	21	17	29	23	11	33	25
U_{mean} (m/s)	1.26	1.22	0.94	1.11	0.58	1.16	1.08	0.94	0.94	0.76	0.46	0.88	0.71
u_* (m/s)	0.16	0.17	0.14	0.162	0.092	0.098	0.086	0.074	0.095	0.094	0.036	0.14	0.075
Fr	0.88	0.81	0.65	0.77	0.45	0.77	0.75	0.73	0.56	0.51	0.44	0.49	0.45
δ^* (m)	0.034	0.043	0.049	0.048	0.038	0.024	0.024	0.014	0.049	0.033	0.008	0.082	0.043
θ (m)	0.019	0.023	0.025	0.027	0.018	0.016	0.017	0.01	0.028	0.022	0.006	0.043	0.028
f (-)	0.123	0.153	0.175	0.168	0.199	0.056	0.050	0.049	0.082	0.123	0.048	0.195	0.090
f' (-)	0.066	0.067	0.061	0.067	0.050	0.047	0.045	0.044	0.040	0.040	0.034	0.052	0.040
f'' (-)	0.057	0.086	0.114	0.101	0.148	0.009	0.005	0.005	0.042	0.083	0.014	0.143	0.049
$f_{Uniform}$ (-)	0.062	0.062	0.063	0.059	0.070	0.057	0.054	0.062	0.056	0.056	0.076	0.053	0.055

Subsequently, the skin friction factor values were evaluated using the following four methods:

a: Applying the friction factor of the equivalent uniform flow and the measured momentum thickness (Equation (20)): In this approach, the momentum thickness must be calculated via velocity profile. The results are comparable to the exact values of the flow characteristics method with equivalent Lamb–Shields parameter (Figure 7a). While there is a general correlation between these values, the deviation is also clear because of the φ parameter.

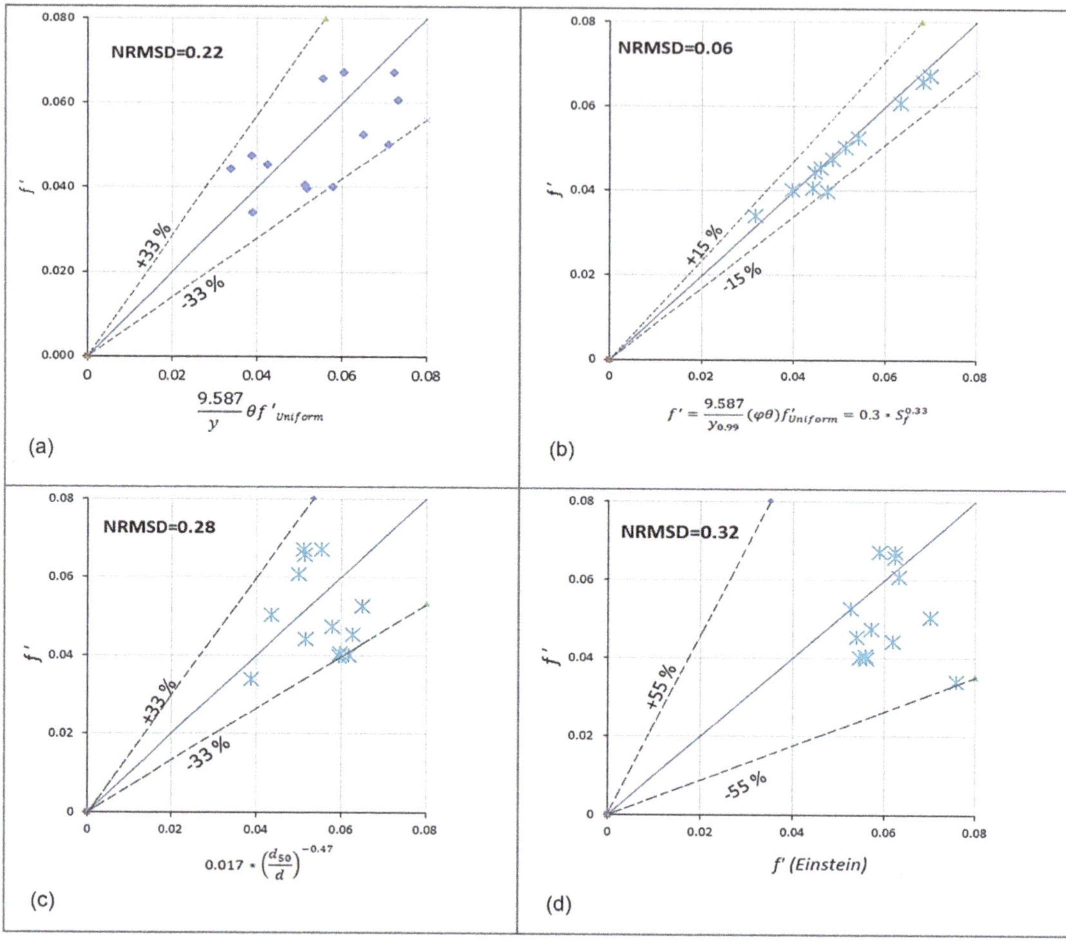

Figure 7. Skin friction factor calculated using Equation (20) (**a**), Equation (21) (**b**), modified Einstein–Strickler Equation (**c**), and classic Einstein–Strickler Equation (**d**), in the Marbor River. The vertical axis shows the skin friction factor calculated with the flow characteristic method. The ranges of deviations are shown with dashed blue lines.

b: Applying the friction factor of the equivalent uniform flow and the modified momentum thickness (Equation (21)): Using the φ parameter in this approach leads to an independent equation that only relies on the energy slope, and the results are very close to the exact values. This approach is independent of velocity profile which makes it a suitable method in engineering works (Figure 7b). The applied equation is shown in Table 2 ($f' = 0.3 * S_f^{0.33}$). The calculated values of φ (using Equation (24)) and the measured values of this parameter are also shown in Table 4. Investigation of the values of φ at the 37 stations of the two gravel-bed rivers (Marbor and Gamasyab) shows that for the majority of the velocimetry stations, $\varphi < 1$. For the centerline of the explored reach of the Marbor River, higher values of φ ($\varphi \geq 1$) are observed in the local contractions (Sec. I and Sec III), while the bed profile seems to have a much lower impact on the values of φ (Figure 8). The higher values of φ in the contractions might be related to the morphological drag of walls. Essentially, wall drag is the main source of scattering of results, particularly in natural channels [33]. However, Lamb et al. (2008) [26] mentioned that the effect of wall drag increases in the high bed slops (S > 0.02). In this study, this phenomenon was observed for two sections, although more data sets are required to evaluate the value of φ for various ranges of contractions. Figure 9 shows the surface waves near the wall of a contraction zone, which clearly shows the development of a horizontal boundary layer.

Table 4. Comparison between measured and calculated φ.

Section	I		II			III			IV			V	
Station	Left	Right	Left	Cent.	Right	Left	Cent.	Right	Left	Cent.	Right	Left	Cent.
Calculated φ	1.21	1.14	0.85	0.95	0.72	1.23	1.07	1.30	0.85	0.90	0.81	0.82	0.68
Measured φ	1.18	1.11	0.83	0.93	0.71	1.22	1.07	1.30	0.78	0.77	0.87	0.81	0.69

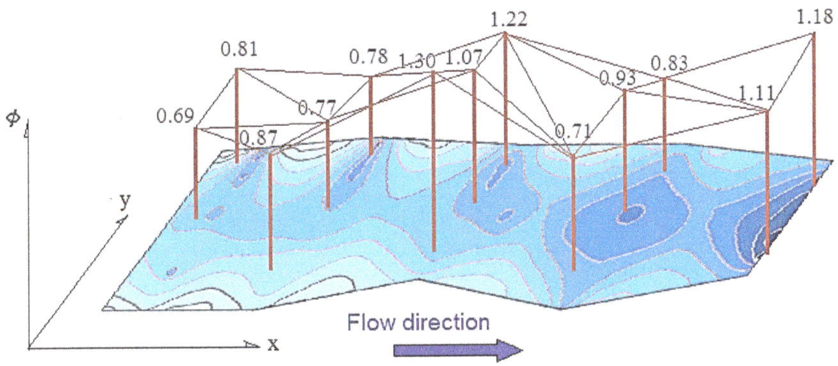

Figure 8. 3D illustration of the measured φ in the investigated reach.

Figure 9. Surface waves and the contraction streamlines, which are visible near the wall of a contraction zone (Marbor River).

c: Applying modified friction factor of the equivalent uniform flow: Modified Einstein—Strickler Equation was used (by considering $K_s = D_{50}$. See Table 2). This method is also independent of the velocity profile. The results contain a higher level of errors in comparison with the two previous methods, although it is less scattered in comparison with the classic Einstein–Strickler Equation (Figure 7c).

d: Applying friction factor of the equivalent uniform flow: The classic Einstein–Strickler Equation according to Equation (9) (Figure 7d).

6. Discussion

The findings of this study indicate that the skin friction factor can be calculated via Equation (21) with high accuracy in comparison with other classic methods. Many studies have shown that the average velocity is not the only relevant velocity scale in determining bed friction, and the effect of flow characteristics needs to be considered [34–38]. Correspondingly, the presented approach does not rely on average velocity as a single variable. Actually, it contains the effects of momentum thickness and its deviations in topographic forms. This method calculates the skin friction factor as an exponential function of energy slope, which means the skin friction factor rises when the energy slope increases. Lamb et al. (2008) [26] explained this phenomenon as a result of backwater effects and an associated pressure differential, which increased the mobility of particles on steeper slopes. This method also eliminated the need for velocity profiles which play a significant role in many fluvial parameters. However, the concept of the φ parameter is not restricted to the measurements where energy slope is available. Applying this concept in the classic Einstein–Strickler method leads to a modified version of the method, evaluating skin friction factor without considering energy slope and velocity profiles.

In order to compare the accuracy of the methods which are developed based on the φ parameter, the relative error ($(f'_{BLCM} - f')/f'_{BLCM}$) was calculated for 100 velocimetry points (Figure 10). This was done for a wide range of energy slopes (0.001–0.1). The results showed that Equation (21) contains less than 5% of error in comparison to the boundary layer flow characteristics method for energy slopes between 0.0014 and 0.056, considering the normal range of friction factor in gravel-bed rivers (Figure 10a). For the high energy slope ($S_f > 0.06$), the error increases linearly. The results also showed that for the conditions where the energy slope is not available, the modified Einstein's Equation (based on the φ parameter's concept) can reduce the error up to 50% on average (Figure 10b). In addition to the Marbor river, for other rivers with different energy slopes, the accuracy of results obtained from Equation (21) can be compared with the results obtained from the Einstein equation (Figure 11). According to the Figure, the Einstein equation is not appropriately accurate in low and high energy slopes. One reason can be attributed to the assumptions taken into account by Einstein. The normal or semi-normal flow regime is not prevalent in rivers with very steep energy slopes because of the tendency of flow to accelerate in such conditions. On the other hand, very low energy slopes are prevalent in the presence of flow blockage or decelerating flow. Despite the Einstein equation, Equation (21) takes energy slope into account. Consequently, the accuracy of the estimated skin friction factor increased on extreme slopes. However, it must be considered that the findings of this research are limited by the number of rivers where flow data are available. As presented in Figure 11a, the accuracy of results decreased sparsely in higher energy slopes. However, the accuracy of the results is still acceptable in comparison with classic methods. Despite the energy slope, relative errors do not comply with a predictable pattern for average velocity and relative roughness (Figure 12). The relative error remains below 6% in most cases. However, large relative errors have been observed in the presence of higher relative roughness. Consequently, it is recommended that future research studies focus on the accuracy of the proposed approach in rivers with higher relative roughness.

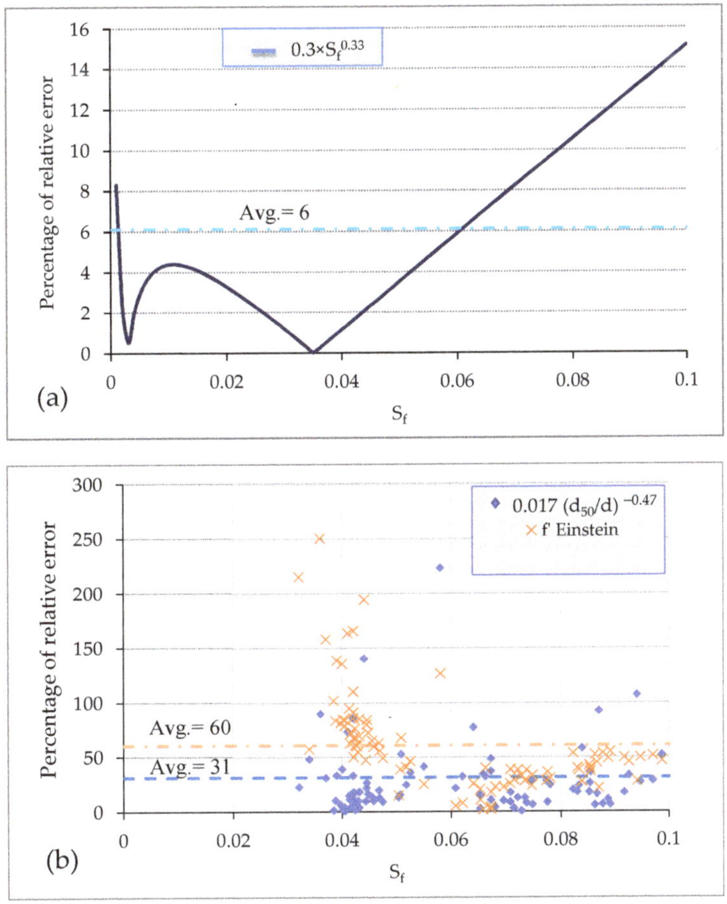

Figure 10. The relative errors of Equation (21) (**a**) and Einstein Equation (**b**) in comparison with BLCM for determining the skin friction factor. The horizontal axis represents the total energy slope.

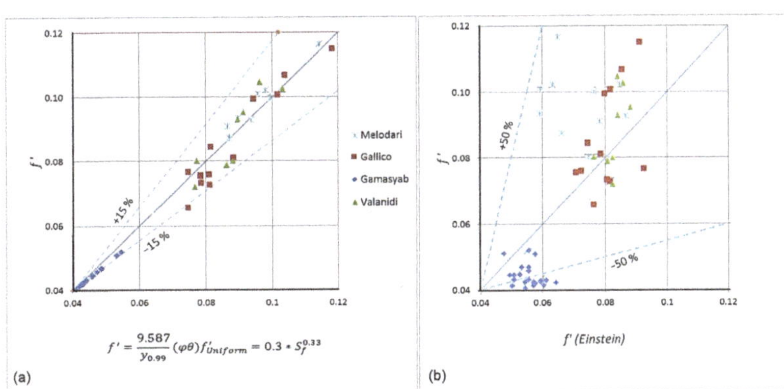

Figure 11. Comparison between the values of skin friction factor calculated via BLCM and Equation (21) (**a**) and Einstein equation (**b**).

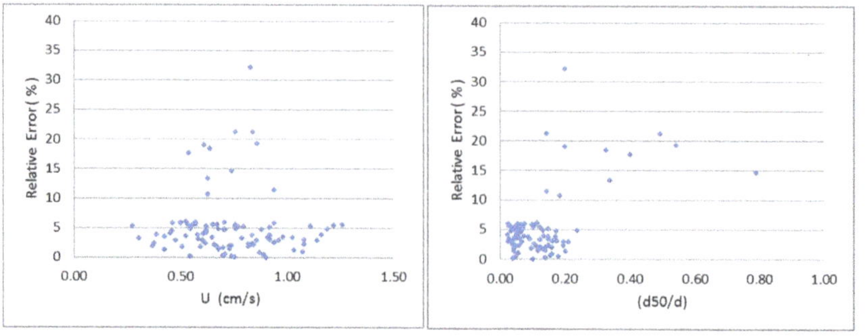

Figure 12. Relative error of skin friction factors calculated using Equation (24) and the BLCM method in comparison with average velocity (**left**) and relative roughness (**right**).

7. Conclusions

The skin drag in open-channel flow is determinable through the modified momentum thickness of the boundary layer. In order to derive a general equation, a 1D force-balance model was applied based on the momentum thickness of the boundary layer. The values of the initial model were compared to standard uniform flow models of skin friction. The deviation of the predicted values (known as φ) was formulated through dimensionless parameters and an optimization procedure that used data sets of eight gravel-bed rivers with 100 velocimetry stations. Applying the φ parameter in the Einstein–Strickler Equation, the skin friction factor reveals an exponential function of the energy slope. The comparison with previous studies (e.g., [26]) shows a similar relation for the total friction factor and bed slope. This method was also considered for a field case study and the results showed a high correlation between the new method and the Keulegan method. The high correlation rate is also expectable for an energy slope range between 0.001 and 0.1, which involves the majority of gravel-bed rivers. The application of φ parameters can lead to engineer-friendly equations which are applicable in engineering works where velocity profile is not available. The results showed that for the conditions where the energy slope is not available, the modified Einstein's Equation (based on the φ parameter's concept) can reduce the error up to 50% on average.

Author Contributions: Conceptualization, H.A.; methodology, H.A.; software, M.K. (Masoud Kazem); validation, M.K. (Masoud Kazem), H.A. and M.N.-S.; resources, H.A.; data curation, M.K. (Masoud Kazem); writing—original draft preparation, M.K. (Masoud Kazem) and M.N.-S.; writing—review and editing, M.N.-S.; visualization, M.K. (Masoud Kazem); supervision, H.A., M.N.-S. and M.K (Moses Karakouzian). All authors have read and agreed to the published version of the manuscript.

Funding: This research received no external funding.

Institutional Review Board Statement: Not applicable.

Informed Consent Statement: Not applicable.

Data Availability Statement: Not applicable.

Conflicts of Interest: The authors declare no conflict of interest.

References

1. Colosimo, C.; Copertino, V.A.; Veltri, M. Friction factor evaluation in gravel-bed rivers. *J. Hydraul. Eng.* **1988**, *114*, 861–876. [CrossRef]
2. Rennie, C.D.; Millar, R.G. Discussion of "Estimation of Gravel-Bed River Flow Resistance" by Colin, D. Rennie and Robert, G. Millar. *J. Hydraul. Eng.* **1999**, *125*, 1317–1319. [CrossRef]
3. Julien, P.Y.; Klaassen, G.J.; Ten Brinke, W.B.M.; Wilbers, A.W.E. Case study: Bed resistance of Rhine River during 1998 flood. *J. Hydraul. Eng.* **2002**, *128*, 1042–1050. [CrossRef]

4. Milukow, H.A.; Binns, A.D.; Adamowski, J.; Bonakdari, H.; Gharabaghi, B. Estimation of the Darcy–Weisbach friction factor for ungauged streams using Gene Expression Programming and Extreme Learning Machines. *J. Hydrol.* **2019**, *568*, 311–321. [CrossRef]
5. Zhenlin Li, M. Direct skin friction measurements and stress partitioning over movable sand ripples. *J. Geophys. Res. Ocean* **1994**, *99*, 791–799. [CrossRef]
6. Hey, R.D. Flow resistance in gravel-bed rivers. *J. Hydraul. Div.* **1979**, *105*, 365–379. [CrossRef]
7. Kazem, M.; Afzalimehr, H.; Sui, J. Formation of Coherent Flow Structures beyond Vegetation Patches in Channel. *Water* **2021**, *13*, 2812. [CrossRef]
8. Kazem, M.; Afzalimehr, H.; Sui, J. Characteristics of Turbulence in the Downstream Region of a Vegetation Patch. *Water* **2021**, *13*, 3468. [CrossRef]
9. Juez, C.; Battisacco, E.; Schleiss, A.J.; Franca, M.J. Assessment of the performance of numerical modeling in reproducing a replenishment of sediments in a water-worked channel. *Adv. Water Resour.* **2016**, *92*, 10–22. [CrossRef]
10. Franca, M.J.; Ferreira, R.M.; Lemmin, U. Parameterization of the logarithmic layer of double-averaged streamwise velocity profiles in gravel-bed river flows. *Adv. Water Resour.* **2008**, *31*, 915–925. [CrossRef]
11. Nikora, V.I.; Smart, G.M. Turbulence characteristics of New Zealand gravel-bed rivers. *J. Hydraul. Eng.* **1997**, *123*, 764–773. [CrossRef]
12. Nikora, V.; Goring, D.; McEwan, I.; Griffiths, G. Spatially averaged open-channel flow over rough bed. *J. Hydraul. Eng.* **2001**, *127*, 123–133. [CrossRef]
13. Yalin, M.S. *Mechanics of Sediment Transport*; Pergamon Press: New York, NY, USA, 1972; Volume 5, pp. 99–100.
14. Afzalimehr, H.; Singh, V.P.; Najafabadi, E.F. Determination of form friction factor. *J. Hydrol. Eng.* **2010**, *15*, 237–243. [CrossRef]
15. Afzalimehr, H.; Anctil, F. Vitesse de frottement associée à un écoulement non uniforme et une rugosité relative intermédiaire. *J. Hydraul. Res.* **2001**, *39*, 181–186. [CrossRef]
16. Schlichting, H.; Gersten, K. Fundamentals of boundary-layer theory. In *Boundary-Layer Theory*; Springer: Berlin/Heidelberg, Germany, 2000; pp. 29–49. [CrossRef]
17. Yen, B.C. (Ed.) *Channel Flow Resistance: Centennial of Manning's Formula*; Water Resources Publication, LLC.: Littleton, CO, USA, 1992.
18. Bertin, S.; Groom, J.; Friedrich, H. Isolating roughness scales of gravel-bed patches. *Water Resour. Res.* **2017**, *53*, 6841–6856. [CrossRef]
19. Mendicino, G.; Colosimo, F. Analysis of Flow Resistance Equations in Gravel-Bed Rivers with Intermittent Regimes: Calabrian fiumare Data Set. *Water Resour. Res.* **2019**, *55*, 7294–7319. [CrossRef]
20. Luo, M.; Wang, X.; Yan, X.; Huang, E. Applying the mixing layer analogy for flow resistance evaluation in gravel-bed streams. *J. Hydrol.* **2020**, *589*, 125119. [CrossRef]
21. Einstein, H.A. Flow on a movable bed. In Proceedings of the Second Hydraulics Conference, Iowa City, IA, USA, 1–4 June 1942; Howe, J.W., Rouse, H., Eds.; University of Iowa: Iowa City, IA, USA, 1943; pp. 332–341, Bulletin 27.
22. Keulegan, G.H. Laws of turbulent flow in open channels. *J. Res. Bur. Stand.* **1938**, *21*, 707–741. [CrossRef]
23. Ackers, P.; White, W.R. Sediment transport: New approach and analysis. *J. Hydraul. Div.* **1973**, *99*, 2041–2059. [CrossRef]
24. van Rijn, L.C. Equivalent roughness of alluvial bed. *J. Hydraul. Div.* **1982**, *108*, 1215–1218. [CrossRef]
25. Hammond, F.D.C.; Heathershaw, A.D.; Langhorne, D.N. A comparison between Shields' threshold criterion and the movement of loosely packed gravel in a tidal channel. *Sedimentology* **1984**, *31*, 51–62. [CrossRef]
26. Lamb, M.P.; Dietrich, W.E.; Venditti, J.G. Is the critical Shields stress for incipient sediment motion dependent on channel-bed slope? *J. Geophys. Res. Earth Surf.* **2008**, *113*, F02008. [CrossRef]
27. Sarma, K.V.; Lakshminarayana, P.; Rao, N.L. Velocity distribution in smooth rectangular open channels. *J. Hydraul. Eng.* **1983**, *109*, 270–289. [CrossRef]
28. Zagarola, M.V.; Perry, A.E.; Smits, A.J. Log laws or power laws: The scaling in the overlap region. *Phys. Fluids* **1997**, *9*, 2094–2100. [CrossRef]
29. Gadbois, J.; Wilkerson, G. Uniform flow development length in a rough laboratory flume. In Proceedings of the World Environmental and Water Resources Congress 2014, Portland, OR, USA, 1–5 June 2014; pp. 1234–1242. [CrossRef]
30. Wolman, M.G. A method of sampling coarse river-bed material. *EOS Trans. Am. Geophys. Union* **1954**, *35*, 951–956. [CrossRef]
31. Juez, C.; Soares-Frazao, S.; Murillo, J.; García-Navarro, P. Experimental and numerical simulation of bed load transport over steep slopes. *J. Hydraul. Res.* **2017**, *55*, 455–469. [CrossRef]
32. Francalanci, S.; Solari, L.; Toffolon, M. Local high-slope effects on sediment transport and fluvial bed form dynamics. *Water Resour. Res.* **2009**, *45*, W05426. [CrossRef]
33. Parker, G.; Wilcock, P.R.; Paola, C.; Dietrich, W.E.; Pitlick, J. Physical basis for quasi-universal relations describing bankfull hydraulic geometry of single-thread gravel bed rivers. *J. Geophys. Res. Earth Surf.* **2007**, *112*. [CrossRef]
34. Papanicolaou, A.N.; Diplas, P.; Evaggelopoulos, N.; Fotopoulos, S. Stochastic incipient motion criterion for spheres under various bed packing conditions. *J. Hydraul. Eng.* **2002**, *128*, 369–380. [CrossRef]
35. Schmeeckle, M.W.; Nelson, J.M. Direct numerical simulation of bedload transport using a local, dynamic boundary condition. *Sedimentology* **2003**, *50*, 279–301. [CrossRef]
36. Zanke, U.C.E. On the influence of turbulence on the initiation of sediment motion. *Int. J. Sediment Res.* **2003**, *18*, 17–31.

37. Wu, F.C.; Yang, K.H. Entrainment probabilities of mixed-size sediment incorporating near-bed coherent flow structures. *J. Hydraul. Eng.* **2004**, *130*, 1187–1197. [CrossRef]
38. Vollmer, S.; Kleinhans, M.G. Predicting incipient motion, including the effect of turbulent pressure fluctuations in the bed. *Water Resour. Res.* **2007**, *43*, W05410. [CrossRef]

hydrology MDPI

Article

Impacts of Climate Change and Variability on Precipitation and Maximum Flows in Devil's Creek, Tacna, Peru

Edwin Pino-Vargas [1], Eduardo Chávarri-Velarde [2,*], Eusebio Ingol-Blanco [2,*], Fabricio Mejía [1], Ana Cruz [1] and Alissa Vera [1]

[1] Department of Civil Engineering, Universidad Nacional Jorge Basadre Grohmann, Tacna 23000, Peru; epinov@unjbg.edu.pe (E.P.-V.); 20210308@lamolina.edu.pe (F.M.); acruzb@unjbg.edu.pe (A.C.); averam@unjbg.edu.pe (A.V.)

[2] Department of Water Resources, Universidad Nacional Agraria La Molina, Lima 15024, Peru

* Correspondence: echavarri@lamolina.edu.pe (E.C.-V.); eingol@lamolina.edu.pe (E.I.-B.)

Abstract: Global projections of climate change indicate negative impacts on hydrological systems, with significant changes in precipitation and temperature in many parts of the world. As a result, floods and droughts are expected. This article discusses the potential effects of climate change and variability on the maximum precipitation, temperature, and hydrological regime in Devil's Creek, Tacna, Peru. The outputs of precipitation and daily temperature of fifteen regional climate models were used for the RCP4.5 and RCP8.5 emission scenarios. The methodology used includes the bias correction and downscaling of meteorological variables using the quintiles mapping technique, hydrological modeling, the evaluation of two emission scenarios, and its effect on the maximum flows of the stream. The results of the multi-model ensemble show that the maximum annual precipitation will probably increase by more than 30% for the RCP4.5 and RCP8.5 scenarios for the 2021–2050 period relative to the 1981–2005 period. Likewise, as expected, the maximum flows could increase by 220% and 154% for the RCP4.5 scenarios for the 2021–2050 and 2051–2080 terms, respectively, and 234% and 484% for the RCP8.5 scenarios and for the 2021–2050 and 2051–2080 terms, respectively, concerning the recorded historical value, increasing the probability of flood events and damage in populations located downstream.

Keywords: maximum precipitation; maximum flow; climate change; hydrological modeling

Citation: Pino-Vargas, E.; Chávarri-Velarde, E.; Ingol-Blanco, E.; Mejía, F.; Cruz, A.; Vera, A. Impacts of Climate Change and Variability on Precipitation and Maximum Flows in Devil's Creek, Tacna, Peru. *Hydrology* **2022**, *9*, 10. https://doi.org/10.3390/hydrology9010010

Academic Editors: Carmelina Costanzo, Tommaso Caloiero and Roberta Padulano

Received: 23 November 2021
Accepted: 30 December 2021
Published: 5 January 2022

1. Introduction

Changes in temperature and precipitation patterns, due to the increase in greenhouse gas concentrations, affect hydrological processes. Consequently, negative impacts are expected on water resources for agriculture, urban uses, mining, industry, aquatic life in rivers and lakes, and hydroelectric power production. Similarly, spatial changes in the intensity and frequency of precipitation can affect the magnitude and frequency of flows, increasing the intensity of floods and droughts, with important impacts on economic activities at the local and regional levels [1,2].

On the other hand, on a global scale, studies show that the temperature increases by more than 3.5 °C, under the RCP8.5 emission scenario. By the end of the 21st century, precipitation is projected to reduce by more than 20% for medium and low latitudes [3]. Similarly, an increase in evaporation and a decrease in soil moisture content and groundwater recharge are expected. Consequently, drought conditions and increased evapotranspiration rates are projected for summer in subtropical regions, as well as medium and low latitudes [3].

In Peru, a national evaluation of climate scenarios carried out by the Peruvian National Meteorological and Hydrological Service (SENAMHI), estimated a progressive increase in the maximum and minimum temperature, by 2.8 °C on average by the end of the 21st century. Similarly, a reduction in precipitation of 40% is projected for winter, and an average increase of 20% during summer and spring [4].

An increase in temperature with a decrease in precipitation will produce a drastic decrease in water availability in rivers and other natural sources, as well as a substantial increase in evapotranspiration rates, meaning greater water consumption, affecting agricultural and municipal uses, and the production of electrical and industrial energy. Additionally, a slow but sustained reduction in glaciers in Peru is expected, which plays an important role in the supply of water to the water systems of communities in the Andean regions.

On the other hand, many regions of the north, center, and south of Peru are frequently seriously affected by floods and huaycos (debris flow and mudslides) as a result of high-intensity precipitation which occurs in the basins, influenced by the El Niño phenomenon. It is estimated that 26 El Niño events have occurred during the 20th century. The most severe events occurred in 1924–1925, 1982–1983, 1997–1998 [5], and most recently in 2016–2017. These events have generated great damages and economic losses in the social, productive, and infrastructure sectors. The total sectoral damages caused by the El Niño phenomena 1982–1983 and 1997–1998 were around USD 3200 million and USD 3500 million, respectively, including direct and indirect damages [6]. In the latest 2016–2017 event, total damages were estimated to be around USD 4 billion [7].

Similarly, at the JORGE BASADRE rain gauge, located in the city of Tacna, precipitation of 5.1 mm was recorded in 24 h in 2017, constituting one of the highest values ever recorded for November. The intensities were variable and covered a large part of the coastal zone of the Tacna region [8].

On the other hand, extreme precipitation events are commonly represented using IDF precipitation curves. Extreme weather events are becoming more severe and frequent, which leads to uncertainties as to how prepared the infrastructure is to face these changes. Infrastructure designs are based on the IDF precipitation curves with the assumption of stationarity, which means that the statistical properties of future events will be similar to those of the past [9,10]. However, climate change is expected to alter climate extremes, a concept called non-stationarity [11].

The main objective of this research is to evaluate the potential impacts of climate change and variability on precipitation and maximum flows in the Devil's Creek, located in the Tacna region, Peru. The Tacna region is located in the northern region of the Atacama Desert. The Atacama Desert is one of the largest hyper-arid deserts in the world [12].

For this purpose, the outputs of the meteorological variables of fifteen regional climate models of the Coupled Model Intercomparison Project Phase 5 [13] were used as inputs for the hydrological model. A bias correction was applied for temperature and precipitation using the quantile mapping method [14–17], which enables bias correction of the regional climate model simulations in comparison with the data observed in rain gauges. The changes in precipitation and maximum flow were evaluated for the period 2021–2050 and the RCP4.5 (intermediate) and RCP8.5 (high) emission scenarios.

The results, in combination with the evaluation of the maximum precipitation and temperature, the analysis of frequencies, and hydrological modeling, will help to answer the research questions such as: What would be the changes of the maximum precipitation and temperature in the sub-basin of the Devil's Creek under historical conditions and climate change scenarios? What changes would the frequency of maximum flows experience in the study area under scenarios of variability and climate change? What are the differences between historical conditions and climate change?

2. Materials and Methods

2.1. Precipitation Data and Historical Temperature

Historical data of daily precipitation and temperature of five (5) rain gauges near the study basin (Figure 1) were collected and analyzed for the period 1966–2020. The completion and extension of information were carried out with the Climatol software, which uses an approach based on the method used by Paulhus and Kohler [16] to complete the missing daily precipitation data. This consists of spatial interpolation of the normal precipitation rate of neighboring rain gauges. This proportion method is extended in the Climatol package with options to use differences and full standardization to normalize the data [18]. However, because the JORGE BASADRE rain gauge only had daily precipitation data since 1993 (Table 1), and because a minimum period of 30 years is necessary, for analysis of the frequency of maximum precipitation, it was considered appropriate to evaluate information from the product named PISCO (Peruvian Interpolated data of the SENAMHI Climatological and hydrological Observations. Precipitation v2.0) [19], available from SENAMHI (National Service of Meteorology and Hydrology of Peru). To use these data, they had to be corrected using the quantile mapping technique and validated for the Devil's Creek area using local rain gauges (JORGE BASADRE, Calana, Calientes, Sama Grande, and Palca) as reference rainfall or observed values. These data have made it possible to complete the daily precipitation dataset for the JORGE BASADRE rain gauge, from 1981–1992.

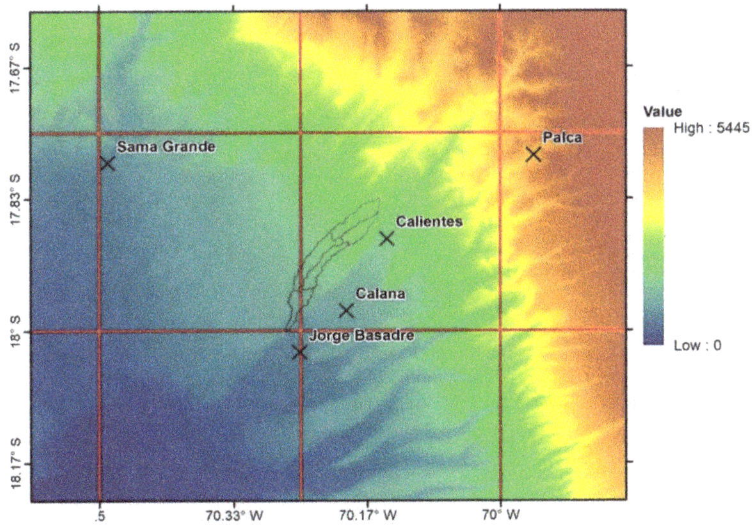

Figure 1. Location of the Devil's Creek with the meteorological rain gauges selected for the downscaling and cells of the general circulation models (GCMs).

Table 1. Availability of precipitation data.

Rain Gauge	No. Values	Start Date	Final Date	% Gaps	Duration (Years)
Calana	19,704	1 January 1966	31 December 2020	2	55
JORGE BASADRE	9952	1 January 1993	31 December 2020	3	28
Calientes	19,298	1 January 1966	31 December 2020	4	55
Sama Grande	19,789	1 January 1966	31December 2020	1	55
Palca	15,866	1 July 1966	31 December 2020	20	55

2.2. General Circulation Models and Scenarios

When considering climate change, one of the challenges water resource managers often face is deciding which general circulation models (GCMs) should be used to assess the impacts of climate change on water resource systems. This is a puzzling question because all GCMs demonstrate uncertainty in the prediction of historical climate variables [20,21]. However, some criteria, such as spatial resolution, the degree of atmospheric–ocean coupling, and the availability of multiple realizations, can be taken into account when selecting a set of GCMs to reduce uncertainties in the predictions of maximum precipitation and floods based on any individual GCM [2,22,23]. Considering the high degree of uncertainty that climate models present in the projection of precipitation, fifteen regional climate models have been deemed convenient to be evaluated in this research.

Currently, there are regionally scaled models (25 km × 25 km) whose outputs still need bias correction and spatial downscaling to improve their performance in a particular region. This research uses 15 general circulation models of the CMIP5 project (the NASA NEX-GDDP dataset–Coupled Model Intercomparison Project Phase 5): ACCESS1-0, bcc-csm1-1, CanESM2, CCSM4, CESM1-BGC, CNRM-CM5, CSIRO-Mk3-6-0, GFDL -CM3, GFDL-ESM2G, GFDL-ESM2M, IPSL-CM5A-LR, MIROC-ESM, MIROC-ESM-CHEM, MIROC5, MPI-ESM-LR, MPI-ESM-MR [24].

On the other hand, climate data from the downscaled emission scenarios RCP4.5 (intermediate emission) and RCP8.5 (high emissions) are used. These scenarios have been selected based on their emission trajectories; medium and high, respectively, for the period 2021–2080.

2.3. Bias Correction and Downscaling

Downscaling can be defined as a technique that increases the resolution of GCMs to obtain the climate at a local scale. There are two fundamental methods for downscaling large-scale data from GCM results: statistical and dynamic downscaling. Their concepts have been discussed in various articles [25–28]. This research uses a statistical downscaling of the outputs of the previously described regional models. Quantile mapping is applied to perform bias correction in regional climate model simulations compared to observed data. This method is designed to fit the distribution of the modeled data, so that they match observed climate data [29,30]; in this case, precipitation and temperature. The following expression was applied and resolved in Rstudio to find the corrected future value:

$$y^f_{corr,i} = F_y^{-1}\left(F_x\left(x_i^f\right)\right) \quad (1)$$

where $y^f_{corr,i}$ is the future value corrected at time i, and $F_x\ (x)$, $F_y\ (y)$ represent the empirical cumulative distribution functions of the model (x) and observations (y). To perform bias correction and the downscaling, this study uses a historical period and a base period simulated by climate models, both from January 1981 to December 2005. With this, the correction for the future period 2021–2080 is applied.

2.4. Climate Model Ensembles

The evaluation of the capacity of a climate model to simulate trends in extreme events is complex due to internal climate variability, whose simulated phases are unique for the realization of each model [31]. On the other hand, we know that, currently, the models have improved a lot in projecting the future climate. However, a high degree of uncertainty still persists, showing great variability between models in mainly projecting precipitation. In this sense, the multi-model ensembles highlight the uncertainty in climate predictions that result from structural differences in global climate models, as well as the uncertainty due to variations in the initial conditions or parameterizations of the model [32]. Similarly, several investigations have shown that a weighted ensemble method, based on the simulation performance of the models, may have better projection abilities than the equal-weighted ensembles [33–36].

In this study, the ensemble of downscaled climate scenarios of the 14 global circulation models (GCMs) applied to the Devil's Creek is given by the weighted average whose weights are determined according to the performance of each model concerning the historical period 1981–2005. The determination of the weights of each GCM is determined by the genetic algorithm [37], whose objective function is to minimize:

$$F_{obj} = min \sum_{j=1981}^{2005} \left(P_{hist,j} - \sum_{i=1}^{14} (w_i * P_{GCM,j}) \right)^2 \quad (2)$$

where P_{hist} is the historical maximum annual precipitation for the Devil's Creek, for the year j, P_{GCM} is the maximum annual precipitation of the GCM of the Devil's Creek, and w is the weight of each GCM. The objective is to minimize the error between the observed value and the simulated value, optimizing the weights that each model would have based on its performance with the history.

The proposed objective function is interpreted as the ensemble of the 14 GCMs in the Devil's Creek for the period 1981–2005 to be identical to the maximum annual precipitation generated in the creek.

The genetic algorithm begins with an initial population subjected to the selection process of the objective function evaluation to later apply the genetic crossing and mutation operators. These are responsible for diversifying the individuals of the initial population [37,38]. A simple ensemble with equal weighting is used for the minimum and maximum temperatures.

2.5. Hydrological Modeling

2.5.1. Frequency Analysis of Total Daily Precipitation for the Northern Area of the City of Tacna

In the Caplina river basin, there are five rain gauges: two main climate rain gauges (La Yarada and Calana), two ordinary rain gauges (Calientes and Palca), and a main agricultural climate rain gauge (JORGE BASADRE) [39]. Due to its proximity to the city of Tacna, the total daily precipitation recorded at the JORGE BASADRE rain gauge was considered representative of the northern area of the city of Tacna. This rain gauge has a continuous record of 28 years (1993 to 2020) (Figure 2). The frequency analysis was performed from the partial duration series, which was adjusted to the Gumbel distribution.

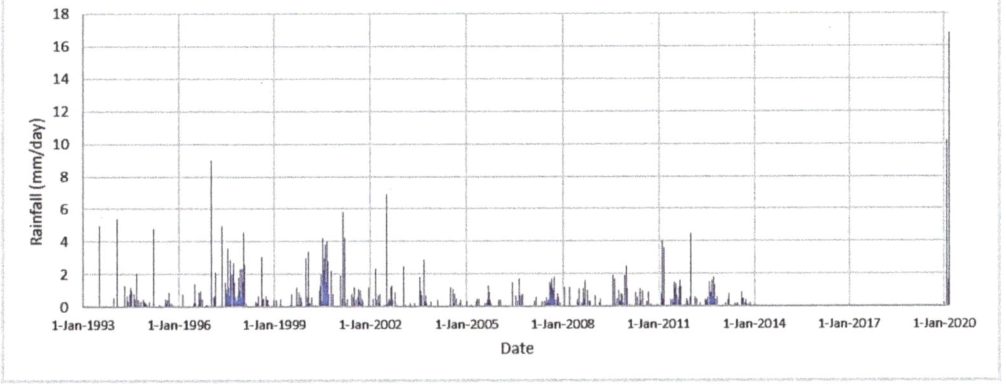

Figure 2. Total daily precipitation—JORGE BASADRE rain gauge.

Figure 3 shows that the precipitation event recorded on 21 February 2020 has a return period of 255.8 years.

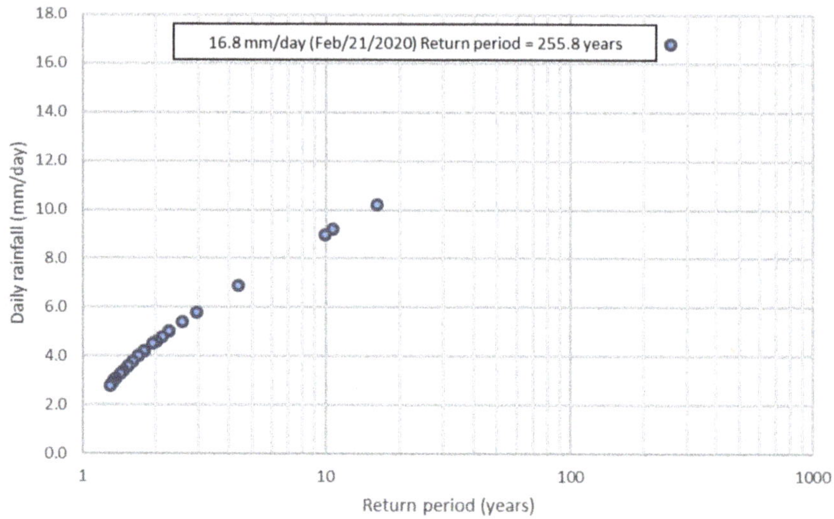

Figure 3. Frequency analysis of the partial series of total daily precipitation at JORGE BASADRE rain gauge (from 1993 to 2020).

2.5.2. Estimated Precipitation Events for the Devil's Creek

Figure 4 shows the comparison of the total daily rainfall discharged from the PISCO product for the coordinates of the JORGE BASADRE rain gauge and the total daily rainfall from the JORGE BASADRE rain gauge, during the period 1 January 1993 to 21 April 2014.

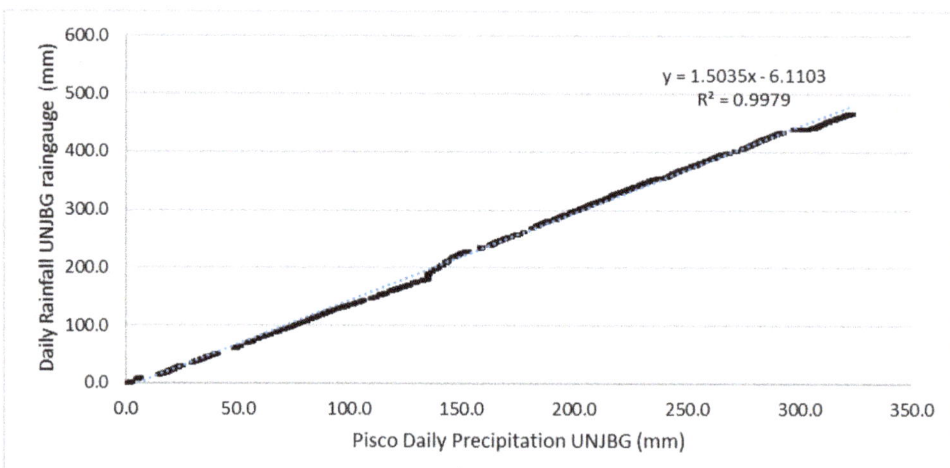

Figure 4. Comparison double-mass between the precipitation data of the PISCO product in JORGE BASADRE rain gauge and the rainfall of the JORGE BASADRE rain gauge, during 7780 days (1 January 1993–21 April 2014).

The efficiency criterion used to evaluate the similarity of the daily rainfall at the JORGE BASADRE rain gauge and the PISCO daily precipitation was the Nash–Sutcliffe model. The efficiency criterion of Nash and Sutcliffe efficiency (E) is defined by the following equation [40].

$$E = 1 - \frac{\sum_1^N (O_i - P_i)^2}{\sum_1^N (O_i - \overline{O})^2} \tag{3}$$

where O_i is the daily rainfall recorded, P_i is the PISCO daily precipitation, $\overline{O}$ is the mean daily rainfall recorded, and N is the number of observations. The ranges of E lay between 0.75 and 1.0 (Very good), 0.65 and 0.75 (Good), 0.5 and 0.65 (Satisfactory), and $\leq$0.5 (Unsatisfactory).

The Nash–Sutcliffe efficiency index of the PISCO product was 0.83, which shows a very good correspondence between the series indicated above [41]. Likewise, it was found that the difference between the total daily precipitation depth discharged from the PISCO product for the coordinates of the JORGE BASADRE rain gauge with respect to the total daily rainfall (depth) from the JORGE BASADRE rain gauge was −31%. Similarly, it was found that the difference between the total daily precipitation depth discharged from the PISCO product for the coordinates of the middle and upper parts of the Devil's Creek with respect to the total daily precipitation depth discharged from the PISCO product for the JORGE BASADRE rain gauge was −6%. Figure 5 shows a comparison of the total daily rainfall discharged from the PISCO product for the coordinates of the middle and upper parts of the Devil's Creek and the total daily rainfall from the JORGE BASADRE rain gauge discharged from the PISCO product, during the period 01 January 1993 to 21 April 2014.

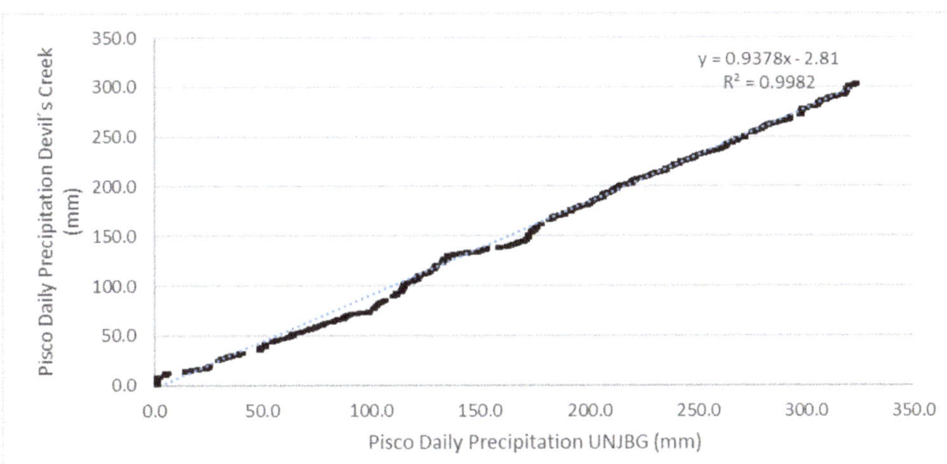

Figure 5. Comparation double-mass between the precipitation data of the PISCO product in the JORGE BASADRE rain gauge versus the precipitation data of the PISCO product in Devil's Creek over 7780 days (1 January 1993–21 April 2014).

On the other hand, the FIAG UNJBG automatic rain gauge has continuously recorded precipitation every 30 min from 7 December 2019 to the present (7 May 2021). Given that the FIAG UNJBG automatic rain gauge is located near the JORGE BASADRE rain gauge, it was assumed that the precipitation which falls every 30 min in the middle and upper parts of the Devil's Creek differed by −38% with respect to FIAG UNJBG automatic rain gauge. The calculations are as follows:

- JORGE BASAGRE rain gauge = 1.31 × PISCO precipitation in JORGE BASADRE;
- PISCO precipitation for the coordinates of the middle and upper parts of the Devil's Creek × 1.06 = PISCO precipitation in JORGE BASADRE;
- The FIAG UNJBG automatic rain gauge is located near the JORGE BASADRE rain gauge; therefore, we assumed that the rainfall data of the FIAG UNJBG automatic rain gauge were the same that the rainfall data of the JORGE BASADRE rain gauge;
- Then, PISCO precipitation for the coordinates of the middle and upper parts of the Devil's Creek × 1.38 = Rainfall data of the FIAG UNJBG automatic rain gauge. The value 1.38 is a result of 1.31 × 1.06.

Considering this, the estimated precipitation for the middle and upper parts of the Devil's Creek for the period 7 December 2019 to 7 May 2021 is shown in Figure 6. In that period, three main events have been identified.

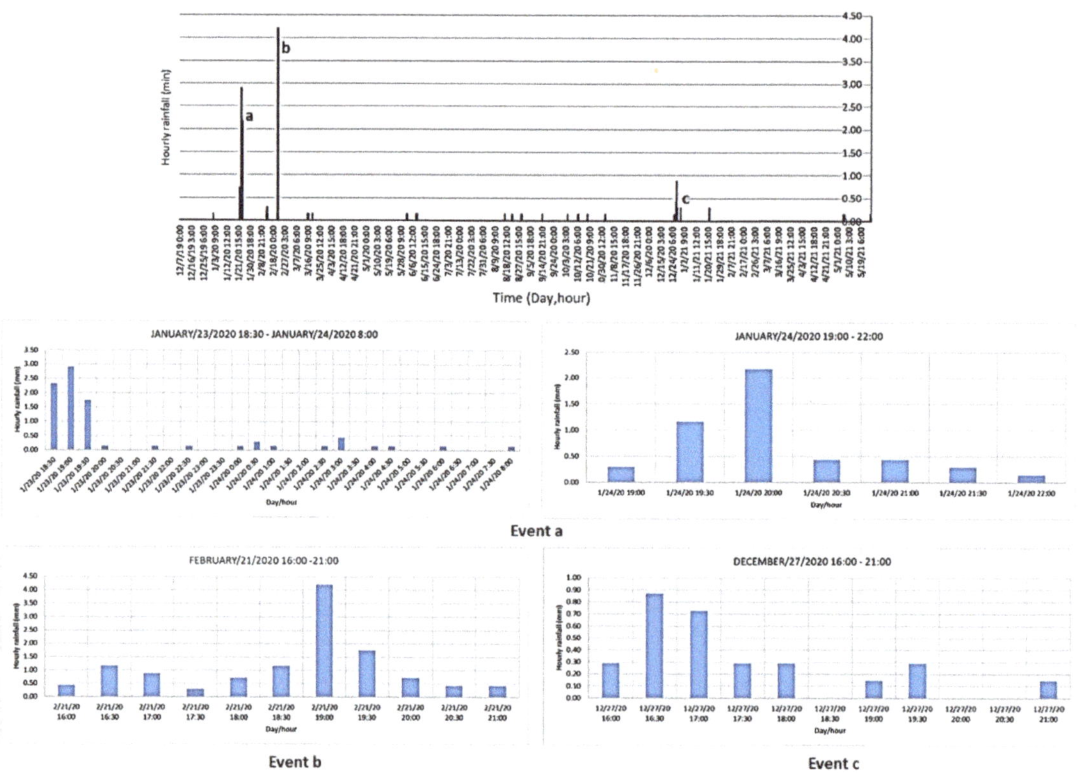

Figure 6. Estimated hourly rainfall for the Devil's Creek. (**a**) Event on 23–24 January 2020; (**b**) event on 21 February 2020; and (**c**) event on 27 December 2020.

In order to corroborate the lower amount of precipitation in the Devil's Creek concerning the FIAG UNJBG automatic rain gauge, a space–time analysis of the information recorded at the JORGE BASADRE rain gauge was carried out with other daily satellite products, such as IMERG and CHIRPS. To do this, two virtual rain gauges were defined to be compared with the JORGE BASADRE rain gauge. Table 2 shows the location of each of the aforementioned rain gauges.

Table 2. Virtual rain gauges and JORGE BASADRE rain gauge.

No	Rain Gauge	Length	Latitude	Elevation (masl)	Source	Record
1	Jorge Basadre G.	−70.2515°	−18.0268°	552	UNJBG	1993–2014, 2017–2021
2	EV01	−70.25°	−17.95°	941	-	-
5	EV02	−70.15°	−17.85°	1560	-	-

Additionally, Figure 7 shows the location of the aforementioned rain gauges. The two virtual rain gauges are located in the lower and upper parts of the basin. It should be noted that the quadrants correspond to the available satellite information.

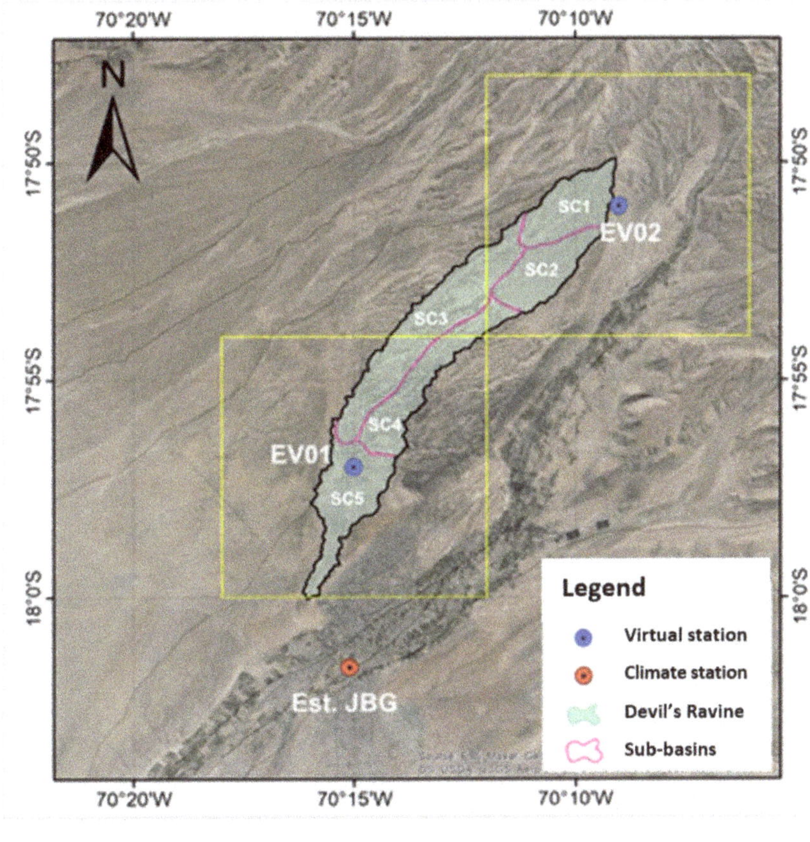

Figure 7. Location of virtual rain gauges and the JORGE BASADRE rain gauge.

For the spatial analysis of precipitation, three satellite products were used, as shown in Table 3. Similarly, for the IMERG product, the early and final versions were used. The information for each product was the total daily precipitation.

In the case of the IMERG product, the information was available from 2000 to date, and in the case of CHIRPS, the information was complete for the entire analysis period.

Figure 8 shows the series of total daily precipitation data from the JORGE BASADRE rain gauge and the series of each satellite product downloaded for the same coordinate from the JORGE BASADRE rain gauge. It can be seen that the data downloaded for the total daily precipitation of the IMERG Early product slightly overestimated the precipitation, and the IMERG Final product underestimated the precipitation information at the JORGE

BASADRE rain gauge. In contrast, the data downloaded for the total daily precipitation of the CHIRPS product were greater than the data for the total daily precipitation of the JORGE BASADRE rain gauge.

Table 3. Satellite and data base products analyzed.

Product	Version	Abbreviation	Source	Resolution	Frequency	Term
Peruvian Interpolated data of SENAMHI's Climatological and Hydrological Observations	V. 2.1	PISCO	SENAMHI	0.1° × 0.1°	Daily	1981–2016
Integrated Multi-satellite Retrievals for GPM	Early V06B	IMERG-F	NASA	0.1° × 0.1°	Daily and 30 min	2000–2021
Integrated Multi-satellite Retrievals for GPM	Final V06B	IMERG-E	NASA	0.1° × 0.1°	Daily and 30 min	2000–2021
Climate Hazards group Infrared Precipitation with Rain gauges	V. 2.0	CHIRPS	UCSB (x)	0.05° × 0.05°	Daily	1981–2021

(x) University of California Santa Bárbara.

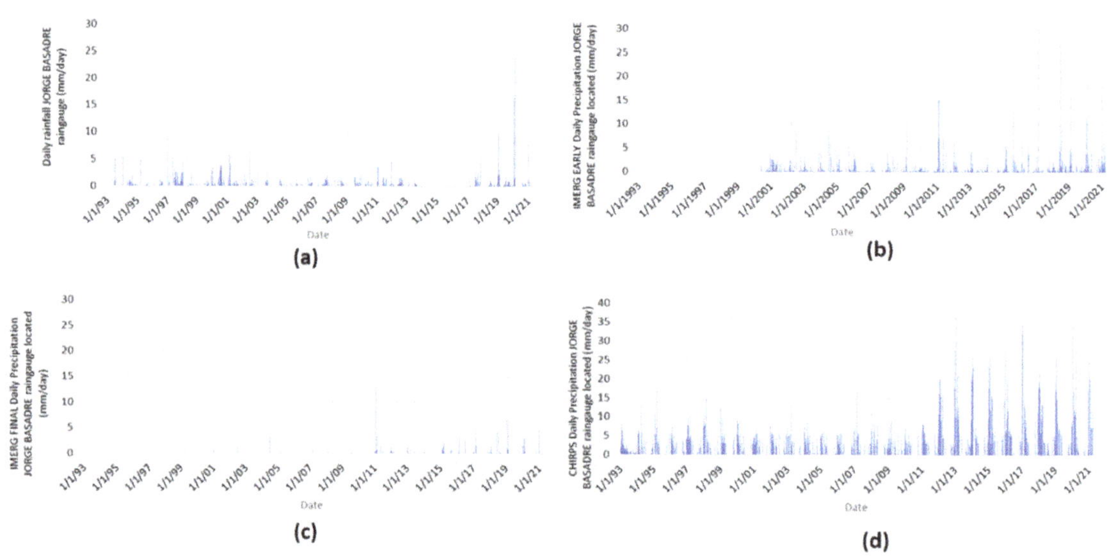

Figure 8. Comparation between (**a**) daily rainfall in JORGE BASADRE rain gauge and daily precipitation from satellites products: (**b**) IMERG Early, (**c**) IMERG Final, and (**d**) CHIRPS.

Although the information from the satellite products used does not have a good correlation with the data observed from the JORGE BASADRE rain gauge, a spatial correlation between them can be determined to establish indicators of spatial variation. The spatial correlation was carried out between the three established rain gauges: JORGE BASADRE rain gauge, Virtual Rain Gauge 01 (lower part of the Creek), and Virtual Rain Gauge 2 (upper part of the creek).

Figure 9 shows the results of the correlations made concerning the JORGE BASADRE rain gauge and the virtual rain gauges EV01 and EV02. With the IMERG Early product, the results obtained were that the precipitation in the EV01 area was 40% lower than the precipitation in the JORGE BASADRE rain gauge, whereas the EV02 rain gauge area had 41% less precipitation than the JORGE BASADRE rain gauge area. Similarly, with the IMERG Final product, the results obtained were that the precipitation in the EV01 area was

33% lower than the precipitation in the JORGE BASADRE rain gauge, whereas the EV02 rain gauge area had 71% less precipitation than the zone of JORGE BASADRE rain gauge.

Concerning the CHIRPS product, the precipitation in the EV01 and EV02 zones was lower by 27% and 45%, respectively, in comparison to the JORGE BASADRE rain gauge. In summary, the previous results corroborate that the amount of precipitation in the Devil's Creek was less than the amount of precipitation in the city of Tacna, during the period of common observation between them.

Table 4 shows a summary table of the linear regression coefficients obtained in the comparison of the data series. It is observed that the area of the EV01 virtual rain gauge, on average, would have 31% less precipitation than the JORGE BASADRE rain gauge, whereas the area of the EV02 virtual rain gauge would have 51.7% less precipitation than the JORGE BASADRE rain gauge.

Table 4. Summary of the regression coefficients and percentage between the JORGE BASADRE rain gauge and the virtual rain gauge (EV01 and EV02).

Virtual Rain Gauge	PISCO JB Rain Gauge	IMERG Early JB Rain Gauge	IMERG Final JB Rain Gauge	CHIRPS JB Rain Gauge	Mean	EV01 and EV02 Have Less Precipitation than JORGE BASADRE Rain Gauge (%)
EV01	0.76	0.60	0.67	0.73	0.69	100% − 69% = 31.0%
EV02	0.51	0.59	0.29	0.54	0.483	100% − 48.3% = 51.7%

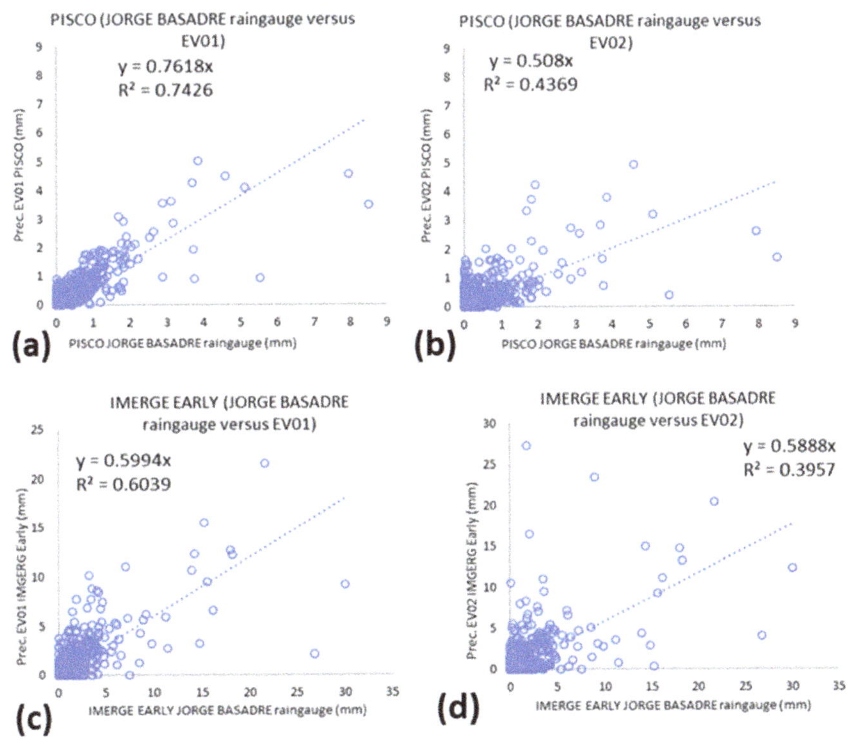

Figure 9. *Cont.*

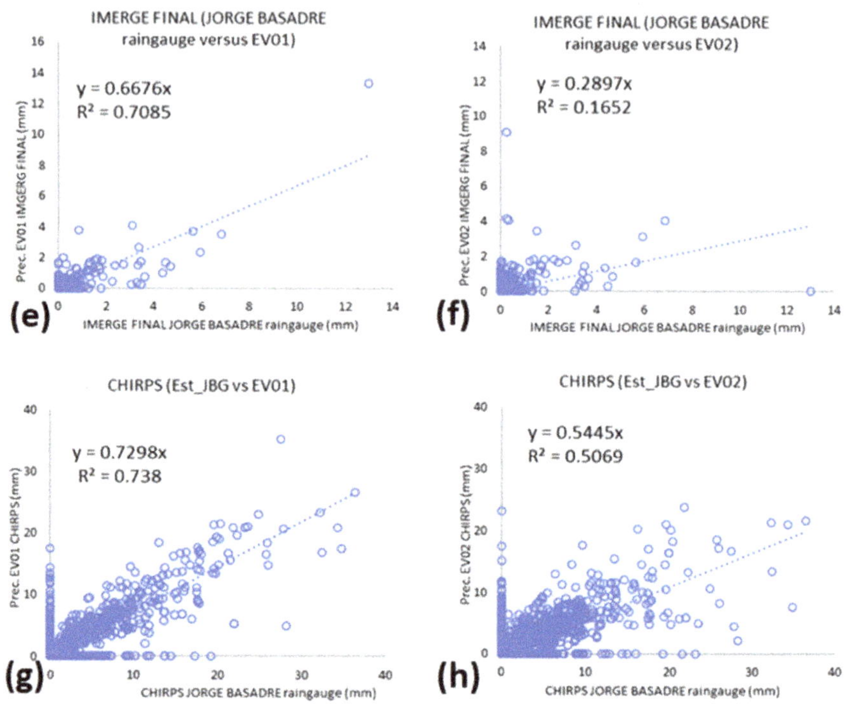

Figure 9. Correlations were made between the total daily precipitation data between the JORGE BASADRE rain gauge and the virtual rain gauges, (EV01 and EV02) using the products: (**a**,**b**) PISCO, (**c**,**d**) IMERGE Early, (**e**,**f**) IMERGE Final, and (**g**,**h**) CHIRPS.

2.5.3. Types of Soils and Infiltration Capacity in the Devil's Creek

Characterization of the soil type was carried out through field samplings conducted on 22 July 2020. Figure 10 shows the places where three soil samplings and the corresponding infiltration tests were carried out. The information on the infiltration tests is presented in Table 5.

Table 5. Location of sampling points, texture, infiltration equation, and hydraulic conductivity at saturation (Ks).

Sampling	UTM Coordinates	Texture	Infiltration Equation F (mm), t (min)	Ks (mm/min)
1	368477E, 8019016N	Clayey silt	$F = 4.0\ t^{0.81}$	0.872
2	368477E, 8019035N	Sandy silt with gravel and clay	$F = 3.67\ t^{0.87}$	1311
3	368328E, 8018959N	Sandy silt with gravels	$F = 7.2\ t^{0.72}$	0.749

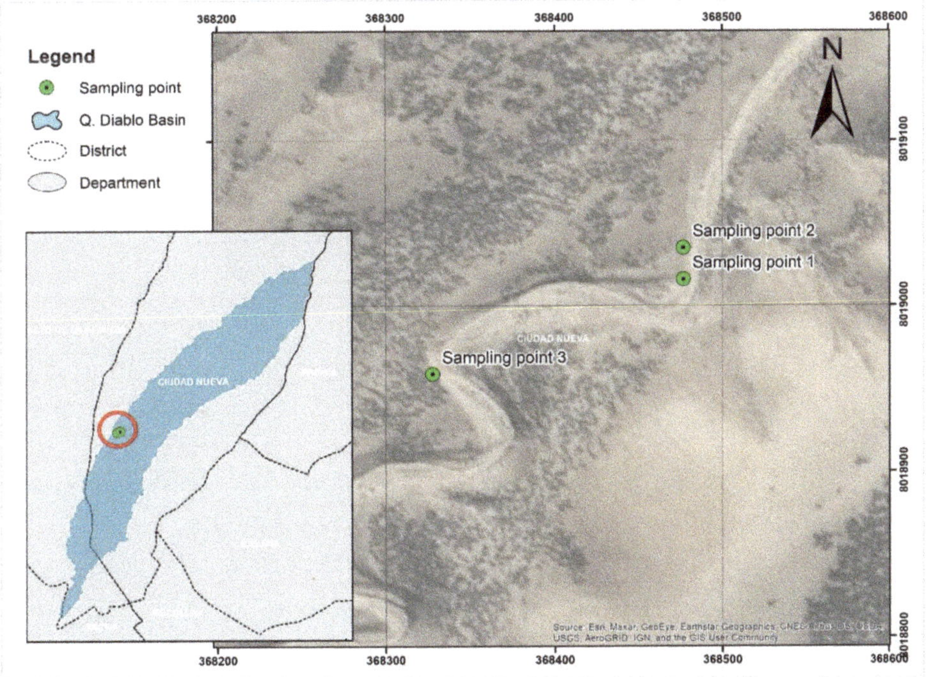

Figure 10. Location of soil sampling points.

2.5.4. Hydrological Model of the Devil's Creek

Hydrological modeling of the maximum flow produced in the Devil's Creek was carried out using the RS MINERVE program.

The RS MINERVE program is open access and widely used in Peru and national institutions such as the National Water Authority (ANA, by its abbreviation in Spanish) and the Potable Water and Sewerage Service of Lima (SENAMHI, by its abbreviation in Spanish) [42].

The RS MINERVE program is a flow simulation program which allows the modeling of complex hydraulic and hydrological networks following a semi-distributed approach. The program is capable of representing not only the main hydrological processes such as snow and glacier melting, surface and sub-surface runoff, but also regulatory infrastructure such as retention dams, spillways, water intakes, turbines, and pumps, among others.

RS MINERVE integrates different hydrological precipitation-runoff models such as GSM, SOCONT, SAC-SMA, GR4J, and HBV. Likewise, it allows the inclusion of hydraulic structures through different models (reservoirs, turbines, spillways, etc.) [43].

The model used for the hydrological modeling of the Devil's Creek was the SOCONT (Soil CONTribution model). The SOCONT and GR4J models are more sensitive in extreme event modeling than the HBV and SAC models [44]. As shown in Figure 11, the SOCONT model procedure consists of the Snow-SD model simulating the evolution of the glacial layer (melting and accumulation) as a function of temperature (T) and precipitation (P), and calculating the equivalent precipitation (Peq). In the case of Devil's Creek, and because it is not a glacial stream, the equivalent precipitation constituted the precipitation. This equivalent precipitation was used as input to the GR3 model that takes into account the potential evapotranspiration (ETP) and generates the net intensity of rainfall for the SWMM model.

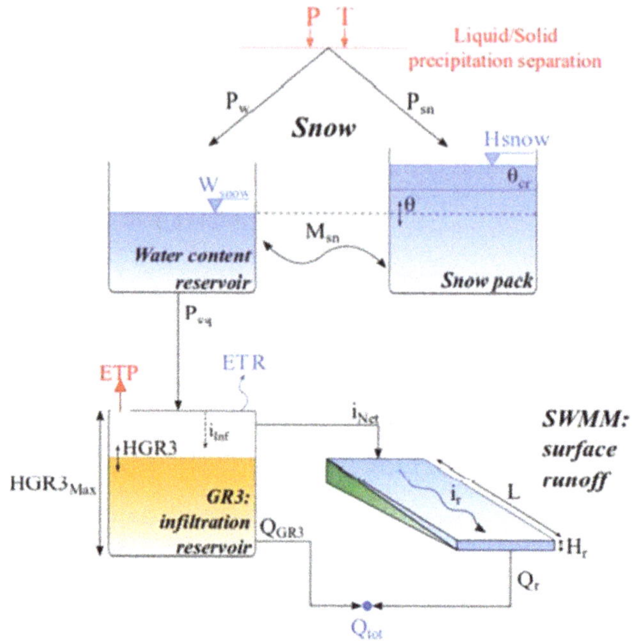

Figure 11. Composition of the SOCONT model.

The flow components of the SOCONT model are explained below:

- P: precipitation (L/T);
- T: Temperature (°C);
- P_w: Liquid precipitation (L/T);
- P_{sn}: Solid precipitation (L/T);
- P_{eq}: Equivalent precipitation (L/T) equal P_w (L/T), because there is no glacial area in the basin;
- i_{Inf}: Infiltration intensity (L/T)

  $i_{Inf} = P_{eq}\ (1 - (H_{GR3}/H_{GR3Max})^2)$, If $H_{GR3} \leq H_{GR3Max}$, where H_{GR3} (L) is the actual capacity of infiltration reservoir (L), and H_{GR3Max} (L) is the maximum capacity of the infiltration reservoir.
  $i_{Inf} = 0$, if $H_{GR3} > H_{GR3Max}$;
- ETR: Real evapotranspiration (L/T)

  $ETR = ETP\ \sqrt{H_{GR3}/H_{GR3Max}}$, If $H_{GR3} \leq H_{GR3Max}$
  $ETR = ETP$, If $H_{GR3} > H_{GR3Max}$;
- i_{Net}: Net intensity (L/T), $i_{Net} = P_{eq} - i_{Inf}$;
- Q_{GR3}: Base discharge (L^3/T)

  $Q_{GR3} = K_{GR3} \cdot H_{GR3} \cdot A$, If $H_{GR3} \leq H_{GR3Max}$, where K_{GR3} is the release coefficient of the infiltration reservoir (1/T) and A is the surface (L^2).
  $Q_{GR3} = K_{GR3} \cdot H_{GR3Max} \cdot A$, if $H_{GR3} > H_{GR3Max}$;
- I_r: Runoff intensity (L/T), $I_r = K_r \sqrt{J_0} H_r^{5/3} \cdot 1/L$, where K_r is the Strickler coefficient ($L^{1/3}$/T), J_0 is the average slope of the plane, H_r is the runoff water level downstream of the surface (L), and L is the length of the plane (L);
- Q_r: Surface runoff (L^3/T), $Q_r = I_r \cdot A$;
- Q_{tot}: Total runoff (L^3/T), $Q_{tot} = Q_{GR3} + Q_r$.

According to Table 6, the total number of parameters necessary for hydrological modeling is 16, of which 7 parameters are typical of the SOCONT model and also take into account 4 initial conditions.

Table 6. SOCONT model parameters and plugins [45].

Object	Name	Units	Description	Regular Range
SOCONT	A	m^2	Surface	>0
	S	mm/°C/d	Reference degree-day snowmelt coefficient	0.5 to 20
	SInt	mm/°C/d	Degree-day snowmelt coefficient	0 to 4
	Smin	mm/°C/d	Minimal degree-day snowmelt coefficient	≥0
	SPh	d	Phase shift of the sinusoidal function	1 to 365
	ThetaCri	-	Critical relative water content of the snow pack	0.1
	bp	d/mm	Melt coefficient due to liquid precipitation	0.0125
	Tcp1	°C	Minimum critical temperature for liquid precipitation	0
	Tcp2	°C	Maximum critical temperature for solid precipitation	4
	Tcf	°C	Critical snowmelt temperature	0
	HGR3Max	m	Maximum height of infiltration reservoir	0 to 2
	KGR3	1/s	Release coefficient of infiltration reservoir	0.00025 to 0.1
	L	m	Length of the plane	>0
	J0	-	Runoff slope	>0
	Kr	$m^{1/3}/s$	Strickler coefficient	0.1 to 90
	CFR	-	Refreezing coefficient	0 to 1
	SWEIni	m	Initial snow water equivalent height	-
	HGR3Ini	m	Initial level in infiltration reservoir	-
	HrIni	m	Initial runoff water level downstream of the surface	-
	ThetaIni	-	Initial relative water content in the snow pack	-

3. Results

3.1. Projected Maximum Annual Precipitation

3.1.1. Period 2021–2050

- RCP4.5 scenario

The maximum annual precipitation projected for the period 2021–2050 under the RCP4.5 emission scenario in the Devil's Creek is presented in Figure 12. The results show a variation range from 0.79 mm to 283 mm with an average of 5.64 mm and a standard deviation of 12.65 mm. The maximum value was projected by the model MPI_ESM_LR (the coupled Max Planck Institute Earth System Model). Likewise, the models bcc_csm1_1, CanESM2, CCSM4, GFDL_ESM2G, GFDL_ESM2M, MPI_ESM_LR projected maximum annual precipitation in a range of 27 mm to 283 mm. The bcc_csm1_1 and GFDL_ESM2M models predicted around 46 mm by 2034.

On the other hand, given the uncertainty in the projections, an arithmetic average of all models has been estimated, as well as an ensemble (dashed blue line) which has been calculated, assigning a greater weight to the model that has better performance concerning the maximum annual rainfall observed (see Equation (2)). The Access1-0 model was discarded from the analysis of maximum annual precipitation because it projected high values greater than 500 mm. The results of the assembly of the models indicate variation in the annual maximum daily precipitation from 2.77 mm to 24.70 mm, with an average of 5.70 mm for the period 2021–2050. The range of uncertainty in the predictions is shown in Figure 13, in which the linear extensions represent the highest and lowest values; the upper, central, and lower limits of the box represent the percentiles of 75%, 50%, and 25%, respectively; and the solid circles represent the outliers, which correspond to the maximum values predicted by each model.

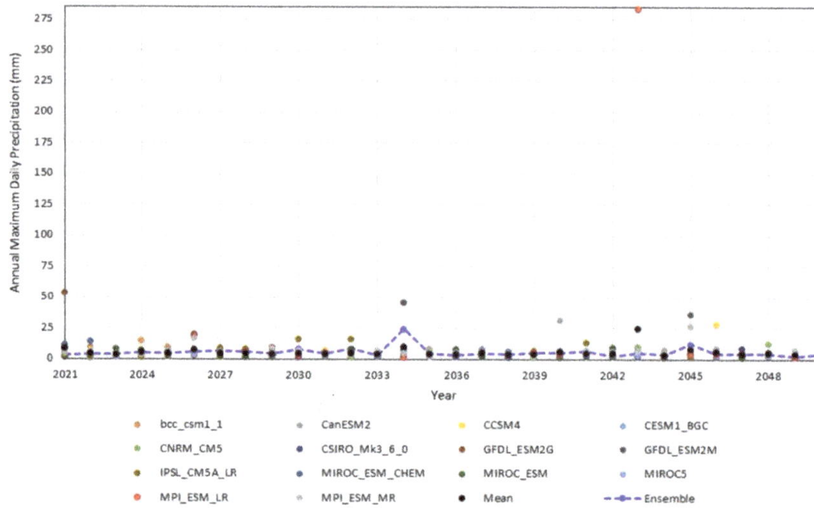

Figure 12. Annual maximum daily precipitation projected by 14 climate models under the RCP4.5 scenario for the Devil's Creek. The dashed blue line corresponds to the ensemble. Period: 2021–2050.

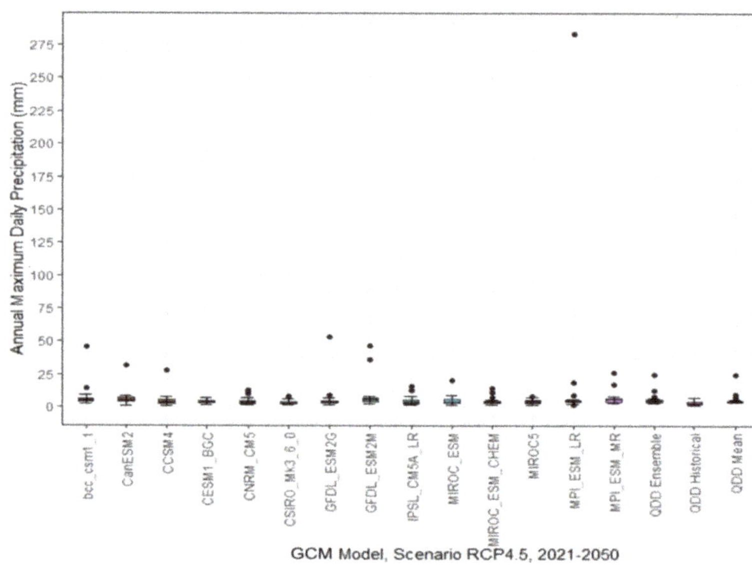

Figure 13. Boxplot for future projections of the annual maximum daily precipitation of 14 GCMs under the RCP4.5 scenario in the Devil's Creek, period 2021–2050. The linear extensions represent the highest and lowest values; the upper, middle and lower limits of the box represent the percentiles of 75%, 50%, and 25%, respectively; and the solid circles represent the outliers.

- RCP8.5 Scenario

The annual maximum daily precipitation projected for the 2021–2050 term under the RCP8.5 emission scenario in the Devil's Creek is shown in Figure 14. A varied range of 1.70 mm to 75.7 mm is evidenced with an average of 5.24 mm and a standard deviation of 3.29 mm for all models, indicating a low dispersion concerning the mean. The maximum value was projected by the GFDL_ESM2M (Fluid Dynamics Laboratory (GFDL) GFDL-ESM2M model).

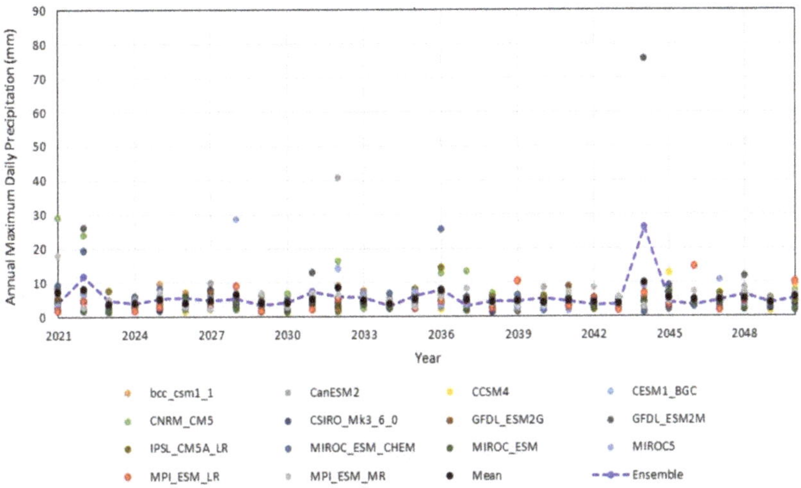

Figure 14. Annual maximum daily precipitation projected by 14 climate models under the RCP8.5 scenario for the Devil's Creek. The dashed blue line corresponds to the ensemble. Period: 2021–2050.

Likewise, the CanESM2, CNRM_CM5, GFDL_ESM2M, MIROC_ESM_CHEM, MIROC5 models projected the highest annual maximum precipitation values in a range of 25.5 mm to 75.7 mm, indicating that, in this period of analysis, heavy precipitation could occur in the Devil's Creek.

On the other hand, the results of the model ensemble indicate a variation in the maximum annual precipitation from 3.14 mm to 25.96 mm, with an average of 5.80 mm for the 2021–2050 term under the RCP8.5 scenario.

The range of uncertainty in the predictions is shown in Figure 15. Similarly, the linear extensions represent the highest and lowest values; the upper, central, and lower limits of the box represent the percentiles of 75%, 50%, and 25%, respectively; and the solid circles represent the outliers, which correspond to the maximum values predicted by each model.

3.1.2. Period 2051–2080

- RCP4.5 Scenario

The maximum annual precipitation projected for the period 2051–2080 under the RCP4.5 emission scenario in the Devil's Creek is shown in Figure 16. The results indicate a variation range from 0.86 mm to 95.46 mm with an average of 5.55 mm and a standard deviation of 5.20 mm for all models, indicating a low dispersion concerning the mean. The maximum value was projected by the CCSM4 model (The Community Climate System Model Version 4). Likewise, of the 14 models evaluated, 7 of them CCSM4, CNRM_CM5, GFDL_ESM2M, IPSL_CM5A_LR, MIROC_ESM_CHEM, MIROC5, MPI_ESM_LR projected the highest annual maximum precipitation values in a range of 21 mm to 96 mm, indicating that heavy precipitation could occur in the Devil's Creek during the analysis period under this emission scenario.

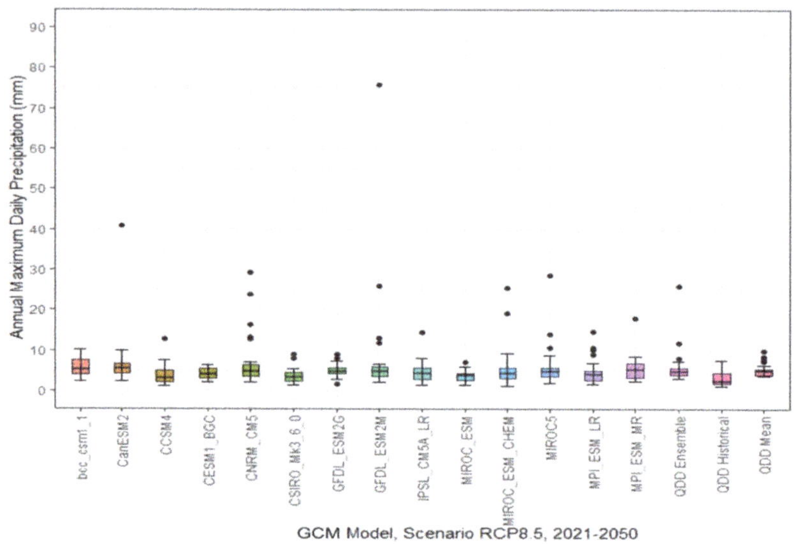

Figure 15. Boxplot for future projections of the annual maximum daily precipitation of 14 GCMs under the RCP8.5 scenario in the Devil's Creek, period 2021–2050. The linear extensions represent the highest and lowest values; the upper, middle and lower limits of the box represent the percentiles of 75%, 50%, and 25%, respectively; and the solid circles represent the outliers.

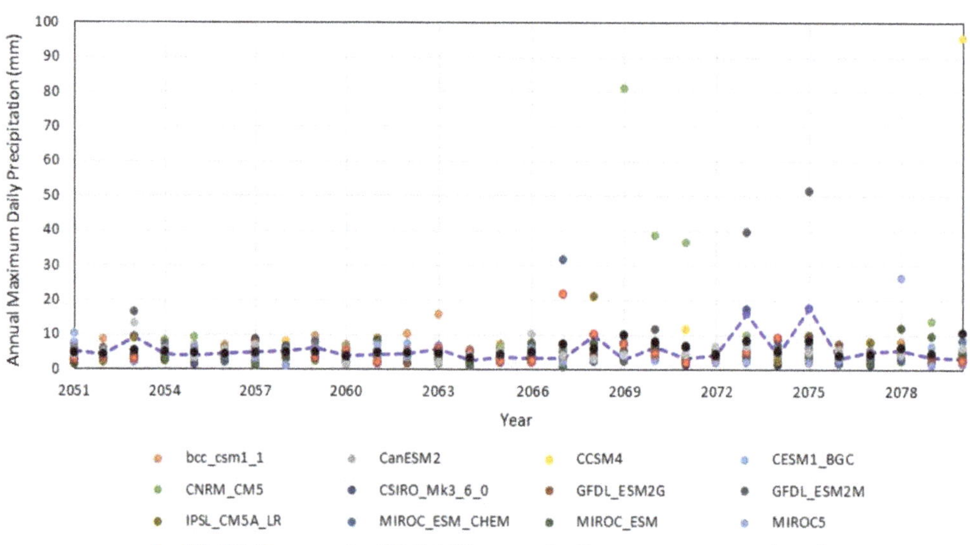

Figure 16. Annual maximum daily precipitation projected by 14 climate models under the RCP4.5 scenario for Devil's Creek. The dashed blue line corresponds to the ensemble. Period: 2051–2080.

On the other hand, the results of the model ensemble indicate a variation in the maximum annual precipitation from 2.77 mm to 17.86 mm, with an average of 5.60 mm for the 2051–2080 term under the RCP4.5 scenario. The range of uncertainty in the predictions is presented in Figure 17.

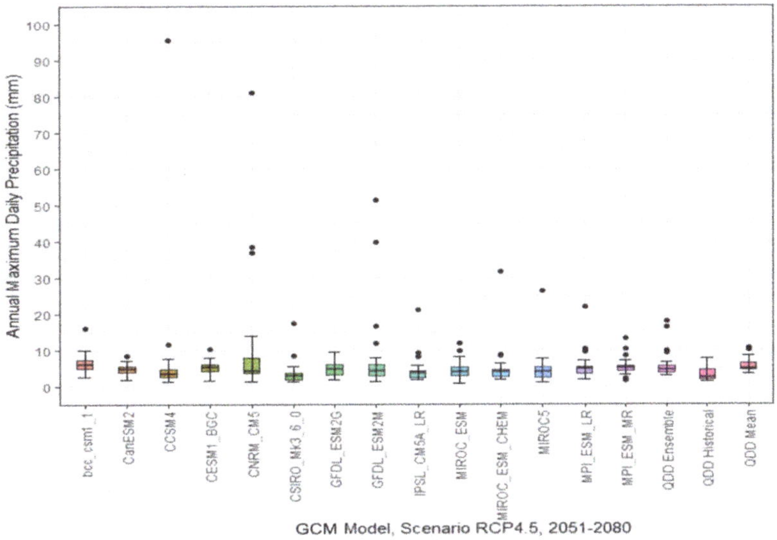

Figure 17. Boxplot for future projections of the annual maximum daily precipitation of 14 GCMs under the RCP4.5 scenario in the Devil's Creek, period 2051–2080. The linear extensions represent the highest and lowest values; the upper, middle and lower limits of the box represent the percentiles of 75%, 50%, and 25%, respectively; and the solid circles represent the outliers.

- RCP8.5 Scenario

The maximum annual precipitation projected for the period 2051–2080 under the RCP8.5 emission scenario in the Devil's Creek is shown in Figure 18. The results indicate a variation range from 0.62 mm to 224.65 mm with an average of 6.0 mm and a standard deviation of 10.50 mm for all models, indicating a high dispersion concerning the mean. The maximum value is projected by the bcc_csm1_1 (the Beijing Climate Center Climate System Model). Likewise, of the 14 models evaluated, 6 of them bcc_csm1_1, CanESM2, CNRM_CM5, GFDL_ESM2G, IPSL_CM5A_LR, MIROC5 projected the highest annual maximum precipitation values in a range from 19.28 mm to 224.65 mm, indicating that heavy rainfall could occur in the stream of the river. Diablo during the analysis period under this broadcast scenario.

On the other hand, the results of the model ensemble indicate a variation in the maximum annual precipitation from 2.92 mm to 49.74 mm, with an average of 7.21 mm for the 2051–2080 term under the RCP8.5 scenario. The range of uncertainty in the predictions is presented in Figure 19. Similarly, the linear extensions represent the highest and lowest values; the upper, middle, and lower limits of the box represent the percentiles of 75%, 50%, and 25%, respectively.

On the other hand, a relative change has been calculated in relation to the historical average for the period 1981–2005, simulated by the corrected and scaled GCM for the Devil's Creek. Under the RCP4.5 scenario, the annual maximum daily precipitation could increase by 32.44%, on average; with a range from −35.80% to +470.77%. Under the RCP8.5 scenario, the maximum annual precipitation will probably increase by 34.64%, on average, during the 2021–2050 term. Values range from −27.24% to 502.37%. Positive and negative values indicate a probable increase and decrease in the annual maximum daily precipitation, respectively. For the 2021–2050 period, under the RCP4.5 scenario, the maximum annual precipitation could increase by 29.79%, on average; the range is from −35.66% to +314.45%. On the other hand, under the RCP8.5 scenario, the maximum annual precipitation could increase by 67.23%, on average. For this scenario, the values range from

−32.24% to 1053.97%. The maximum percentage value corresponds to a positive anomaly of 45.43 mm concerning the historical average simulated by the GCM.

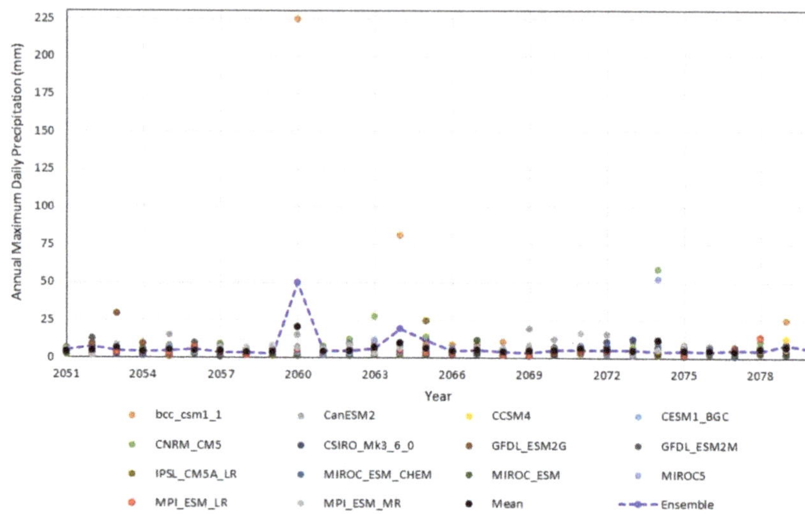

Figure 18. Annual maximum daily precipitation projected by 14 climate models under the RCP8.5 scenario for the Devil's Creek. The dashed blue line corresponds to the ensemble. Period: 2051–2080.

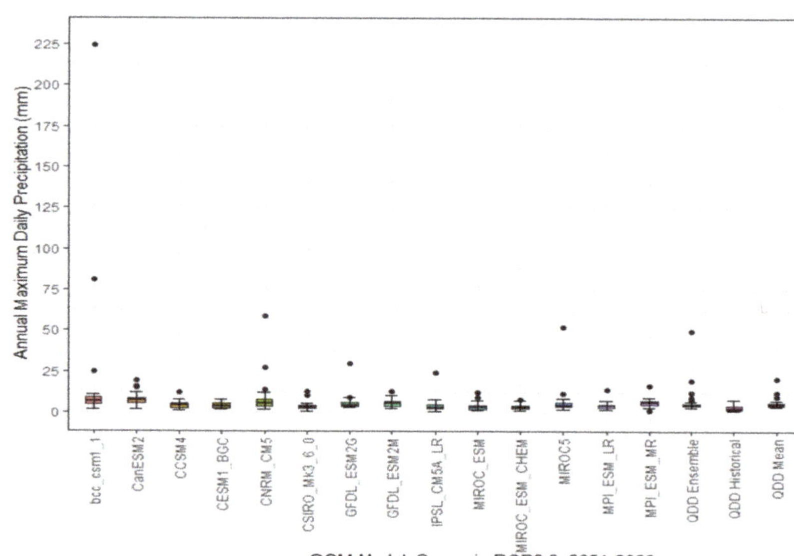

Figure 19. Boxplot for future projections of the annual maximum daily precipitation of 14 GCMs under the RCP8.5 scenario in the Devil's Creek, 2051–2080 term. The linear extensions represent the highest and lowest values; the upper, middle, and lower limits of the box represent the percentiles of 75%, 50%, and 25%, respectively; and the solid circles represent the outliers.

3.2. Projected Temperature

3.2.1. Annual Average Temperature

The temperature under climate change is another important variable that needs to be evaluated and projected to determine the potential impact on the hydrological response of the basin under study. Figures 20 and 21 present the minimum average annual temperature simulated by regional climate models, after bias correction and scaling for the study area, under RCP4.5 and RCP8.5 scenarios, for the 1981–2100 term. The black line corresponds to the averaged time series of all models evaluated. Similarly, as evidenced by other researchers, climate models agree in projecting a positive trend in temperature. However, from 2050 onwards, greater variability and a decrease in the trend are observed under the RCP4.5 scenario. On the other hand, as expected, the RCP8.5 high-emissions scenario projected the highest values. In both scenarios, the MIRO-ESM-CHEM model projects the highest minimum temperature values. Figures 22 and 23 show the average annual maximum temperature projected by the regional climate models for the study area, under the RCP4.5 and RCP8.5 scenarios, 1981–2100 term. Similarly, the black line corresponds to the averaged time series of all the evaluated models and for both scenarios, the MIRO-ESM-CHEM model projects the highest values of maximum temperature.

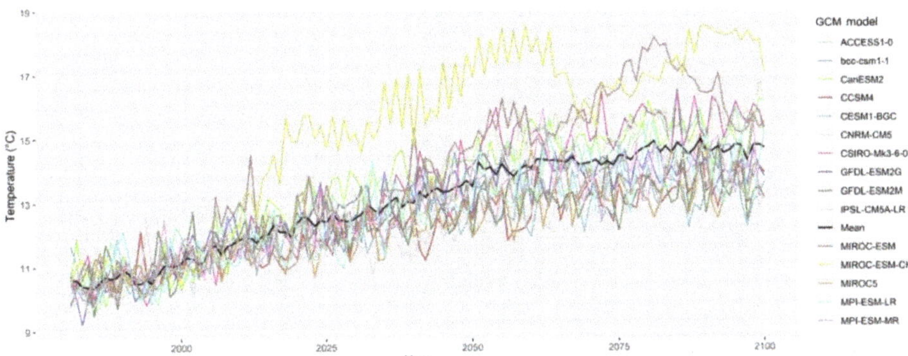

Figure 20. Minimum annual average temperature simulated by climate models, corrected and scaled for the Devil's Creek, 1981–2100 term, RCP4.5 emission scenario. The black line represents the averaged ensemble of 15 GCMs.

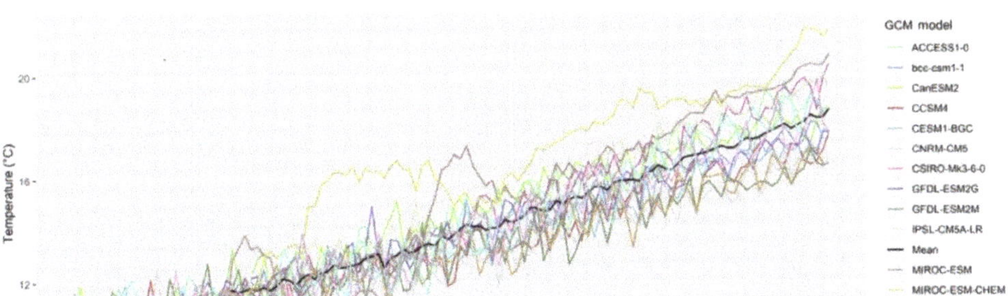

Figure 21. Minimum annual average temperature simulated by climate models, corrected and scaled for the Devil's Creek, 1981–2100 term, RCP8.5 emission scenario. The black line represents the averaged ensemble of the 15 GCMs.

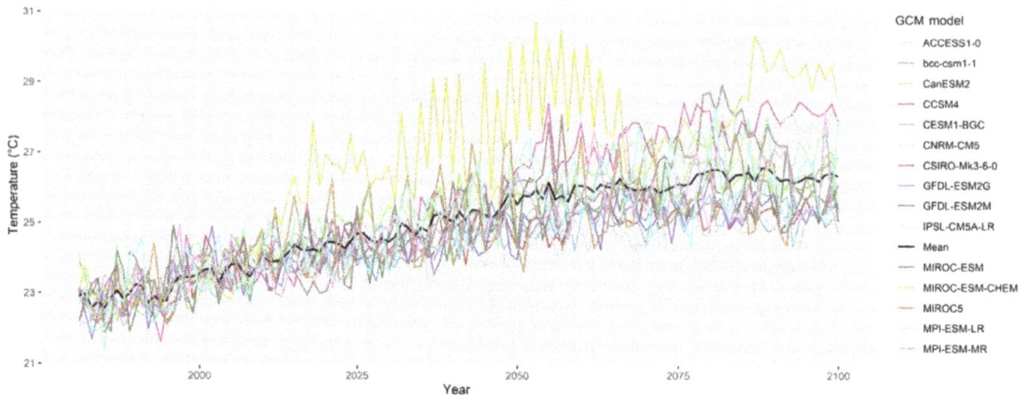

Figure 22. Maximum annual average temperature simulated by climate models, corrected and scaled for the Devil's Creek, 1981–2100 term, RCP4.5 emission scenario. The black line represents the averaged ensemble of 15 GCMs.

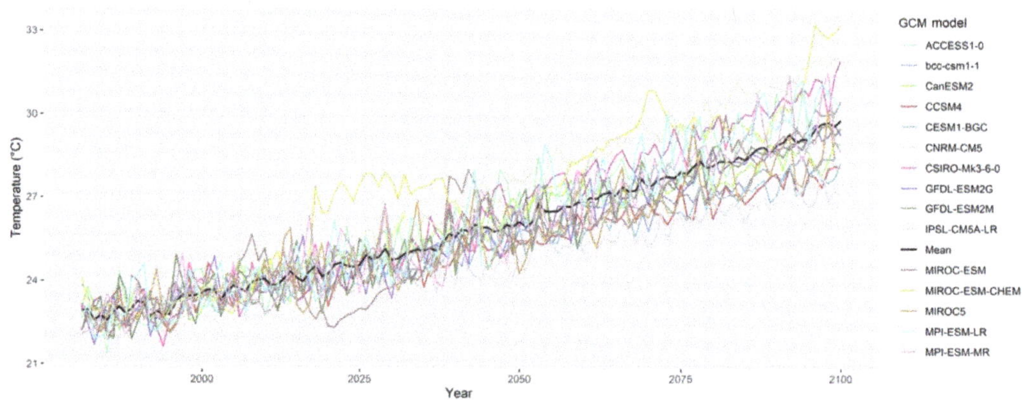

Figure 23. Maximum annual average temperature simulated by regional climate models, corrected and scaled for the Devil's Creek, 1981–2100 term, RCP8.5 emission scenario. The black line represents the averaged ensemble of 15 GCMs.

3.2.2. Average Monthly Temperature

Figure 24 presents the minimum monthly average temperature under the RCP4.5 and RCP8.5 scenarios for the Devil's Creek, 2021–2050 and 2051–2080 periods for the average multi-model ensemble of 15 regional climate models. Likewise, the historical period of 1981–2005 simulated by the GCM is observed. Monthly increases of 2.13 °C and 3.45 °C on average are projected for the minimum temperature under the RCP4.5 scenario, and of 2.62 °C and 4.90 °C under the RCP8.5 scenario, for the 2021–2050 and 2051–2080 terms, respectively. All of them corresponded with the 1981–2005 term. Likewise, the months of May, June, July, and August, as well as the period from January to March, project the largest increases in the minimum temperature for both scenarios. Figure 25 presents the maximum monthly average temperatures for the RCP4.5 and RCP8.5 scenarios. The results indicate an increase in the monthly average maximum temperature of 1.79 and 2.85 °C under the RCP4.5 scenario for the 2021–2050 and 2051–2080 terms, both relative to the 1985–2005 term. On the other hand, under the RCP8.5 scenario, the average increase is 2.12 °C and 4.06 °C for the periods previously described. For both scenarios, June, July, and August show the largest increases, meaning that there will probably be greater warming during winter in the coming decades. Likewise, the minimum temperatures tend to increase slightly in relation to the maximum temperatures.

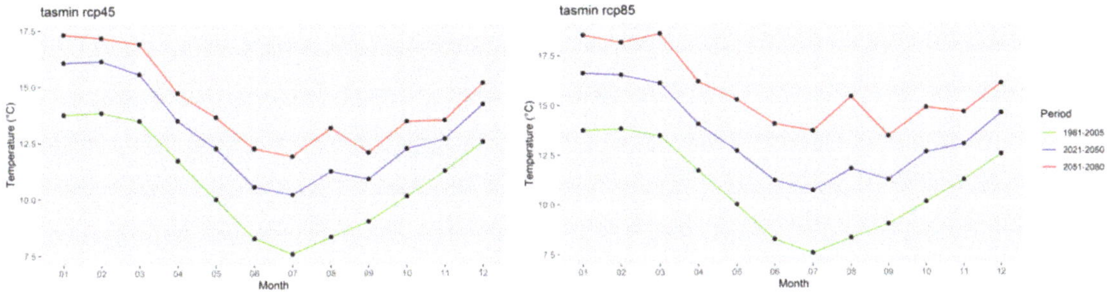

Figure 24. Average monthly temperature (minimum) under RCP4.5 and RCP8.5 scenarios for the Devil's Creek, 2021–2050 and 2051–2080 periods. The multi-model ensemble average of 15 GCMs.

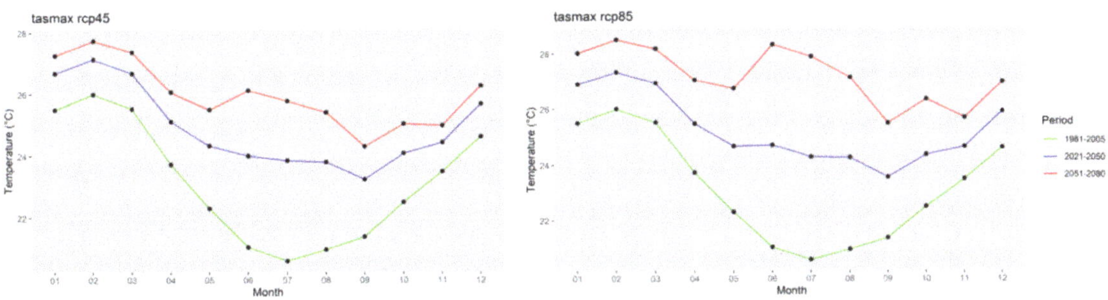

Figure 25. Average monthly temperature (maximum) under RCP4.5 and RCP8.5 scenarios for the Devil's Creek, 2021–2050 and 2051–2080 periods. The multi-model ensemble average of 15 GCMs.

3.2.3. Monthly Changes

In this study, we have provided average temperature changes for two periods: 2020–2050 and 2051–2080, relative to the period 1981–2005. Figure 26 shows the monthly average temperature changes (minimum) under RCP4.5 and RCP8.5 scenarios for the Devil's Creek, for the multi-model average ensemble of 15 GCMs corrected and downscaled. For both scenarios and the two study periods, changes are positives; consequently, minimum temperature would increase in the next decades under climate change. Under the RCP4.5 scenario, the largest positive changes are projected in July and August: 2.7 °C on average for the period 2021–2050. Additionally, this is a change of 4.4 °C, on average, for winter (JJA) during the period 2051–2080. The lowest minimum temperature changes are projected for spring (SON) with 1.8 °C and 2.9 °C for both periods, respectively. On the other hand, under the RCP8.5 scenario, the highest values of increase in the minimum temperature are projected for winter (JJA), more than 3.2 °C and 6. 3 °C for both periods.

Figure 27 shows the monthly average temperature change (maximum) under the RCP4.5 and RCP8.5 scenarios. Under the RCP4.5 scenario, it is evident that the greatest positive change in the maximum temperature would occur in winter (JJA), more than 3.0 °C and 4.9 °C for both periods. Similarly, for the RCP8.5 high-emissions scenario, in winter, the highest increases are projected, ranging from 3.5 to 6.9 °C, on average, for the periods 2021–2050 and 2051–2080, respectively.

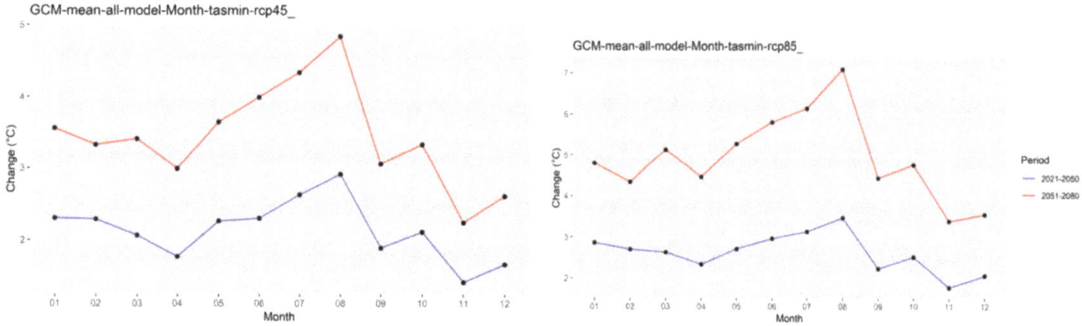

Figure 26. Average monthly temperature change (minimum) under RCP4.5 and RCP8.5 scenarios for the Devil's Creek, 2021–2050 and 2051–2080 periods. The multi-model ensemble average of 15 GCMs.

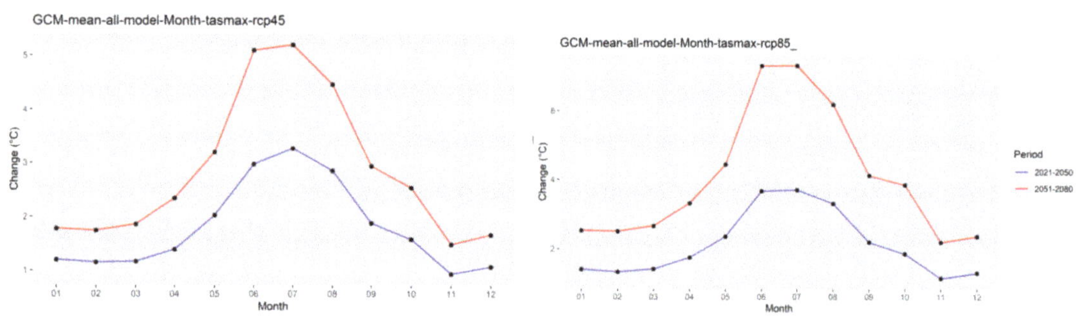

Figure 27. Average monthly temperature change (maximum) under RCP4.5 and RCP8.5 scenarios for the Devil's Creek, 2021–2050 and 2051–2080 periods. The multi-model ensemble average of 15 GCMs.

3.3. Calibration and Generation of Maximum Flows in the Devil's Creek

3.3.1. Under Historical Conditions

As a precedent to the generation of the maximum flows produced in the Devil's Creek, it should be mentioned that the channel of this creek was interrupted by an informal embankment used as a trail to access farms in the area of the Alto de la Alianza hill. The collapse of this embankment, located 2.0 km from the town of La Florida (City of Tacna), caused the debris flow on 21 February 2020. Figure 28 shows the site before and after the collapse of the informal embankment called Paso Camiara.

The calibration of the hydrological model was carried out by modeling the maximum flow produced in the Devil's Creek, dated 21 February 2020, as a result of the event of maximum rain and rupture of the Paso Camiara informal embankment. Figure 29 shows the topology of the Devil's Creek, generated in the RS Minerve model.

As seen in Figure 7, the Devil's Creek was subdivided into five sub-basins: two in the upper part (SC1 and SC2), two in the middle part (SC3 and SC4), and one in the lower part (SC5). Each of the sub-basins is linked to a virtual precipitation rain gauge. The precipitation events for each sub-basin are shown in Table 7.

The rain event of 21 February 2020, attributable to the positive anomaly of the sea surface temperature between +1 °C and +2 °C, produced off the coast of Tacna between 22 January 2020 and 22 February 2020 (Figure 30) [12].

Riverbed of the Devil's Creek obstructed by an informal embankment as an access trail to farms on the Alto de la Alianza hill. Source INGEMMET (2016).

The place of the informal embankment collapsed by the flood of 22 February 2020 and that caused the flood towards the city of Tacna.

Figure 28. Before and after the Devil's Creek channel was obstructed by an informal embankment as an access trail to farms in the Alto de la Alianza hill.

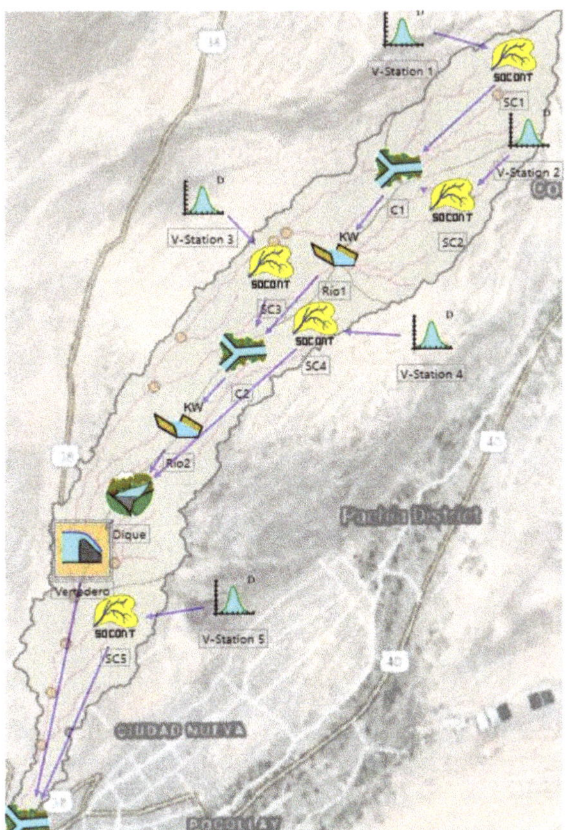

Figure 29. Structure of the RS Minerve Model for simulation of the 21 February 2020 event.

Table 7. The intensity of precipitation over each sub-basin (mm/h).

Date Hour	SC5	SC4 and SC3	SC2 and SC1
21 February 2020 15:00	0.00	0.00	0.00
21 February 2020 16:00	0.43	0.39	0.34
21 February 2020 17:00	2.03	1.83	1.62
21 February 2020 18:00	1.01	0.91	0.81
21 February 2020 19:00	5.36	4.82	4.29
21 February 2020 20:00	2.46	2.21	1.97
21 February 2020 21:00	0.86	0.77	0.69
21 February 2020 22:00	0.00	0.00	0.00

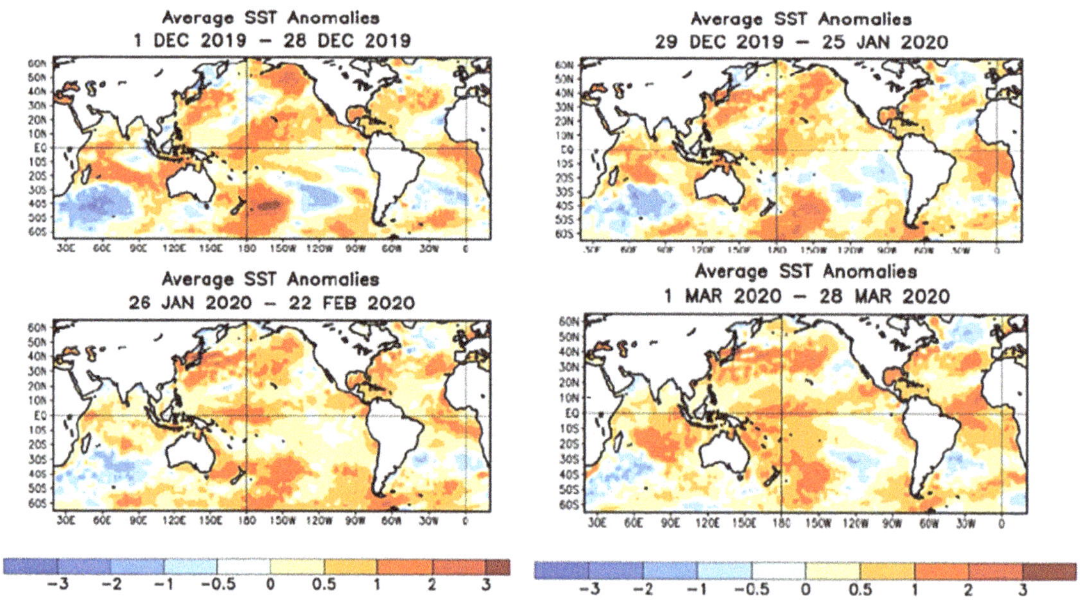

Figure 30. Sea surface temperature anomalies between 1 December 2019 and 28 March 2020.

Sub-basins 1 and 2 discharge their waters to River 1, and this flow joins the runoff produced by Sub-basin 3. This discharge feeds River 2, which joins the production of run-off from Sub-basin 4, before entering the informal Paso Camiara embankment. This structure served to model the dam break effect. Finally, the abrupt discharge from the dam break joins the runoff produced in Sub-basin 5. The parameters and initial conditions for each of the sub-basins are shown in Table 8. Likewise, the parameters and initial conditions for each river, are shown in Table 9.

In order to enter the RS Minerve, utilizing the bathymetry information of the Paso Camiara informal embankment (height vs. volume), the calculations were made using information from the topographic survey (Figure 31).

To simulate the dam break effect, it was assumed that this would occur when the dam's maximum height of 13.0 m was reached.

A necessary aspect to carry out the calculation of the dam break is the determination of the width of the breach.

To calculate the width of the rectangular breach (b), the formula proposed by Froehlich (2008) can be used, obtained by processing 69 cases of dam failures [46]:

$$b = 0.27\, k_0\, V^{0.32}\, H_b^{0.040} \quad (4)$$

where b (m) is the width of the breach, k_0 (Adim) is 1.3 (Overflow failures) and 1.0 (For other cases), V (m^3) is the volume of water stored above the bottom of the breach, and H_b (m) is the height of the breach.

Table 8. SOCONT model parameters for each sub-basin.

Sub-Basins		SC1	SC2	SC3	SC4	SC5
		SOCONT Model Parameters				
A	m^2	8,446,871	7,775,105	16,991,236	8,542,694	11,086,674
bp	d/mm	0.0125	0.0125	0.0125	0.0125	0.0125
CFR	-	1	1	1	1	1
HGR3Max	m^2	0.1	0.1	0.1	0.2	0.5
J0	-	0.102	0.060	0.028	0.036	0.047
KGR3	1/s	0.001	0.001	0.001	0.001	0.001
Kr	$m^{1/3}/s$	2	2	2	2	2
L	m	1489.2	1301.7	1514.2	1008.1	1613.8
S	mm/°C/d	5	5	5	5	5
Sint	mm/°C/d	0	0	0	0	0
Smin	mm/°C/d	0	0	0	0	0
SPh	d	80	80	80	80	80
Tcf	°C	0	0	0	0	0
Tcp1	°C	0	0	0	0	0
Tcp2	°C	4	4	4	4	4
ThetaCri	-	0.1	0.1	0.1	0.1	0.1
		Initial conditions				
SWEIni	m	0	0	0	0	0
ThetaIni	-	0	0	0	0	0
HGR3Ini	m	0.1	0.1	0.1	0.1	0.1
HrIni	m	0	0	0	0	0

Table 9. Riverbed model parameters by cinematic approximation.

Riverbed		River 1	River 2
	Parameters		
L	m	12,541.4	8622.1
B0	m	5	12
m	-	1	1
J0	-	0.03	0.0335
K	$m^{1/3}/s$	30	30
N	-	1	1
	Initial conditions		
Qini	m^3/s	0	0

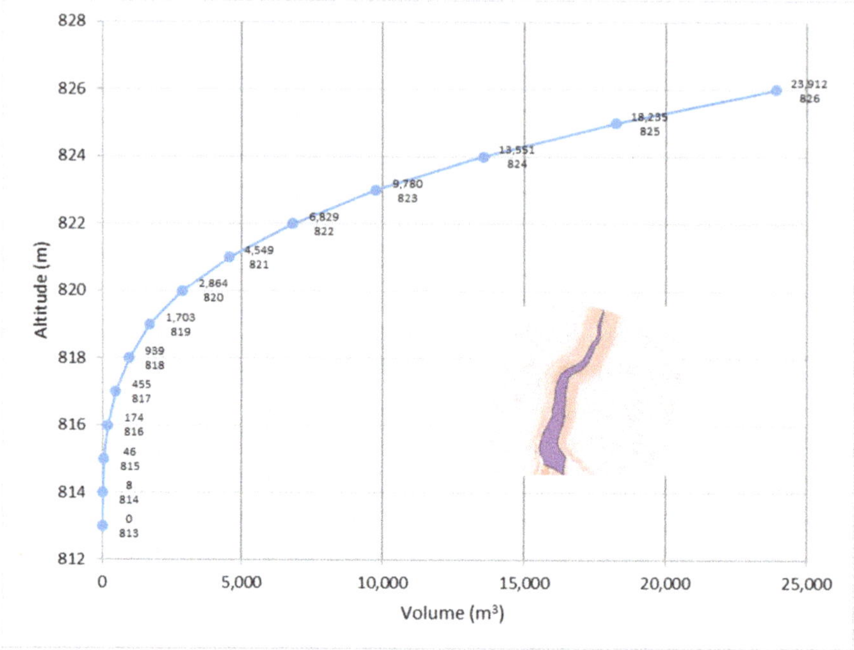

Figure 31. Bathymetry of the informal Paso Camiara embankment.

If we consider $k_0 = 1.3$, the volume of water stored before the dam break was 23,911.5 m^3, and if the total water height above the breach is 13.0 m, there will be a breach width of 9.8 m.

Likewise, if the breach is considered as a spillway, with a coefficient of 1.5, the discharge flow at the instant of the total rupture of the dam would be 688.7 m^3/s.

$$Q = 1.5 \times b \times Hmax^{1.5} = 1.5 \times 9.8 \times 13^{1.5} = 688.7\ m^3/s$$

The results of these calculations are shown in Table 10.

Table 10. Instantaneous discharge flow due to breach of the Paso Camiara informal embankment.

	Hc (m)	V (Hm^3)	b (m)	Qp (m^3/s)
Paso Camiara Dam	13	0.0239115	9.8	688.7

The simulated hydrographs produced by each of the sub-basins are shown in Figure 32.

The entrance hydrograph to the Paso Camiara informal embankment is shown in Figure 33. A maximum inflow flow of 10.72 m^3/s produced at 22:00 h was calculated. It should be mentioned that the Devil's Creek does not have a gauging station. The maximum flow of 10.72 m^3/s was contrasted with the water footprints left by the event in the riverbed of the Devil's Creek.

Figure 34 shows the height and flow hydrographs in the Paso Camiara dam. The results of the simulation show that from 4:00 p.m. on 21 February 2020, the filling of the Paso Camiara informal embankment began, breaking at a maximum water height of 15.7 m and discharging a maximum flow of 2550.8 m^3/s. Figure 35 shows the debris flow which produced floods that caused the loss of three human lives as well as great economic losses in Tacna city.

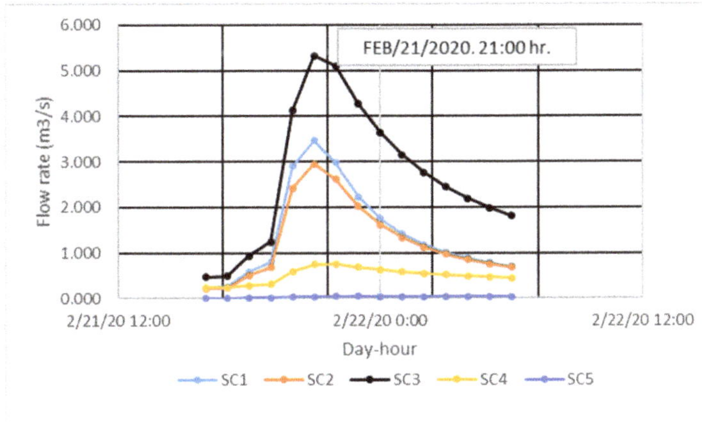

Figure 32. Hydrographs generated by the sub-basins.

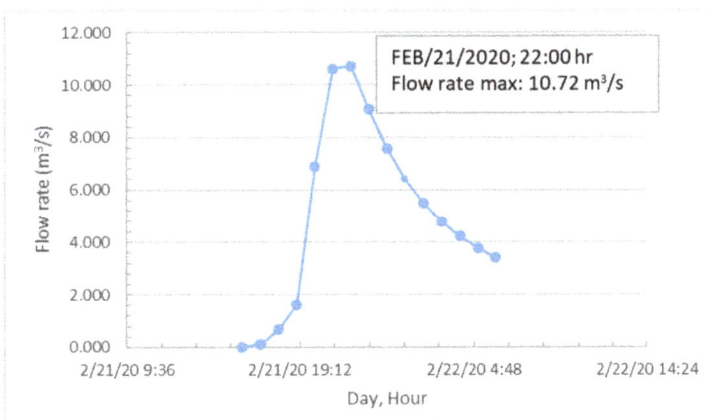

Figure 33. Hydrograph of entry to the Paso Camiara informal embankment.

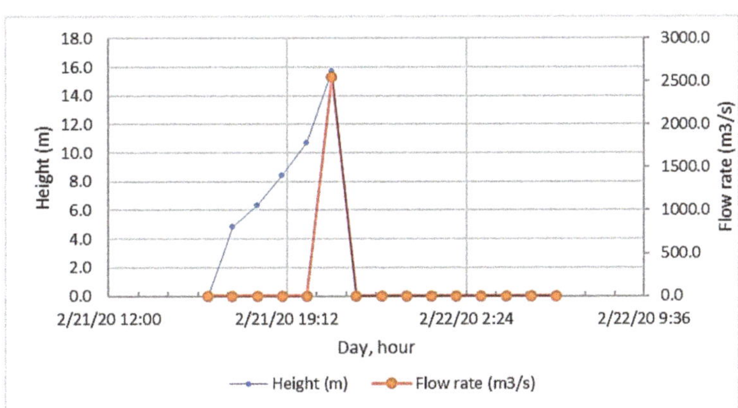

Figure 34. Hydrographs of water height in the dam and flow discharged due to the collapse of the Paso Camiara dam.

Figure 35. Gestion newspaper reports: debris flow in Tacna left three people dead, 22 February 2020.

3.3.2. Under Climate Change Conditions

To carry out the hydrological modeling of the maximum flow in the Devil's Creek, produced by maximum rainfall events generated by climate change models, a topology was used without considering the Paso Camiara informal embankment. Therefore, Sub-basins 1 and 2 discharge their waters to River 1, and this flow joins the runoff produced by Sub-basin 3. This discharge feeds River 2, which joins the production of run-off from Sub-basin 4, which joins the runoff produced in Sub-basin 5. Likewise, the parameters and initial conditions for each of the sub-basins were maintained. Figure 36 shows the structure of the RS Minerve model.

The modeling of the maximum flow product of precipitation intensities over each sub-basin (mm/h) for the RCP4.5 and 8.5 scenarios, and for 2021–2050 and 2051–2080 terms (Table 11), are shown in Table 12 and Figure 37, respectively.

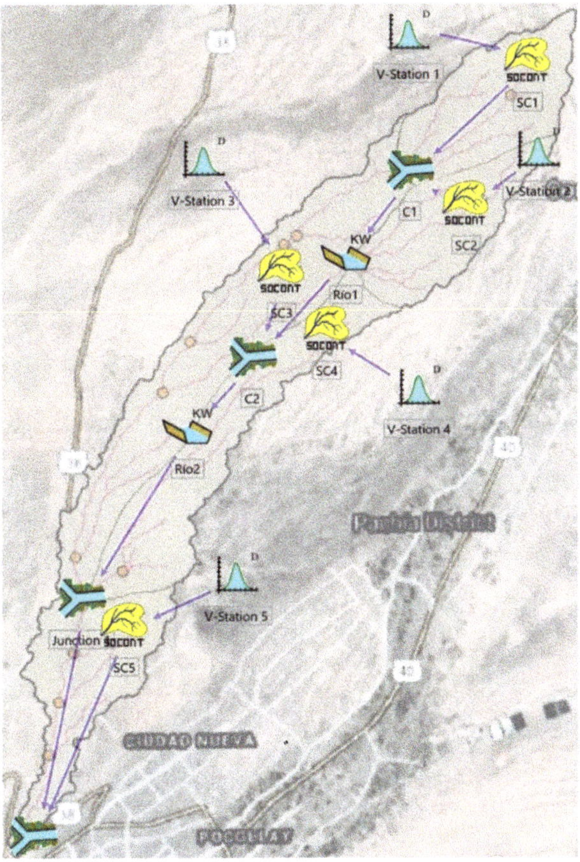

Figure 36. Structure of the RS Minerve Model for the simulations of events with different climate change scenarios.

Table 11. Precipitation intensity over each sub-basin (mm/h) for RCP4.5 and 8.5 scenarios, from 2021 to 2050 and from 2051 to 2080.

	2021–2050 (RCP4.5)			2051–2080 (RCP4.5)			2021–2050 (RCP8.5)			2051–2080 (RCP8.5)		
Hours	**SC5**	**SC4 and SC3**	**SC2 and SC1**	**SC5**	**SC4 and SC3**	**SC2 and SC1**	**SC5**	**SC4 and SC3**	**SC2 and SC1**	**SC5**	**SC4 and SC3**	**SC2 and SC1**
0	0.00	0.00	0.00	0.00	0.00	0.00	0.00	0.00	0.00	0.00	0.00	0.00
1	0.87	0.78	0.70	0.63	0.57	0.51	0.92	0.83	0.73	1.76	1.58	1.41
2	4.11	3.70	3.29	2.98	2.69	2.39	4.34	3.90	3.47	8.31	7.48	6.65
3	2.04	1.84	1.64	1.48	1.34	1.19	2.16	1.94	1.73	4.13	3.72	3.31
4	10.85	9.77	8.68	7.88	7.09	6.30	11.45	10.31	9.16	21.94	19.75	17.55
5	4.98	4.48	3.98	3.62	3.25	2.89	5.26	4.73	4.20	10.07	9.06	8.06
6	1.74	1.57	1.39	1.26	1.14	1.01	1.84	1.65	1.47	3.52	3.17	2.82
7	0.00	0.00	0.00	0.00	0.00	0.00	0.00	0.00	0.00	0.00	0.00	0.00
Total in 6 h	24.60	22.14	19.68	17.86	16.07	14.29	25.96	23.36	20.77	49.74	44.77	39.79

Table 12. Flood hydrographs for RCP4.5 and RCP8.5 scenarios from 2021 to 2050 and from 2051 to 2080.

Time (Hour)	Years 2021–2050 RCP4.5	Years 2021–2050 RCP8.5	Years 2051–2080 RCP4.5	Years 2051–2080 RCP8.5
1.0	0.0	0.0	0.0	0.0
2.0	0.1	0.1	0.1	0.2
3.0	2.2	2.5	1.2	14.0
4.0	4.9	5.5	2.9	16.6
5.0	23.7	26.0	13.5	76.4
6.0	30.4	32.9	19.1	78.6
7.0	26.0	27.7	17.6	55.7
8.0	19.1	20.1	13.8	34.9
9.0	14.4	15.1	10.9	23.9
10.0	11.3	11.7	8.9	17.5
11.0	9.1	9.5	7.4	13.4
12.0	7.6	7.8	6.3	10.7
13.0	6.4	6.6	5.4	8.7
14.0	5.5	5.7	4.7	7.3
15.0	4.9	5.0	4.2	6.3

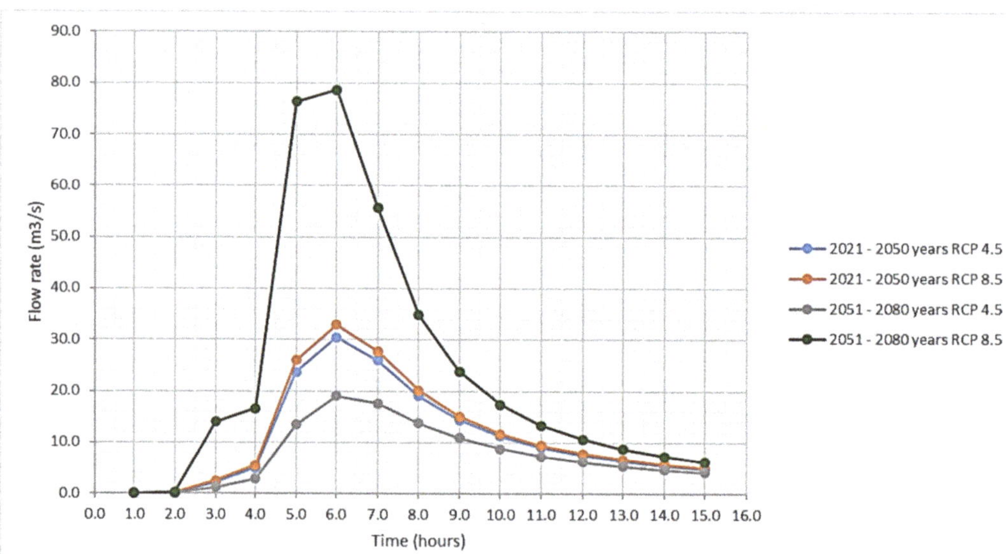

Figure 37. Flood hydrographs for RCP4.5 and RCP8.5 scenarios for 2021–2050 and 2051–2080 terms.

4. Discussion

This research addressed the impacts of natural variability and climate change on the maximum precipitation and maximum flows in the Devil's Creek, Tacna, Peru. For the projection of the future maximum annual precipitation, maximum temperature, and minimum temperature, the analysis is based on the daily output of 15 general circulation models (GCMs) of the CMIP5 project (Coupled Model Intercomparison Project Phase 5) and considers two emission scenarios: RCP4.5 and RCP8.5. The Access1-0 model projection was discarded due to a high overestimation of maximum precipitation. Furthermore, given the uncertainty in the projections by the GCMs [31], an ensemble of 14 GCMs was assessed. In general, although most models do not agree in projecting similar trends in precipitation as evidenced in other investigations [21,45,46], the projections indicate that the future pattern of maximum annual precipitation will experience significant changes, with increases for the two periods evaluated and for both scenarios.

Compared with historical conditions, heavy precipitation events are intensified, especially for the high-emissions scenario, according to the multi-model ensemble of 14 GCMs. Under the RCP4.5 scenario, the maximum annual precipitation could increase by more than 32%, on average, whereas under the RCP8.5 scenario, it could increase by 35%, on average, during the 2021–2050 period relative to the historical period of 1981–2005. On the other hand, for the 2051–2080 period, the maximum annual precipitation projected for the RCP4.5 scenario tends to decrease slightly. However, significant positive change is projected under the RCP8.5 scenario relative to the 1981–2005 period.

Regarding future temperature changes, as evidenced in other investigations, the GCMs agree in projecting a positive temperature trend [47,48]. However, as of 2050, greater variability and a decrease in the low trend are observed under the RCP4.5 scenario for the Devil's Creek area. Additionally, as expected, the RCP8.5 scenario projected the highest temperature values. For both scenarios, the values in June, July, and August denote a high range of positive changes of the minimum temperature. The behavior is similar for the monthly average maximum temperature during the 2021–2050 term for both scenarios. In contrast, it occurs during the 2050–2080 term under the RCP8.5 scenario that indicates significant warming during June, July, and August; similarly, it happens for January, February, and March.

Regarding the historical rainfall event of 21 February 2020, in the Devil's Creek, it can be affirmed that this is attributable to the positive anomaly of the sea surface temperature off the coast of Tacna during the days before that date.

Likewise, the lowest estimated rainfall sheet for the event of 21 February 2020, in the Devil's Creek, concerning the rainfall sheet recorded in the city of Tacna, is attributed to the shorter distance from the city of Tacna to the coastal area. This was demonstrated, in the absence of registered information on rainfall on the Devil's Creek, through the spatio-temporal analysis of the information registered in the city of Tacna (JORGE BASADRE rain gauge) and satellite products such as IMERG and CHIRPS.

Regarding the debris flow produced as a result of the rainfall event of 21 February 2020, which caused a loss of human lives, this was the consequence of the collapse of an informal embankment built as a trail, located 2.0 km upstream from the town center named La Florida (City of Tacna).

Regarding the historical and future hydrological modeling, these were carried out through the RS Minerve program. The historical hydrological modeling was calibrated by estimating the footprints of maximum flows produced in the middle section of the Devil's Creek and the evidence and testimonies recorded in videos of the magnitude of the disaster caused by the debris flow of 21 February 2020. The future hydrological modeling was carried out for two time periods: 2021 to 2050 and 2051 to 2080, for RCP4.5 and RCP8.5 scenarios, respectively. The results will be used for the design of studies and alternatives for the protection of the population in the area of influence of the Devil's Creek.

5. Conclusions

Climate models agree in projecting a positive trend in surface temperature. However, from 2050 onwards, greater variability and a decrease in the trend were observed under the RCP4.5 scenario. On the other hand, as expected, the RCP8.5 high-emissions scenario projected the highest values. For both scenarios and for both periods, minimum and maximum temperature would increase under climate change. Higher positive monthly changes are projected in June, July, and August, meaning that winters would be warmer.

The result of the model ensemble, under the RCP4.5 scenario, indicated a variation in the maximum annual precipitation from 2.77 mm to 24.70 mm, with an average of 5.70 mm for the 2021–2050 term. Additionally, this was from 3.14 mm to 25.96 mm, with an average of 5.80 mm under the RCP8.5 scenario. For the 2051–2080 term, the multi-model ensemble of 14 GCMs indicated a variation in the maximum annual precipitation from 2.77 mm to 17.86 mm, with an average of 5.60 mm under the RCP4.5 scenario and from 2.92 mm to 49.74 mm, averaging 7.21 mm under the RCP8.5 scenario.

Regarding the relative change, under the RCP4.5 scenario, the maximum annual precipitation could increase by 32%, on average. Under the RCP8.5 scenario, it would probably increase by 35%, on average, during the 2021–2050 term. On the other hand, for the 2050–2080 term, the maximum annual precipitation could increase by 30% under the RCP4.5 scenario and 65% under the RCP8.5 scenario.

The rainfall event of 21 February 2020, in the Devil's Creek, is attributable to the positive anomaly of the sea surface temperature between +1 °C and +2 °C, produced off the coast of Tacna between 22 January 2020, and 22 February 2020.

The debris flow produced as a result of the rainfall event of 21 February 2020 was the consequence of the collapse of an informal embankment built as a trail, located 2.0 km upstream from the town of La Florida (City of Tacna).

Historical hydrological modeling using the RS Minerve model was calibrated by estimating the footprints of maximum flows produced in the middle section of the Devil's Creek and the evidence and testimonies recorded in videos of the magnitude of the disaster caused by the debris flow of 21 February 2020.

The maximum flood volume in the Devil's Creek could increase by 220% and 154% for the RCP4.5 scenario, for the 2021–2050 and 2051–2080 terms, respectively.

The maximum flood volume in the Devil's Creek could increase by 234% and 484% for the RCP8.5 scenario for the 2021–2050 and 2051–2080 terms, respectively.

Author Contributions: E.I.-B. designed, collected observed data, wrote the first draft, and edited the document. E.C.-V. designed, collected observed data, wrote the first draft, and edited the document. E.P.-V. guided writing and review. F.M., A.C. and A.V. collected and processed information. All authors have read and agreed to the published version of the manuscript.

Funding: Universidad Nacional Jorge Basadre Grohmann, Tacna, Perú.

Institutional Review Board Statement: Not applicable.

Informed Consent Statement: Not applicable.

Data Availability Statement: Not applicable. The data corresponds to reports from Peruvian public institution that under agreement provided such information with academic purposes.

Acknowledgments: The authors would like to thank the National Water Authority (ANA) and the National Meteorology and Hydrology Service (SENAMHI) for providing the valuable information used in this work.

Conflicts of Interest: The authors declare no conflict of interest.

References

1. Ingol, E.M.; Mckinney, D.C. *Modeling Climate Change Impacts on Hydrology and Water Resources: Case Study Rio Conchos Basin*; Center for Research in Water Resources Bureau of Engineering Research, The University of Texas at Austin: Austin, TX, USA, 2011.
2. Stagl, J.; Mayr, E.; Koch, H.; Hattermann, F.; Huang, S.; Judith, S. *Managing Protected Areas in Central and Eastern Europe Under Climate Change, Effects of Climate Change on the Hydrological Cycle in Central and Eastern Europe*; Springer Open: Berlin/Heidelberg, Germany, 2013; pp. 31–43.
3. IPCC. *Climate Change and Water. Contribution to the Fourth Assessment Report of the Intergovernmental Panel on Climate Change*; Bates, B.C., Kundzewicz, Z.W., Wu, S., Palutikof, J.P., Eds.; IPCC Technical Paper VI; IPCC: Geneva, Switzerland, 2008.
4. SENAMHI. Escenarios Climáticos en el Perú para el Año 2030. 2009. Available online: https://idesep.senamhi.gob.pe/portalidesep/files/tematica/cambio_climatico/Escenarios_climaticos_en_el_Peru_para_el_ano_2030.pdf (accessed on 21 November 2021).

5. Changnon, S.A. *El Niño 1997–1998: The Climate Event of the Century*; Oxford University Press: New York, NY, USA, 2000; p. 36. ISBN 978-0-19-803096-6. Available online: https://oxford.universitypressscholarship.com/view/10.1093/oso/9780195135510.001.0001/isbn-9780195135510 (accessed on 21 November 2021).
6. Comunidad Andina de Fomento (CAF). El Fenomeno El Niño 1997–1998 Memoria; Retos Y Soluciones. 2000. Available online: https://scioteca.caf.com/bitstream/handle/123456789/676/Las%20lecciones%20de%20El%20Ni%C3%B1o.Per%C3%BA.pdf (accessed on 21 November 2021).
7. El Niño Costero. *Las Inundaciones de 2017 en el Perú*; ISET-International, Soluciones Prácticas y el Programa de Resiliencia a las Inundaciones de Zúrich: La Paz, Bolivia, 2017.
8. Pino, E.; Ramos, L.; Avalos, O.; Tacora, P.; Chávarri, E.; Angulo, O.; Ascensios, D.; Mejía, J. Effect of Environmental and Geological Characteristics on Water Quality in the Caplina River Basin, Tacna, Peru. *Tecnol. Cienc. Agua* **2017**, *8*, 77–99. [CrossRef]
9. Cheng, L.; AghaKouchak, A. Nonstationary Precipitation Intensity-Duration-Frequency Curves for Infrastructure Design in a Changing Climate. *Sci. Rep.* **2014**, *4*, 7093. [CrossRef]
10. Sarhadi, A.; Soulis, E.D. Time-varying extreme rainfall intensity-duration-frequency curves in a changing climate. *Geophys. Res. Lett.* **2017**, *44*, 2454–2463. [CrossRef]
11. Hu, H.; Ayyub, B. Extreme Precipitation Analysis and Prediction for a Changing Climate. *ASCE-ASME J. Risk Uncertain. Eng. Syst. Part A Civ. Eng.* **2018**, *4*, 04018029. [CrossRef]
12. Pino, E.; Chávarri, E. Evidence of climate change in the hyper-arid region of the southern coast of Peru, head of the Atacama Desert. *Tecnol. Cienc. Agua* **2022**, *3*, 1. [CrossRef]
13. CMIP5. Coupled Model Intercomparison Project Phase 5. 2013. Available online: https://pcmdi.llnl.gov/mips/cmip5/ (accessed on 21 November 2021).
14. Han, Z.Y.; Tong, Y.; Gao, X.J. Correction based on quantile mapping for temperature simulated by the RegCM4. *Clim. Change Res.* **2018**, *14*, 331–340.
15. Yang, X.; Wood, E.F.; Sheffield, J.; Ren, L.; Zhang, M.; Wang, Y. Bias correction of historical and future simulations of precipitation and temperature for china from CMIP5 models. *J. Hydrometeorol.* **2018**, *19*, 609–623. [CrossRef]
16. Tong, Y.; Gao, X.; Han, Z.; Xu, Y.; Xu, Y.; Giorgi, F. Bias correction of temperature and precipitation over China for RCM simulations using the QM and QDM methods. *Clim. Dyn.* **2021**, *57*, 1425–1443. [CrossRef]
17. Paulhus, J.L.H.; Kohler, M.A. Interpolation of missing precipitation records. *Mon. Weather. Rev.* **1952**, *80*, 129–133. [CrossRef]
18. Guijarro, J.A. Homogenization of climatic series with Climatol. In *Reporte Técnico State Meteorological Agency (AEMET)*; Balearic Islands Office: Madrid, Spain, 2018.
19. Aybar, C.; Lavado-Casimiro, W.; Huerta, A.; Fernández, C.; Vega, F.; Sabino, E.; Felipe-Obando, O. Uso del Producto Grillado "PISCO" de Precipitación en Estudios, Investigaciones y Sisitemas Operacionales de Monitoreo y Pronóstico Hidrometeorológico. 2017. Available online: https://www.senamhi.gob.pe/load/file/01402SENA-8.pdf (accessed on 21 November 2021).
20. Warren, W.; Parkinson, C. *An Introduction to Three-Dimensional Climate Modeling*, 2nd ed.; University Science Books; National Center for Atmospheric Research and NASA Goddard Space Flight Center: California, VA, USA, 2005; p. 368.
21. IPCC. *Climate Change 2013: The Physical Science Basis. Contribution of Working Group I to the Fifth Assessment Report of the Intergovernmental Panel on Climate Change*; Stocker, T.F.D., Qin, G.-K., Plattner, M., Tignor, S.K., Allen, J., Boschung, A., Nauels, Y., Xia, V.B., Midgley, P.M., Eds.; Cambridge University Press: Cambridge, UK, 2013; p. 1535.
22. Flato, G.J.; Marotzke, B.; Abiodun, P.; Braconnot, S.C.; Chou, W.; Collins, P.; Cox, F.; Driouech, S.; Emori, V.; Eyring, C.; et al. Evaluation of climate models. In *Climate Change 2013: The Physical Science Basis. Contribution of Working Group I to the Fifth Assessment Report of the Intergovernmental Panel on Climate Change*; Cambridge University Press: Cambridge, UK.
23. Overland, J.E.; Wang, M.; Bond, N.A.; Walsh, J.E.; Kattsov, V.M.; Chapman, W.L. Considerations in the Selection of Global Climate Models for Regional Climate Projections: The Arctic as a Case Study. *J. Clim.* **2011**, *24*, 1583–1597. [CrossRef]
24. NASA. NEX-GDDP: NASA Earth Exchange Global Daily Downscaled Climate Projections, the Coupled Model Intercomparison Project Phase 5. 2012. Available online: https://www.nccs.nasa.gov/sites/default/files/NEX-GDDP-CMIP6-Tech_Note.pdf (accessed on 21 November 2021).
25. Wilby, R.L.; Wigley, T.M.L. Downscaling General Circulation Model Ouput: Review of Methods and Limitations. *Prog. Phys. Geogr.* **1997**, *21*, 530–548. [CrossRef]
26. Chong-Yu, X. From GCMs to River Flow: A Review of Downscaling Methods and Hydrologic Modeling Approach. *Prog. Phys. Geogr.* **1999**, *23*, 229–249.
27. Wilby, R.L.; Charles, S.P.; Zorita, E.; Timbal, B.; Whetton, P.; Mearns, L.O. Guidelines for Use of Climate Scenarios Developed From Statistical Downscaling Methods. 2004. Available online: https://www.ipcc-data.org/guidelines/dgm_no2_v1_09_2004.pdf (accessed on 21 November 2021).
28. Fowler, H.J.; Blenkinsop, S.; Tebaldi, C. Linking climate change modeling to impacts studies: Recent advances in downscaling techniques for hydrological modeling. *Int. J. Clim.* **2007**, *27*, 1547–1578. [CrossRef]
29. Teutschbein, C.; Seiber, J. Bias correction of regional climate model simulations for hydrological climate-change impact studies: Review and evaluation of different methods. *J. Hydrol.* **2012**, *456–457*, 12–29. [CrossRef]
30. Maraun, G. Bias Correction, Quantile Mapping, and Downscaling: Revisiting the Inflation Issue, Helmholtz Centre for Ocean Research Kiel, Kiel, Germany. *J. Clim.* **2013**, *26*, 2137–2143. [CrossRef]

31. Perkins, S.E.; Fischer, E.M. The usefulness of different realizations for the model evaluation of regional trends in heat waves. *Geophys. Res. Lett.* **2013**, *40*, 5793–5797. [CrossRef]
32. Semenov, M.; Stratonovitch, P. Use of multi-model ensembles from global climate models for assessment of climate change impacts. *Clim. Res.* **2010**, *41*, 1–14. [CrossRef]
33. Bishop, C.H.; Abramowitz, G. Climate model dependence and the replicate Earth paradigm. *Clim. Dyn.* **2013**, *41*, 885–900. [CrossRef]
34. LeDuc, M.; Mailhot, A.; Frigon, A.; Martel, J.-L.; Ludwig, R.; Brietzke, G.B.; Giguère, M.; Brissette, F.; Turcotte, R.; Braun, M.; et al. The ClimEx Project: A 50-Member Ensemble of Climate Change Projections at 12-km Resolution over Europe and Northeastern North America with the Canadian Regional Climate Model (CRCM5). *J. Appl. Meteorol. Clim.* **2019**, *58*, 663–693. [CrossRef]
35. Annan, J.D.; Hargreaves, J.C. On the meaning of independence in climate science. *Earth Syst. Dyn.* **2017**, *8*, 211–224. [CrossRef]
36. Wang, B.; Zheng, L.; Liu, D.L.; Ji, F.; Clark, A.; Yu, Q. Using multi-model ensembles of CMIP5 global climate models to reproduce observed monthly rainfall and temperature with machine learning methods in Australia. *Int. J. Clim.* **2018**, *38*, 4891–4902. [CrossRef]
37. Holland, J.H. Genetic algorithms. *Sci. Am.* **1992**, *267*, 44–50. [CrossRef]
38. Kramer, O. Genetic algorithms. In *Genetic Algorithm Essentials*; Springer: Berlin/Heidelberg, Germany, 2017; pp. 11–19.
39. SENAMHI. Boletín Hidroclimático Mensual Dirección Zonal 7. 2021. Available online: https://repositorio.senamhi.gob.pe/handle/20.500.12542/1209 (accessed on 21 November 2021).
40. Krause, P.; Boyle, D.P.; Bäse, F. Comparison of different efficiency criteria for hydrological model assessment. *Adv. Geosci.* **2005**, *5*, 89–97. [CrossRef]
41. Moriasi, D.N.; Gitau, M.W.; Pai, N.; Daggupati, P. Hydrologic and Water Quality Models: Performance Measures and Evaluation Criteria. *Trans. ASABE* **2015**, *58*, 1763–1785. [CrossRef]
42. Muñoz, R.; Huggel, C.; Drenkhan, F.; Vis, M.; Viviroli, D. Comparing model complexity for glacio-hydrological simulation in the data-scarce Peruvian Andes. *J. Hydrol. Reg. Stud.* **2021**, *37*, 100932. [CrossRef]
43. Hidalgo, I.G.; Paredes-Arquiola, J.; Andreu, J.; Lerma-Elvira, N.; Lopes, J.E.G.; Cioffi, F. Hydropower generation in future climate scenarios. *Energy Sustain. Dev.* **2020**, *59*, 180–188. [CrossRef]
44. Astorayme, M.; García, J.; Suarez, W.; Felipe, O.; Huggel, C.; Molina, W. Modelización hidrológica con un enfoque semidistribuido en la cuenca del río Chillón, Perú. *Rev. Peru. Geo Atmosférica RGPA* **2015**, *4*, 109–124. Available online: http://www.senamhi.gob.pe/rpga/pdf/2015_vol04/paper8.pdf (accessed on 21 November 2021).
45. Garcia, J.; Foehn, A.; Fluixa, J.; Roquire, B.; Brauchli, T.; Paredes, J.; Cesare, G.; Minerve, R.S. *Technical Manual*; Centre de Recherche sur l'Environnement Alpin: Sion, Switzerland, 2020.
46. Campos, D. Modelado empírico simple del rompimiento de presas pequeñas de tierra (hidrograma de salida). *Ing. Investig. Tecnol.* **2013**, *14*, 377–388. [CrossRef]
47. Dosio, A.; Jones, R.G.; Jack, C.; Lennard, C.; Nikulin, G.; Hewitson, B. What can we know about future precipitation in Africa? Robustness, significance and added value of projections from a large ensemble of regional climate models. *Clim. Dyn.* **2019**, *53*, 5833–5858. [CrossRef]
48. Ortega, G.; Arias, P.A.; Villegas, J.C.; Marquet, P.A.; Nobre, P. Present-day and future climate over central and South America according to CMIP5/CMIP6 models. *Int. J. Clim.* **2021**, *41*, 6713–6735. [CrossRef]

hydrology MDPI

Article

Spatiotemporal Trend Analysis of Temperature and Rainfall over Ziway Lake Basin, Ethiopia

Aster Tesfaye Hordofa [1,2,*], Olkeba Tolessa Leta [3], Tane Alamirew [4] and Abebe Demissie Chukalla [5]

1 Africa Centre of Excellence for Water Management, Addis Ababa University, Addis Ababa 1176, Ethiopia
2 Faculty of Water Resources and Irrigation Engineering, Arba Minch University Water Technology Institute (AWTI), Arba Minch 4400, Ethiopia
3 Bureau of Watershed Management and Modeling, St. Johns River Water Management District, 4049 Reid Street, Palatka, FL 32177, USA; OLeta@sjrwmd.com
4 Ethiopian Institute of Water Resources, Addis Ababa University, Addis Ababa 1176, Ethiopia; tena.a@wlrc-eth.org
5 The Department of Land and Water Management, IHE Delft Institute for Water Education, 2611 AX Delft, The Netherlands; a.chukalla@un-ihe.org
* Correspondence: aster.tesfaye@aau.edu.et

Abstract: Rainfall and temperature trends detection is vital for water resources management and decision support systems in agro-hydrology. This study assessed the historical (1983–2005) and future (2026–2100) rainfall, maximum temperature (T_{max}), and minimum temperature (T_{min}) trends of the Ziway Lake Basin (Ethiopia). The daily observed rainfall and temperature data at eleven stations were obtained from the National Meteorological Agency (NMA) of Ethiopia, while simulated historical and future climate data were obtained from the Coupled Model Intercomparison Project 5 (CMIP5) datasets under Representative Concentration Pathways (RCP) of 4.5 and 8.5. The CMIP5 datasets were statistically downscaled by using the climate model data for hydrologic modeling (CMhyd) tool and bias corrected using the distribution mapping method available in the CMhyd tool. The performance of simulated rainfall, T_{max}, and T_{min} of the CMIP5 models were statistically evaluated using observation datasets at eleven stations. The results showed that the selected CMIP5 models can reasonably simulate the monthly rainfall, T_{max}, and T_{min} at the majority of the stations. Modified Mann–Kendall trend test were applied to estimate the trends of annual rainfall, T_{max}, and T_{min} in the historical and future periods. We found that rainfall experienced no clear trends, while T_{max}, and T_{min} showed consistently significant increasing trends under both RCP 4.5 and 8.5 scenarios. However, the warming is expected to be greater under RCP 8.5 than RCP 4.5 by the end of the 21st century, resulting in an increasing trend of T_{max} and T_{min} at all stations. The greatest warming occurred in the central part of the basin, with statistically significant increases largely seen by the end of the 21st century, which is expected to exacerbate the evapotranspiration demand of the area that could negatively affect the freshwater availability within the basin. This study increases our understanding of historic trends and projected future change effects on rainfall- and evapotranspiration-related climate variables, which can be used to inform adaptive water resource management strategies.

Keywords: climate change; representative concentration pathways 4.5 and 8.5; modified Mann–Kendall trend test; Ziway Lake Basin; Rift Valley; Ethiopia

Citation: Hordofa, A.T.; Leta, O.T.; Alamirew, T.; Chukalla, A.D. Spatiotemporal Trend Analysis of Temperature and Rainfall over Ziway Lake Basin, Ethiopia. *Hydrology* **2022**, 9, 2. https://doi.org/10.3390/hydrology9010002

Academic Editors: Tommaso Caloiero, Carmelina Costanzo, Roberta Padulano and Alain Dezetter

Received: 3 November 2021
Accepted: 16 December 2021
Published: 22 December 2021

1. Introduction

Climate change may have significant consequences on temperature and rainfall patterns, which are the most vital climatic elements used in the decision-making process for integrated water resources management [1,2]. For example, integrated water resources management models utilize temperature and rainfall as important input data for assessing freshwater availability [3]. The evidence of rapid global climate change includes an overall increase in temperature and decrease in rainfall that could result in declining freshwater availability [4–6], shrinking ice sheets, and rising sea level, among others [7,8]. Climate change impacts are more prominent in the African continent, especially countries that are predominantly dependent on a rain-fed agricultural system [9]. For example, several studies on temperature and rainfall changes have shown that most African countries have experienced warmer temperatures by more than 1 °C, frequent heat waves, and an increase in the number of hot days since 1901 [10]. In addition, both seasonal and annual rainfall conditions of the African countries are highly variable [11,12].

Ethiopia, located in the eastern part of the African horn, strongly depends on rain-fed agriculture that in turn contributes a large portion to the country's economy. However, due to climate change and limited water resources availability, the country's rain-fed and irrigated agricultural productivities are probably at risk, leading to negative consequences on food security [13,14]. This is projected to be accelerated in the future, as reported in many studies [15–18].

The Central Rift Valley (CRV) Lakes Basin of Ethiopia, located in the eastern part of the country, has experienced considerable seasonal and annual rainfall variability, with a prominent increase in temperature [19]. Some studies have already documented the hydroclimate conditions and climate change impacts in the CRV, including partially in the Ziway Basin (Meki and Katar sub-basins, the two major sub-basins draining into the Ziway Lake) [20,21]. For example, Musie et al. [21] have reported climate variability has negatively impacted the monthly streamflow of the Ziway Lake Basin (ZLB). Abraham et al. [20] indicated future annual decrement in runoff depth from the Katar and Maki rivers. They also documented that reduction in runoff has been caused by decrease in rainfall and increase in temperature. Furthermore, other studies found that an annual decline in Ziway Lake water level is due to a decrease in runoff from both the Katar and Meki rivers, including increase in evapotranspiration from the basin [20,21]. These studies show that water resources of the ZLB are highly exposed to a rapidly changing global climate [22]. The changes in rainfall and temperature in particular are important drivers affecting the runoff and evapotranspiration and thus the lake water level [23]. A good description and understanding of rainfall and temperature trends is thus crucial for many studies related to irrigation (agriculture), climatology, and hydrology. Long-term trend analyses in rainfall and temperature are also important for rain-fed and irrigated agricultural areas, particularly for the ZLB, which is dominated by agricultural land [24]. It has been reported that both rainfall and temperature can impact water availability (Ziway Lake water level), irrigation water demand, water use efficiency, and productivity of the ZLB [25–28].

Previous studies have documented hydrological responses of the ZLB to climate change with some studies based on only the historical period [21] and others based on projected areal rainfall and temperature over the large area [20]. While the previous studies focused on impact of climate change on hydrological responses of the basin [20,21,24], rainfall and temperature trends of the basin have not been analyzed in-depth. Therefore, in-depth characterization, understanding, and trend analysis of rainfall and temperature both spatially and temporally are a vital procedure in assessing climate change impacts on freshwater resources availability, irrigation water uses, and water resources planning [29–36]. This study conducted detailed rainfall and temperature trends analysis at spatial and temporal explicit scale, which is important for the ZLB with its significant topography and experiences with climate variations. The current study also introduced climate model selection criteria relevant to the case study that includes availability of the most commonly used RCP 4.5 and 8.5 scenarios, availability of high spatial resolution and ensemble models, and recommendations from the past studies [37,38]. The study further evaluated the performance of the selected climate models and identified the best performing models for the study area.

The primary aim of this study was to analyze the spatial pattern and temporal trends in annual rainfall and temperature for historical (1983–2005) and future (2026–2100) periods. The future period was split into near-term (2026–2049), middle-term (2050–2072), and far-term (2073–2100) time series. The study utilized rainfall, maximum temperature (T_{max}), and minimum temperature (T_{min}) from eleven stations together with Coupled Model

Intercomparison Project 5 (CMIP5) under two representative concentration pathways (RCP 4.5 and RCP 8.5). In this study, we first selected some CMIP5 models based on the abovementioned selection criteria and downscaled and bias corrected the climate variables of the selected CMIP5 models. We then evaluated the performance of the CMIP5 models' data against observations at eleven stations that are spatially fairly distributed over the ZLB. Finally, we analyzed the annual trends of historical and future temperature and rainfall datasets.

2. Study Area Description

The Ziway Lake Basin (ZLB) is located within 38°00′–39°30′ East and 7°00′–8°30′ North in the Adami Tullu–Jiddo Kombolcha Woreda of the East Shewa Zone, Oromia region, Ethiopia. The basin is about 150 km south of the capital city, Addis Ababa. The town of Ziway (recently named Batu) is situated on the lake's western shore. The altitude of Ziway Lake is approximately 1636 m above mean sea level (amsl), and has a maximum water depth of 4 m, a total basin drainage area of about 7300 km^2 (Figure 1), and a lake volume of 1.5 million cubic meters [29]. Most of the basin is characterized by low to moderately undulating topography but bounded by a steep slope and abrupt faults in the eastern and southeastern escarpments, ranging from 4200 to 1600 m (Figure 1). ZLB experiences a monsoon agro-climate and has a tropical climate with no uniform spatial and temporal climatic conditions. The rainfall patterns are generally affected by the annual oscillation of the intertropical convergence zone that forms a wet summer from June to September [30]. The mean annual rainfall of the basin spatially varies from 417 to 1012 mm, with a noticeable temporal variation at a monthly time scale. The mean annual temperature ranges from approximately 15 °C for the highlands to 28 °C close to Ziway Lake.

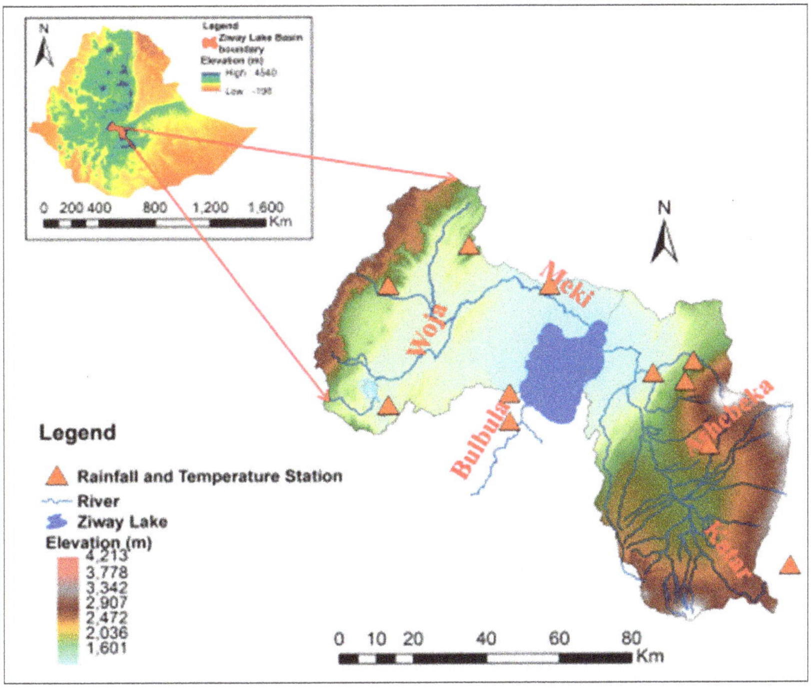

Figure 1. Topographic map of the Ziway Lake Basin.

3. Data and Methods

3.1. Observed Data

Daily maximum and minimum temperature (T_{max} and T_{min}) and rainfall (RF) data of the gauged stations, which were used for historical and future climate downscaling and bias correction, were obtained from the National Meteorological Agency (NMA) of Ethiopia. The dataset was obtained from eleven climate stations distributed over the Ziway Lake Basin for the period from 1983 to 2005.

3.2. Historical and Future Climate Data

We accessed the CMPI5 model data through the four Earth System Gird Federation (ESGF) gateways that manage, analyze, and distribute the model output and observation data [31]. At the time of this study, the CMIP5 version is the most up-to-date set of widely used climate models [31]. In addition, compared to CMIP3, the updated CMIP5 models produce higher resolution projections and use an updated set of greenhouse gas emission scenarios [32]. Among the four representative concentration pathways (RCPs), this study selected the two most commonly used emission scenarios of the future climate, RCP 4.5 and RCP 8.5, which are, respectively, an intermediate and a very high greenhouse gas (GHG) emission [32] scenario. We obtained the CMIP5 (RCP 4.5 and RCP 8.5) data from https://esgf-node.llnl.gov/search/cmip5/ (accessed from 10 January 2021 to 18 January 2021). The CMIP5 output provides global climate metadata that stores the data in the form of network command data form (NetCDF).

3.3. CMIP5 Selection Criteria

A range of Global Climate Models (GCMs) are available to access the future minimum and maximum temperature and rainfall data [33–36]. Three criteria were set to select suitable GCMs that capture climate of the stations in the study area [33]. The criteria are RCP 4.5 and 8.5 scenario availability, model resolutions and ensemble, and past studies in the study area [21,37,38]. Based on these criteria, three CMIP5 GCMs were selected and their corresponding output data, such as rainfall, maximum temperature (T_{max}), and minimum temperature (T_{min}) were downloaded under the two RCP 4.5 and RCP 8.5 scenarios. Table 1 summarizes the characteristics of the selected GCM.

Table 1. CMIP5 global climate models considered in this study.

Modeling Center	Model	Resolution in Degrees	Institute	Reference
CNRM-CERFACS	CNRM-CM5	1.4 × 1.4	Centre National de Recherches Meteorologiques/Centre Europeen de Recherche et Formation Avancees en Calcul Scientifique	[24,38]
MPI-M	MPI-ESM-LR	1.9 × 1.9	Max Planck Institute for Meteorology (MPI-M)	[33,34,38]
CSIRO-QCCCE	CSIRO-MK3.6	1.8 × 1.8	Commonwealth Scientific and Industrial Research Organization in collaboration with the Queensland Climate Change Centre of Excellence	[37,38]

3.4. Data Extraction, Downscaling, and Bias Correction

Downscaling the GCM–CMIP5 outputs to finer spatial resolutions is necessary for a reliable assessment of the regional impact of climate change on rainfall and temperature. The statistical downscaling method, which is the most widely used due to its quick and effective assessments of local climate change impacts [39,40], is applied to downscale and generate GCM-CMIP5 data (rainfall and temperature) at individual stations. We extracted the large-scale climate variables obtained from CMPI5 model for historical, RCP 4.5 and 8.5 scenarios and downscaled using the climate model data for hydrologic modeling (CMhyd) tool [41]. We made a bias correction of the rainfall and temperature data using the distribution mapping method, which is recommended in the previous studies, and also available in the CMhyd software [38]. We used the eleven stations' observed temperature and rainfall data for the bias correction and downscaling of the climate data from the three selected GCM–CMIP5 models (Table 1).

3.5. Climate Models Performance Evaluation

We evaluated the performance of the downscaled and bias-corrected CMIP5 models over the Ziway Lake Basin using monthly observations of rainfall, minimum temperature, and maximum temperature from 1983 to 2005 at eleven stations. We applied three statistical evaluation metrics that include percent bias (PBIAS), root mean square error (RMSE), and correlation coefficient (r) [42].

3.6. Rainfall and Temperature Trend Analysis

3.6.1. Mann–Kendall Test

The foremost measures of time series data including climate change are trend analysis, which identifies the general increasing or decreasing tendencies of the climate variables [43]. There are two main standard types of trend analysis: nonparametric and parametric methods. Parametric analysis is the most preferred method, but it is applicable for stationarity and normally distributed and serial-independent data, which is not possible in most hydro-climatological time series records [44]. In contrast, the nonparametric trend test requires time series data that are independent and less sensitive to outliers [45]. In this study, we used the nonparametric methods of Mann–Kendall (MK) to detect the rainfall and temperature trends for both historical and future periods: historical (1985–2005), near-term (2026–2049), middle-term (2050–2072), and far-term (2073–2100) time series. Because of its reliability for detecting monotonic trends in climate time series data, the nonparametric method of MK has been widely applied to discovering trends in hydrometeorological time series [45–48].

For comparison purpose, we also produced the spatial distribution maps for both mean annual temperature (T_{max} and T_{min}) and rainfall. For this, we used the known-distance-based interpolation method called inverse distance weighting [49]. Then, we applied ArcGIS technique to generate maps showing the spatial distribution for rainfall and temperature (T_{max} and T_{min}) over the ZLB.

3.6.2. Modified Mann–Kendall Test

The Mann–Kendall trend (MK) test is a nonparametric test used to reliably estimate trends of time series data. The original Mann–Kendall trend test does not consider serial correlation and seasonality effect [48]. In the actual world, the time series data are autocorrelated and this autocorrelation produces a misconception of trend test results [50,51]. To solve this problem, researchers and scholars proposed the modified Mann–Kendall tests, such as the Hamed and Rao modified MK Test [51], Yue and Wang modified MK Test [45], and modified MK test using the whitening method [52]. The modified MK (MMK) test, as proposed by Hamed and Rao [51] considers all the significant autocorrelation coefficients in a time series. Consequently, in this study, we used the widely applied MMK method [51]. We identified the trends of the selected variables (T_{max}, T_{min}, and rainfall) using the MMK and the Sen's slope test for annual time scales. We performed the analyses using the Python package called PyMannKendall in Jupyter Notebook [53]. The presence of a statistically significant trend is determined using the MK value. This statistic is used to test the null hypothesis of no trend exists against the alternative hypothesis that a trend does exist. A positive MK value indicates an increasing trend in the time series, while a negative MK value indicates a decreasing trend. The MMK trend test uses a two-tailed test method for evaluating and determining trend significance by simultaneously running three confidence intervals (90%, 95%, and 99%) [54].

4. Results and Discussion

4.1. *Performance Evaluation of CMIP5 Models*

The three CMIP5 models (CNMR-CM5, MIP-ESM, and CSIRO-MK3.6) show a positive correlation coefficient for monthly observed rainfall with r values ranging from 0.4 to 1 at the eleven stations (Table A1, Appendix A). The CNMR-CM5 model showed the highest r-value of 1 at three stations (Kulumsa, Meraro, and Sagure) and a medium r-value of 0.6 at Bui station, suggesting higher (medium) agreements with the observed dataset. On average, the spatial correlation coefficients of the selected models range from 0.5 to 0.73. The results suggest that from the three models, the CNRM-CM5 model showed relatively a good agreement with the observed data at all stations (0.73). This may be due to its high horizontal spatial resolution compared to the other models, including its resolution improvements that increased both in the atmosphere (from 2.8° to 1.4°) and in the ocean (from 2° to 1°) [55–57].

All models showed good performance in simulating the monthly T_{max} with r values ranging from 0.5 to 0.7, which indicates acceptable agreement against the observed data (Table A2, Appendix A). In addition, all models showed lower magnitude of RMSE (0.3 to 1.6 °C), indicating the simulated monthly T_{max} values have good agreement with the observed monthly T_{max} values of the eleven stations in the basin. The PBIAS values are in the range −15.5% to 40.1% for T_{max}, which indicates that the selected models showed both underestimation and overestimation of the observed T_{max} values.

Similar to T_{max}, r values of the three models for the T_{min} range from 0.5 to 0.7, while the RMSE values range from 0.8 to 1.6 °C (Table A3, Appendix A). This indicates that the monthly simulated T_{min} values by the three climate models reasonably represented the observed T_{min} values at all stations. The values of PBIAS are in the range −10.8% to 37.3% for T_{min}, confirming the selected models both underestimated and overestimated T_{min} at different stations. Overall, the selected CMIP5 models can reasonably simulate the monthly rainfall and temperature (T_{max} and T_{min}) at the majority of the eleven stations (see Tables A1–A3, Appendix A).

4.2. *Historical Annual Rainfall and Temperature Trends*

4.2.1. Temporal Trends of Annual Historical Rainfall and Temperature

The modified Mann–Kendall trend test and Sen's slope estimators for historical annual rainfall are presented in Table 2. The table shows increasing trends in rainfall for some stations while the other stations show decreasing trends under certain conditions of the three selected climate models (CNMR-CM5, CSIRO-MK3.6, and MIP-ESM-LR). The results generally indicate insignificant increasing or decreasing trends at almost all stations and for all the three models, which in turn highlights that rainfall trends varied spatially over the study area. However, a significant increasing trend in annual rainfall is shown at Arata, at 5% significance level for CNRM-CM5 model output.

Table 2. Mann–Kendall trend and Sen's slope estimator value for annual baseline (observed) and historical rainfall within the Ziway Lake Basin from 1983–2005.

Station Test	CNMR-CM5		CSIRO-MK3.6		MIP-ESM-LR	
	MK Trend	Sen's Slope	MK Trend	Sen's Slope	MK Trend	Sen's Slope
Ziway	−2.14	−0.52	0.16	0.08	0.79	0.33
Meki	0.42	0.10	0.01	0.06	0.79	0.34
Arata	2.17 *	0.56	−0.02	−0.02	−0.42	−0.14
Butajira	−1.21	−0.31	0.63	0.27	0.63	0.23
Tora	−0.50	−0.19	−0.26	−0.19	−0.58	−0.30
Bui	−0.66	−0.23	0.63	0.29	0.63	0.31
Kulumsa	0.90	0.37	−0.53	−0.07	0.05	0.07
Assela	0.85	0.37	−0.16	−0.03	0.05	0.10
Sagure	1.06	0.36	0.48	0.18	1.02	0.01
Meraro	0.48	0.18	−0.90	−0.42	−0.58	−0.21
Adamitulu	1.11	0.35	0.11	0.05	0.90	0.20

Table A4 in Appendix A summarizes the modified Mann–Kendall test results for T_{max}. The table clearly indicates that historical (model simulated) T_{max} data have experienced statistically significant upward trends.

Table A5 of Appendix A summarizes the modified Mann–Kendall trend test results for T_{min}. Similar to T_{max}, the table indicates that historical (simulated) T_{min} data have also experienced statistically significant upward trends. Results show that there are significant warming trends for annual T_{max} and T_{min} from 1983 to 2005 for all stations at different significance levels (see Tables A4 and A5, Appendix A). Overall, both stations and GCM-CMIP5 datasets showed similar upward slopes, highlighting the suitability of downscaled T_{max} and T_{min} data for trend analysis.

4.2.2. Spatial Distribution of Historical Mean Annual Rainfall and Temperature

Although this study conducted historical spatial trend analysis for the three selected GCM-CMIP5 models, we only present graphical results for CNMR-CM5 model that showed the best performance with observed data as compared to the other models. However, the statistical performance of the other models is summarized in Appendix A. The spatial distribution of mean annual rainfall and temperature over the Ziway Lake Basin for the period 1983 to 2005 are shown in Figure 2 for both observed (stations) and historical (GCM-CMIP5 simulated) data. The spatial distribution of stations (observed) and historical (simulated) mean annual rainfall and temperature over ZLB showed similar patterns. Figure 2a clearly shows that the western part of Ziway Lake Basin (in the part of Meki River) had received more rainfall than the eastern part of the Basin (in the part of Katar River), which is consistent with model simulation outputs (Figure 2b).

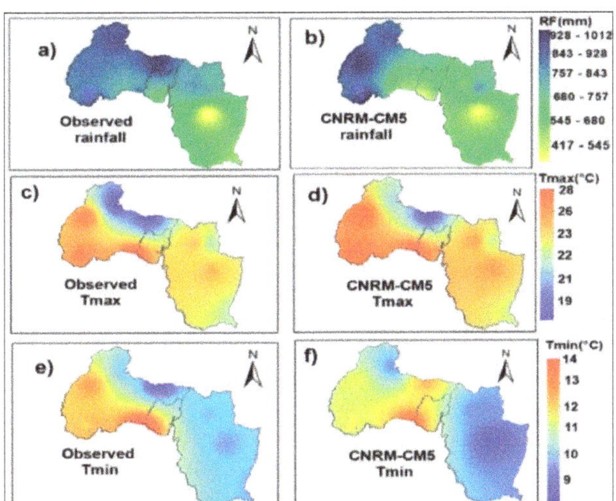

Figure 2. Spatial distribution of observed and CNRM-CM5 model annual mean rainfall (mm/year) and annual mean temperatures (T_{max} and T_{min} (°C)) for (**a**) observed and (**b**) CNRM-CM5 rainfall (**c**) observed T_{max}, (**d**) CNRM-CM5 T_{max}, (**e**) observed T_{min}, and (**f**) CNRM-CM5 T_{min} over the ZLB (1983–2005).

Figure 2 also shows the spatial distribution of mean annual T_{max} and T_{min} over ZLB for the period from 1983 to 2005. Both observed and simulated temperatures over the study area indicated similar spatial distribution; T_{max} ranges from 28 to 19 °C (Figure 2c,d) and T_{min} ranges from 14 to 9 °C (Figure 2e,f).

4.3. Future Annual Rainfall and Temperature Trends

4.3.1. Temporal Trends of Annual Future Rainfall and Temperature

Annual Rainfall Trend

Table 3 shows the rainfall trend analysis for the future period (2026–2049) under RCP 4.5 and RCP 8.5 scenarios. The annual average rainfall showed nonsignificant decreasing trends under the RCP 4.5 emission scenario at most of the stations. However, under the RCP 8.5 emission scenario, annual rainfall data showed insignificant increasing or decreasing trends except at Meki and Arata stations, which showed significantly increasing trends at 5% significance level.

Table 3. Projected Mann–Kendall trend and Sen's slope estimator results average rainfall for eleven stations within the Ziway Lake Basin for the period 2026–2049.

Model	CNMR-CM5				CSIRO-MK3.6				MIP-ESM-LR			
Station Test	**MK RCP 4.5**	**Sen's RCP 4.5**	**MK RCP 8.5**	**Sen's RCP 8.5**	**MK RCP 4.5**	**Sen's RCP 4.5**	**MK RCP 8.5**	**Sen's RCP 8.5**	**MK RCP 4.5**	**Sen's RCP 4.5**	**MK RCP 8.5**	**Sen's RCP 8.5**
Ziway	0.71	0.52	0.38	0.23	0.94	0.06	1.61	1.34	−0.55	0.22	−1.41	−2.35
Meki	−0.54	−0.11	1.07	0.29	0.12	0.24	2.06 *	2.37	0.38	−0.03	0.07	0.04
Arata	−0.53	−0.09	1.41	0.53	0.02	0.14	2.01 *	1.35	0.24	0.12	−0.27	−0.15
Butajira	0.48	0.08	0.57	0.20	−0.02	−0.12	1.46	1.53	−0.38	0.08	−0.05	−0.01
Tora	0.28	0.05	0.07	0.04	0.67	0.53	1.61	1.92	−0.34	−0.02	−0.12	−0.10
Bui	0.17	0.03	0.37	0.15	0.62	0.60	1.36	0.93	0.05	−0.10	0.07	0.02
Kulumsa	−0.42	−0.07	0.47	0.25	−0.07	−0.05	1.60	1.14	−0.55	−0.01	0.22	0.05
Assela	−0.53	−0.07	0.62	2.90	−0.02	−0.13	1.70	1.05	−0.25	0.06	0.32	0.07
Sagure	−0.18	−0.03	0.57	0.15	−0.07	−0.23	1.51	0.69	−0.67	−0.06	1.66	1.05
Meraro	−0.34	−0.05	0.52	0.07	−0.27	−0.19	1.31	0.71	−0.38	−0.08	0.17	0.08
Adamitulu	−0.54	−0.09	0.39	0.23	1.07	0.41	−0.06	−0.01	0.70	0.03	−0.17	−0.16

During the mid-term period (2050–2072), the annual rainfall is expected to show increasing trends at Bui and Tora stations at the 5% significance level under the RCP 8.5 scenario for the CNRM-CM5 model, including the Ziway station under the MIP-ESM-LR model (Table A6, Appendix A). Significant increasing trends are also projected at Arata, Meki, and Ziway stations under the RCP 4.5 for the CSIRO-MK3.6 model. During the far-term period (2073–2100), the annual rainfall under the RCP 8.5 scenario is projected to have insignificant positive trends for all models (Table A7). In general, the projected rainfall showed insignificant increase and decrease trends during the entire 21st century under both RCP 4.5 and RCP 8.5 (Figure 3).

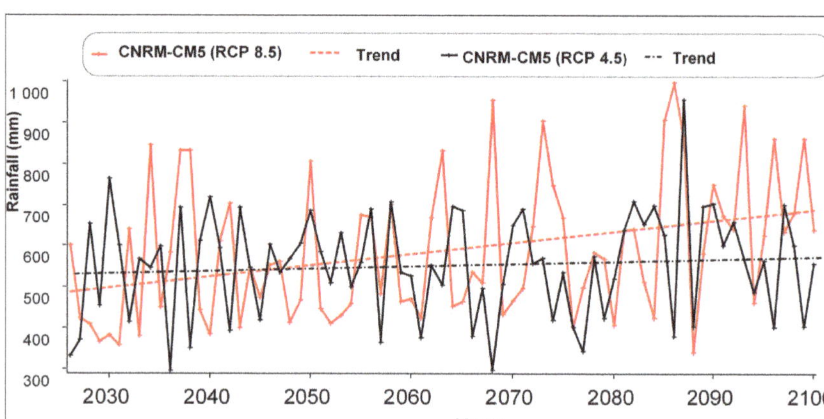

Figure 3. The annual rainfall from 2026 to 2100 for CNRM-CM5 model under RCP 4.5 and 8.5 scenarios.

The Intergovernmental Panel on Climate Change (IPCC) [58] reported frequent and intensive occurrence of extreme events such as droughts, under both RCP 4.5 and RCP 8.5. For example, IPCC [58] forecasted (with 66% certainty) that droughts will become more frequent, longer, and intensive by the end of the twenty-first century due to an overall decrease in rainfall amounts and increase in temperature. However, our findings indicate that the general trend in rainfall is unpredictable for the Ziway Lake Basin (ZLB), which may cause an increase or decrease in water availability that could potentially affect the farming practices in the ZLB. The results of prior studies on hydrological responses to climate change also illustrated fluctuation in rainfall [20,21]. Such trends are expected to affect the agricultural production and irrigation water requirements in the basin [59–61].

Annual Future Maximum Temperature (T_{max}) and Minimum Temperature (T_{min}) Trends

Table 4 presents the near-term (2026–2049) T_{max} Mann–Kendall trend analysis for the three selected models (CNRM-CM, CSIRO-MK3.6. and MIP-ESM-LR) for both RCP 4.5 and RCP 8.5 scenarios. Maximum temperature showed significantly increasing trends at 0.1%, 1%, and 5% significance levels under both RCP 4.5 and 8.5 scenarios. However, each model indicated a slightly different increasing level of maximum temperature.

Table 4. Projected Mann–Kendall trend and Sen's slope estimator results for annual mean T_{max} for eleven stations within the Ziway Lake Basin for the period 2026–2049.

Model	CNMR-CM5				CSIRO-MK3.6				MIP-ESM-LR			
Station Test	**MK RCP 4.5**	**Sen's RCP 4.5**	**MK RCP 8.5**	**Sen's RCP 8.5**	**MK RCP 4.5**	**Sen's RCP 4.5**	**MK RCP 8.5**	**Sen's RCP 8.5**	**MK RCP 4.5**	**Sen's RCP 4.5**	**MK RCP 8.5**	**Sen's RCP 8.5**
Ziway	2.00 *	0.04	1.86	0.03	2.43 *	0.03	3.35 ***	0.05	1.51	0.02	2.66 **	0.03
Meki	2.21 *	0.04	2.51 *	0.05	2.95 **	0.05	4.14 ***	0.05	1.51	0.03	1.71	0.03
Arata	2.21 *	0.04	2.85 **	0.07	3.35 ***	0.07	3.35 ***	0.05	1.56	0.03	1.66	0.03
Butajira	2.21 *	0.04	1.17	0.02	3.20 ***	0.04	3.34 ***	0.05	1.56	0.02	1.66	0.02
Tora	1.56	0.03	2.00 *	0.04	1.56	0.02	3.64 ***	0.06	1.66	0.02	2.98 **	0.03
Bui	2.20 *	0.04	0.57	0.01	0.60	0.01	3.69 ***	0.06	2.71 **	0.02	2.01 *	0.03
Kulumsa	2.20 *	0.04	2.41 *	0.04	2.90 **	0.04	3.35 ***	0.05	1.51	0.02	1.61	0.02
Assela	2.9 **	0.04	2.80 **	0.04	3.17 ***	0.05	3.34 ***	0.05	2.32 *	0.02	1.69	0.03
Sagure	2.65 **	0.04	2.65 **	0.04	3.62 ***	0.07	3.35 **	0.05	1.56	0.02	3.66 **	0.02
Meraro	2.80	0.05	2.80 **	0.05	3.37 ***	0.04	3.32 ***	0.05	1.61	0.02	1.71	0.03
Adamitulu	2.50 **	0.07	3.25 **	0.07	3.52 ***	0.04	2.12 *	0.04	0.87	0.01	3.96 **	0.02

As shown in Table A8, regardless of the applied scenarios, the mid-term (2050–2072) maximum temperature is expected to be higher than the near-term (2026–2049). In addition, as it should be expected, the RCP 8.5 scenario is projected to cause more warming than the RCP 4.5 scenario by the end of this century (Figure 4). We noticed a significant increasing trend of 0.02 to 0.03 °C/year using the CNRM-CM5 model. The CSIRO-MK3.6 model showed a larger significant increasing rate of 0.08 to 0.09 °C/year within the basin (Table A9).

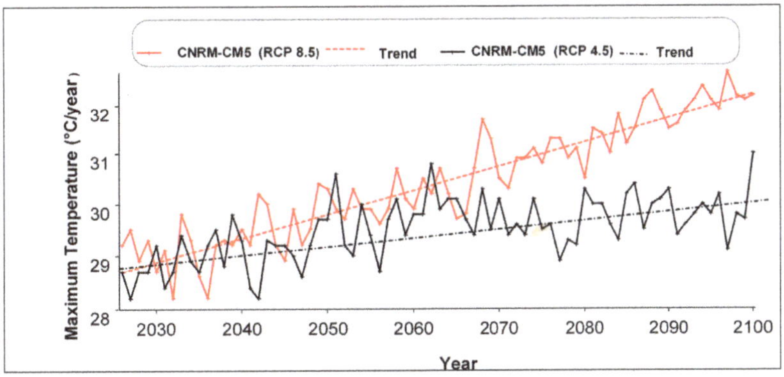

Figure 4. Annual average T_{max} trends from 2026 to 2100 under RCP (4.5 and 8.5) for CNRM-CM5.

Overall, the annual maximum temperature at the eleven stations of the basin is generally projected to increase at an increasing rate (Figure 4). More importantly, the RCP 8.5 scenario is expected to cause a higher increase in maximum temperature as compared to the RCP 4.5 scenario (Figure 4). This could be due to a continuous increase in radiative forcing emissions under the RCP 8.5 scenario [45].

Similar to the maximum temperature, the minimum temperature is expected to increase at increasing rate at all stations for both RCP 4.5 and 8.5 scenarios (Tables A10–A12, Appendix A). As compared to the RCP 4.5 scenario, the RCP 8.5 scenario is predicted to cause more warming by the end of the 21st century. For example, the majority of the stations showed a significant annual minimum temperature change of 0.1 degree by the end of this century under RCP 8.5 and CSIROM-MK3.6 model (Table A12).

In general, irrespective of the two RCP 4.5 and 8.5 scenarios, the future annual mean temperatures (T_{max} and T_{min}) have shown increasing trends relative to the historical trends. This indicates a warmer temperature in the near-, mid-, and far-terms [62]. In addition, RCP 8.5 is predicted to cause a larger increase in temperature than that of RCP 4.5 at all stations, especially by the end of the 21st century. This is consistent with the conclusions drawn by the IPCC Fifth Assessment Report (AR5) [62], which indicates an increase in evapotranspiration within the area and crop water demand [63,64]. Additionally, the increase in an increasing rate of future temperature will likely lead to a negative effect on freshwater availability, crop production, and the phenological days for crops [13,14,65,66].

4.3.2. Spatial Distribution of Future Mean Annual Rainfall and Temperature

Figure 5 shows the spatial distribution of projected mean annual rainfall using the CNRM-CM5 model under both RCP 4.5 and 8.5 scenarios within the ZLB. The figure depicts a slight difference in rainfall distribution when compared to the spatial distribution of historical rainfall (see Figure 2). The future rainfall projections indicate wetter conditions over the basin under both RCP 4.5 and 8.5 scenarios except during the near-term of RCP 4.5 and mid-term of RCP 8.5 scenario (Figure 5a,d), which exhibit lower annual rainfall related to the other periods.

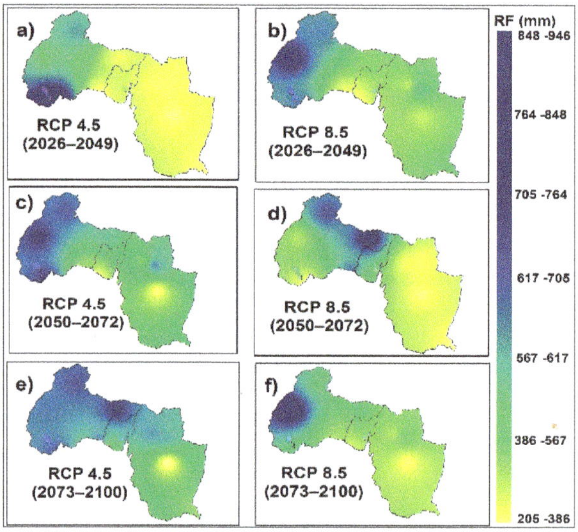

Figure 5. Spatial distribution of CNRM-CM5 model mean annual rainfall (mm/year) within the ZLB for RCP 4.5 and 8.5 scenarios and three future periods: (**a**) near-term RCP 4.5, (**b**) near-term RCP 8.5, (**c**) mid-term RCP 4.5, (**d**) mid-term RCP 8.5, (**e**) far-term RCP 4.5, and (**f**) far-term RCP 8.5.

The rainfall pattern is unevenly distributed. The distribution showed lower values of rainfall corresponding to the lower elevation (in the central part of the basin) and higher values corresponding to the higher elevations, particularly, in the southern part. The average rainfall depth at each gauge station during the analyzed period of 2026–2049, 2050–2072, and 2073–2100 under both scenarios exhibit different distribution, which may be due to change in climatic conditions and topographic and physiographic factors.

The rainfall pattern shows a slight decrease within the major parts of the basin (west, north, and east) under both RCP 4.5 and 8.5 scenarios of the near-term (Figure 5a) and mid-term (Figure 5d), respectively. The eastern and central parts of the ZLB are expected to receive lower rainfall values as compared with the western part of the basin (Figure 5). Overall, as compared to the RCP 4.5 scenario, the RCP 8.5 scenario is expected to cause a larger decrease in rainfall amounts by the end of the 21st century (Figure 5f).

The spatial distributions of projected mean annual T_{max} and T_{min} of CNRM-CM5 for the two RCP (4.5 and 8.5) scenarios are shown in Figures 6 and 7, respectively. The future minimum and maximum temperatures appear to have similar spatial distribution with the historical (1983–2005) temperature (see Figure 2). However, relative to the historical values, both the future T_{max} and T_{min} values are predicted to be noticeably increasing over the basin under both RCP 4.5 and 8.5 scenarios.

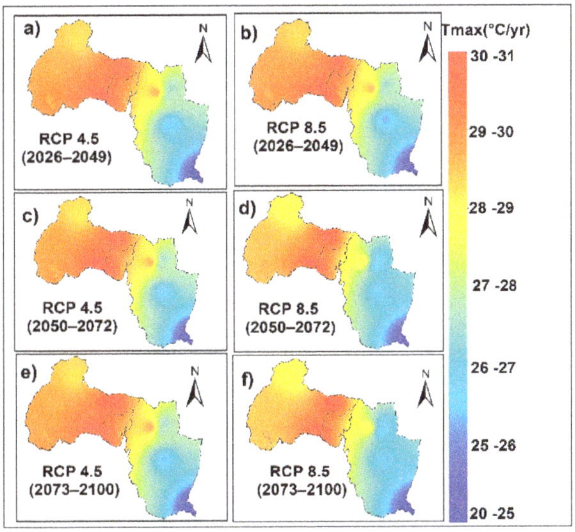

Figure 6. Spatial distribution of CNRM-CM5 model mean annual T_{max} (°C) within the ZLB for RCP 4.5 and 8.5 scenarios and three future periods: (**a**) near-term RCP 4.5, (**b**) near-term RCP 8.5, (**c**) mid-term RCP 4.5, (**d**) mid-term RCP 8.5, (**e**) far-term RCP 4.5, and (**f**) far-term RCP 8.5.

The central part of the basin will experience higher T_{max} and T_{min} values that range from 20 to 31 °C (Figure 6) and 11 to 19 °C (Figure 7), respectively. Apart from the western (in the part of Meki River) and central (around Ziway Lake), some small portions of the eastern part (in the part of Katar River) are expected to experience a cooler temperature.

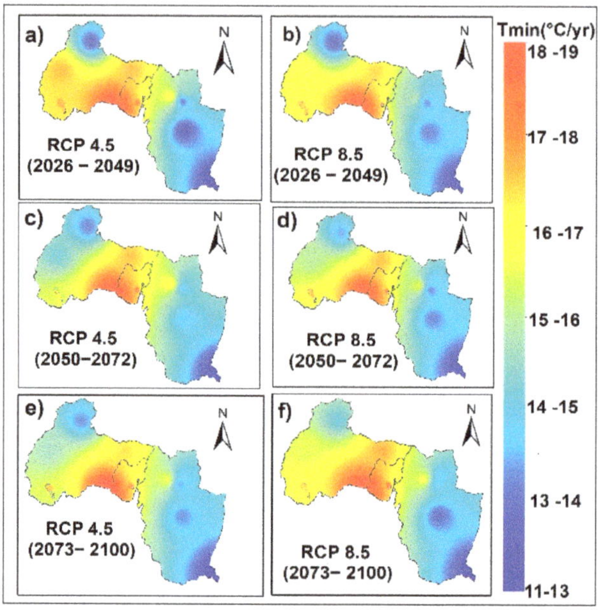

Figure 7. Spatial distribution of CNRM-CM5 model mean annual T_{min} (°C) within the ZLB for RCP 4.5 and 8.5 scenarios and three future periods: (**a**) near-term RCP 4.5, (**b**) near-term RCP 8.5, (**c**) mid-term RCP 4.5, (**d**) mid-term RCP 8.5, (**e**) far-term RCP 4.5, and (**f**) far-term RCP 8.5.

5. Conclusions

Understanding spatial and temporal trends in annual rainfall and temperature is of interest in a basin's water resource management and decision-making processes. Such studies are more important in basins where agricultural development is strongly dependent on rain-fed agriculture and irrigation, such as the Ziway Lake Basin (ZLB), Ethiopia. This study analyzed the historical (1983 to 2005) and future (2026–2100) trends of annual rainfall, minimum temperature (T_{min}), and maximum temperature (T_{max}) for the ZLB under RCP 4.5 and RCP 8.5 scenarios.

The results in this study revealed that the climate models reproduced well the spatial and temporal patterns of the annual mean rainfall and temperature of the study area.

The annual rainfall at the eleven stations of the ZLB showed both negative and positive insignificant trend. As the agriculture system of the basin highly depends on rain-fed agriculture and extensive irrigation, future climate change may have negative implications on the freshwater availability and agricultural productivity of the basin. We found that annual T_{max} and T_{min} of the ZLB are projected to increase at an increasing rate for both RCP 4.5 and 8.5. However, the RCP 8.5 is predicted to cause higher warming by the end of the 21st century. Such change is expected to increase the evapotranspiration of the basin, the key component of the hydrologic cycle, and thus negatively affects the freshwater availability and ecosystem functioning of the basin. The increase in temperature further increases the crop water requirement, which exacerbate the water supply–demand gap.

The findings of this study offer in-depth and useful information for better understanding and managing water resources and implementing mitigation measures to climate change in the Ziway Lake Basin. In addition, policy makers and relevant stakeholders such as farmer unions in the basin could also benefit from this study to minimize the negative impacts of climate change on freshwater availability and agricultural productivity.

As this study analyzed the annual rainfall and temperature trends only, future research work focuses on analyzing the spatiotemporal trends of seasonal rainfall and temperature. Furthermore, future studies should also compare the CMIP5 and CMIP6 models datasets and assess their ability to simulate the spatiotemporal distribution of rainfall and temperature.

Author Contributions: Conceptualization, A.T.H., O.T.L. and A.D.C.; data curation and analysis, A.T.H.; formula analysis and methodology, A.T.H.; writing—original draft, A.T.H.; literature review, A.T.H.; editing, A.T.H., O.T.L. and A.D.C.; supervision O.T.L., T.A. and A.D.C. All authors have read and agreed to the published version of the manuscript.

Funding: This work was financially supported by the Africa Center of Excellence for Water Management, Addis Ababa University, Ethiopia.

Institutional Review Board Statement: Not applicable.

Informed Consent Statement: Not applicable.

Data Availability Statement: The data used in this study are available at the World Climate Research Program (WCRP) Climate Explorer https://esgf-node.llnl.gov/search/cmip5/ (accessed from 10 January 2021 to 18 January 2021), daily CMIP5 project data of historical and future precipitation and maximum and minimum temperature under RCP 4.5 and RCP 8.5 emission scenarios.

Acknowledgments: The authors would like to thank the National Meteorological Agency (NMA) of Ethiopia for providing the gauged climate data. We would also like to express our sincere gratitude to the Africa Center of Excellence for Water Management, Addis Ababa University, for the support to conduct this research.

Conflicts of Interest: The authors declare no conflict of interest.

Appendix A

Table A1. Statistical performance indices of the GCMs–CMIP5 for monthly rainfall of 1983 to 2005.

Model	CNMR-CM5			CSIRO-MK3.6			MIP-ESM-LR		
Station	RMSE (mm)	PBIAS (%)	r	RMSE (mm)	PBIAS (%)	r	RMSE (mm)	PBIAS (%)	r
Ziway	27.4	11.2	0.8	22.0	29.9	0.4	40.7	29.6	0.5
Meki	27.9	2.8	0.8	31.2	12.5	0.5	28.2	12.8	0.6
Arata	31.8	−10.8	0.9	40.6	−11.6	0.5	37.2	−20.3	0.6
Butajira	39.7	35.0	0.7	28.4	−3.5	0.4	43.8	5.3	0.5
Tora	33.5	1.5	0.8	39.0	−1.2	0.4	46.3	−2.3	0.5
Bui	38.2	−4.2	0.6	34.6	2.8	0.5	32.4	2.4	0.6
Kulumsa	9.1	−9.0	1.0	16.1	−1.6	0.4	44.9	−2.8	0.5
Assela	26.1	7.8	0.9	27.4	12.3	0.5	47.5	12.0	0.5
Sagure	16.9	12.2	1.0	10.6	12.2	1.0	39.6	15.0	0.6
Merero	8.5	15.8	1.0	19.2	16.7	0.5	23.4	33.1	0.6
Adamitulu	16.2	0.0	0.9	48.6	−9.7	0.4	35.8	−9.5	0.5

Table A2. Statistical indices values of GCMs–CMIP5 output monthly maximum temperature during 1983–2005 over Ziway Lake Basin.

Model	CNMR-CM5			CSIRO-MK3.6			MIP-ESM-LR		
Station	RSME (C°)	PBIAS (%)	r	RSME (C°)	PBIAS (%)	r	RSME (C°)	PBIAS (%)	r
Ziway	1.0	0.3	0.7	1.1	0.4	0.7	1.2	0.4	0.6
Meki	1.3	40.1	0.6	1.3	28.0	0.6	1.2	50.2	0.6
Arata	1.0	22.3	0.7	1.1	22.4	0.7	1.0	22.3	0.7
Butajira	0.9	−1.7	0.6	1.0	−1.6	0.6	1.0	−1.6	0.5
Tora	0.9	0.8	0.7	1.0	0.8	0.7	1.0	0.8	0.6
Bui	0.9	15.1	0.6	0.3	1.0	0.6	1.1	10.7	0.6
Kulumsa	1.1	−0.9	0.6	1.1	−0.8	0.6	1.2	−0.9	0.5
Assela	1.1	−8.8	0.7	1.1	−8.8	0.7	1.2	−8.8	0.6
Sagure	1.3	−8.6	0.7	1.0	−22.9	0.6	1.4	−15.6	0.5
Merero	1.0	−15.5	0.7	1.1	−7.4	0.7	1.1	−15.5	0.6
Adamitulu	1.2	0.0	0.7	1.2	6.7	0.7	1.2	0.0	0.6

Table A3. Statistical indices values of GCMs–CMIP5 output monthly minimum temperature during 1983–2005 over Ziway Lake Basin.

Station Test	CNRM-CM5		CSIRO-MK3.6		MIP-ESM-LR	
	MK Trend	Sen's Slope	MK Trend	Sen's Slope	MK Trend	Sen's Slope
Ziway	2.01 *	0.03	2.15 *	0.03	2.32 **	0.04
Meki	2.21 *	0.03	2.07 *	0.03	2.32 **	0.03
Arata	2.31 *	0.03	2.29 *	0.03	2.43 **	0.03
Butajira	2.1 *	0.03	2.20 *	0.02	2.69 **	0.05
Tora	1.56	0.02	2.12 *	0.02	2.69 **	0.05
Bui	2.20 *	0.03	2.15 *	0.02	2.27 **	0.03
Kulumsa	2.20 *	0.03	2.91 **	0.05	2.32 **	0.03
Assela	2.9 **	0.05	2.05 *	0.02	2.38 **	0.03
Sagure	2.65 **	0.04	2.07 *	0.02	2.27 **	0.03
Merero	2.80 **	0.05	2.16 *	0.02	2.51 **	0.03
Adamitulu	2.50 **	0.05	2.20 *	0.02	2.32 **	0.03

Table A4. Mann–Kendall trend and Sen's slope estimator value for annual baseline and historical T_{max} at Ziway Lake Basin from 1983 to 2005.

Model Station	CNMR-CM5			CSIRO-MK3.6			MIP-ESM-LR		
	RSME (C°)	PBIAS (%)	r	RSME (C°)	PBIAS (%)	r	RSME (C°)	PBIAS (%)	r
Ziway	1.1	21.3	0.8	1.3	21.3	0.7	1.2	21.3	0.7
Meki	1.4	37.3	0.7	1.5	37.1	0.6	1.8	12.2	0.4
Arata	1.2	−2.1	0.7	1.2	−2.1	0.6	1.0	−2.1	0.7
Butajira	0.9	−7.8	0.6	0.8	−18.3	0.7	1.0	−18.3	0.6
Tora	0.8	2.2	0.7	0.8	2.2	0.7	1.0	2.2	0.4
Bui	1.2	−2.3	0.7	0.8	2.2	0.7	1.0	−2.2	0.7
Kulumsa	1.1	−10.8	0.6	1.5	−7.4	0.5	1.1	12.1	0.7
Assela	1.0	1.1	0.7	1.0	−20.4	0.7	1.2	−11.8	0.8
Sagure	0.9	−7.4	0.7	1.5	−43.9	0.5	1.0	−27.1	0.6
Merero	0.9	−3.2	0.7	1.4	−44.6	0.6	1.0	−44.8	0.6
Adamitulu	1.0	0.0	0.8	1.0	0.0	0.8	1.1	0.0	0.6

Table A5. Mann–Kendall trend and Sen's slope estimator value for annual baseline and historical T_{min} at Ziway Lake Basin from 1983 to 2005.

Station Test	CNRM-CM5		CSIRO-MK3.6		MIP-ESM-LR	
	MK Trend	Sen's Slope	MK Trend	Sen's Slope	MK Trend	Sen's Slope
Ziway	2.85 **	0.02	2.01 *	0.03	2.32 *	0.03
Meki	1.74	0.02	1.95	0.02	2.17 *	0.03
Arata	2.32 *	0.03	2.11 *	0.02	2.3 *	0.03
Butajira	2.93 **	0.02	3.16 **	0.04	2.75 **	0.04
Tora	2.48 *	0.02	2.69 **	0.03	2.67 **	0.04
Bui	2.48 *	0.02	3.06 **	0.04	2.43 *	0.03
Kulumsa	1.69	0.01	2.01 *	0.02	1.42	0.01
Assela	1.58	0.01	2.11 *	0.02	2.34 *	0.03
Sagure	1.69	0.01	2.91 **	0.04	2.32 *	0.03
Merero	1.74	0.01	2.64 **	0.03	2.32 *	0.03
Adamitulu	1.85	0.01	2.06 *	0.03	2.32 *	0.03

Table A6. Projected Mann–Kendall trend and Sen's slope estimator results average rainfall for 11 stations over Ziway Lake Basin from the period 2050–2072.

Model	CNMR-CM5				CSIRO-MK3.6				MIP-ESM-LR			
Station Test	**MK RCP 4.5**	**Sen's RCP 4.5**	**MK RCP 8.5**	**Sen's RCP 8.5**	**MK RCP 4.5**	**Sen's RCP 4.5**	**MK RCP 8.5**	**Sen's RCP 8.5**	**MK RCP 4.5**	**Sen's RCP 4.5**	**MK RCP 8.5**	**Sen's RCP 8.5**
Ziway	0.68	0.35	1.56	0.38	2.69 **	2.16	1.42	1.91	−0.32	−0.08	2.17 *	0.51
Meki	0.30	0.46	1.70	0.71	2.59 **	2.33	1.47	3.29	−1.17	−0.27	0.82	0.53
Arata	0.49	0.47	1.61	0.64	2.64 **	2.73	1.14	1.56	−0.87	−0.19	1.10	0.64
Butajira	0.35	0.43	1.47	0.48	2.17 *	1.74	1.19	1.05	−0.77	−0.23	1.00	0.46
Tora	−0.12	−0.10	2.17 *	1.06	1.89	1.75	1.65	1.98	−1.17	−0.26	0.63	0.33
Bui	−0.07	−0.23	2.03 *	1.04	1.99	3.60	1.33	1.54	−0.82	−0.30	0.58	0.53
Kulumsa	−0.30	−0.30	1.75	0.76	2.17 *	2.23	1.24	1.23	−0.77	−0.17	0.72	0.34
Assela	0.30	0.41	1.33	1.20	2.36 *	1.89	1.19	1.20	−1.41	−0.36	0.72	0.45
Sagure	0.16	0.26	1.33	0.55	2.17 *	1.76	0.82	0.44	−0.92	−0.25	1.19	1.20
Meraro	0.30	0.20	1.28	0.30	1.89	1.33	1.33	1.37	−1.27	−0.22	0.82	0.46
Adamitulu	0.21	0.16	0.49	0.36	2.73 **	1.86	−0.03	−0.01	−0.07	−0.05	0.86	0.55

Table A7. Projected Mann–Kendall trend and Sen's slope estimator results of annual average rainfall for 11 stations over Ziway Lake Basin from the period 2073–2100.

Model	CNMR-CM5				CSIROM-MK3.6				MIP-ESM-LR			
Station Test	**MK RCP 4.5**	**Sen's RCP 4.5**	**MK RCP 8.5**	**Sen's RCP 8.5**	**MK RCP 4.5**	**Sen's RCP 4.5**	**MK RCP 8.5**	**Sen's RCP 8.5**	**MK RCP 4.5**	**Sen's RCP 4.5**	**MK RCP 8.5**	**Sen's RCP 8.5**
Ziway	−1.45	−1.03	1.28	0.58	1.28	0.91	0.62	1.12	0.35	0.07	1.50	0.40
Meki	−1.63	−1.51	1.06	0.71	0.88	1.02	1.15	2.52	−0.93	−0.33	1.63	0.88
Arata	−1.50	−1.86	1.10	0.96	1.37	2.77	0.62	0.76	0.09	0.05	1.59	0.99
Butajira	−1.32	−1.08	1.23	0.52	1.19	1.06	1.85	2.14	0.35	0.10	1.72	0.81
Tora	−1.45	−2.06	0.93	0.68	1.54	1.12	1.63	2.66	−0.40	−0.15	1.16	1.03
Bui	−1.81	−3.14	0.84	0.65	1.76	2.27	0.62	1.00	−2.47	−1.10	1.45	0.99
Kulumsa	−1.90	−2.53	0.75	0.58	0.88	0.72	0.48	0.73	−2.60	−0.92	1.67	0.85
Assela	−0.79	−0.57	0.84	0.44	1.90	1.08	0.35	0.41	0.02	0.01	1.63	0.76
Sagure	−0.97	−0.79	1.28	0.63	1.28	0.95	0.04	0.03	−0.79	−0.25	0.35	0.41
Meraro	−0.66	−0.37	1.06	0.46	0.62	0.41	1.37	0.93	−0.88	−0.26	1.32	0.52
Adamitulu	−0.93	−0.33	0.11	0.12	1.68	1.38	0.01	−0.01	−0.71	−0.15	0.04	0.03

Table A8. Projected Mann–Kendall trend and Sen's slope estimator results of annual mean T_{max} for 11 stations over Ziway Lake Basin from the period 2050–2072.

Model	CNMR-CM5				CSIROM-MK3.6				MIP-ESM-LR			
Station Test	**MK RCP 4.5**	**Sen's RCP 4.5**	**MK RCP 8.5**	**Sen's RCP 8.5**	**MK RCP 4.5**	**Sen's RCP 4.5**	**MK RCP 8.5**	**Sen's RCP 8.5**	**MK RCP 4.5**	**Sen's RCP 4.5**	**MK RCP 8.5**	**Sen's RCP 8.5**
Ziway	2.45 *	0.04	2.15 *	0.03	4.52 ***	0.09	4.28 ***	0.09	0.02	0.01	3.00 **	0.04
Meki	2.07 *	0.03	2.07 *	0.03	4.56 ***	0.09	4.09 ***	0.09	0.03	0.01	2.91 **	0.04
Arata	2.08 *	0.03	2.29 *	0.03	4.54 ***	0.09	4.27 ***	0.09	0.03	0.01	2.95 **	0.04
Butajira	2.08 *	0.03	2.20 *	0.02	4.57 ***	0.09	4.27 ***	0.09	0.02	0.01	3.04 **	0.04
Tora	2.36 *	0.04	2.12 *	0.02	4.67 ***	0.08	4.40 ***	0.08	0.02	0.01	2.86 **	0.04
Bui	2.07 *	0.03	2.15 *	0.02	4.63 ***	0.08	4.4 ***	0.08	0.02	0.01	2.73 **	0.03
Kulumsa	2.08 *	0.03	2.91 **	0.03	4.52 **	0.09	4.28 ***	0.09	0.02	0.01	2.87 **	0.04
Assela	4.09 ***	0.05	2.05 *	0.02	4.50 **	0.09	4.27 ***	0.09	0.02	0.01	3.04 **	0.04
Sagure	4.09 ***	0.05	2.07 *	0.02	4.32 ***	0.08	4.26 ***	0.09	0.02	0.01	2.95 **	0.04
Meraro	4.09 ***	0.04	2.16 *	0.02	4.40 ***	0.09	4.28 ***	0.09	0.02	0.01	3.09 **	0.04
Adamitulu	4.13 ***	0.05	2.20 *	0.02	4.63 ***	0.09	4.14 ***	0.08	0.01	0.01	3.37 ***	0.07

Table A9. Projected Mann–Kendall trend and Sen's slope estimator results of annual mean T_{max} for 11 stations over Ziway Lake Basin from the period 2073–2100.

Model	CNMR-CM5				CSIROM-MK3.6				MIP-ESM-LR			
Station Test	**MK RCP 4.5**	**Sen's RCP 4.5**	**MK RCP 8.5**	**Sen's RCP 8.5**	**MK RCP 4.5**	**Sen's RCP 4.5**	**MK RCP 8.5**	**Sen's RCP 8.5**	**MK RCP 4.5**	**Sen's RCP 4.5**	**MK RCP 8.5**	**Sen's RCP 8.5**
Ziway	3.67 ***	0.07	2.78 **	0.03	4.63 ***	0.07	4.63 ***	0.07	1.51	0.02	1.90	0.03
Meki	3.62 ***	0.07	2.73 **	0.03	4.63 ***	0.07	4.63 ***	0.07	1.51	0.03	1.90	0.03
Arata	3.71 ***	0.07	2.5 *	0.03	4.63 ***	0.07	4.63 ***	0.07	1.56	0.03	1.81	0.03
Butajira	3.71 ***	0.06	2.47 *	0.03	4.63 ***	0.07	4.63 ***	0.07	1.56	0.02	1.85	0.03
Tora	3.20 ***	0.06	3.17 **	0.04	5.47 ***	0.06	4.72 ***	0.07	1.66	0.02	2.25	0.04
Bui	3.29 ***	0.05	3.31 ***	0.04	5.47 ***	0.07	4.63 ***	0.07	1.71	0.02	2.20	0.03
Kulumsa	3.34 ***	0.06	3.22 **	0.04	4.63 ***	0.07	4.63 ***	0.07	1.51	0.02	1.98	0.03
Assela	3.67 ***	0.06	2.30 *	0.03	4.63 ***	0.07	4.63 ***	0.07	1.51	0.02	1.90	0.03
Sagure	3.67 ***	0.06	2.47 *	0.03	4.63 ***	0.07	4.63 ***	0.07	1.56	0.02	1.85	0.03
Meraro	3.62 ***	0.07	2.78 **	0.03	4.63 ***	0.07	4.38 ***	0.06	1.61	0.02	0.17	0.00
Adamitulu	3.62 ***	0.08	3.57 ***	0.05	4.38 ***	0.06	4.47 **	0.07	0.87	0.01	2.42*	0.05

Table A10. Projected Mann–Kendall trend and Sen's slope estimator results of annual mean T_{min} for 11 stations over Ziway Lake Basin from the period 2026–2049.

Model	CNMR-CM5				CSIROM-MK3.6				MIP-ESM-LR			
Station Test	**MK RCP 4.5**	**Sen's RCP 4.5**	**MK RCP 8.5**	**Sen's RCP 8.5**	**MK RCP 4.5**	**Sen's RCP 4.5**	**MK RCP 8.5**	**Sen's RCP 8.5**	**MK RCP 4.5**	**Sen's RCP 4.5**	**MK RCP 8.5**	**Sen's RCP 8.5**
Ziway	3.89 ***	0.07	3.45 ***	0.05	4.14 ***	0.08	4.14 ***	0.08	2.46 *	0.04	1.96	0.04
Meki	4.29 ***	0.07	3.89 ***	0.06	4.12 ***	0.08	4.14 ***	0.08	2.36 *	0.04	1.96	0.04
Arata	2.26 *	0.04	2.85 **	0.03	4.14 ***	0.08	4.09 ***	0.08	2.31 *	0.04	2.01 *	0.04
Butajira	1.51	0.03	1.17	0.02	4.14 ***	0.08	4.14 ***	0.08	2.36 *	0.04	2.01 *	0.04
Tora	0.07	0.00	2.00 *	0.01	4.78 ***	0.07	4.79 ***	0.07	2.86 **	0.04	1.71	0.04
Bui	0.05	0.00	1.17	0.01	4.78 ***	0.07	4.09 ***	0.08	2.80 **	0.04	1.76	0.04
Kulumsa	1.61	0.04	2.41 *	0.02	4.14 ***	0.08	4.14 ***	0.08	2.46 *	0.04	1.91	0.04
Assela	1.12	0.02	2.80 **	0.02	4.14 ***	0.08	4.14 ***	0.08	2.41 *	0.04	1.96	0.04
Sagure	1.17	0.03	2.65 **	0.02	4.09 ***	0.08	4.14 ***	0.08	2.42 *	0.04	2.01 *	0.04
Meraro	1.12	0.02	2.80 **	0.03	4.14 ***	0.08	4.39 ***	0.07	2.31 *	0.04	1.28	0.04
Adamitulu	2.25 *	0.05	3.25 **	0.04	4.39 ***	0.07	4.79 ***	0.07	3.20 **	0.06	2.21 *	0.06

Table A11. Projected Mann–Kendall trend and Sen's slope estimator results of annual mean T_{min} for 11 stations over Ziway Lake Basin from the period 2050–2072.

Model	CNMR-CM5				CSIROM-MK3.6				MIP-ESM-LR			
Station Test	**MK RCP 4.5**	**Sen's RCP 4.5**	**MK RCP 8.5**	**Sen's RCP 8.5**	**MK RCP 4.5**	**Sen's RCP 4.5**	**MK RCP 8.5**	**Sen's RCP 8.5**	**MK RCP 4.5**	**Sen's RCP 4.5**	**MK RCP 8.5**	**Sen's RCP 8.5**
Ziway	1.28	0.02	4.12 ***	0.02	4.58 ***	0.10	4.20 ***	0.09	1.14	0.02	3.61 ***	0.05
Meki	1.00	0.02	4.30 ***	0.02	4.63 ***	0.10	4.6 ***	0.10	1.14	0.02	3.60 ***	0.05
Arata	1.03	0.02	4.52 ***	0.02	4.65 ***	0.09	4.18 ***	0.09	1.12	0.02	3.53 ***	0.05
Butajira	1.00	0.02	4.54 ***	0.02	4.56 ***	0.10	4.18 ***	0.09	1.05	0.01	3.57 ***	0.05
Tora	1.47	0.03	4.24 ***	0.02	4.14 ***	0.09	4.14 ***	0.09	0.72	0.01	3.26 **	0.05
Bui	0.98	0.02	4.14 ***	0.02	4.14 ***	0.09	4.14 ***	0.09	0.82	0.01	3.35 ***	0.05
Kulumsa	1.03	0.02	4.46 ***	0.03	4.56 ***	0.10	4.18 ***	0.09	1.14	0.01	3.48 ***	0.04
Assela	1.75	0.03	4.58 ***	0.02	4.58 ***	0.10	4.19 ***	0.09	1.10	0.01	3.61 ***	0.04
Sagure	1.75	0.03	4.09 **	0.02	4.09 **	0.08	4.17 ***	0.09	1.10	0.02	3.57 ***	0.04
Meraro	1.70	0.03	4.54 ***	0.02	4.54 ***	0.10	4.18 ***	0.09	1.10	0.01	3.61 ***	0.05
Adamitulu	1.56	0.03	4.62 ***	0.02	4.62 ***	0.10	4.06 ***	0.08	1.56	0.02	4.18 ***	0.05

Table A12. Projected Mann–Kendall trend and Sen's slope estimator results of annual mean T_{min} for 11 stations over Ziway Lake Basin from the period 2073–2100.

Model	CNMR-CM5				CSIROM-MK3.6				MIP-ESM-LR			
Station Test	**MK RCP 4.5**	**Sen's RCP 4.5**	**MK RCP 8.5**	**Sen's RCP 8.5**	**MK RCP 4.5**	**Sen's RCP 4.5**	**MK RCP 8.5**	**Sen's RCP 8.5**	**MK RCP 4.5**	**Sen's RCP 4.5**	**MK RCP 8.5**	**Sen's RCP 8.5**
Ziway	2.63 **	0.08	2.59 **	0.06	4.43 ***	0.03	4.76 ***	0.10	1.51	0.02	2.78 **	0.05
Meki	2.63 **	0.08	2.59 **	0.06	3.95 ***	0.05	4.76 ***	0.10	1.51	0.03	2.78 **	0.05
Arata	2.57 *	0.08	2.64 **	0.06	3.45 ***	0.07	4.76 ***	0.10	1.56	0.03	2.73 **	0.05
Butajira	2.63 **	0.08	2.69 **	0.06	3.20 **	0.04	4.76 ***	0.10	1.56	0.02	2.68 **	0.04
Tora	2.69 **	0.09	2.36 *	0.06	1.56	0.02	4.89 ***	0.09	1.66	0.02	2.82 **	0.06
Bui	2.69 **	0.07	2.31 *	0.05	1.60	0.01	4.76 ***	0.10	1.71	0.02	2.60 **	0.05
Kulumsa	2.63 **	0.09	2.31 *	0.06	2.90 **	0.04	4.76 ***	0.10	1.51	0.02	2.73 **	0.05
Assela	2.6 **	0.08	2.69 **	0.06	3.1 **	0.05	4.76 ***	0.10	1.51	0.02	2.60 **	0.04
Sagure	2.6 **	0.08	2.69 **	0.06	3.62 ***	0.07	4.76 ***	0.10	1.56	0.02	2.64 **	0.04
Meraro	2.6 **	0.08	2.64 **	0.06	3.37 ***	0.04	4.98 ***	0.08	1.61	0.02	1.03	0.00
Adamitulu	2.63 **	0.10	2.54	0.07	2.52 **	0.04	4.89 ***	0.09	0.87	0.01	3.44 ***	0.07

Mann–Kendall test of significance levels: 90% (*), 95% (**), 99% (***).

References

1. Wilby, R.L.; Dawson, C.W.; Barrow, E.M. SDSM—A decision support tool for the assessment of regional climate change impacts. *Environ. Model. Softw.* **2002**, *17*, 145–157. [CrossRef]
2. Bonebrake, T.C.; Mastrandrea, M.D. Tolerance adaptation and precipitation changes complicate latitudinal patterns of climate change impacts. *Proc. Natl. Acad. Sci. USA* **2010**, *107*, 12581–12586. [CrossRef] [PubMed]
3. Hanjra, M.A.; Ferede, T.; Gutta, D.G. Reducing poverty in sub-Saharan Africa through investments in water and other priorities. *Agric. Water Manag.* **2009**, *96*, 1062–1070. [CrossRef]
4. Carpenter, S.R.; Fisher, S.G.; Grimm, N.B.; Kitchell, J.F. Global Change and Freshwater Ecosystems. *Annu. Rev. Ecol. Syst.* **1992**, *23*, 119–139. [CrossRef]
5. Kundzewicz, Z.W.; Mata, L.J.; Arnell, N.W.; Döll, P.; Jimenez, B.; Miller, K.; Oki, T.; Şen, Z.; Shiklomanov, I. The implications of projected climate change for freshwater resources and their management. *Hydrol. Sci. J.* **2008**, *53*, 3–10. [CrossRef]
6. Leta, O.T.; El-Kadi, A.I.; Dulai, H. Implications of Climate Change on Water Budgets and Reservoir Water Harvesting of Nuuanu Area Watersheds, Oahu, Hawaii. *J. Water Resour. Plan. Manag.* **2017**, *143*, 05017013. [CrossRef]
7. Kundu, A.; Dwivedi, S.; Chandra, V. Precipitation Trend Analysis over Eastern Region of India Using Cmip5 Based Climatic Models. *ISPRS-Int. Arch. Photogramm. Remote Sens. Spat. Inf. Sci.* **2014**, *XL-8*, 1437–1442. [CrossRef]

8. Bamber, J.L.; Oppenheimer, M.; Kopp, R.E.; Aspinall, W.P.; Cooke, R.M. Ice sheet contributions to future sea-level rise from structured expert judgment. *Proc. Natl. Acad. Sci. USA* **2019**, *116*, 11195–11200. [CrossRef] [PubMed]
9. Schilling, J.; Hertig, E.; Tramblay, Y.; Scheffran, J. Climate change vulnerability, water resources and social implications in North Africa. *Reg. Environ. Chang.* **2020**, *20*, 15. [CrossRef]
10. Kotir, J.H. Climate change and variability in Sub-Saharan Africa: A review of current and future trends and impacts on agriculture and food security. *Environ. Dev. Sustain.* **2010**, *13*, 587–605. [CrossRef]
11. Challinor, A.; Wheeler, T.; Garforth, C.; Craufurd, P.; Kassam, A. Assessing the vulnerability of food crop systems in Africa to climate change. *Clim. Chang.* **2007**, *83*, 381–399. [CrossRef]
12. Collins, J.M. Temperature Variability over Africa. *J. Clim.* **2011**, *24*, 3649–3666. [CrossRef]
13. Alemayehu, A.; Bewket, W. Local climate variability and crop production in the central highlands of Ethiopia. *Environ. Dev.* **2016**, *19*, 36–48. [CrossRef]
14. Block, P.J.; Strzepek, K.; Rosegrant, M.W.; Diao, X. Impacts of considering climate variability on investment decisions in Ethiopia. *Agric. Econ.* **2008**, *39*, 171–181. [CrossRef]
15. Evangelista, P.; Young, N.; Burnett, J. How will climate change spatially affect agriculture production in Ethiopia? Case studies of important cereal crops. *Clim. Chang.* **2013**, *119*, 855–873. [CrossRef]
16. Baker, T.J.; Miller, S.N. Using the Soil and Water Assessment Tool (SWAT) to assess land use impact on water resources in an East African watershed. *J. Hydrol.* **2013**, *486*, 100–111. [CrossRef]
17. Gummadi, S.; Rao, K.P.C.; Seid, J.; Legesse, G.; Kadiyala, M.D.M.; Takele, R.; Amede, T.; Whitbread, A. Spatio-temporal variability and trends of precipitation and extreme rainfall events in Ethiopia in 1980–2010. *Theor. Appl. Clim.* **2018**, *134*, 1315–1328. [CrossRef]
18. Esayas, B.; Simane, B.; Teferi, E.; Ongoma, V.; Tefera, N. Trends in Extreme Climate Events over Three Agroecological Zones of Southern Ethiopia. *Adv. Meteorol.* **2018**, *2018*, 7354157. [CrossRef]
19. Wagesho, N.; Jain, M.K.; Goel, N.K. Effect of Climate Change on Runoff Generation: Application to Rift Valley Lakes Basin of Ethiopia. *J. Hydrol. Eng.* **2013**, *18*, 1048–1063. [CrossRef]
20. Abraham, A.T.; Woldemicheala, A.; Muluneha, A.; Abateb, B. Hydrological Responses of Climate Change on Lake Ziway Catchment, Central Rift Valley of Ethiopia. *J. Earth Sci. Clim. Change* **2018**, *9*, 6. [CrossRef]
21. Musie, M.; Sen, S.; Chaubey, I. Hydrologic Responses to Climate Variability and Human Activities in Lake Ziway Basin, Ethiopia. *Water* **2020**, *12*, 164. [CrossRef]
22. Belihu, M.; Abate, B.; Tekleab, S.; Bewket, W. Hydro-meteorological trends in the Gidabo catchment of the Rift Valley Lakes Basin of Ethiopia. *Phys. Chem. Earth, Parts A/B/C* **2018**, *104*, 84–101. [CrossRef]
23. Rosa, L.; Chiarelli, D.D.; Rulli, M.C.; Dell'Angelo, J.; D'Odorico, P. Global agricultural economic water scarcity. *Sci. Adv.* **2020**, *6*, eaaz6031. [CrossRef] [PubMed]
24. Musie, M.; Momblanch, A.; Sen, S. Exploring future global change-induced water imbalances in the Central Rift Valley Basin, Ethiopia. *Clim. Chang.* **2021**, *164*, 47. [CrossRef]
25. Fernández García, I.; Lecina, S.; Ruiz-Sánchez, M.C.; Vera, J.; Conejero, W.; Conesa, M.R.; Domínguez, A.; Pardo, J.J.; Léllis, B.C.; Montesinos, P. Trends and Challenges in Irrigation Scheduling in the Semi-Arid Area of Spain. *Water* **2020**, *12*, 785. [CrossRef]
26. Greaves, G.E.; Wang, Y.-M. Effect of regulated deficit irrigation scheduling on water use of corn in southern Taiwan tropical environment. *Agric. Water Manag.* **2017**, *188*, 115–125. [CrossRef]
27. Cancela, J.J.; Trigo-Córdoba, E.; Martínez, E.M.; Rey, B.J.; Bouzas-Cid, Y.; Fandiño, M.; Mirás-Avalos, J.M. Effects of climate variability on irrigation scheduling in white varieties of *Vitis vinifera* (L.) of NW Spain. *Agric. Water Manag.* **2016**, *170*, 99–109. [CrossRef]
28. Eshete, D.G.; Sinshaw, B.G.; Legesse, K.G. Critical review on improving irrigation water use efficiency: Advances, challenges, and opportunities in the Ethiopia context. *Water-Energy Nexus* **2020**, *3*, 143–154. [CrossRef]
29. Legesse, D.; Ayenew, T. Effect of improper water and land resource utilization on the central Main Ethiopian Rift lakes. *Quat. Int.* **2006**, *148*, 8–18. [CrossRef]
30. Alhamshry, A.; Fenta, A.A.; Yasuda, H.; Kimura, R.; Shimizu, K. Seasonal Rainfall Variability in Ethiopia and Its Long-Term Link to Global Sea Surface Temperatures. *Water* **2020**, *12*, 55. [CrossRef]
31. Abdulla, G.M. *Annual Earth System Grid Federation 2019 Progress Report*; Lawrence Livermore National Lab. (LLNL): Livermore, CA, USA, 2019. [CrossRef]
32. Lamarque, J.-F.; Kyle, G.P.; Meinshausen, M.; Riahi, K.; Smith, S.J.; van Vuuren, D.P.; Conley, A.J.; Vitt, F. Global and regional evolution of short-lived radiatively-active gases and aerosols in the Representative Concentration Pathways. *Clim. Change* **2011**, *109*, 191–212. [CrossRef]
33. Basheer, A.K.; Lu, H.; Omer, A.; Ali, A.B.; Abdelgader, A.M.S. Impacts of climate change under CMIP5 RCP scenarios on the streamflow in the Dinder River and ecosystem habitats in Dinder National Park, Sudan. *Hydrol. Earth Syst. Sci.* **2016**, *20*, 1331–1353. [CrossRef]
34. McSweeney, C.F.; Jones, R.G.; Lee, R.W.; Rowell, D.P. Selecting CMIP5 GCMs for downscaling over multiple regions. *Climate Dynamics.* **2015**, *44*, 3237–3260. [CrossRef]
35. Lutz, A.F.; ter Maat, H.W.; Biemans, H.; Shrestha, A.B.; Wester, P.; Immerzeel, W.W. Selecting representative climate models for climate change impact studies: An advanced envelope-based selection approach. *Int. J. Climatol.* **2016**, *36*, 3988–4005. [CrossRef]

36. Emami, F.; Koch, M. Agricultural Water Productivity-Based Hydro-Economic Modeling for Optimal Crop Pattern and Water Resources Planning in the Zarrine River Basin, Iran, in the Wake of Climate Change. *Sustainability* **2018**, *10*, 3953. [CrossRef]
37. Abraham, T.; Abate, B.; Woldemicheal, A.; Muluneh, A. Impacts of Climate Change under CMIP5 RCP Scenarios on the Hydrology of Lake Ziway Catchment, Central Rift Valley of Ethiopia. *J. Environ. Earth Sci.* **2018**, *8*, 81–90.
38. Musie, M.; Sen, S.; Srivastava, P. Application of CORDEX-AFRICA and NEX-GDDP datasets for hydrologic projections under climate change in Lake Ziway sub-basin, Ethiopia. *J. Hydrol. Reg. Stud.* **2020**, *31*, 100721. [CrossRef]
39. Gudmundsson, L.; Bremnes, J.B.; Haugen, J.E.; Engen-Skaugen, T. Downscaling RCM precipitation to the station scale using statistical transformations—a comparison of methods. *Hydrol. Earth Syst. Sci.* **2012**, *16*, 3383–3390. [CrossRef]
40. Smid, M.; Costa, A.C. Climate projections and downscaling techniques: A discussion for impact studies in urban systems. *Int. J. Urban Sci.* **2018**, *22*, 277–307. [CrossRef]
41. Rathjens, H.; Bieger, K.; Srinivasan, R.; Chaubey, I.; Arnold, J.G. *CMhyd User Manual: Documentation for Preparing Simulated Climate Change Data for Hydrologic Impact Studies*; SWAT: Garland, TX, USA, 2016. Available online: https://swat.tamu.edu/media/115265/bias_cor_man.pdf (accessed on 2 November 2021).
42. Hordofa, A.T.; Leta, O.T.; Alamirew, T.; Kawo, N.S.; Chukalla, A.D. Performance Evaluation and Comparison of Satellite-Derived Rainfall Datasets over the Ziway Lake Basin, Ethiopia. *Climate* **2021**, *9*, 113. [CrossRef]
43. Hoffmann, P.; Spekat, A. Identification of possible dynamical drivers for long-term changes in temperature and rainfall patterns over Europe. *Theor. Appl. Clim.* **2021**, *143*, 177–191. [CrossRef]
44. Mohorji, A.M.; Şen, Z.; Almazroui, M. Trend Analyses Revision and Global Monthly Temperature Innovative Multi-Duration Analysis. *Earth Syst. Environ.* **2017**, *1*, 9. [CrossRef]
45. Yue, S.; Pilon, P.; Cavadias, G. Power of the Mann–Kendall and Spearman's rho tests for detecting monotonic trends in hydrological series. *J. Hydrol.* **2002**, *259*, 254–271. [CrossRef]
46. Mann, H.B. Nonparametric tests against trend. *Econometrica J. Econom. Soc.* **1945**, *1*, 245–259. [CrossRef]
47. Serinaldi, F.; Kilsby, C.G.; Lombardo, F. Untenable nonstationarity: An assessment of the fitness for purpose of trend tests in hydrology. *Adv. Water Resour.* **2018**, *111*, 132–155. [CrossRef]
48. Bari, S.H.; Rahman, M.T.U.; Hoque, M.A.; Hussain, M.M. Analysis of seasonal and annual rainfall trends in the northern region of Bangladesh. *Atmospheric Res.* **2016**, *176–177*, 148–158. [CrossRef]
49. De Lima, M.I.P.; Carvalho, S.; De Lima, J.L.M.P.; Coelho, M.F.E.S. Trends in precipitation: Analysis of long annual and monthly time series from mainland Portugal. *Adv. Geosci.* **2010**, *25*, 155–160. [CrossRef]
50. Hu, Y.; Maskey, S.; Uhlenbrook, S. Expected changes in future temperature extremes and their elevation dependency over the Yellow River source region. *Hydrol. Earth Syst. Sci.* **2013**, *17*, 2501–2514. [CrossRef]
51. Hu, Y.; Maskey, S.; Uhlenbrook, S.; Zhao, H. Streamflow trends and climate linkages in the source region of the Yellow River, China. *Hydrol. Process.* **2011**, *25*, 3399–3411. [CrossRef]
52. Yue, S.; Pilon, P.; Phinney, B.; Cavadias, G. The influence of autocorrelation on the ability to detect trend in hydrological series. *Hydrol. Process.* **2002**, *16*, 1807–1829. [CrossRef]
53. Hussain, M.M.; Mahmud, I. pyMannKendall: A python package for non-parametric Mann Kendall family of trend tests. *J. Open Source Softw.* **2019**, *4*, 1556. [CrossRef]
54. Cox, D.R.; Stuart, A. Some Quick Sign Tests for Trend in Location and Dispersion. *Biometrika* **1955**, *42*, 80–95. [CrossRef]
55. Tehrani, E.N.; Sahour, H.; Booij, M.J. Trend analysis of hydro-climatic variables in the north of Iran. *Theor. Appl. Clim.* **2019**, *136*, 85–97. [CrossRef]
56. Kamworapan, S.; Surussavadee, C. Evaluation of CMIP5 Global Climate Models for Simulating Climatological Temperature and Precipitation for Southeast Asia. *Adv. Meteorol.* **2019**, *2019*, 1067365. [CrossRef]
57. Voldoire, A.; Sanchezgomez, E.; Mélia, D.S.Y.; Decharme, B.; Cassou, C.; Senesi, S.; Valcke, S.; Beau, I.; Alias, A.; Chevallier, M.; et al. The CNRM-CM5.1 global climate model: Description and basic evaluation. *Clim. Dyn.* **2013**, *40*, 2091–2121. [CrossRef]
58. Rogelj, J.; Meinshausen, M.; Knutti, R. Global warming under old and new scenarios using IPCC climate sensitivity range estimates. *Nat. Clim. Change* **2012**, *2*, 248–253. [CrossRef]
59. Gurara, M.A.; Jilo, N.B.; Tolche, A.D. Impact of climate change on potential evapotranspiration and crop water requirement in Upper Wabe Bridge watershed, Wabe Shebele River Basin, Ethiopia. *J. Afr. Earth Sci.* **2021**, *180*, 104223. [CrossRef]
60. Deressa, T.T.; Hassan, R.M. Economic Impact of Climate Change on Crop Production in Ethiopia: Evidence from Cross-section Measures. *J. Afr. Econ.* **2009**, *18*, 529–554. [CrossRef]
61. Gadissa, T.; Nyadawa, M.; Behulu, F.; Mutua, B. The Effect of Climate Change on Loss of Lake Volume: Case of Sedimentation in Central Rift Valley Basin, Ethiopia. *Hydrology* **2018**, *5*, 67. [CrossRef]
62. Intergovernmental Panel on Climate Change (IPCC). Summary for Policymakers. Climate Change 2014: Part of the Working Group III Contribution to the Fifth Assessment Report of the IPCC, Geneva, Switzerland. Available online: http://pure.iiasa.ac.at/11125 (accessed on 14 September 2021).
63. Asare-Nuamah, P.; Botchway, E. Understanding climate variability and change: Analysis of temperature and rainfall across agroecological zones in Ghana. *Heliyon* **2019**, *5*, e02654. [CrossRef] [PubMed]
64. Kelkar, S.M.; Kulkarni, A.; Rao, K.K. Impact of Climate Variability and Change on Crop Production in Maharashtra, India. *Curr. Sci.* **2020**, *118*, 1235–1245. [CrossRef]

65. Gebrehiwot, T.; van der Veen, A. Assessing the evidence of climate variability in the northern part of Ethiopia. *J. Dev. Agric. Econ.* **2013**, *5*, 104–119. [CrossRef]
66. Tesfaye, K.; Aggarwal, P.K.; Mequanint, F.; Shirsath, P.B.; Stirling, C.M.; Khatri-Chhetri, A.; Rahut, D.B. Climate Variability and Change in Bihar, India: Challenges and Opportunities for Sustainable Crop Production. *Sustainability* **2017**, *9*, 1998. [CrossRef]

hydrology MDPI

Article

Improving Hillslope Link Model Performance from Non-Linear Representation of Natural and Artificially Drained Subsurface Flows

Nicolás Velásquez [1,*], Ricardo Mantilla [2], Witold Krajewski [1], Morgan Fonley [3] and Felipe Quintero [1]

1 IIHR Hydroscience & Engineering, Iowa Flood Center, The University of Iowa, Iowa City, IA 52240, USA; witold-krajewski@uiowa.edu (W.K.); felipe-quintero@uiowa.edu (F.Q.)
2 Civil Engineering Department, University of Manitoba, Winnipeg, MB R3T 2N2, Canada; ricardo.mantilla@umanitoba.ca
3 Department of Mathematics and Computer Science, Alma College, Alma, MI 48801, USA; fonleymr@alma.edu
* Correspondence: nicolas-giron@uiowa.edu

Abstract: This study evaluates the potential for a newly proposed non-linear subsurface flux equation to improve the performance of the hydrological Hillslope Link Model (HLM). The equation contains parameters that are functionally related to the hillslope steepness and the presence of tile drainage. As a result, the equation provides better representation of hydrograph recession curves, hydrograph timing, and total runoff volume. The authors explore the new parameterization's potential by comparing a set of diagnostic and prognostic setups in HLM. In the diagnostic approach, they configure 12 different scenarios with spatially uniform parameters over the state of Iowa. In the prognostic case, they use information from topographical maps and known locations of tile drainage to distribute parameter values. To assess performance improvements, they compare simulation results to streamflow observations during a 17-year period (2002–2018) at 140 U.S. Geological Survey (USGS) gauging stations. The operational setup of the HLM model used at the Iowa Flood Center (IFC) serves as a benchmark to quantify the overall improvement of the model. In particular, the new equation provides better representation of recession curves and the total streamflow volumes. However, when comparing the diagnostic and prognostic setups, the authors found discrepancies in the spatial distribution of hillslope scale parameters. The results suggest that more work is required when using maps of physical attributes to parameterize hydrological models. The findings also demonstrate that the diagnostic approach is a useful strategy to evaluate models and assess changes in their formulations.

Keywords: hydrology; tiling; subsurface flow; Iowa

Citation: Velásquez, N.; Mantilla, R.; Krajewski, W.; Fonley, M.; Quintero, F. Improving Hillslope Link Model Performance from Non-Linear Representation of Natural and Artificially Drained Subsurface Flows. *Hydrology* **2021**, *8*, 187. https://doi.org/10.3390/hydrology8040187

Academic Editors: Carmelina Costanzo, Tommaso Caloiero and Roberta Padulano

Received: 30 November 2021
Accepted: 17 December 2021
Published: 20 December 2021

1. Introduction

Flood forecasts that are calculated using regional distributed hydrological models are becoming more common and relevant because they also provide information about internal watershed processes in large domains, along with predicted hydrographs for all streams in the river network. These forecasts are expected to be accurate at the region's ungauged watersheds [1] as a consequence of appropriate spatial representation of processes and parameters in the model.

Current hydrological models correctly identify many aspects of the streamflow hydrographs, thereby providing acceptable forecasts. However, they still struggle to reproduce the hydrograph recession. According to [2], modelers need to pay more attention to storm runoff's slow flow, which is a crucial component of the recession. For regional models, recession becomes more challenging because its non-linearity increases with the spatial scale [3–5]. Landscape properties such as topography, soil, and the stream network seem to be involved in the recession variability [6–8]. Additionally, human landscape and land-use interventions, such as tile drainage, can help to restore river health [9]. Nevertheless, these interventions also have the potential to alter streamflow and its recession [10,11]. A good representation of the recession may influence the estimation of the flood duration, and flood occurrence during successive events.

1.1. Issues with the Hillslope Link Model (HLM) in Iowa

The Iowa Flood Center (IFC) produces flood forecasts for the state of Iowa using the Hillslope Link Model (HLM) [12–15]. The operational HLM represents the hillslope subsurface flux, using a linear-reservoir equation. According to [15], the current HLM configuration accurately estimates peak flows with an overall acceptable performance in Iowa. However, the model has some limitations in capturing the hydrograph recessions and the total runoff volume at some locations, as reported in [14]. The discrepancies between simulated and observed recessions are more common in watersheds with tiling drainage. Sample streamflow simulation results, using the IFC HLM operational model for three Iowa watersheds are presented in Figure 1a–c (in red). The model performance is described in terms of Kling Gupta Efficiency [16], which is defined as $\mathrm{KGE} = 1 - \sqrt{(\rho-1)^2 + (\alpha-1)^2 + (\beta-1)^2}$, where ρ, α, and β denote correlation, the ratio of standard deviation (σ_s/σ_o), and the ratio of the mean (μ_s/μ_o) between simulated and observed streamflow, respectively. The model's limitations are most evident in the watersheds located in the north and west regions of Iowa, where the model has low performance in terms of the Kling Gupta Efficiency (KGE) index (Figure 1d). We associate the model's poor performance in the region of north-central Iowa, known as the Des Moines Lobe, with the widespread use of artificial subsurface drainage (known as tile drains) in the region [11].

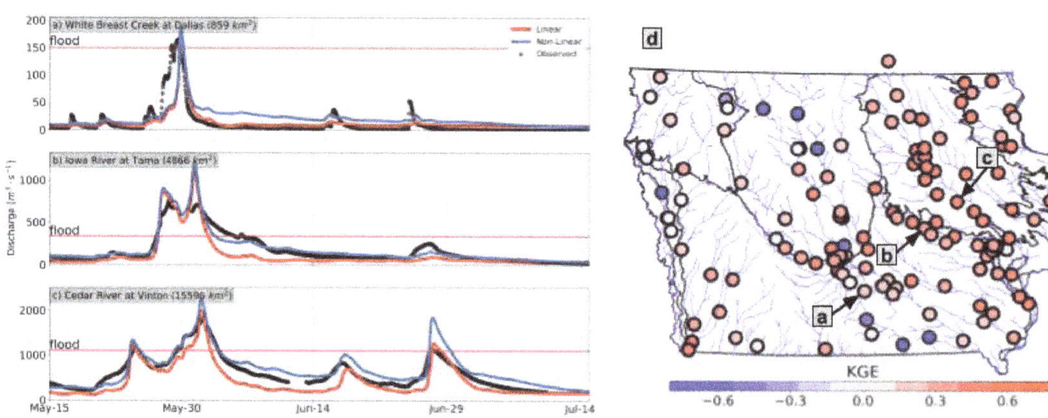

Figure 1. (**a–c**) Observed (black) and simulated streamflows by the linear (red) and the non-linear (blue) setups at three USGS gauged stations. (**d**) Mean annual KGE performance of the HLM linear setup for Iowa between 2002 and 2018.

To address these issues, the authors of [17] developed a subsurface non-linear equation that can represent subsurface flow from hillslopes with different steepness and soil conductivities, as well as the presence of tile drainage. The blue lines in Figure 1a–c show the resulting hydrographs using the non-linear equation with parameters corresponding to no tile and a steepness of 2% [17]. Compared with the linear equation of the operational HLM, the non-linear equation tends to improve the total streamflow volume and the simulated recession shapes. However, we still observe discrepancies (Figure 1a,b) that are attributed to issues with the parameter values and spatial representation of processes. The model proposed in [15] serves as motivation for this study, to evaluate the performance of such model modifications.

1.2. The Diagnostic-Prognostic Approach

According to the authors of [18], the development of a hydrological model is subject to the hypothesis-testing process. This process evaluates, rejects, and replaces model components. We performed a diagnostic–prognostic analysis of the model at 140 USGS gauges in Iowa to test the utility of the non-linear equation to represent the hillslope subsurface flux. In this case, we adapted the diagnostic-prognostic approach developed in studies on evapotranspiration [19–21]. Our diagnostic setups have simplified, spatially uniform parameter values, while the prognostic scenarios use maps to determine parameter values. The diagnostic-prognostic approach offers complementary information about the model [22] and the required independence to perform model comparisons [23].

According to [15], an insightful way to improve models starts with model performance verification, followed by structure modification. We expanded on this approach by using the diagnostic-prognostic analysis to add tools to verify the model's processes and required parameters. Our objective is to identify the non-linear model parameters for Iowa, their uncertainties, and the model limitations.

The paper is organized as follows: we first describe the HLM model and the equations governing the hillslope processes, including the linear equation and non-linear equations to represent subsurface and drainage tile fluxes. Next, we describe the diagnostic and prognostic setups. Then, we compare the diagnostic and prognostic approach results using 140 USGS stations and we analyze the parameters' influence on the model performance. Finally, we provide conclusions based on our experiment results.

2. Materials and Methods

2.1. Model Description

The Hillslope Link Model (HLM) represents the hydrological processes at the hillslope scale (Figure 2a,b) and routes the streamflow through the channel network (Figure 2c). At the hillslopes, HLM has three storages, ponded surface (S_p [m]), topsoil (S_T [m]), and subsurface storage (S_s [m]). The water from the ponded storage can either infiltrate the topsoil (q_{pT} $[\text{m} \cdot \text{min}^{-1}]$) or flow as runoff to the channel link (q_{pL} $[\text{m} \cdot \text{min}^{-1}]$). The water in the topsoil percolates (q_{Ts} $[\text{m} \cdot \text{min}^{-1}]$) to the soil storage. Finally, the water in the soil storage seeps into the channel link as subsurface runoff (q_{sL} $[\text{m} \cdot \text{min}^{-1}]$). Evaporation occurs from the three storages as a removal of volume from the model. Once in the river network, HLM transports the channel water (q $[\text{m}^3 \cdot \text{s}^{-1}]$) downstream. A detailed description of the hillslope and stream routing process can be found in [11,14].

The surface runoff, infiltration, and percolation rates are linked through the reference speed v_r and the shape of the hillslope. Each hillslope has a parameter $k_2[\text{min}^{-1}]$ (Equation (1)) that depends on the hillslope link length ($L_i[\text{m}]$) and area ($A_h[\text{m}^2]$), along with the reference velocity v_r. The parameter k_2 is the inverse of the runoff residence time in the hillslope. The runoff q_{pL} and the infiltration q_{pT} are linked to k_2 through Equations (2) and (3), respectively. Moreover, the infiltration also depends on the topsoil depth (T_l) assumed to be equal to $0.1[\text{m}]$. Additionally, the percolation rate q_{Ts} is computed as a proportion of k_2, expressed by k_i. Usually, k_i is 2% of k_2; however, its value may change depending on the soil and topographical properties.

$$k_2 = v_r \cdot \left(\frac{L_i}{A_h}\right) \cdot 60 \tag{1}$$

$$q_{pL} = k_2 \cdot S_p \tag{2}$$

$$q_{pT} = k_2 \cdot S_p \cdot 99 \cdot (1 - S_T/T_l)^3 \tag{3}$$

$$q_{Ts} = k_2 \cdot S_T \cdot k_i \tag{4}$$

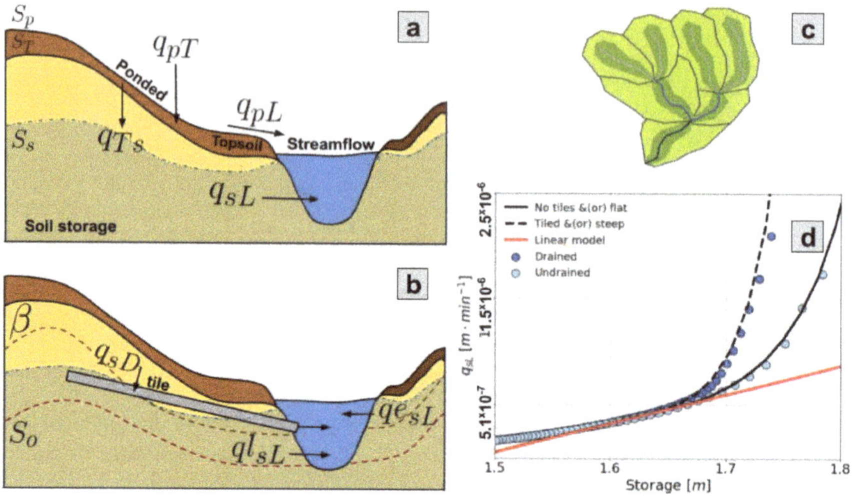

Figure 2. Hillslope Link Model spatial discretization and schematic of the storages and processes represented at the hillslope scale: (**a**) HLM hillslope process using the linear subsurface flux equation; (**b**) hillslope process including the active layer (β), the exponential flux (qe_{sL}), and the tile drainage flux (q_{sD}); (**c**) watershed decomposition into hillslopes and channel links; and (**d**) functional form of the subsurface flux in the function of the soil storage (S_s) after [17].

The current HLM setup represents the subsurface flux to the channels (q_{sL} $[\mathrm{m} \cdot \mathrm{min}^{-1}]$) with a linear equation (red line on Figure 2d). The equation releases water to the channel at a rate m, when S_s is greater than the no-flow threshold (S_o), as follows,

$$q_{sL} = m \cdot (S_s - S_o) \tag{5}$$

Ref. [17] developed a set of parameterizations for ordinary differential equations that adds a non-linear component to Equation (5) when S_s is above threshold storage. The following exponential equation (continuous line on Figure 2d) is added to Equation (5) if S_s is greater than the activation threshold β [m],

$$qe_{sL} = \alpha(S_s - \beta)e^{17(S_s - \beta)} \tag{6}$$

where α is a parameter that depends on the hillslope properties, such as its steepness and the soil conductivity. In [17], the authors also developed an exponential equation that applies when the hillslope has tiles. The following equation (dashed line on Figure 2d) is added when S_s is greater than the tile relative depth D_d [m],

$$q_{sD} = c(S_s - D_d)e^{35(S_s - D_d)} \tag{7}$$

In the described scheme, the subsurface flux becomes a set of equations that HLM activates, depending on the value of S_s relative to the thresholds S_o, β, and D_d. The segmented subsurface runoff is as follows,

$$q_sL = \begin{cases} ql_{sL} & if \quad S_s < \beta \\ ql_{sL} + qe_{sL} & if \quad S_s > \beta \\ ql_{sL} + qe_{sL} + q_sD & if \quad S_s > D_d \end{cases} \tag{8}$$

The relative tile depth (D_d) is independent of β, so either could be larger depending on the tile configuration and the hillslope properties. Moreover, if there are no tiles, Equation (8) is limited to its two first expressions. More details on the subsurface equation development can be found in [17].

2.2. Model Setup and Data

We used both diagnostic and prognostic approaches to test the performance using the non-linear equation. We used the river network for the state of Iowa derived from a DEM of 90 m and decomposed into about 420,000 individual hillslopes, following the approach presented in [11]. The precipitation force corresponds to hourly Stage IV QPEs [24]. We forced the evapotranspiration using the mean annual monthly values derived from MODIS [25] for the region.

Equation (8) offers a formulation for the subsurface flux that we want to validate for the Iowa domain. In this process, we can fix parameters uniformly over the space or distribute them spatially. A uniform setup assumes that each hillslope in the region uses the same model parameters, while a distributed setup assumes parameter variability as a function of landscape properties. Neither approach is without error because the parameters are only approximate and could depend upon unknown factors that are variable in space. The fixed setup is unrealistic, and the distributed setup may be subject to spatial errors. However, both approaches are complementary. Fixed setups could help assess the ability of Equation (8) to improve the accuracy of the simulated streamflow fluctuations. In contrast, a distributed setup helps to validate the parameter description given by the map(s). Considering this, we used both approaches to validate the new q_{sL} equation and to explore the limits of the so-called predefined setups. In the distribute parameters case, we use the steepness of the hillslopes (Figure 3a) and the tiles localization according to the Iowa Department of Natural Resources (DNR) (Figure 3b).

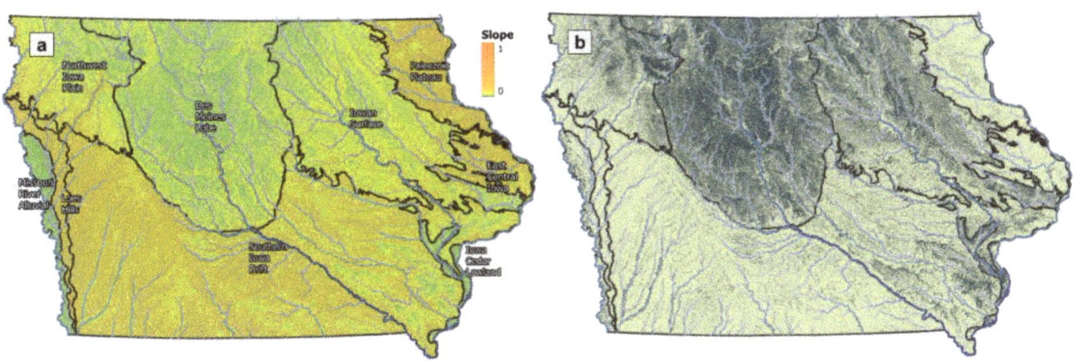

Figure 3. Maps of the hillslope steepness (**a**) and tile drainage localization according to the Iowa DNR (**b**).

The model validation consists of comparing fixed (diagnostic) and distributed (prognostic) HLM setups (Figure 4). The diagnostic setup (Figure 4a) shows how different formulations could significantly improve the model across the region. On the other hand, the prognostic setups (Figure 4b) show the improvements and limitations derived from the application of "known" spatial variables.

The formulation of Equation (8) relies on the percolation rate because the non-linear formulation depends upon the amount of water in the subsurface storage. The described dependence increases the relevance of the percolation parameter (k_i). The distribution of k_i can be derived from maps of the soil profile properties. However, using an additional map may increase the errors affecting the comparison of both setups. For this reason, we choose to fix three different percolation rates for the diagnostic and prognostic setups (Figure 4c).

Moreover, we used the same values for S_0, β, and D_d in both setups. In [17] S_0 oscillates between 1.4 and 1.55 in function of the slope. Nevertheless, additional features of the hillslope such as the bedrock depth, and soil type determine the value of S_0. To avoid adding additional uncertainty sources, we fixed $S_0 = 1.48$. On the other hand, for β and D_d we used the values described by [17] of 1.67 and 1.635, respectively. In Table 1 we summarize the described diagnostic and prognostic setups.

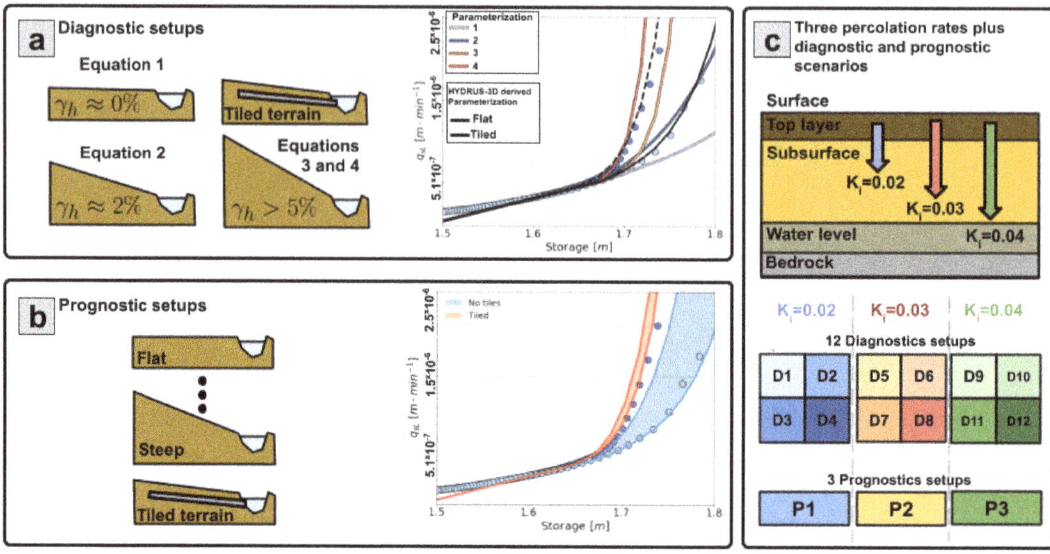

Figure 4. Diagnostic and prognostic experiment setup: (**a**) diagnostic case with four non-linear subsurface flux equations fixed for the domain of Iowa after [17] (**b**) prognostic case, with equations varying with the hillslopes steepness (blue scenario) and the presence of tiles (red scenario); and (**c**) percolation rates fixed for the different scenarios. Their combination gives us 12 diagnostic scenarios and three prognostic scenarios.

Table 1. HLM diagnostic and prognostic setups summary.

Identifier	Type	Slope	Tiled	k_i
D1	Diagnostic	0%	False	0.02
D2	Diagnostic	2%	False	0.02
D3	Diagnostic	5%	False	0.02
D4	Diagnostic	2%	True	0.02
D5	Diagnostic	0%	False	0.03
D6	Diagnostic	2%	False	0.03
D7	Diagnostic	5%	False	0.03
D8	Diagnostic	2%	True	0.03
D9	Diagnostic	0%	False	0.04
D10	Diagnostic	2%	False	0.04
D11	Diagnostic	5%	False	0.04
D12	Diagnostic	2%	True	0.04
P1	Prognostic	Variable	Variable	0.02
P2	Prognostic	Variable	Variable	0.03
P3	Prognostic	Variable	Variable	0.04

2.2.1. Diagnostic Setups

In the diagnostic setup (Figure 4a), we created four base parametrizations, using Equation (8) for the Iowa domain. The parametrizations range from flat hillslopes (light blue line on Figure 4a) to steep or tiled hillslopes (red line on Figure 4a). In the tiled case, we used a steepness of 2% under the assumption that tiles are usually installed in flat terrains. Then, we combined the four parameterizations and the three k_i rates to obtain 12 diagnostic (D) scenarios (D1 to D12 in Figure 4c). In the scenarios, D1 to D4 use $k_i = 0.02$; D5 to D8 use $k_i = 0.03$; and D9 to D12 use $k_i = 0.04$. A summary of the diagnostic scenarios is provided in Table 1.

2.2.2. Prognostic Setups

In the prognostic setup, we distributed parameter values in function of the hillslopes steepness and the Iowa DNR map describing tile presence (Figure 4b). According to [17] the coefficient α of Equation (6) changes in function of the hillslope steepness (γ_h) with a linear equation. Using the following equation, we assigned the parameter α to each hillslope, obtaining variable $q_s L$ curves that oscillate between the blue bands shown in Figure 4b:

$$\alpha = \gamma_h \left(8.5 * 10^{-8}\right) + 9.48 * 10^{-7} \tag{9}$$

Additionally, we include Equation (8) for tiled terrain. To do it, we assigned tiles to the hillslopes based in the map in Figure 4b. In the tile drainage case, we changed c in function of the slope using the following equation:

$$c = 3.4 * 10^{-8} \cdot log(\gamma_h) + 5.7 * 10^{-7} \tag{10}$$

Then, we developed the prognostic (P) scenarios P1, P2, and P3 using the distributed parameters α and c with the percolation rates k_i of 0.02, 0.03 and 0.04, respectively (distributed setups in Figure 4c). A summary of the prognostic scenarios is provided in Table 1.

3. Results

3.1. Insights from a Diagnostic-Prognostic Approach

The diagnostic and prognostic setups produced significant differences between the model outputs. In Figure 5, we present the simulated hydrographs at three watersheds simulated by the diagnostic scenario D4 in blue and the prognostic scenario P1 in red. In this case, the diagnostic setup assumes that all the hillslopes have tiles or are steep. On the other hand, the prognostic setup assumes there to be tiles only at some hillslopes and that the parameter α of Equation (6) varies with the steepness. In these three cases, the diagnostic (or fixed) setup produces a longer recession curve than the one obtained by the prognostic setup. The diagnostic case has a better match to the Iowa River at Tama (Figure 5b), while the prognostic setup exhibits a better match to the White Breast Creek (Figure 5a) and at the Cedar River (Figure 5c). Figure 5 gives a brief description of the expected differences between the setups. Additionally, it shows that Equation (8) can improve the streamflow representation, given the correct set of parameters that are obtained.

According to Figure 5, the non-linear model can produce a good representation of the hydrograph falling limb and early recession, depending on the parameters. Considering the described sensitivity, we compare the event-based KGE of the non-linear setups and the linear model (Figure 6). The KGE equation summarizes the correlation (γ), the mean ratio (μ), and the deviation ratio (σ). Our results suggest that the KGE performance depends heavily on the percolation rate (k_i). With $k_i = 0.02$ (first row of Figure 6), all the non-linear setups tend to improve the linear model, with a significant performance decrease in some events. Conversely, values of k_i equal to 0.03 and 0.04 do not exhibit a significant KGE change (second and third rows of Figure 6). Cases such as D5 and D11 exhibited a performance such as that obtained by the linear model. Other cases, such as D9, tended toward a general decrease in performance. D6, D8, and P2 exhibited a slight performance increase. The described results highlight the relevance of the percolation rate and the subsurface parameters. The comparison with the linear model shows that Equation (8) can significantly improve the model performance, depending on the parameters.

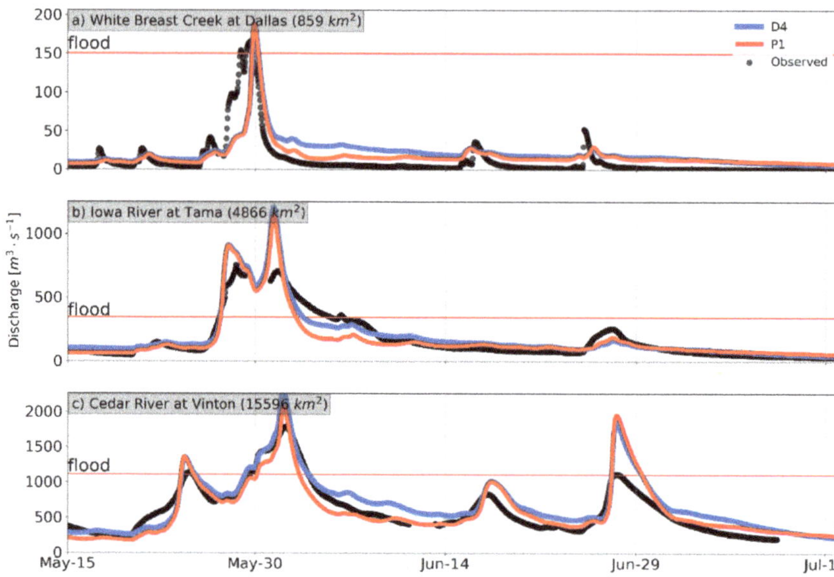

Figure 5. (**a**–**c**) Diagnostic (blue) and prognostic (red) simulations done for the year 2013. Results are compared with the USGS observations (black).

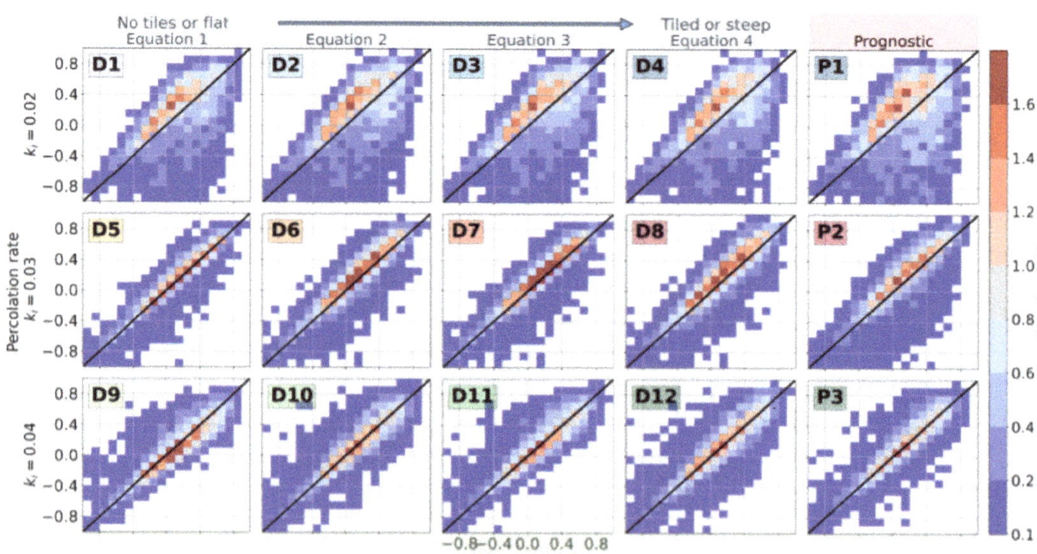

Figure 6. Event-based KGEs comparison between the diagnostics setups and the linear model. Each row corresponds to a fixed percolation rate. Columns correspond to the four fixed equations. The color bar shows the percentage of events that fall at each bin of the 2D histogram.

Differences among the scenarios are highlighted by comparing the performance gauge by gauge. First, we choose the diagnostic (D) and prognostic (P) setup with the best performance at each gauge. For this, we used the KGE to select the setup that outperformed the others for most of the events. In Figure 7, we present the KGE distribution and the percentage of time each scenario was chosen. We found similarities between the diagnostic and prognostic chosen setups when grouped by the percolation rate values (k_i). D4 and P1 ($k_i = 0.02$) have a similar KGE distribution, as do D8 and P2 ($k_i = 0.03$) and the group that includes D9, D11, D12, and P3 ($k_i = 0.04$). The similarities among the described groups highlights the relevance of k_i. Moreover, some differences also highlight the relevance of Equation (8) parameters.

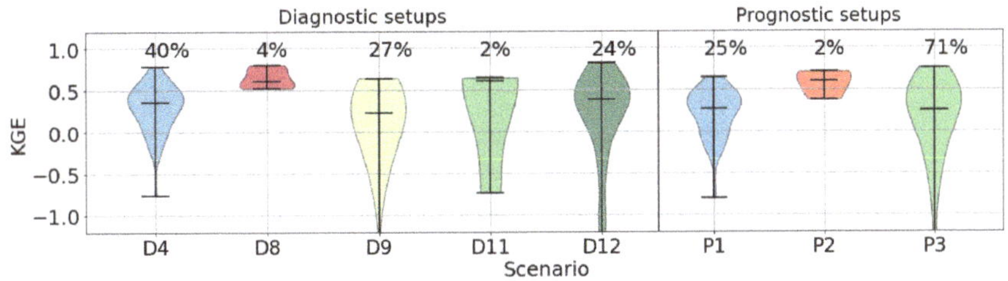

Figure 7. Event-based KGE distribution for the selected scenarios at each station.

The results presented in Figure 7 follow a spatial distribution. Figure 8 shows each USGS gauge colored by the diagnostic (Figure 8a) and prognostic (Figure 8b) setups with the best performance. In both cases, the percolation rate defines the spatial distribution. We can identify how the chosen setups (Figure 8) follow the Iowa landforms to some extent in the diagnostic case (see Figure 3a). Scenario D12 is recurrent over the Des Moines Lobe and the Northwest Iowa Plain. D9 recurs over the Missouri River Alluvial and Loess Hills landforms. We found that D4 dominates over the Southern Iowa Drift area. In the remaining regions, we see a mix of scenarios. The spatial distribution is similar among the chosen prognostic scenarios (Figure 8b) and seems to be highly influenced by the percolation rates, represented here by tones of blue ($k_i = 0.02$), red ($k_i = 0.03$), and green ($k_i = 0.04$).

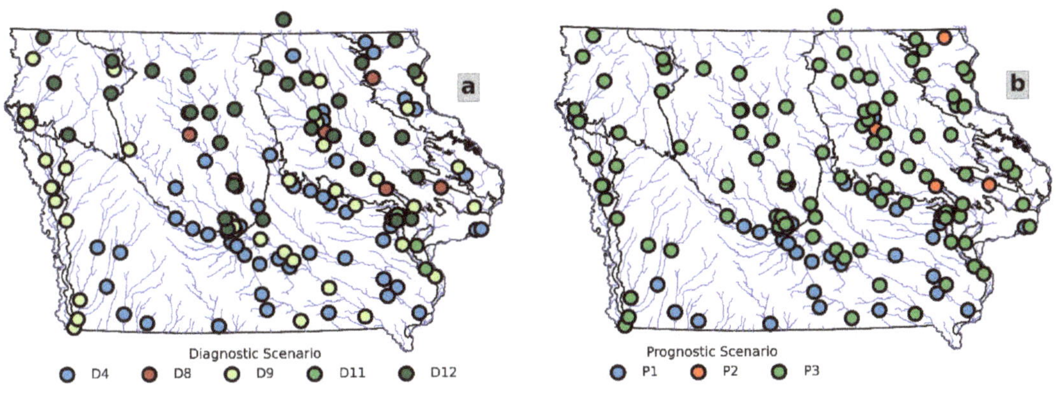

Figure 8. Spatial distribution of the scenarios with best KGE performance at each USGS station: (**a**) results obtained from the diagnostic scenarios; and (**b**) results obtained from the prognostic scenarios. The green, red, and blue gauge colors correspond to the percolation rates of 0.02, 0.03, and 0.04, respectively.

According to Figure 8, the chosen diagnostic and prognostic scenarios share percolation rates. However, differences exist in the spatial performance improvement distribution (Figure 9). Figure 9a,b show the diagnostic and prognostic scenarios of KGE improvement with respect to the linear model. With only two cases of negative KGE differences (red dots on Figure 9a), the diagnostic scenarios outperform the linear model at almost all the USGS gauges. Alternatively, in the prognostic case (Figure 9b), the count of negative KGE differences increases to 13, while the number of gauges decreases in cases where the improvement is more significant than 0.1 (yellow). We attribute the decrease in prognostic case performance to the parameter's spatial distribution.

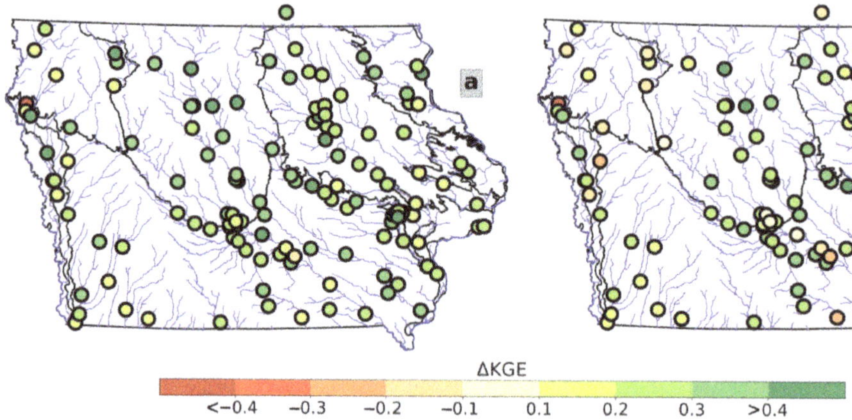

Figure 9. Mean KGE spatial difference of the diagnostic and prognostic scenarios with respect to the linear model: (**a**) diagnostic KGE minus linear model KGE; and (**b**) prognostic KGE minus linear model KGE.

The prognostic scenario performance decrease occurs mostly over the east and west regions of Iowa. The most significant decrease is observed for the Northwest Iowa Plains landform (Figure 9b). In this region, the chosen diagnostic setups were D12 and D9 (Figure 8a), suggesting a mix between tiled terrain and flat hillslopes. Over the Southern Iowan Drift landform area, the k_i value is the same for the diagnostic and prognostic scenarios. However, the prognostic scenario performance declines at several stations in this region. On the other hand, the Iowa Surface region exhibits more k_i discrepancies between both scenarios, as well as a higher number of performance differences.

The described results suggest a level of heterogeneity in the parameters shown by the diagnostic and prognostic scenarios. This heterogeneity creates difficulties when choosing the most adequate regional parameterization for the model, regardless of whether it is fixed (diagnostic) or distributed (prognostic). To address this issue, we compare the KGE (upper diagonal in Figure 10) and the mean ratio μ (lower diagonal in Figure 10) of the chosen scenarios. We estimate μ as the ratio between the simulated (μ_{sim}) and observed (μ_{obs}) flows with values near indicating a perfect match. According to Figure 10, the KGE and mean ratio of scenarios D4 and P1 outperform almost all the scenarios. Additionally, both scenarios have the highest percentage of events with KGE values above 0.4 (blue bars in Figure 10 histograms). Compared with the linear model, the D4 and P1 mean ratio correction is significant. In both plots (Linear-D4 and Linear-P1), there are almost no events where the linear setup outperformed D4 and P1. The scenarios D4 and P1 have the same k_i (0.02) value; however, their subsurface parameters are different.

The parameters of D4 are fixed for all the domains following line 4 of Figure 4a. This parameterization represents highly conductive soils or the presence of tiles. On the other hand, P1 parameters follow the hillslope steepness with Equation (9), and the presence of tiles described by the map in Figure 4b. The described differences in the parameters seem to develop slight dissimilarities in performance. According to panel D4-P1 in Figure 10, the KGE performance is similar in both, although D4 has a better performance in some events. Moreover, the panel P1-D4 shows that the mean ratio description of both setups is similar. Considering that D4 assumes tiles everywhere, our results suggest a high presence of tile-like signatures.

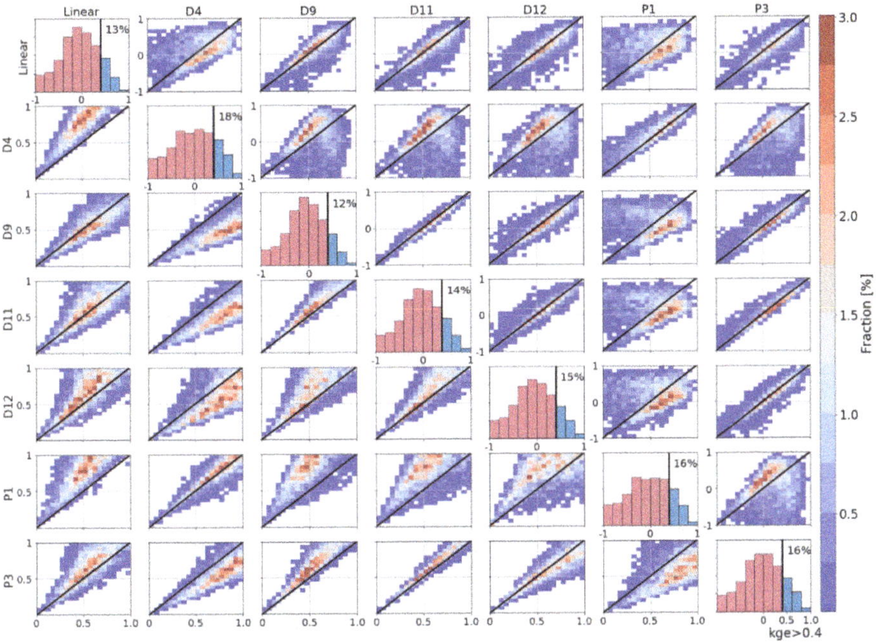

Figure 10. An event-based KGE comparison of the diagnostic and prognostic dominant scenarios. Each row compares a scenario against the others. The upper diagonal panels correspond to the KGE histogram of the scenarios. Over the diagonal the KGE histogram of each setup, a coloring in of the percentage of events with blue is visible, with a KGE of above 0.4. The lower diagonal compares the event-based mean-ratio error.

3.2. Extended Metrics

According to the diagnostic and prognostic KGE comparisons, the performance differences between the two scenarios are relatively small. However, the KGE is subject to three parameters that do not necessarily reflect all the relevant changes in the simulated streamflows. With this in mind, we also compared the NSE (Nash Sutcliffe efficiency), the hit rate, and the lags (Figure 11a–c, respectively). The NSE contrasts the simulated data prediction skill with the mean value of the observations. An NSE of below 0 indicates that the mean value performs better than the model, and an NSE of 1 indicates a perfect simulation. The hits rate is the percentage of time that is shared by observations and simulations during floods. A hit rate of zero corresponds to missing all the floods, and a hit rate of one corresponds to a perfect match. The lag is a measure of the displacements applied over the simulated data to maximize the correlation. We made hourly displacements from −48 h to 48 h. Negative lag values correspond to cases in which the simulated data exhibit responses earlier than the observed, and positive values correspond to the opposite behavior. A simulated series with good performance must have lags near zero.

Results from Figure 11 show that D4 and P1 have some similarities and some relevant differences. Regarding the NSE (Figure 11a), D4 has a slightly better score. Regarding the hits rate (Figure 11b), it is hard to tell which scenario has a better performance. Furthermore, P1 has a higher fraction of hit rates approaching one, but it also has higher frequencies at some lower intervals. The number of lags is also similar (Figure 11c). Nevertheless, P1 tends toward negative lag values more than D4 does, representing more frequent early peak estimations.

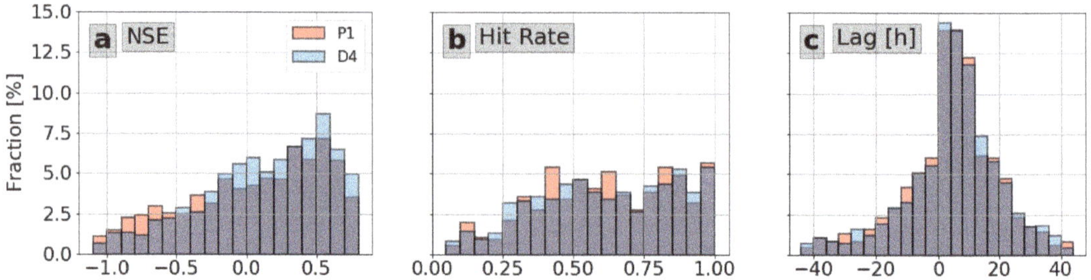

Figure 11. Histograms of performance metrics for the scenarios D4 and P1. Panel a compares the NSE. Panel (**b**) compares the Hit Rate considering (**a**) Hit when the simulated and observed streamflow are above the flood level of the gauge. Panel (**c**) compares the Lags (in hours) required to maximize the correlation between observations and simulations.

In addition to the indexes, we compare the simulated peaks of the chosen diagnostic and prognostic scenarios. Because the gauged watersheds areas range between 40 and 18,000 km^2, we performed a scale-independent comparison. To obtain scale-independent peaks (Z), we divided the peaks Q_p $[\mathrm{m}^3 \cdot \mathrm{s}^{-1}]$ by the mean annual peak $\overline{Q_p}$ $[\mathrm{m}^3 \cdot \mathrm{s}^{-1}]$. Then, for each event of each link, we computed the difference between the simulated (Z_s) and the observed (Z_o) standardized peaks. The peak flow estimation of D4 and P1 exhibited a similar performance, with D4 being superior. The D4 scenario reaches a fraction of 32% for differences near zero (Figure 12), while in P1 this value drops to 28% (Figure 12). Additionally, P1 has a higher fraction of errors equal or greater than one.

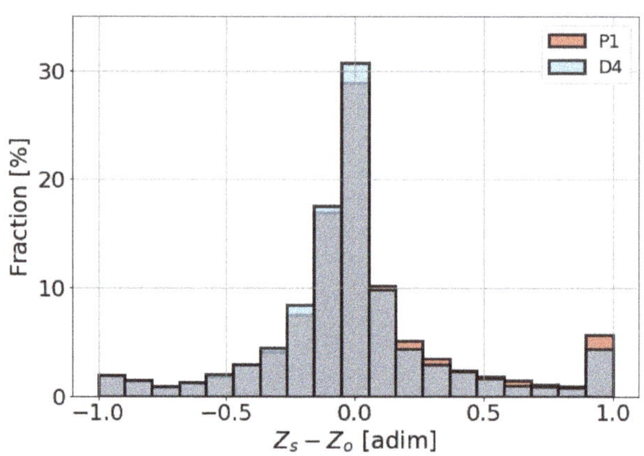

Figure 12. Histograms of the standardized peak flows difference for the D4 (blue) and P1(red) scenarios.

We expected the diagnostic superiority because, in the prognostic case, we impose restrictions based on maps. The resulting differences among simulations emphasize the parameters' relevance and the need for their correct representation. Contrasted with the diagnostic scenarios, the prognostic scenarios tend to reduce the performance. The differences between D4 and P1 suggest that the landscape descriptors could have errors that lead to decreases in the modeling performance. Additionally, our results suggest that there may be more tiles than the ones represented by the map in Figure 4b.

3.3. Analysis of Parameter Values

The diagnostic and prognostic scenarios offer different ways to determine the values of parameters in space. In the diagnostic cases, we identified the best fixed-parameter combination for each gauged watershed. In the prognostic cases, we pre-defined a set of distributed parameters based on the available information. In a previous step, we defined the best diagnostic and prognostic setup for each gauge (Figure 8a,b, respectively). According to our results, a spatial distribution of the parameters seems to be explained by k_i and the parameters of Equation (8). Additionally, the chosen diagnostic scenarios outperform the chosen prognostic ones (Figure 8a,b). In some gauges, the performance differences are small; however, in others, the difference is relatively large. This is an interesting result because the only difference between both cases is the parameterization of Equation (8). Considering the described performance differences, we explore in more detail how they are related to the parameterizations of the diagnostic and prognostic setups.

We compare Equation (8) setup for the diagnostic and prognostic scenarios to evaluate whether variations in the parameters explain the observed performance differences. We made the comparison at each gauged watershed. For the comparison, the prognostic setup has a set of curves $q_{sL}(P)$ for a given watershed (light blue lines in Figure 13), and there is one diagnostic curve $q_{sL}(D)$ for the same watershed (dark blue line in Figure 13). Using Equation (11), we compare $q_{sL}(D)$ with the 50th percentile of $q_{sL}(P)$ for storages between 1.6 and 1.7 [m] (green region in Figure 13). We choose the 1.6–1.7 range because it describes the shift the linear to the exponential expression. Besides, the hillslopes subsurface storage (S_s) is usually in this range during storm events.

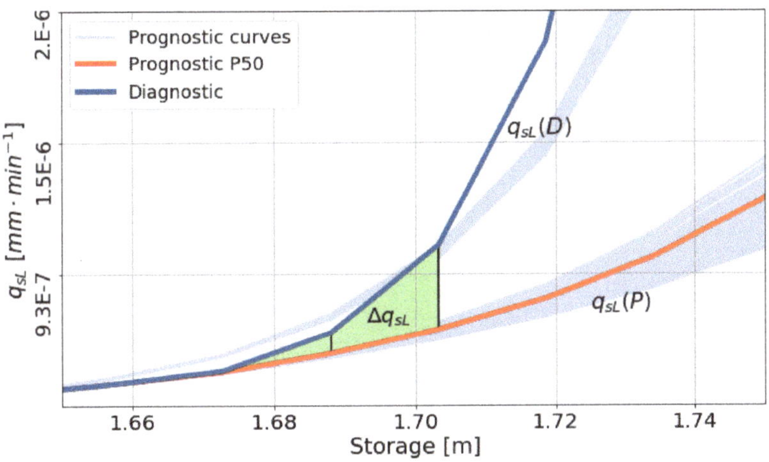

Figure 13. Example of the q_{sL} parameterization comparison between the diagnostic and prognostic scenarios. The light blue lines are q_{sL} curves of the prognostic scenario for a given watershed. The red line corresponds to the 50th percentile of the prognostic q_{sL} curves. The dark blue line is the q_{sL} curve of the diagnostic scenario. We used the green region to perform a comparison between the scenarios.

$$\Delta q_{sL} = \frac{1}{N} \cdot \sum_{i}^{N} \frac{q_{sL}(D)_i - P50(q_{sL}(P))_i}{q_{sL}(D)_i} \quad (11)$$

According to Figure 14, the differences of the parameters (Δq_{sL}) do not adequately explain the performance differences between the diagnostic and the prognostic scenarios. In some cases, low absolute values of Δq_{sL} are linked to low KGE differences. However, the described behavior does not apply for large absolute values of Δq_{sL}. Figure 14 shows many watersheds in which the differences of the absolute parameters are larger than 10% (x-axis), while the KGE absolute differences are low. Additionally, some cases with low absolute Δq_{sL} exhibit large KGE differences. On the other hand, according to the colors of Figure 14 (non-absolute Δq_{sL}), positive values of Δq_{sL} are related to low KGE differences; and negative values of Δq_{sL} correspond to high KGE differences.

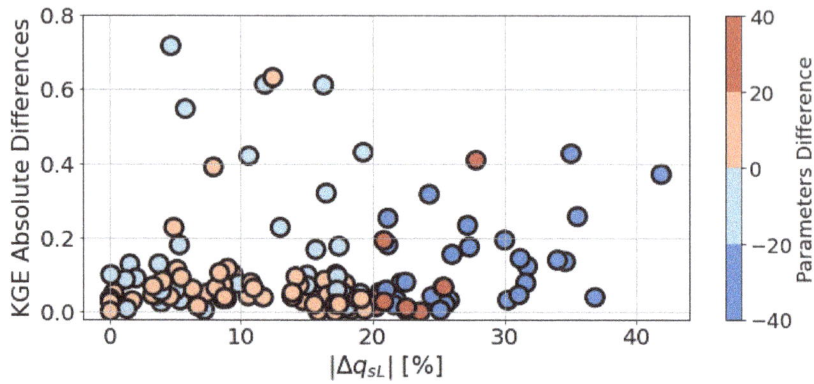

Figure 14. Parameters' absolute differences vs. KGE differences at the USGS gauges. The colors correspond to non-absolute differences in the parameters.

We also compared the Δq_{sL} and the KGE differences in space. According to Figure 15, the coincidences between the KGE and Δq_{sL} do not show a strong regional pattern. We observe some similarities only in the Des Moines Lobe and the Northwest Iowa Plains regions. In both cases, some significant KGE differences correspond with large absolute Δq_{sL} values. Additionally, there is a match between low KGE and Δq_{sL} differences in the Iowan Surface region, with some exceptions.

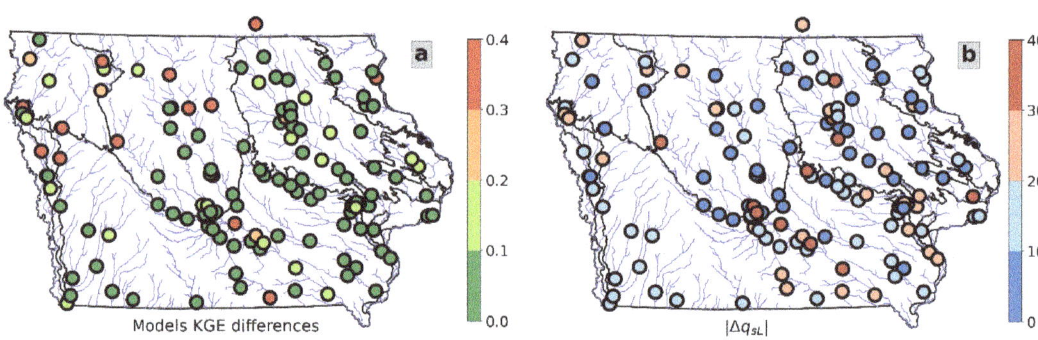

Figure 15. Maps showing KGE differences (**a**), and Δq_{sL} (**b**), in USGS gauges.

It is hard to establish a relationship between the diagnostic and prognostic parameters and their performance differences. We attribute this lack of correlation to the model's non-linear transformations at the hillslope scale and throughout the network. It is expected that the parameters would alter the model's output. However, our results show that a pre-defined distribution of the parameters could lead to modeling errors that are hard to identify.

4. Conclusions

The Iowa Flood Center (IFC) has been making operational flood forecasts for the state of Iowa since 2010. IFC forecasters use the hydrological Hillslope Link Model (HLM), along with rainfall data. The HLM has been accurate in forecasting streamflow fluctuations at several scales [14]. However, the model has limitations in its representation of the recession curve, and it underestimates the total streamflow volume. Moreover, the limitations seem to increase in a tiled landscape. The authors of [17] attributed these limitations to the linear equation HLM uses to represent the subsurface flux and the lack of an equation representing the tiled terrain. To address this issue, [15] developed an exponential equation that can be parameterized to represent the function of the hillslope steepness and the presence of tiles.

This paper evaluated the exponential equation proposed by [17], which represents subsurface hillslope–link interaction in HLM. The equation can represent hillslopes with and without tile drainage. We performed the equation evaluation at 140 USGS gauges in Iowa. The analysis used hourly records between 2002 and 2018. In the evaluation, we compared the exponential equation with a linear equation. The comparison used a diagnostic and a prognostic approach to establish the parameters. In the diagnostic setup, we implemented 12 fixed parameter scenarios, while in the prognostic setup, we distributed the parameters with consideration of the hillslope steepness and presence of tiles. In both cases, we considered three fixed percolation rates. Results from this study indicate the following:

- Compared with the linear equation, the exponential equation corrects the volume bias on the simulated streamflow. We attribute the correction to the active layer threshold on the exponential equation and the significant outflow increase once the storage is above this threshold. In contrast, in the linear equation, the water remains in the soil for extended periods because of the described absence of these processes.
- Depending on the parameters, the exponential equation could improve the performance of HLM. We found that the exponential equation outperforms the linear equation for several parameter combinations with changes in the shape of the hydrograph, the simulated peaks, and timing. We also found significant differences using different combinations of the equation parameters and the percolation rate.
- The percolation rate plays a significant role in the representation of the subsurface flux from the described combinations. We found spatial coincidences in the percolation rates when choosing the best diagnostic and prognostic scenarios. Additionally, the percolation rate induces changes comparable to those produced by the exponential equation's parameters.
- Determining the distributed parameters of HLM remains challenging. In this paper, we used the diagnostic and prognostic approaches to analyze the parameters of HLM. The diagnostic approach assumes unknown conditions and fixed parameters over the space. On the other hand, the prognostic method is the more classical approach, in which the parameters are derived from maps of the landscape. In our experiments, the diagnostic setups tended to outperform the prognostic setups. Additionally, had difficulty in identifying a link between the diagnostic and prognostic parameters and their respective performances.

In the current work, we showed that a better representation of the processes and the correct parameters can improve a hydrological model. The improvement is supported by comparisons performed at 140 USGS gauges. Moreover, the differences between the diagnostic and prognostic setups suggest that identifying the parameters is still challenging. Despite the limitation related to the number of gauges, the diagnostic approach reveals the parameters' potential spatial distribution.

Two main factors may explain the differences in parameters and performance between the diagnostic and prognostic setups: errors in the landscape description and unrepresented processes in HLM. Uncertainties exist in the tile localization maps; likewise, limitations exist in the representation of the average steepness at the hillslope scale. On the other hand, we unrepresented processes in some regions of Iowa, such as potholes over the northwest and agricultural terraces in the west. It is difficult to identify which of these factors is more relevant to the implementation of a hydrological model. However, according to our results, the use of maps as landscape descriptors may lead to the detection of errors that are usually hidden in a posterior calibration process. Moreover, we found it difficult to identify the errors caused by the prescribed distributed parameters. Both issues could be addressed using diagnostic setups that help to identify the uncertainties derived from the parameters and their possible regional distributions.

Author Contributions: Conceptualization, N.V., R.M., W.K. and M.F.; Methodology, N.V., R.M., W.K. and M.F.; Writing—original draft, N.V., R.M., M.F. and F.Q. All authors have read and agreed to the published version of the manuscript.

Funding: This research was funded by MID-America Transportation Center (Grant number: 69A3551747107), the Iowa Highway Research Board and Iowa Department of Transportation (Contract number: TR-699).

Data Availability Statement: The data supporting the results of this work can be found in the following repository.

Acknowledgments: This work was completed with support from the Iowa Flood Center, MID-America Transportation Center (Grant number: 69A3551747107), the Iowa Highway Research Board and Iowa Department of Transportation (Contract number: TR-699).

Conflicts of Interest: The authors declare no conflict of interest.

References

1. Samaniego, L.; Kumar, R.; Attinger, S. Multiscale parameter regionalization of a grid-based hydrologic model at the mesoscale. *Water Resour. Res.* **2010**, *46*, 1–25. [CrossRef]
2. Mandeville, A.N. Insights gained from four component hydrograph separation. *Hydrol. Res.* **2016**, *47*, 606–618. [CrossRef]
3. Chen, B.; Krajewski, W.F. Recession analysis across scales: The impact of both random and nonrandom spatial variability on aggregated hydrologic response. *J. Hydrol.* **2015**, *523*, 97–106. [CrossRef]
4. Clark, M.P.; Rupp, D.E.; Woods, R.A.; Tromp-van Meerveld, H.J.; Peters, N.E.; Freer, J.E. Consistency between hydrological models and field observations: Linking processes at the hillslope scale to hydrological responses at the watershed scale. *Hydrol. Process.* **2009**, *23*, 311–319. [CrossRef]
5. Harman, C.J.; Sivapalan, M.; Kumar, P. Power law catchment-scale recessions arising from heterogeneous linear small-scale dynamics. *Water Resour. Res.* **2009**, *45*, 1–13. [CrossRef]
6. Tallaksen, L.M. A review of baseflow recession analysis. *J. Hydrol.* **1995**, *165*, 349–370. [CrossRef]
7. Biswal, B.; Marani, M. Geomorphological origin of recession curves. *Geophys. Res. Lett.* **2010**, *37*, 1–5. [CrossRef]
8. Shaw, S.B.; Riha, S.J. Examining individual recession events instead of a data cloud: Using a modified interpretation of dQ/dt-Q streamflow recession in glaciated watersheds to better inform models of low flow. *J. Hydrol.* **2012**, *434–435*, 46–54. [CrossRef]
9. Zhang, Z.; Li, Y.; Wang, X.; Li, H.; Zheng, F.; Liao, Y.; Tang, N.; Chen, G.; Yang, C. Assessment of river health based on a novel multidimensional similarity cloud model in the Lhasa River, Qinghai-Tibet Plateau. *J. Hydrol.* **2021**, *603*, 127100. [CrossRef]
10. Schilling, K.E.; Gassman, P.W.; Arenas-Amado, A.; Jones, C.S.; Arnold, J. Quantifying the contribution of tile drainage to basin-scale water yield using analytical and numerical models. *Sci. Total Environ.* **2019**, *657*, 297–309. [CrossRef] [PubMed]
11. Schilling, K.E.; Helmers, M. Effects of subsurface drainage tiles on streamflow in Iowa agricultural watersheds: Exploratory hydrograph analysis. *Hydrol. Processes Int. J.* **2008**, *4506*, 4497–4506. [CrossRef]
12. Mantilla, R.; Gupta, V.K. A GIS Numerical Framework to Study the Process Basis of Scaling Statistics in River Networks. *October* **2005**, *2*, 404–408. [CrossRef]
13. Demir, I.; Krajewski, W.F. Towards an integrated Flood Information System: Centralized data access, analysis, and visualization. *Environ. Model. Softw.* **2013**, *50*, 77–84. [CrossRef]
14. Krajewski, W.F.; Ceynar, D.; Demir, I.; Goska, R.; Kruger, A.; Langel, C.; Mantilllla, R.; Niemeier, J.; Quintero, F.; Seo, B.C.; et al. Real-time flood forecasting and information system for the state of Iowa. *Bull. Am. Meteorol. Soc.* **2017**, *98*, 539–554. [CrossRef]
15. Quintero, F.; Krajewski, W.F.; Seo, B.C.; Mantilla, R. Improvement and evaluation of the Iowa Flood Center Hillslope Link Model (HLM) by calibration-free approach. *J. Hydrol.* **2020**, *584*, 124686. [CrossRef]
16. Gupta, H.V.; Kling, H.; Yilmaz, K.K.; Martinez, G.F. Decomposition of the mean squared error and NSE performance criteria: Implications for improving hydrological modelling. *J. Hydrol.* **2009**, *377*, 80–91. [CrossRef]
17. Fonley, M.R.; Qiu, K.; Velásquez, N.; Haut, N.K.; Mantilla, R. Development and Evaluation of an ODE Representation of 3D Subsurface Tile Drainage Flow Using the HLM Flood Forecasting System. *Water Resour. Res.* **2021**, *57*, e2020WR028177. [CrossRef]
18. Clark, M.P.; Kavetski, D.; Fenicia, F. Pursuing the method of multiple working hypotheses for hydrological modeling. *Water Resour. Res.* **2011**, *47*, 1–16. [CrossRef]
19. Sur, C.; Park, S.Y.; Kim, J.S.; Lee, J.H. Prognostic and diagnostic assessment of hydrological drought using water and energy budget-based indices. *J. Hydrol.* **2020**, *591*, 125549. [CrossRef]
20. Kalma, J.D.; McVicar, T.R.; McCabe, M.F. Estimating land surface evaporation: A review of methods using remotely sensed surface temperature data. *Surv. Geophys.* **2008**, *29*, 421–469. [CrossRef]
21. Allen, R.G.; Pereira, L.S.; Howell, T.A.; Jensen, M.E. Evapotranspiration information reporting: I. Factors governing measurement accuracy. *Agric. Water Manag.* **2011**, *98*, 899–920. [CrossRef]
22. Yilmaz, M.T.; Anderson, M.C.; Zaitchik, B.; Hain, C.R.; Crow, W.T.; Ozdogan, M.; Chun, J.A.; Evans, J. Comparison of prognostic and diagnostic surface flux modeling approaches over the Nile River basin. *Water Resour. Res.* **2014**, *50*, 386–408. [CrossRef]

23. Crow, W.T.; Li, F.; Kustas, W.P. Intercomparison of spatially distributed models for predicting surface energy flux patterns during SMACEX. *J. Hydrometeorol.* **2005**, *6*, 941–953. [CrossRef]
24. Lin, Y. GCIP/EOP Surface: Precipitation NCEP/EMC 4KM Gridded Data (GRIB) Stage IV Data. UCAR/NCAR-Earth Observing Laboratory. Available online: https://data.eol.ucar.edu/dataset/21.006 (accessed on 30 November 2021).
25. Running, S.; Mu, Q.; Zhao, M. MOD16A2 MODIS/Terra Net Evapotranspiration 8-Day L4 Global 500 m SIN Grid V006. 2017. Available online: https://lpdaac.usgs.gov/products/mod16a2v006/ (accessed on 30 November 2021).

hydrology MDPI

Article

Improving Operational Short- to Medium-Range (SR2MR) Streamflow Forecasts in the Upper Zambezi Basin and Its Sub-Basins Using Variational Ensemble Forecasting

Rodrigo Valdés-Pineda [1,*], Juan B. Valdés [1], Sungwook Wi [2], Aleix Serrat-Capdevila [3] and Tirthankar Roy [4]

[1] Department of Hydrology and Atmospheric Sciences, University of Arizona, J W Harshbarger Bldg., 1133 James E. Rogers Way #122, Tucson, AZ 85721, USA; jvaldes@arizona.edu
[2] Department of Civil and Environmental Engineering, University of Massachusetts, Amherst, MA 01003, USA; sungwookwi@ecs.umass.edu
[3] Global Water Practice, The World Bank, Washington, DC 20433, USA; aserratcapdevila@worldbank.org
[4] Department of Civil and Environmental Engineering, College of Engineering, University of Nebraska, 900 N. 16th St., Lincoln, NE 68588, USA; roy@unl.edu
* Correspondence: rvaldes@arizona.edu

Abstract: The combination of Hydrological Models and high-resolution Satellite Precipitation Products (SPPs) or regional Climatological Models (RCMs), has provided the means to establish baselines for the quantification, propagation, and reduction in hydrological uncertainty when generating streamflow forecasts. This study aimed to improve operational real-time streamflow forecasts for the Upper Zambezi River Basin (UZRB), in Africa, utilizing the novel Variational Ensemble Forecasting (VEF) approach. In this regard, we describe and discuss the main steps required to implement, calibrate, and validate an operational hydrologic forecasting system (HFS) using VEF and Hydrologic Processing Strategies (HPS). The operational HFS was constructed to monitor daily streamflow and forecast them up to eight days in the future. The forecasting process called short- to medium-range (SR2MR) streamflow forecasting was implemented using real-time rainfall data from three Satellite Precipitation Products or SPPs (The real-time TRMM Multisatellite Precipitation Analysis TMPA-RT, the NOAA CPC Morphing Technique CMORPH, and the Precipitation Estimation from Remotely Sensed data using Artificial Neural Networks, PERSIANN) and rainfall forecasts from the Global Forecasting System (GFS). The hydrologic preprocessing (HPR) strategy considered using all raw and bias corrected rainfall estimates to calibrate three distributed hydrological models (HYMOD_DS, HBV_DS, and VIC 4.2.b). The hydrologic processing (HP) strategy considered using all optimal parameter sets estimated during the calibration process to increase the number of ensembles available for operational forecasting. Finally, inference-based approaches were evaluated during the application of a hydrological postprocessing (HPP) strategy. The final evaluation and reduction in uncertainty from multiple sources, i.e., multiple precipitation products, hydrologic models, and optimal parameter sets, was significantly achieved through a fully operational implementation of VEF combined with several HPS. Finally, the main challenges and opportunities associated with operational SR2MR streamflow forecasting using VEF are evaluated and discussed.

Keywords: variational ensemble forecasting; hydrologic processing strategies or hypotheses; SR2MR streamflow forecasting; real-time hydrologic forecasting system; satellite precipitation products; multi models; best streamflow prediction; inverse variance weighting; inverse probability weighting

Citation: Valdés-Pineda, R.; Valdés, J.B.; Wi, S.; Serrat-Capdevila, A.; Roy, T. Improving Operational Short- to Medium-Range (SR2MR) Streamflow Forecasts in the Upper Zambezi Basin and Its Sub-Basins Using Variational Ensemble Forecasting. *Hydrology* **2021**, *8*, 188. https://doi.org/10.3390/hydrology8040188

Academic Editor: Andrea Petroselli

Received: 2 November 2021
Accepted: 15 December 2021
Published: 20 December 2021

1. Introduction

1.1. Decisions and Limitations of Hydrologic Forecasting

At any spatiotemporal scale, critical decisions about the design, functionality, and operability of a Hydrologic Forecasting System (HFS) need to be made to reduce the total hydrologic uncertainty (THU) propagated from different components of a hydrologic modelling paradigm (HMP). In fact, reducing the total hydrological uncertainty is key to developing reliable Integrated Water Resources Management (IWRM) strategies across spatial and temporal scales. For river basins across the world, the allocation of water resources largely relies on accurate streamflow forecasts. In Africa, for instance, the waters of the Upper Zambezi River Basin (UZRB) are shared by eight countries: Angola, Namibia, Zambia, Botswana, Malawi, Tanzania, Zimbabwe, and Mozambique. However, the administrative complexities created by the transnational nature of the Zambezi Basin (Figure 1) result in inconsistencies in the operation and maintenance of the rain gauges and stream gauges (see Table 1), and consequently a lack of reliable hydrologic data for the implementation of an HFS. For example, rainfall or streamflow time series with missing records can undermine the effectiveness of calibration and validation schemes, consequently increasing the propagation of total hydrological (meteorological) uncertainty for the establishment of hydrologic processing strategies. Therefore, an appropriate identification and quantification of uncertainty (at any level) can help reducing the THU for the final development of streamflow forecast products.

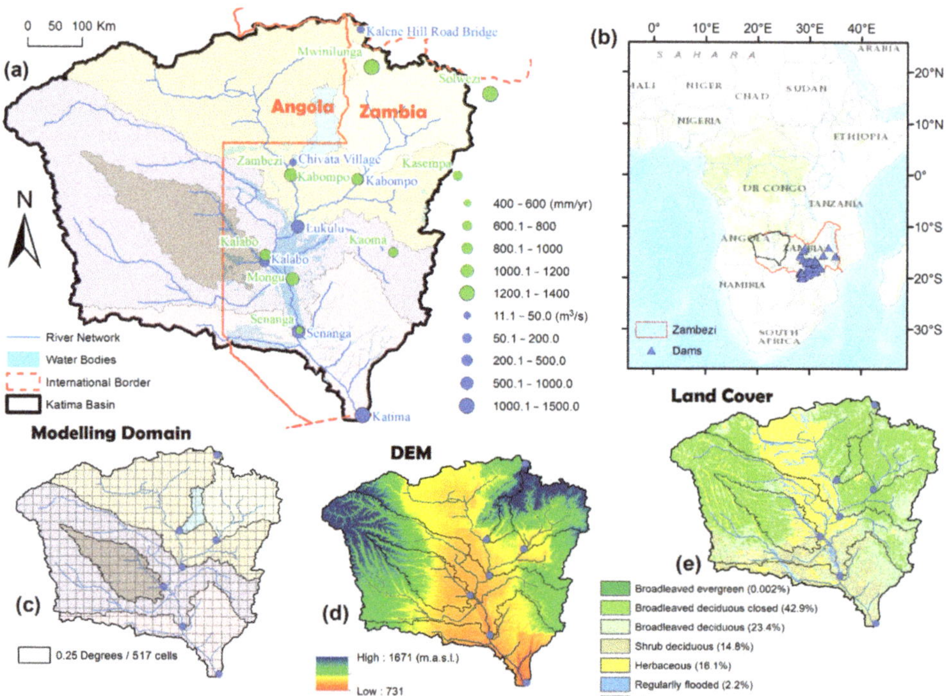

Figure 1. (**a**) Upper Zambezi River Basin (UZRB) delineated above the Katima Mulilo streamgauge. The green markers represent 9 rain gauges available in the basin. The blue markers represent the streamgauges used in this study; (**b**) Location of the Zambezi Basin in the African continent. The map also shows the location of major hydropower and water storage projects; (**c**) Modelling domain selected to implement the real-time HFS (RT-HFS). The modelling domain was set up using grid cells at 0.25° of spatial resolution; (**d**) Land cover map based on [1]. The basin is dominated by broadleaved trees (~66%), herbaceous (16.1%), and shrubs (14.8%), whereas only a little (~0.6%) of the area is managed or represents agricultural; and (**e**) Digital Elevation Model (DEM) based on Hydrosheds (90 m resolution). The spatial distribution of the vegetation types is consistent with the elevational pattern of the basin, which ranges between approximately 731 and 1671 m above sea level [2].

1.2. Knowledge Gaps and Justification of the Study

The first HFS for Zambezi was applied in 2011, where the use of the Kalman filter was combined with a simple two-layer conceptual hydrological model to forecast streamflow in three sub-basins [3]. Satellite-based soil moisture estimates were used to calibrate the aggregated hydrologic model, which was able to generate daily streamflow forecasts up to 40 days into the future. Meier [4] argued that the spatial resolution of the satellite data needed to be improved together with the implementation of more sophisticated hydrologic models. In 2014, the daily floodplain behavior of the Zambezi was simulated by [5] applying a modified reservoir approach for the SWAT model [6]. Their results showed that the modified version of SWAT improved the simulation of daily streamflow and floodplain development in the Zambezi basin. Several other hydrological models have been satisfactorily calibrated and validated in other sub-basins of the Zambezi with available records (i.e., [7–12]). Despite all these modelling efforts in the Zambezi Basin, an operational HFS for the undisturbed flows of the poorly gauged UZRB has yet to be established. The primary objective of an operational HFS is to generate an accurate short- to medium-range (SR2MR) streamflow forecast that can inform water distribution schemes at the relevant spatial and temporal scales. The forecasts can be obtained from multiple ensembles constructed from Variational Ensembles Forecasting (VEF) approaches.

1.3. Variational Ensemble Forecasting (VEF) to Improve Operational Streamflow Forecasts

An operational HFS can only be implemented if the realtime and short-term forecasts of the input data (e.g., rainfall, temperature, etc.) are readily available. Therefore, the use of multiple satellite precipitation products (or regional climate models), combined with multiple conceptual and physic-based hydrologic models, can provide important insights about the practical and scientific aspects of implementing operational HFS in poorly gauged basins [13,14]). In this regard, many Variational Ensemble Forecasting (VEF) algorithms have been proposed to improve the representation of the components included in a Hydrologic Forecasting System (HFS). Previous studies have applied VEF based on multiproduct, multimodel, or multi-initialization schemes (see for example [13–21]). However, the evaluation of VEF including an additional dimension with multiple optimal parameter sets has not been explored in simulation schemes or operational forecasting yet' neither has the role of Hydrologic Processing Strategies to improve the assimilation of VEF applications in an operational HFS context.

1.4. Purpose of This Paper

The main motivation of this study is to describe, analyze, and discuss the main steps required to design and implement an HFS, aimed to improve SR2MR streamflow forecasts in the UZRB and its sub-basins during the operational stage. The technical functionality and operability of the HFS is assessed using Hydrologic Processing Strategies (HPS) within the context of a Variational Ensemble Forecasting (VEF) hydrologic modelling paradigm [22], i.e., optimal combination of multiple precipitation products, multiple hydrologic models, and multiple parameters sets. The THU propagated by the HFS is evaluated and compared according to different sources of uncertainty, i.e., satellite-based, or model-based rainfall estimates, hydrological models, and optimal parameters. As detailed later in this paper, a more comprehensive overview of the whole modeling paradigm used to implement an operational HFS can help significantly to generate an improved streamflow forecast products that are closer to the hydrological conditions of the basin under study.

2. Methods

2.1. The Upper Zambezi River Basin (UZRB) Domain

The operational HFS implementation for the UZRB and its sub-basins with records (Table 1), considered the drainage area (~339,521 km^2) delineated above the Katima Mulilo streamgauge (Table 1 and Figure 1a). The mean annual streamflow at the UZRB (measured at Victoria Falls) represents about 25% of the mean annual streamflow measured at the

Zambezi Delta outlet (~4200 m^3/s), the largest contribution of all tributaries within the whole Zambezi Basin [23,24]. The historic mean daily flows (1942–2017) measured at Katima Mulilo stream gauge were about 1174 m^3/s (Table 1); however, during extreme episodes the maximum daily streamflows can exceed more than six times the mean daily streamflows. The basin is dominated by broadleaved trees (~66%), herbaceous (16.1%), and shrubs (14.8%), whereas only a little (~0.6%) of the area is managed or represents agriculture (Figure 1c). The spatial distribution of these vegetation types is consistent with the elevational pattern of the basin, which ranges between approximately 731 and 1671 m above sea level (Figure 1d). The UZRB domain for the operational HFS implementation was set up using grid cells at 0.25° of spatial resolution, approximately 25 km at the Equator (Figure 1e). This area above Katima Mulilo was selected, because it is the unique portion of the Zambezi that does not have ongoing hydropower or water storage infrastructure projects (see Figure 1b). This is an advantage for HFS implementation since the presence of dams or any other anthropogenic water regulations can be an important limitation when the primary objective is to simulate natural streams.

Table 1. Description of streamgauges use in this study. The HFS column shows those streamgauges (sub-basins) used for the calibration of hydrologic models.

Country	Streamgauge	South Latitude	East Longitude	Area (km^2)	Altitude (m.a.s.l.)	Average Flow (m^3/s)	Period	Missing (%)	HFS
Zambia	Kalene Hill Road Bridge	−11.13	24.25	764	1261	12.3	1977–2004	34.81	No
Zambia	Chivata Village	−13.33	23.15	3354	1065	17.4	1962–2004	23.32	Yes
Zambia	Luanginga-Kalabo	−14.96	22.68	34,621	1021	59.0	1958–2004	8.94	Yes
Zambia	Kabompo Pontoon	−13.60	24.21	42,740	1029	252.2	1990–2005	51.04	Yes
Zambia	Lukulu	−14.38	23.233	206,531	1012	772.0	1950–2004	12.44	Yes
Zambia	Senanga	−16.11	23.25	284,538	992	972.6	1947–2004	8.40	Yes
Namibia	Katima Mulilo	−17.48	24.3	339,521	746	1174.5	1942–2017	13.54	Yes

2.2. *Forecasting Timescales and Water Management Activities*

An HFS can be implemented for either of the three (or a combination thereof) principal forecasting timescales: (1) Realtime Monitoring or Short- to Medium-Range Forecasting (SR2MR), (2) Subseasonal to Seasonal or Short- to Long-Range Forecasting (SR2LR), or (3) Climate Change Predictions. The choice of forecasting timescales is partly determined by the relevant water management goals, and the needs of end users of the forecasts. In this study, the need for better water management schemes for flood warning and water allocation in the lower Zambezi Basin prompted the implementation of an HFS for realtime streamflow monitoring and short- to medium-range (SR2MR) forecasting in the UZRB and its sub-basins. As mentioned in the previous section, the HFS was implemented to forecast streamflow in the UZRB because it is the only part of the whole Zambezi River Basin without water regulation infrastructure, i.e., natural streamflow patterns can be observed in the UZRB. The SR2MR scheme allows streamflow forecasting from realtime up to eight days into the future (Figure 2).

2.3. *The Operational Context of a Hydrologic Modeling Paradigm*

A functional HFS requires the design of a hydrologic modelling paradigm under an operational scheme. The design can range from a simple propagation of meteorological forecasts using a single hydrological model to more advanced techniques which can include multiparameter, multimodel, or multi-initialization schemes (Figure 3). The use of a simple propagation scheme does not allow for the quantification of the hydrological uncertainty propagated from the model structure or from the model parameters. However, by adding more parameter sets for a single hydrological model (i.e., [15]), by adding more hydrological models [13,14,16–18], or simply by performing more initializations of the initial starting conditions of a single model [19,20], the hydrological uncertainty either from the parameters, model assumptions, or initial conditions can be quantified.

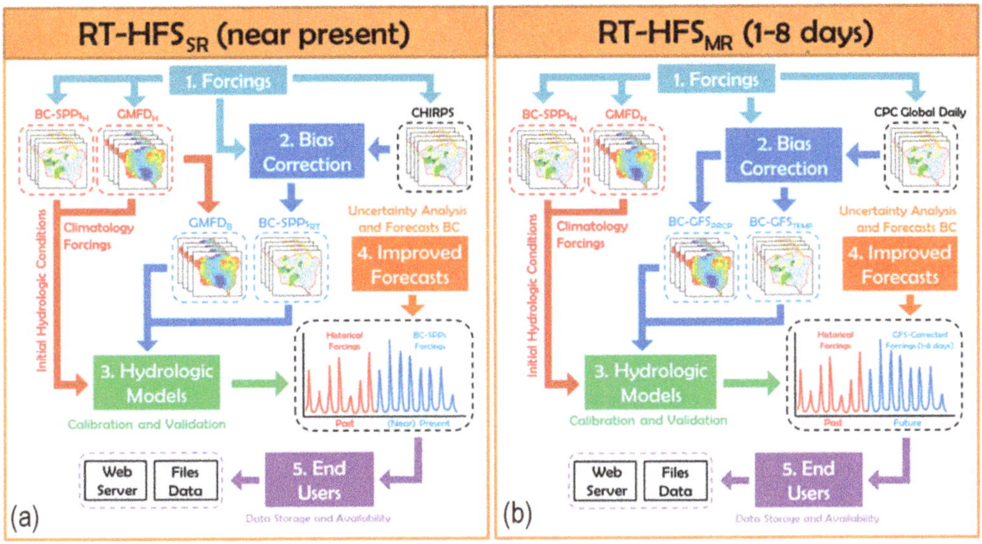

Figure 2. Structure and major modules required to implement an operational Realtime Hydrologic Forecasting System (RT-HFS) at (**a**) short-range timescales RT-HFS$_{SR}$ and at (**b**) medium-range timescales RT-HFS$_{MR}$.

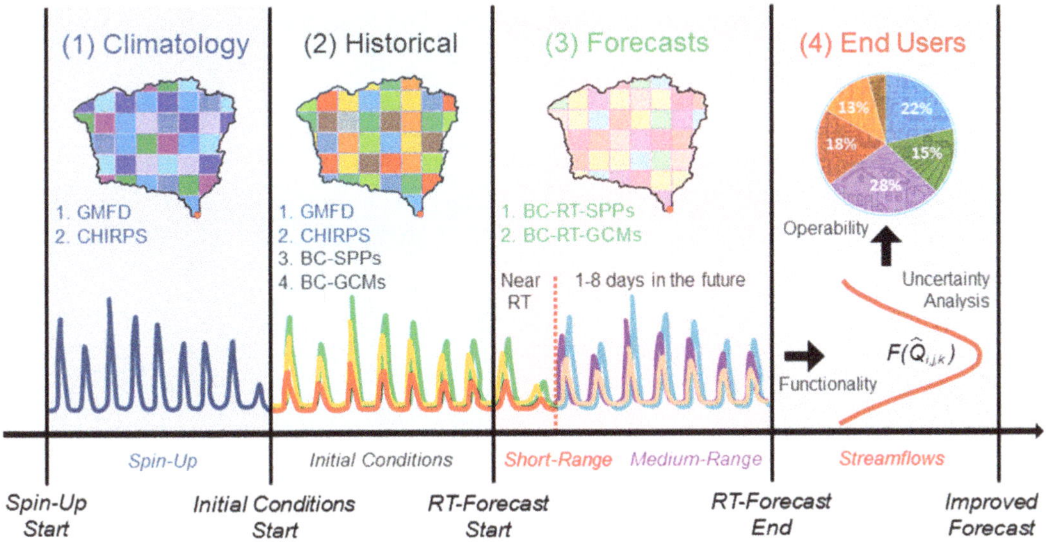

Figure 3. Simplified hydrologic modelling paradigm (HMP) used in the operational HFS implementation for the UZRB and its sub basins. Each of the hydrologic processing strategies (HPS) can help quantifying the propagation of total hydrological uncertainty (THU) from different sources, i.e., input data, model structures and parameters, modelling assumptions, and initial conditions of the models, etc.

In the UZRB, the operational HFS is designed as a VEF modelling paradigm [22] in which the best SR2MR forecasts are derived from the combination of multiple precipitation products, multiple hydrologic models, and multiple parameters sets (Figure 3). This technique makes it possible to increase the range of possible streamflow forecast ensembles that can be used to evaluate and select the best representations of the hydrological states and fluxes for the UZRB and its sub-basins.

2.4. Selection of Hydro Climatological Forcings

The input data can have significant impacts on the propagation of meteorological and total hydrological uncertainty, and consequently, on the final streamflow forecasts. In this regard, it is well known that rainfall is the most important variable for streamflow simulation. At this stage, either point-based instrumental records or gridded-based climatology products can be used to establish baseline conditions and to correct SPPs or RCMs. The UZRB is a poorly gauged basin that lacks a consistent hydrometric network with continuous records (only nine rain gauges with discontinuous records were available for the HFS implementation). Therefore, the hydrological preprocessing hypothesis (also known as meteorological postprocessing hypothesis) [22] was approximated using rainfall climatology provided by CHIRPS (Climate Hazards Group InfraRed Precipitation with Station data) [25,26], and temperature climatology provided by the Global Meteorological Forcing Dataset (GMFD) [27].

To minimize the probability of errors propagated during the implementation of the operational HFS, the quality of the climatological forecast products used in an HFS needs to be evaluated and validated before they can be reliably used for hydrological applications. In the UZRB, three SPPs were evaluated, corrected, and then used to provide short-range (realtime) rainfall estimates (1) TMPA-RT [28]; (2) CMORPH [29]; and (3) PERSIANN [30,31]. To complete the SR2MR scheme, medium-range rainfall forecasts from the Global Forecasting System (GFS) [32] were used. The original GFS product provides rainfall forecasts in six-hour intervals (00, 06, 12, and 18 UTC) up to 16 days into the future. However, since the quantity of missing records between 9 and 16 days in the archives is larger (at least in this region), only rainfall forecasts provided at 00 UTC and up to eight days into the future were selected for operational implementation. (Table 2).

Table 2. Climatology data used to correct (preprocess) rainfall forecasts from SPPs and RCMs.

Product	Institution	Spatial Resolution	Temporal Resolution	Global Coverage	Period
CHIRPS [1] **[26]**	UCSB	0.25° × 0.25°	Daily	50° N–50° S 180° E–180° W	1981 to present
GMFD [2] **[27]**	Princeton	0.25° × 0.25°	Daily	50° N–50° S 180° E–180° W	1981–2012
PERSIANN-CCS [3] **[31]**	UCI	0.25° × 0.25°	3-hourly	37.8° N–40.6° S 28.0° W–56.2° E	1998 to present
CMORPH [4] **[29]**	NOAA-CPC	0.25° × 0.25°	3-hourly	60° N–60° S 180° E–180° W	1998 to present
TMPA-RT [5] **[28]**	NASA GES DISC	0.25° × 0.25°	3-hourly	50° N–50° S 180° E–180° W	1998 to present
GFS [6] **[32]**	NOAA-NCEI	0.25° × 0.25°	3-hourly	90° N–90° S 180° E–180° W	2014 to present
Katima Mulilo	GRDC	Streamgauge	Daily	17.48° S–24.3° W	1942 to present

[1] Climate Hazards Infrared Precipitation with Station data (CHIRPS). [2] Global Meteorological Forcing Dataset (GMFD) [3] Precipitation Estimation from Remotely Sensed Information using Artificial Neural Networks-Cloud Classification System (PERSIANN) [4] Climate Prediction Center morphing method (CMORPH) [5] Tropical Rainfall Measuring Mission (TRMM) Multi-satellite Precipitation Analysis-Real Time [6] Global Forecast System (GFS).

2.5. Hydrologic Models for Operational HFS

The selection of a model structure to be included in the hydrologic modelling paradigm (HMP) and ultimately in the operational HFS is as important as the selection of the model parameters to be used in the calibration process [16]. For an operational HFS, many HMP options can be implemented, including a single selection or an optimal combination of multiple precipitation or climate products, multiple hydrological models, multiple model state initializations, and/or multiple parameters sets. For example, the operational VEF approach implemented in the UZRB, and its sub-basins utilized three distributed hydrologic models (at 0.25° of spatial resolution): (1) HBV_DS, (2) HYMOD_DS, and (3) VIC 4.2.b (Figure 4). The first two models are distributed versions of two traditional well-known hydrological models: HBV [33,34] and HYMOD [35,36]. The VIC model used in this study is a modified version of the well-known Variable Infiltration Capacity (VIC) land surface model [37] that can resolve both the water and the energy balances. The modification allows postprocessing of VIC model outputs with the Lohmann's model for routing [38]; a Gamma distribution to represent the catchment's unit hydrograph; and the linearized version of the Saint-Venant Equations for final river routing. Additional details of the model states, fluxes, and parameters used in this study are provided in the Appendix A.

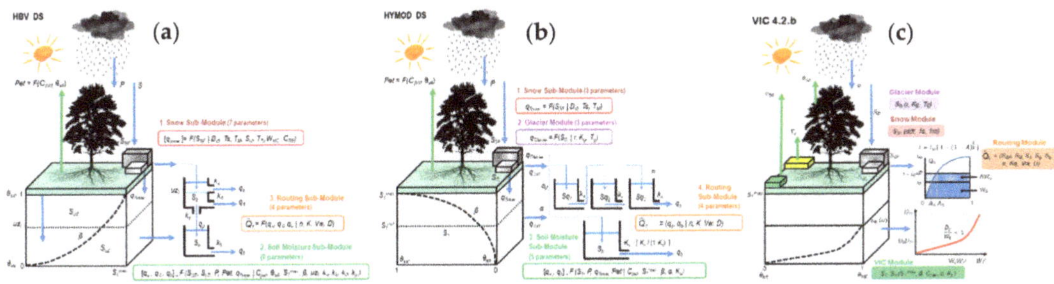

Figure 4. Modelling structures used in the operational HFS implementation for the UZRB. (**a**) HBV_DS, (**b**) HYMOD_DS, and (**c**) VIC 4.2.b. Details about model states, fluxes, and parameters are provided in the Appendix A.

2.6. Calibration of Models Included in the HFS

The models described in the previous section were calibrated and validated for the whole UZRB and its sub-basins at 0.25 degrees of spatial resolution (approximately 25 km at the Equator). Historical available daily streamflow records for the UZRB and its sub-basins were used as reference data (Figures 5 and 6). The genetic algorithm [39] was used to optimize the parameter sets of the three models. However, at this stage, any suitable optimization scheme can be implemented (e.g., [39–43]; and many others) based on the availability of time, resources, and expertise. In this study, a daily pooled calibration considering all observed daily records in the UZRB, and its sub-catchments (Figure 5) was applied at the Massachusetts Green High-Performance Computing Center (MGHPCC). The algorithm was run in parallel processing mode using 100,000 iterations, with population sizes ranging between 100 and 1000 generations. With this approach, the evaluation of the most appropriate population of parameter sets within each generation could be conducted in a more efficient manner. The Kling-Gupta efficiency [44] and the Nash-Sutcliffe efficiency [45], among other measures, were used to evaluate the degree of agreement between observed and simulated streamflow (Figure 6).

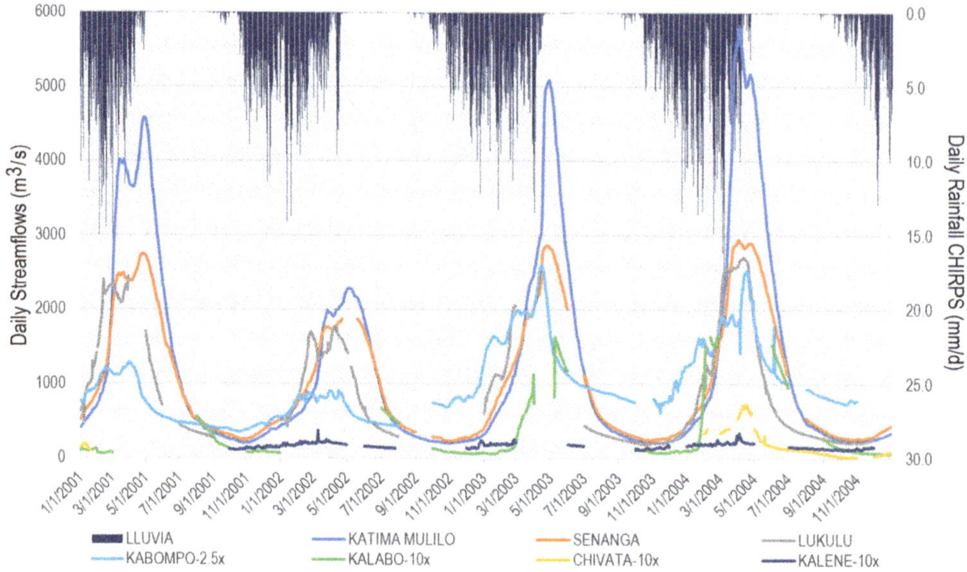

Figure 5. Daily streamflow records for Katima Mulilo, Senanga, Lukulu, Kabompo, Kalabo, Chivata, and Kalene sub-catchments. Daily average rainfall data from CHIRPS are also included. The numbers with "x" next to the names represent an amplification along the *y*-axis for a better visualization of the hydrographs.

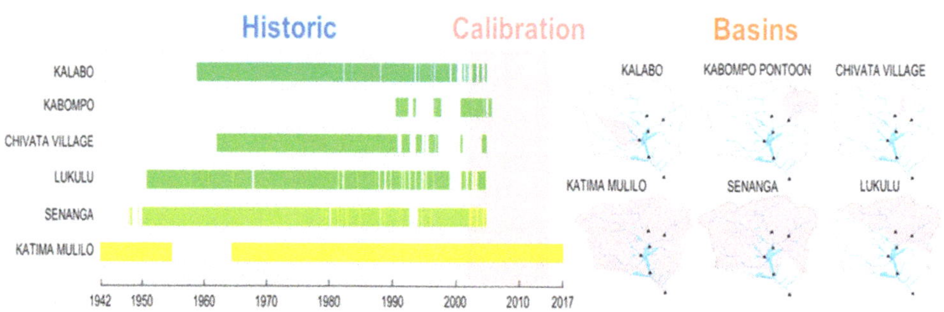

Model	Product	Calibration Period	KGE
HYMOD_DS	CHIRPS	01/01/2002–09/30/2015	0.95
	CMORPH	01/01/2002–09/30/2015	0.73
	TMPA	01/01/2002–09/30/2015	0.82
	PERSIANN	01/01/2002–09/30/2015	0.52
HBV_DS	CHIRPS	01/01/2002–12/31/2015	0.91
	CMORPH	01/01/2002–09/30/2015	0.82
	TMPA	01/01/2002–09/30/2015	0.8
	PERSIANN	01/01/2002–09/30/2015	0.64
VIC_4.2.b	CHIRPS	01/01/2002–09/30/2015	0.83
	CMORPH	01/01/2002–09/30/2015	0.82
	TMPA	01/01/2002–09/30/2015	0.86
	PERSIANN	01/01/2002–09/30/2015	0.77

Figure 6. (**Top**) UZRB and sub-catchments utilized during the daily calibration process. (**Bottom**) Calibration performances for Katima Mulilo Streamgauge (2002–2015).

2.7. Operational Variational Ensemble Forecasting (VEF)

A group of i hydroclimatological inputs, j hydrologic models, and k calibrated model parameters can be used to establish simple or variational ensemble forecasts. A VEF approach can provide a larger number of forecasts than a simple assembling approach, because it evaluates all possible modelling chain sequences that can be arranged to construct an HFS (see Figure 7). At this stage, data assimilation techniques to support the "perfect model assumption" can be also applied to generate deterministic streamflow forecasts. Ref. [46] reviewed and evaluated many available methods to perform deterministic forecasts. However, in an operational context, the experience of streamflow forecasters suggests that uncertainty bands can better support decision making for water management schemes [47–52]. The intention of a VEF approach is to use all available weighted or non-weighted components of a modelling chain (HFS) to generate streamflow forecasts. The final hypothesis about the probability distribution of streamflow forecasts can be approximated using several procedures applied to the Multiproduct, Multimodel, and Multiparameter sets, through the implementation of a VEF approach that can be trained during T_W and used during T_F as:

$$\mathcal{M}(z^{i,j,k}_{y_{T_F}} \mid \hbar^{i}_{u_{T_F}}, \hbar^{j}_{p_{T_W}}, \hbar^{k}_{\theta_{T_W}}) = \mathcal{H}\left(\hbar(z^{i,j,k}_{y_{T_F}} \mid \hbar^{i}_{u_{T_F}}, \hbar^{j}_{p_{T_W}}, \hbar^{k}_{\theta_{T_W}})\right) \forall\, T_W \in \{t_0, \ldots, t\} \text{ and } \forall\, T_F \in \left\{t, \ldots, t_f\right\} \tag{1}$$

$$\hbar^{i}_{u_{T_F}} = \hbar(u^{i}_{T_F} \mid z^{i}_{u_{T_W}}, z^{i}_{u_{T_F}}) \tag{2}$$

$$\hbar^{j}_{p_{T_W}} = \hbar(S^{j}_{T_W} \mid \hbar^{k}_{\theta_{T_W}}) \tag{3}$$

$$\hbar^{k}_{\theta_{T_W}} = \hbar(\theta^{k}_{T_W} \mid z^{k}_{\theta_{T_W}}) \tag{4}$$

$$\hbar^{i,j,k}_{y_{T_F}} = \hbar(z^{i,j,k}_{y_{T_F}} \mid \hbar^{i}_{u_{T_F}}, \hbar^{j}_{p_{T_W}}, \hbar^{k}_{\theta_{T_W}}) \tag{5}$$

- where, $\mathcal{M}$ is a Multiproduct, Multimodel, and Multiparameter Variational Ensemble Streamflow Forecast for the forecast period T_F.
- $\mathcal{H}$ is a family of hypothetical ensemble components for the warmup period T_W and used to forecast the period T_F.
- $\hbar^{i,j,k}_{y_{T_F}}$ is a hypothesis of the hydrologic process from a family of input data i, hydrologic model j, and parameter set k, about the HFS for the forecast period T_F.
- $z^{i,j,k}_{y_{T_F}}$ is the streamflow prediction i, j, k about the HFS response for the forecast period T_F.
- $\hbar^{i}_{u_{T_F}}$ is a hypothesis of the input data from a family of input data i about the HFS for the forecast period T_F.
- $\hbar^{j}_{p_{T_W}}$ is a hypothesis of the hydrologic process from a family of hydrologic models j and for the warmup period T_W.
- $\hbar^{k}_{\theta_{T_W}}$ is a hypothesis of the parameter sets from a family of parameter sets k for the warmup period T_W.
- $u^{i}_{T_F}$ is the control variable for the forecast period T_F.
- $z^{i}_{u_{T_F}}$ is a family of input data i about the HFS for the forecast period T_F.
- $z^{i}_{u_{T_W}}$ is a family of input data i about the HFS for the warmup period T_W.
- $S^{j}_{T_W}$ is the model structure j for the warmup period T_W.
- $\theta^{k}_{T_W}$ is the parameter set k for the warmup period T_W.
- $z^{k}_{\theta_{T_W}}$ is a family of parameter sets k for the warmup period T_W.

The VEF approach (Figure 7) makes it possible to explicitly represent how different components of each model ensemble (modelling chain) vary according to all possible combinations of biogeophysical representations of the climate system and the hydrologic system (e.g., [53]). In addition, one of the novel aspects of this research is introduction of the assumption that the forecast skill of hydrologic models to represent underlying hydrologic processes can be better captured by assessing the generalization capabilities of multiple optimal parameters sets.

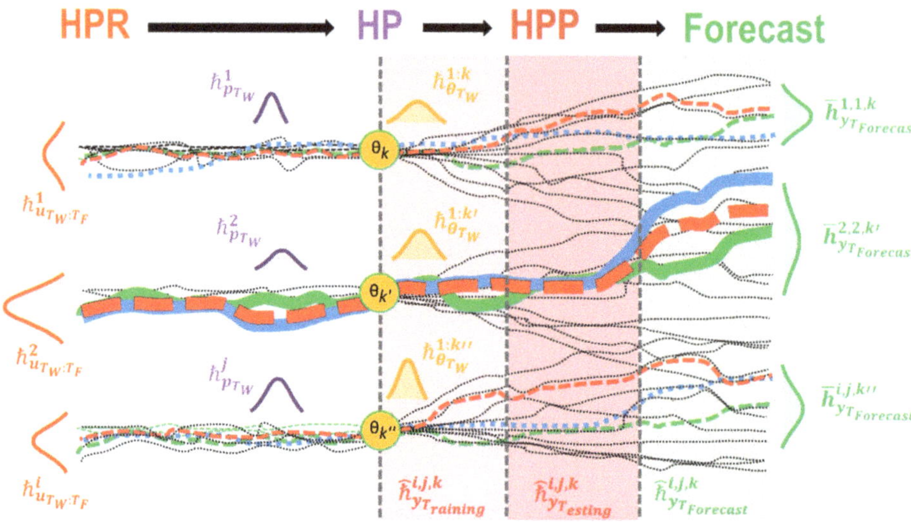

Figure 7. Multi-Input, Multimodel, and Multiparameter Variational Ensemble Forecasting (VEF). Each Hydrologic Processing Strategy (HPS) is shown in the context of VEF.

2.8. Strategies to Reduce Uncertainty and Improve VEF in an Operational Environment

In the development of an operational streamflow forecasting paradigm, it must be determined how total hydrological uncertainty will be reduced to improve streamflow forecasts. This topic is still a matter of discussion among many hydrologists and researchers around the world [47–52]. For instance, standardized processes to decompose, quantify, or evaluate the meteorological or the total hydrological uncertainty propagated from a modelling paradigm are required (see, i.e., Figure 8). The implementation of new techniques for the decomposition of uncertainty can take advantages from VEF approaches to explore all available sources of climate data and physical representations (i.e., model structures and parameters) for hydrologic modelling. Taking this into account, a combination of VEF and hydrological processing hypotheses can help us better understand how the propagation of errors occurs. For example, from multiple climate products the amount of meteorological uncertainty propagated through the modelling chain can be identified; then comparisons can be made to estimate the amount of total hydrological uncertainty (THU) propagated from the same system. This more systematic method to identify uncertainties can be useful to inform additional pre- or postprocessing of THU.

Assuming that an operational HFS can be evaluated as a VEF approach, three main strategies (Figure 8) to establish hydrological processing hypotheses can be applied to evaluate and improve the forecast skill of any hydrological model: (1) Hydrological Preprocessing (HPR) (or Meteorological Postprocessing), which can include the application of bias correction techniques (i.e., physical or statistical improvements) to reduce the propagation of errors or the application of Bayesian or non-Bayesian approaches to estimate an optimal weighted combination of precipitation products; (2) Hydrological Processing (HP), which can include improvements in the models' structure, states, or parameters; and (3) Hydrological Postprocessing (HPP), which can include bias correction techniques applied over the streamflow forecasts or the use of Bayesian approaches to estimate an optimal weighted combination of streamflow forecasts.

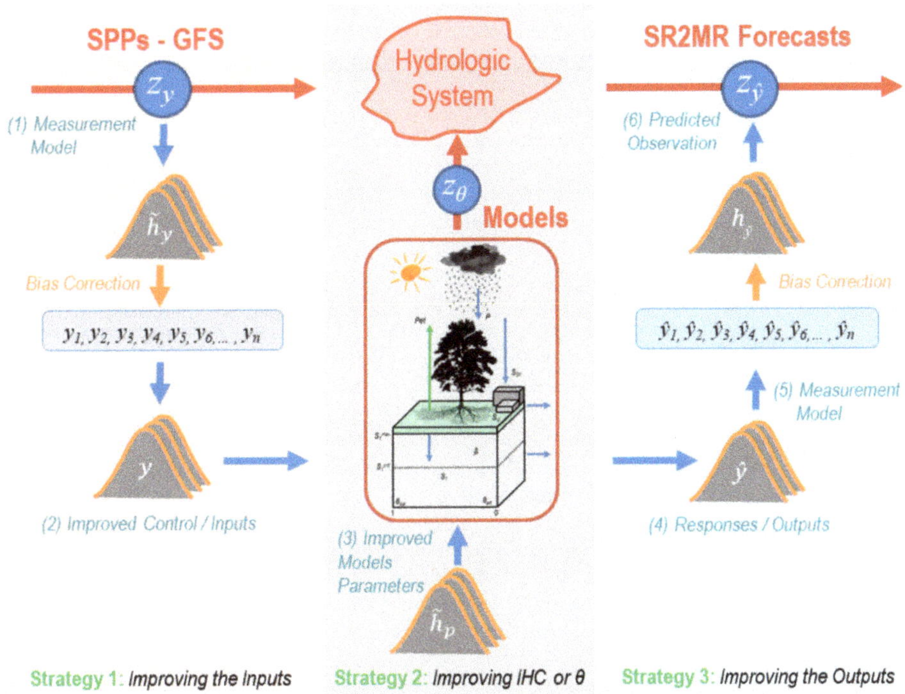

Figure 8. Modelling paradigm implemented in an operational HFS context in the UZRB. The hydrological processing hypotheses that can be implemented to improve streamflow forecasts are displayed as sources of uncertainty (h) propagated from the input data (y_i), and/or from the applied hydrological preprocessing and/or processing techniques. The output data ($\hat{y}_i$) can also propagate uncertainty, which can be minimized through the implementation of hydrological postprocessing techniques.

3. Results

The three main strategies for the evaluation of hydrological hypotheses (HPR, HP, and HPR) were applied to evaluate and improve the forecast skill of operational streamflow forecasts in the UZRB. The implementation of one or another strategy can have significant impacts on the final forecasting products. For example, precipitation bias correction methods (HPR strategies) can dramatically perturb the volume of water entering to the system, in the HP strategies the most sensitive model parameters can significantly modify water routing though the model structure, and HPP strategies can have a large impact by directly perturbing the forecasts to adjust scaling and fitting issues derived from a poor model representation. Details on how these three processing strategies can impact operational streamflow forecasts are discussed in the following sections of this paper.

3.1. Strategy 1: Hydrological Pre-Processing (HPR)

Rainfall forecasts or any other climate forecasts derived from satellites or climatological models are prone to errors that must be corrected. The propagation of these errors is more significant when the forecasts are biased, especially those for rainfall, since this is the most important variable for hydrological modeling. Corrections applied over rainfall records allow for the identification of sensitivities or gaps associated to the improvement of satellite data, for the calibration and validation of variational hydrological models, and for the identification or selection of the best ensembles for any operational implementation of an HFS.

The corrections correspond to the application of any selected HPR hypothesis, and they contribute to the reductions of the propagated meteorological uncertainties through the VEF implementation. One of the most popular HPR strategies is Quantile Mapping [54–57], a technique that has been previously evaluated in the UZRB and compared to Principal Components Analysis (PCA) [13]. In the operational context, the Quantile Mapping (QM) technique is applied at a daily time scale with the assumption that the probability density functions (PDFs) of the rainfall observations and forecasts follow Gamma PDFs. The key idea behind this technique is to swap the quantiles of the simulated data with the quantiles of the observed data. The application of this technique has shown that daily estimates from SPPs and RCMs can be satisfactorily corrected at the catchment scale or at more regional scales (i.e., [54–60]). The previous studies agree with our findings for the UZRB where precipitation estimates from three satellites were significantly improved after the application of the Quantile Mapping (QM) method, used with a Gamma Probability Distribution Function (PDF). The results showed that all raw SPPs (Figure 9a–c) could be satisfactorily corrected at daily time scales (Figure 9d–f). Here, it is important to notice that many missing and false detections of rainfall can be corrected (see all raw to corrected scatters in Figure 9); however, the selection and fitting of a fixed PDF to the whole rainfall dataset can also reduce the performance of rainfall forecasts in some areas if (1) the selected PDF is not a good representation of the local climate conditions; if (2) the application of QM is exclusively tied to regional parameters instead of cell-by-cell parameters or vice versa, and if (3) only fixed temporal parameters are used instead of temporally varying parameters. All these factors can have result on successful or inadequate rainfall corrections that can have a significant impact on the next steps related to model calibration, final structural design, and generation of final operational streamflow forecasts products.

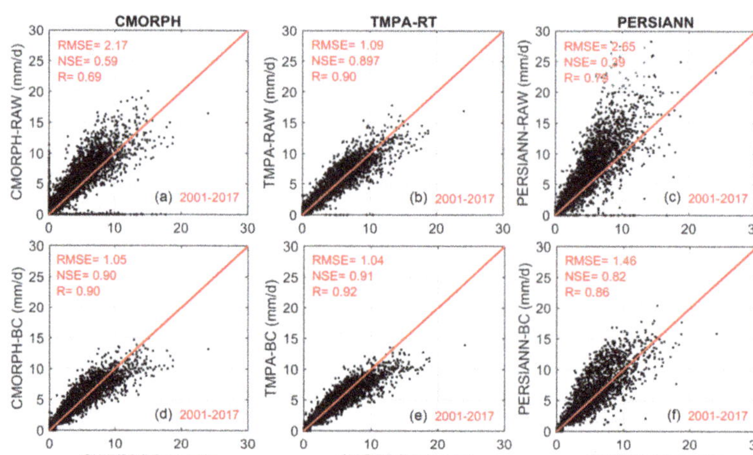

Figure 9. Scatter plots for daily average rainfall in the UZRB. The observed daily rainfall records from CHIRPS (with drizzle effect removed for rainfall $\leq$ 0.1 mm) are compared to raw (**a–c**) and corrected (**d–f**) satellite-based rainfall estimates from CMORPH, TMPA-RT, and PERSIANN for the period 2001–2017. Three error measures are included for comparison: the Root Mean Squared Error (RMSE), the Nash-Sutcliffe Efficiency (NSE), and the Correlation Coefficient (R).

3.2. Strategy 2: Hydrological Processing (HP)

The VEF hydrologic modeling paradigm requires bridging science and engineering for the design of functional and operational HFS. Therefore, all available precipitation datasets utilized during HPR, must be used to evaluate the propagation of meteorological uncertainty into final SR2MR streamflow forecasts (see Figure 10a_1–a_7). Generally, the objective of HP is to constrain the range of valid model outcomes for the application of HPP strategies and for the generation of forecast products. Multimodel Ensemble approaches are a popular alternative to propagate uncertainty into future forecasts; however, they can yield boundless errors in inference, which can produce unbounded uncertainty bands [61]. Although this is an acceptable argument, the science and engineering of streamflow forecasting can play an important role in defining the best alternatives for the management of total hydrologic uncertainty (THU). In doing this, Variational Ensemble Forecasting (VEF) approaches have the advantage that they can be implemented using all available sources of input data, hydrologic models, and optimal parameter sets, to improve the assimilation of forecasts. VEF can also be coupled with regularization techniques to constrain the forecasting range based on the classification and evaluation of historic events to define the best ensembles for HPP.

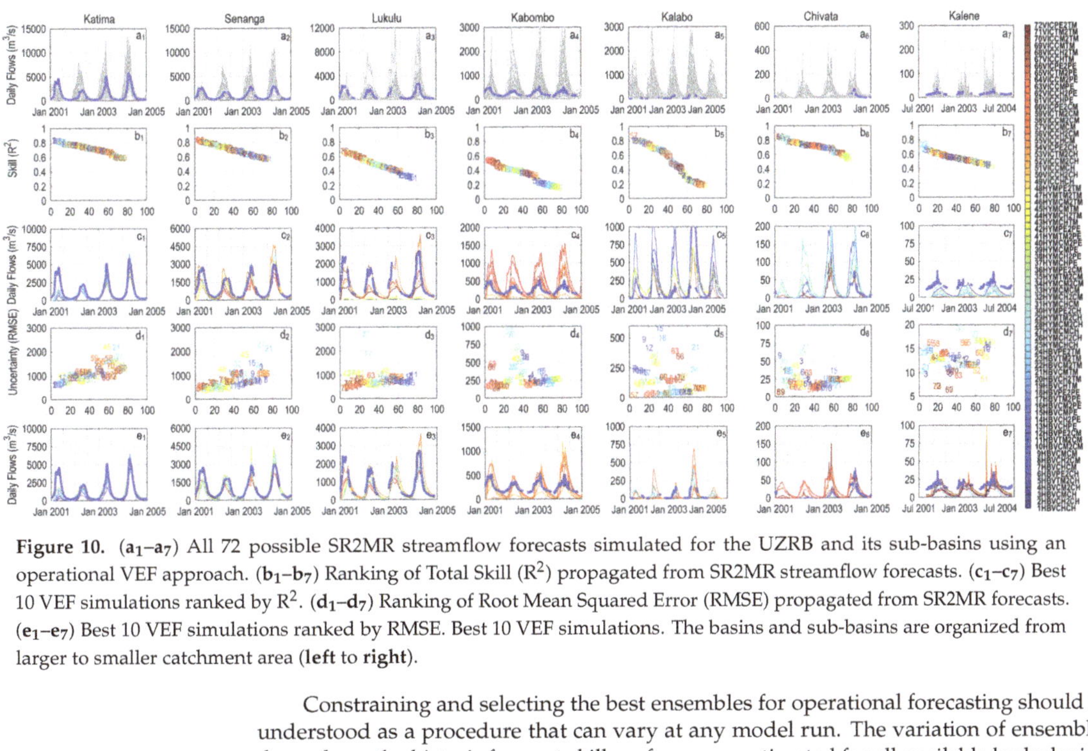

Figure 10. (**a_1–a_7**) All 72 possible SR2MR streamflow forecasts simulated for the UZRB and its sub-basins using an operational VEF approach. (**b_1–b_7**) Ranking of Total Skill (R^2) propagated from SR2MR streamflow forecasts. (**c_1–c_7**) Best 10 VEF simulations ranked by R^2. (**d_1–d_7**) Ranking of Root Mean Squared Error (RMSE) propagated from SR2MR forecasts. (**e_1–e_7**) Best 10 VEF simulations ranked by RMSE. Best 10 VEF simulations. The basins and sub-basins are organized from larger to smaller catchment area (**left** to **right**).

Constraining and selecting the best ensembles for operational forecasting should be understood as a procedure that can vary at any model run. The variation of ensembles depends on the historic forecast skill performance estimated for all available hydrological events and their classified characteristic responses. In the UZRB and its sub-basins, the implementation of an operational VEF approach was used to generate SR2MR streamflow forecasts derived from all possible combinations of SPPs, hydrologic models, and optimal parameter sets (Figure 10a_1–a_7). The propagation of meteorological uncertainty can be quite large in the resulting streamflow traces when the whole input-model-parameter space is mapped and used for operational forecasts. To avoid unbounded uncertainty bands, the VEF approach allows improving the accuracy of streamflow forecasts through a ranking evaluation and posterior identification of the best hydrologic ensembles for the UZRB (Figure 10b_1–b_7 for forecasts ranked using skill analysis of R^2 and Figure 10d_1–d_7 for forecasts ranked using total uncertainty defined as RMSE). Then the best ten raw streamflow forecasts are ranked using skill and/or uncertainty measures (Figure 10c_1–c_7,e_7–e_7). The application of specific or combined verification techniques on streamflow forecasts usually leads to an improvement of the HPP hypotheses. This multiple evaluation is helpful to identify how skill and uncertainty perform over space and time but also to evaluate how the spatial resolution of precipitation products can have a large effect on the operational forecasts. For instance, the smallest sub-basins inside the UZRB (see Kabombo, Kalabo, Chivata, and Kalene in Figure 10) resulted in higher total uncertainty compared to large basins (see Katima, Senanga, and Lukulu in Figure 10). This reduction in performance in the smallest sub-basins is probably associated with the smallest number of available precipitation grid cells, which produces larger averaged forecasting errors if missing or false detections are present in the forecasts.

3.3. *Evaluating Pre-Operational SR2MR Streamflow Forecasts*

All 72 possible SR2MR forecasts that were generated in the UZRB by the combination of multiple rainfall products, hydrological models, and optimal parameter sets, enabled the identification of differences between the satellite precipitation products (SPPs) and their HPR strategies (or bias corrections) but also between the hydrological models and their optimal parameter sets (or HP strategy). From the VEF approach, the best raw model ensembles (Figure 10) were retained (20 out of 72 ensembles were selected for this study) and used for the generation of final streamflow forecasts products and reports for end users.

In general, an expected outcome is to have climatology products with better forecast skill (see for example Figure 11a,b) because of the instrumental corrections applied during the Hydrological Preprocessing (HPR) stage. However, one disadvantage of the climatology products is that they are not available in the SR2MR domain. For this reason, SPPs and RCMs are still needed for SR2MR streamflow forecasts in the UZRB or any other basins around the World. Therefore, the selection of final streamflow forecasts is also operationally based on SR2MR records. Furthermore, the quantification and propagation of retrospective meteorological uncertainties might be required for the release of final forecasts depending on the needs of end users. In the UZRB, multiples ensembles for deterministic streamflow forecasts and the spread of uncertainty by means of a probability density function were established for each daily SR2MR forecast. These forecasts were provided for Namibian Hydrological Services, one of the relevant African institutions that manages water resources in the Upper Zambezi River.

3.4. *Strategy 3: Hydrologic Post-Processing (HPP) for Raw Streamflow Forecasts*

Our inability to generate exact physical representations of natural hydrologic systems creates the need for streamflow forecasts that can quantify and reduce the total hydrologic uncertainty (THU) propagated from a hydrologic modelling paradigm (HMP). Hydrological Postprocessing (HPP) hypotheses focus on establishing standardized methods to quantify and propagate total hydrological uncertainty (which is the sum of all uncertainties i.e., input, parameter, or structural uncertainties propagated into the final streamflow forecasts). Similar to HPR and HP, different methods can be applied for HPP hypotheses, i.e., stochastic, Bayesian, or machine learning methods can be used as post-processors for the final ensemble of SR2MR streamflow forecasts. The objective at this stage is to minimize the error of the deterministic forecast but also the spread of total hydrological uncertainty around the raw streamflow forecasts. The deterministic forecast can be obtained as an optimal weighted SR2MR streamflow forecast, which is used to propagate the total hydrological uncertainty (Figure 12). In the UZRB, a Multivariate Combinatorial Linear Regression (MCLR) approach was applied as a regularization technique to inform the best selection of hydrologic ensembles that minimized the spread of total hydrologic uncertainty. The MCLR was also used as a first level hydrologic postprocessor to resolve scaling issues of the raw streamflow forecasts (Figure 12). The final ensemble of raw and corrected SR2MR streamflow forecasts was then evaluated using two inference-based hydrologic postprocessors: (1) Inverse-Variance Weighting (IVW), and (2) Inverse-Probability Weighting (Figure 12). Both approaches were compared to the best streamflow prediction (BSP) and the Average Streamflow Prediction (AVSP) using the retrospective total hydrologic uncertainty quantified using the root mean square error in millimeters per day (Figure 12). The application of hydrologic postprocessors (MCLR, IVW, and IWP) revealed an improved efficiency of both the forecasts and the propagation of THU in the UZRB and its sub-basins (Figure 12).

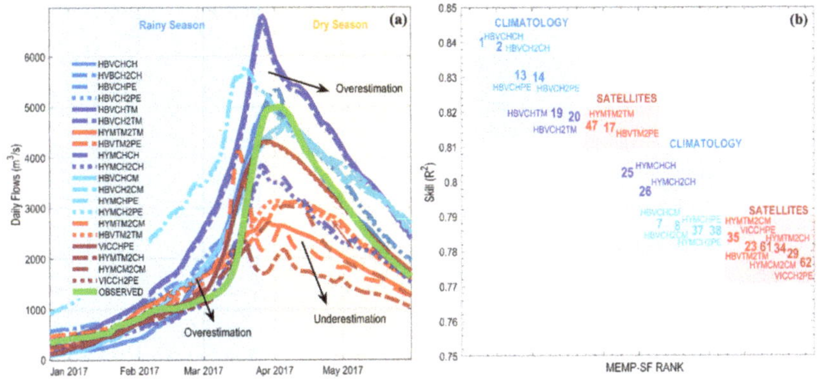

Figure 11. (**a**) Peak streamflow hydrograph (January to June of 2017) for the UZRB at Katima Mulilo, and (**b**) Ranked Predictive Skill (R^2) for the best 20 simulations obtained from the operational VEF approach. The acronyms represent the Hydrological Models (HYM for Hymod; HBV for HBV; and VIC for VIC); the Satellite Precipitation Products or Climatology (CH for CHIRPS and CH2 for drizzle removed effect; CM for CMORPH and CM2 for its bias corrected version; TM for TMPA and TM2 its bias corrected version; PE for PERSIANN and PE2 its bias corrected version); and the utilized parameter set (CH is CHIRPS parameter set; CM is CMORPH parameter set; TM is TMPA parameter set; and PE is PERSIANN parameter set).

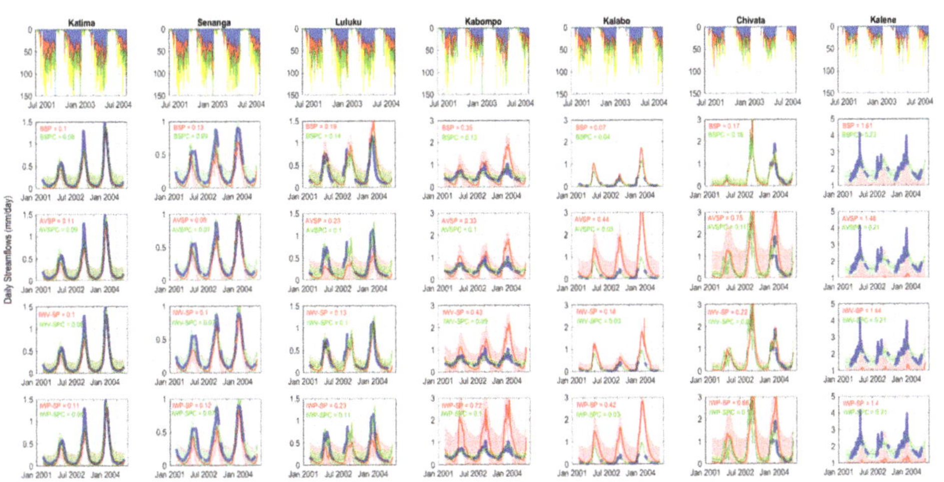

Figure 12. (top) Catchment-average satellite-based precipitation for the UZRB and its sub-basins. SR2MR streamflow forecasts for Katima Mulilo, and its sub-basins Senanga, Lukulu, Kabompo, Kalabo, Chivata, and Kalene (organized from left to right according to their size). The y-axis represents the daily streamflow forecasts, and the x-axis represents the validation and testing periods (2002–2004). The initial streamflow forecasts with their respective uncertainty bands (RMSE in mm) are shown in red. The hydrologically postprocessed (HPP) forecast is shown in light green. BSP is the best streamflow forecast; AVSP is the average streamflow forecast; IVW-SP is the Inverse-Variance Weighting streamflow forecast; IVP-SP is the Inverse-Probability Weighting streamflow forecast. The letter "c" at the end of each acronym represents the regularization applied by combining Multivariate Combinatorial Linear Regression (MCLR) and Inference-Based methods for SR2MR daily streamflow forecasting.

The combination of HPP with short and long-term memory windows (for periods ranging between 5 and 180 days before the event) also proved to be more efficient in improving the performance of SR2MR streamflow forecasts (see Appendix A Figure A3 for details). The effect of HP for operational streamflow forecasting was mainly observed in two aspects: (1) a better performance of the deterministic streamflow forecasts, i.e., correction of scaling issues, and (2) narrower total hydrologic uncertainty bands around the deterministic forecast, i.e., more practical and realistic uncertainty bands for decision makers (see details in Figure 12).

4. Discussion

To establish operational hydrological forecasting systems (HFS) it is important to first define who will be the main and potential end users of the streamflow forecasts products within the basin under consideration. In the UZRB (Figure 1), the need for better water management schemes for flood warning and water allocation schemes has required the generation of daily SR2MR streamflow forecasts (Figure 2). Several water users and water authorities of the countries sharing waters from the UZRB can take advantage of the benefits of this operational HFS implementation, which can largely help developing sustainable water management and allocation activities for this transnational basin. For example, the water users of streamflow forecasts in the UZRB can be informed about the main water management decisions that need to be taken care of for water supply, reservoir management, hydraulic design, etc. These decisions are generally complex; therefore, under such a scenario, the best way to inform the local authorities and end users in the UZRB can be obtained if we had 'perfect' weather forecasts, or climatic predictions, that could be combined with 'perfect' hydrologic models to generate "almost" perfect streamflow forecasts. If this hypothesis were still true only one weather or climatic prediction (model), and one hydrologic model would be needed for streamflow forecasting. For example, the 'perfect' climate estimates of the climate model could be combined with one 'perfect' hydrological model, and the streamflow prediction should be close enough to inform both managers and users, so they can apply the right decisions for water management. The problem of this hypothesis is the fact that so far, hydrologists and meteorologists has not been able to create or establish perfect model representations of climate and hydrology, i.e., models are simplified representations of the hydroclimatological processes occurring across spatiotemporal scales. Therefore, we know and assume that during the operational implementation of the HFS in the UZRB there were countless sources of uncertainty, and different tools and schemes were established in this study to quantify how meteorological and hydrological uncertainty propagate through an operational VEF approach. One of the first adopted techniques that emerged was the utilization of single model realizations as those provided in many previous studies (see for example [3–12]). Then, multi-ensembles of climate predictions were used to quantify and propagate the meteorological uncertainty through the streamflow forecasts. The main issue of this technique is that it did not allow quantifying the uncertainty propagated from the hydrological model either from the model structure or from its parameters. To resolve this problem [15], proposed a way to quantify the hydrological uncertainty from a single model with multiple parameters that can be obtained from the calibration of multiple climate products. The final set of outputs is obtained from a multiproduct and multiparameter scheme. This technique allows quantifying the hydrological uncertainty propagated from multiple parameters sets of a single hydrological model; however, it does not allow quantifying how the uncertainty propagates and varies as a function of the structure of the selected hydrological model. This latter conceptualization of hydrological uncertainty is tied to the modelling assumptions required to establish the physical representation of the hydrologic system. Obviously, these assumptions also vary as a function of the selected hydrological model and its structure. In this context, hydrologists have opted to promote more comprehensive modelling schemes that combine multiproducts and multiple hydrological models [14,16–18]. With this technique, it has been possible to quantify the meteorological uncertainty propagated from the climate products and the hydrological uncertainty propagated from the hydrologic models and their corresponding structures. On the other hand, recent studies [62–64] have argued that the performance of a single hydrological model can be improved together with the streamflow forecasts if multiple climate products are combined with a model that is perturbed by generating multiple initializations with different initial conditions, i.e., changes in surface storage, groundwater storage, or snow storage, among others. The main objective behind this approach is to capture the current hydrologic condition of a catchment to assimilate streamflow. For example, in this regard it has been traditionally assumed that the warmup period is required for hydrologic modelling; however, this approach can be bypassed given that the initial states are adjusted beforehand, taking in consideration the streamflow assimilation. This scheme has also stablished a new source for the quantification of hydrologic uncertainty just by changing the initial conditions of the hydrological model. With this technique, a new research niche has been recognized by hydrologists and now is also applied to quantify total hydrological uncertainty (THU). Having said that, it is also important to add that all the schemes mentioned above have allowed quantifying the hydrological uncertainty in separated procedures, either from multiple climate products, from multiple hydrological models, from multiple parameters sets, or from multiple initializations. In fact, all these hydrologic modeling paradigms (HMP) are still applied in a systematic manner to identify and quantify different sources of hydrological uncertainty.

Despite the existence of all these HMP previously mentioned, none of them have proposed a combined technique to quantify both meteorological and hydrological uncertainties propagated from different sources of a VEF implementation (see Figures 7 and 8), i.e., sources as climate forecasts, modelling assumptions, and optimal parameter sets, that can be evaluated for any operational hydrological forecasting system (HFS) implementation with VEF. To accomplish this issue, we have implemented a VEF approach [22] based on multiple satellite precipitation products (and GFS precipitation forecasts), multiple hydrologic models, and multiple optimal parameters sets for SR2MR daily streamflow forecasting. The VEF approach implemented in the UZRB (Figure 10) has allowed increasing the number of possible hydrologic ensembles available for streamflow forecasting, together with an improvement of the streamflow assimilation (observed versus predicted). It also provides a more comprehensive and systematic framework to identify and propagate the spread of total hydrologic uncertainty in an operational hydrologic forecasting system. Now, the natural question is how much room is left to define new modelling paradigms or techniques that can be used to quantify or minimize the propagation of total hydrological uncertainty? To answer this question, we need to differentiate between what we can do to define the best integrated implementation of an HFS, and what adjustments are required at each separated component (i.e., inputs, models, or outputs) of the HFS. The operational HFS implementation in the UZRB, identified and considered three general hydrologic processing strategies (HPS) that can be applied to any VEF approach. If we consider that any hydrologic modelling paradigm (HMP) can either include multiple inputs; hydrologic models, parameters, initializations; and outputs, then, we can hypothesize that at each component of the HMP it is possible to apply additional techniques or methods to improve streamflow forecasting. These strategies were conceptualized as: (1) Hydrological Preprocessing (HPR), (2) Hydrological Processing (HP), and (3) Hydrological Postprocessing (HPP). Taking advantage of the strategies proposed for the establishment of hydrologic processing strategies or hypotheses, standardized methods for HPR (Figure 9) and HPP (Figure 12) were applied in the UZRB, showing that the performance of raw VEF streamflow forecasts can be significantly improved, and the spread of uncertainty can be better constrained by applying regularization processes that combine the strength of Multivariate Combinatorial Linear Regression (MCLR) and Inference-Based approaches (Figure 12).

The science and engineering of future operational streamflow forecasting in the UZRB will continue concentrating efforts in improving the forecasts; however, new needs from end users might also require improving the physical representation at the catchment scale.

For this, it will be necessary to establish the role of mathematical, statistical, and/or machine learning methods that can be used to correct and propagate the hydrologic uncertainty from different components of a VEF approach, i.e., Bayesian, stochastic, pattern, or inference-based learning methods, etc. The performance of different rainfall products and methods also needs to be evaluated across different catchment sizes and forecast timescales to determine the space–time variability that propagates total hydrologic uncertainty. This evaluation must also be extended to determine the dependence of the accuracy of the streamflow forecasts and the propagation of hydrological uncertainty from physically based models. All the above will indicate the applicability limits of the VEF approach based on multiple precipitation or climate products, multiple hydrologic models, and multiple optimal parameters sets. All of the above will allow the establishment of new robust theoretical and hypothetical paradigms to quantify, evaluate, reduce, and manage the propagation of total hydrologic uncertainty using VEF approaches.

5. Conclusions

This paper described the main stages and processes required to implement and improve an operational hydrologic forecasting system (HFS) in the UZRB and its sub-basins. The process of implementation is very complex, and important decisions needed to be made about the input data (precipitation from satellites or climate products), the hydrologic models to be included along with their optimal parameter sets, and the timescales required for the generation of streamflow forecasts.

Once the HFS was completely operational in the UZRB, additional improvements to the forecasts were required to improve its performance and reduce the spread of total hydrologic uncertainty into the final streamflow forecast products. In this regard, three general strategies to improve the performance of VEF approach were proposed: from Hydrological Preprocessing to Postprocessing techniques that can improve the input data, the hydrologic models (or their structures), the optimal parameter sets, and the raw streamflow forecasts. The whole range of available techniques for operational HFS will require more detailed and standardized conceptualizations. In this regard, bias corrections or preprocessing (HPR) techniques applied over the input data will still play an important role in operational hydrological forecasts for the UZRB. The operational implementation of the VEF combined with regularization and inference-based methods improved the performance of streamflow forecasts as the primary need from end users in the UZRB; however, new alternatives to improve the physical understanding of the basin are still a pending task. Finally, it is important to add that the science of Hydrological Postprocessing is still under an early stage of development, and it still lacks the standardized methods that can be used for these purposes. Emerging methods will need to be evaluated to establish the real boundary between physical and statistical needs in operational streamflow forecasting. This discipline will also require a natural merging of science and engineering (practical applications) in a real-world context to establish baseline conditions for streamflow forecasting and hydrological uncertainty quantification and propagation.

Author Contributions: Conceptualization, R.V.-P. and J.B.V.; methodology, R.V.-P.; software, R.V.-P., S.W. and T.R.; validation, J.B.V. and A.S.-C.; formal analysis, R.V.-P.; investigation, R.V.-P.; resources, J.B.V., A.S.-C. and S.W.; data curation, R.V.-P.; writing—original draft preparation, R.V.-P.; writing—review and editing, R.V.-P., J.B.V., T.R., S.W. and A.S.-C.; visualization, R.V.-P.; supervision, J.B.V.; project administration, A.S.-C.; funding acquisition, J.B.V. and A.S.-C. All authors have read and agreed to the published version of the manuscript.

Funding: This research was supported by NASA-USAID SERVIR Program, Award 11-SERVIR11-58.

Acknowledgments: This research was supported by the NASA-USAID SERVIR Program (Award 11-SERVIR11-58). The first author of this paper was supported by the National Research and Development Agency of Chile (ANID), through the National Doctoral Degree Scholarship Competition of the Formation of Advanced Human Capital Program. The Research Computing Department at the University of Massachusetts provided support on the calibration of the hydrological models. All these contributions are gratefully acknowledged.

Conflicts of Interest: The authors declare no conflict of interest.

Appendix A. Hydrologic Models Used for SR2MR Streamflow Forecasting in the UZRB

The HBV_DS model is a modified distributed version of the Hydrologiska Byrans Vattenbalansavdelning (HBV) model (see details in Bergström, 1976; Seibert and Vis, 2012; Yang and Wi, 2018), and it simulates catchment discharge on a daily time step, based on time series of precipitation and air temperature. The implementation of HBV_DS (Figure A1) requires the calibration of 20 parameters (see Table A1). The Potential Evapotranspiration (PET) is computed as a function of daily mean temperature and hours of daylight using the Hamon Method (Hamon, 1961). In the snow routine, the snow accumulation and snowmelt are computed by a degree-day method (see Moore, 1993; Rango and Martinec, 1995). The actual evaporation and the groundwater recharge are simulated as a function of the actual soil water storage. The surface runoff, the interflow, and the percolation are simulated using a single linear reservoir with three outlets, and the groundwater routing is represented by a single linear reservoir. The sum of these outflows is then routed using the diffusive wave approximation of the linearized Saint-Venant equation (Lohmann et al., 1998).

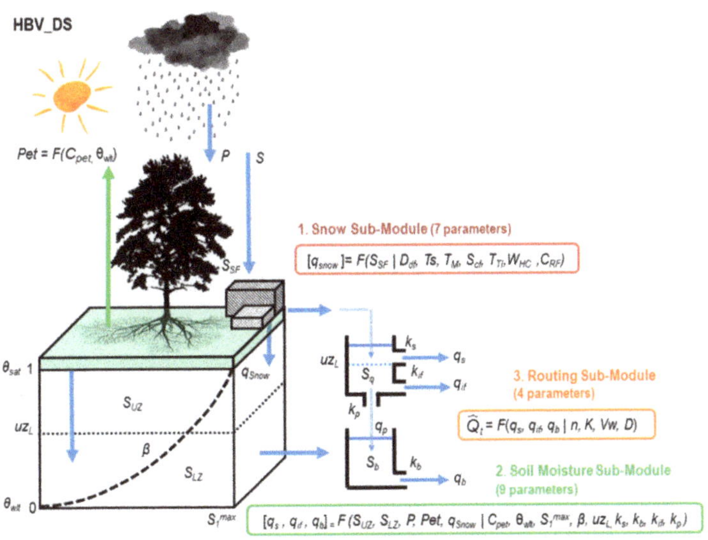

Figure A1. Hydrologiska Byrans Vattenbalansavdelning (HBV) Model Structure (states, fluxes, and parameters).

The HYMOD_DS model (Wi et al., 2015) is a modified version of the original HyMod Hydrological Model (see details in Moore, 1985; Boyle et al., 2000; Gonzalez-Leiva et al., 2016; Valdés-Pineda et al., 2016). The modified distributed version (Figure A2) simulates streamflows on a daily time step and requires daily precipitation and mean temperature as input variables. The implementation of HYMOD_DS requires the calibration of 15 parameters (see Table A1). The model is based on the probability-distributed storage capacity concept (proposed by Moore, 1985) to represent the soil moisture accounting component. Estimates of potential evaporation rates are calculated using the Hamon Method (Hamon, 1961). The rate of change in snow and glacier volume is expressed by the degree day factor (DDF) mass balance model (see Moore, 1993; Stahl et al., 2008). The direct runoff is characterized by an instantaneous unit hydrograph (IUH) (Nash, 1957), in which the catchment is represented as a series of "n" linear reservoirs. The groundwater routing is simplified as a single linear reservoir. Finally, similar to the HBV_DS model, the transport of water in the channel system is described using the diffusive wave approximation of the linearized Saint-Venant equation (Lohmann et al., 1998).

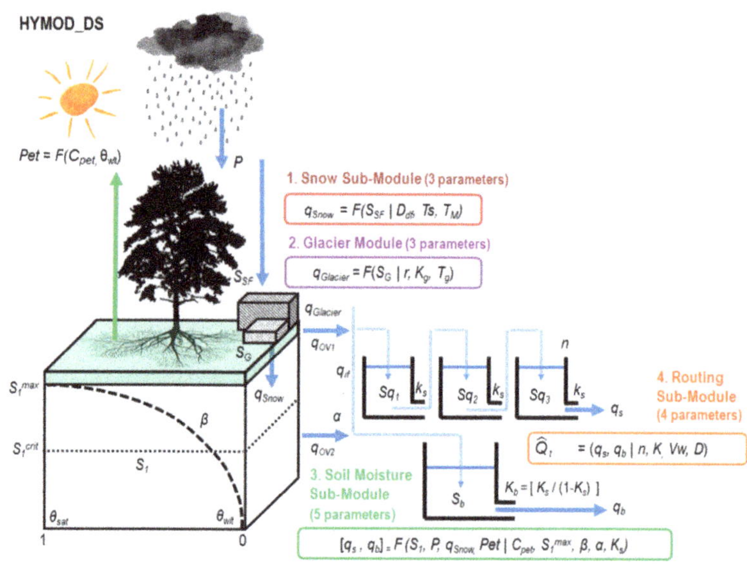

Figure A2. HyMod Hydrologic Model Structure (states, fluxes, and parameters).

The VIC (Variable Infiltration Capacity) Model (Liang et al., 1994, 1996; Cherkauer et al., 2003; Bowling et al., 2004; Bowling and Lettenmaier, 2009) is a large-scale semi-distributed hydrologic model (Figure A3). VIC simulates streamflows on a sub-daily or daily time step and requires daily precipitation, mean daily temperature, and/or mean wind speed as input variables. The VIC model has about 50 parameters; however, its implementation requires the calibration of 5 parameters (Table A1). The model balances both the water and surface energy budgets within the grid cell; and its sub-grid variations are captured statistically. The total evapotranspiration over a grid cell is computed as the sum of three types of evaporation: evaporation from the canopy layer of each vegetation tile, transpiration from each of the vegetation tiles, and evaporation from the bare soil (Liang et al. 1994). The snow model in VIC represents the snowpack as a two-layer medium and solves for energy and mass balance for the ground surface snowpack in a manner similar to other cold land processes models (Anderson, 1976; Wigmosta et al., 1994; Tarboton et al., 1995). The VIC model uses the variable infiltration curve (Zhao et al., 1980) to account for the spatial heterogeneity of runoff generation. It assumes that surface runoff from the upper two soil layers is generated by those areas for which precipitation, when added to soil moisture storage at the end of the previous time step, exceeds the storage capacity of the soil. The formulation of subsurface runoff follows the Arno model conceptualization (Franchini and Pacciani, 1991; Todini, 1996). To finally simulate streamflow, VIC results are postprocessed with a separate routing model (Lohmann, et al., 1996; 1998a; b), based on a linear transfer function to simulate the streamflow.

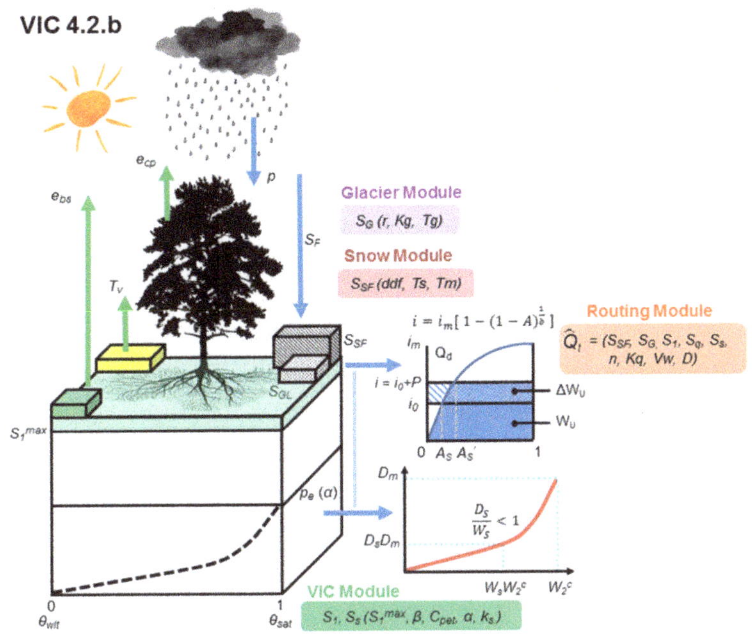

Figure A3. Variable Infiltration Capacity (VIC) Model Structure (states, fluxes, and parameters).

Table A1. Parameters included in the calibration of the HBV_DS, HYMOD_DS, and VIC models.

Module	Parameters	Description	Range	Units	Model
Soil Moisture	C_{pet}	Proportionality Coefficient of Hamon Potential Evapotranspiration	0.1–2	non-dim	HVB–HYMOD
	S_1^{max}	Maximum storage capacity of soil moisture accounting tank	5–1500	(mm)	HVB–HYMOD
	β	Shape parameter of the storage capacity distribution function	0.01–1.99	non-dim	HVB–HYMOD
	α	Split parameter for quick and slow components	0.01–0.99	non-dim	HYMOD
	θ_{wlt}	Soil Permanent Wilting Point (limiting soil moisture for PET occurrence)	0.1–1	non-dim	HBV
	uz_L	Upper reservoir water level for quick runoff occurrence	0–1000	mm	HBV
	K_s	Recession constant for quickflow in the upper reservoir	0.01–0.99	day^{-1}	HVB–HYMOD
	K_b	Recession constant for slowflow in the lower reservoir	0.0001–0.99	day^{-1}	HVB–HYMOD
	K_{if}	Recession constant for interflow in the upper reservoir	0.001–0.15	day^{-1}	HBV
	K_p	Flow rate for percolation between the upper and lower reservoir	0–3	mm day^{-1}	HBV
	b_i	Shape parameter of the Variable Infiltration Capacity curve	0–0.4	non-dim	VIC
	D_2	Second Soil Layer Thickness	0.1–1.5	m	VIC
	D_3	Third Soil Layer Thickness	0.1–1.5	m	VIC
	D_{Smax}	Maximum Baseflow Velocity	0–30	mm day^{-1}	VIC
	D_S	Fraction of Maximum Baseflow Velocity	0–1	non-dim	VIC
	W_S	Fraction of Maximum Soil Moisture content of the third soil layer	0–1	non-dim	VIC

Table A1. *Cont.*

Module	Parameters	Description	Range	Units	Model
Snow	D_{df}	Degree-Day Factor	0.001–10.0	mm °C day^{-1}	HVB–HYMOD
	S_{cf}	Snowfall Correction Factor	0.4–1	non-dim	HBV
	T_S	Temperature threshold for snow falling	0–5	°C	HVB–HYMOD
	T_M	Temperature threshold for snowmelt	0–5	°C	HVB–HYMOD
	T_{TI}	Temperature interval for mixture of snow and rain	0–5	°C	HBV
	W_{HC}	Liquid water holding capacity of the snowpack	0–0.2	non-dim	HBV
	C_{RF}	Refreezing coefficient of the liquid water in snow	0–1	non-dim	HBV
Glacier	r	Glacier melt factor	1–2	non-dim	HYMOD
	K_g	Glacier reservoir release coefficient	0.01–0.99	non-dim	HYMOD
	T_g	Glacier melt temperature threshold	0–10	°C	HYMOD
Routing	n	Grid Unit Hydrograph parameter (number of linear reservoirs)	1–99	non-dim	HVB–HYMOD
	K_q	Grid Unit Hydrograph parameter (reservoir storage constant)	0.01–0.99	day^{-1}	HVB–HYMOD
	Vw	Wave velocity in the linearized Saint-Venant equation	0.5–5.0	m s^{-1}	HYMOD
	D	Diffusivity in the linearized Saint-Venant equation	200–4000	m^2 s^{-1}	HYMOD

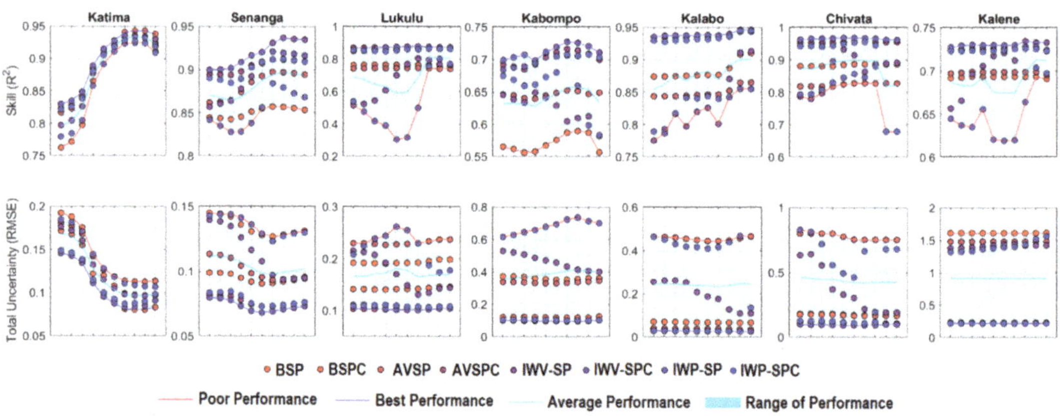

Figure A4. Effect of short and long-term memory (moving window) on the performance of daily streamflow forecasts for the UZRB and its sub-basins. From left to right each plot represents a memory window ranging between 5 and 180 days. The following windows were used: 5, 8, 15, 30, 45, 60, 90, 120, 150, and 180 days. Reddish colors represent aggregated streamflow forecasts and blueish colors represent weighted streamflow forecasts.

References

1. Bartholomé, E.; Belward, A.S. GLC2000: A new approach to global land cover mapping from Earth observation data. *Int. J. Remote Sens.* **2005**, *26*, 1959–1977. [CrossRef]
2. Lehner, B.; Verdin, K.; Jarvis, A. New global hydrography derived from spaceborne elevation data. *Eos* **2011**, *89*, 93–94. [CrossRef]
3. Meier, P.; Frömelt, A.; Kinzelbach, W. Hydrological real-time modelling in the Zambezi river basin using satellite-based soil moisture and rainfall data. *Hydrol. Earth Syst. Sci.* **2011**, *15*, 999–1008. [CrossRef]
4. Meier, P. Real-Time Hydrologic Modelling and Floodplain Modelling in the Kafue River Basin, Zambia. Ph.D. Thesis, ETH Zurich, Zurich, Switerland, 2012.
5. Liechti, C.T.; Matos, J.P.; Ferràs Segura, D.; Boillat, J.L.; Schleiss, A.J. Hydrological modelling of the Zambezi River Basin taking into account floodplain behaviour by a modified reservoir approach. *Int. J. River Basin Manag.* **2014**, *12*, 29–41. [CrossRef]
6. Arnold, J.G.; Moriasi, D.N.; Gassman, P.W.; Abbaspour, K.C.; White, M.J.; Srinivasan, R.; Kannan, N. SWAT: Model use, calibration, and validation. *Trans. ASABE* **2012**, *55*, 1491–1508. [CrossRef]

7. Beck, L.; Bernauer, T. How will combined changes in water demand and climate affect water availability in the Zambezi River basin? *Glob. Environ. Chang.* **2011**, *21*, 1061–1072. [CrossRef]
8. Michailovsky, C.I.; McEnnis, S.; Berry, P.A.M.; Smith, R.; Bauer-Gottwein, P. River monitoring from satellite radar altimetry in the Zambezi River basin. *Hydrol. Earth Syst. Sci.* **2012**, *16*, 2181–2192. [CrossRef]
9. Nones, M.; Ronco, P.; Di Silvio, G. Modelling the impact of large impoundments on the Lower Zambezi River. *Int. J. River Basin Manag.* **2013**, *11*, 221–236. [CrossRef]
10. Gumindoga, W.; Makurira, H.; Phiri, M.; Nhapi, I. Estimating runoff from ungauged catchments for reservoir water balance in the Lower Middle Zambezi Basin. *Water SA* **2016**, *42*, 641–649. [CrossRef]
11. Ndhlovu, G.Z.; Woyessa, Y.E. Evaluation of Streamflow under Climate Change in the Zambezi River Basin of Southern Africa. *Water* **2021**, *13*, 3114. [CrossRef]
12. Gumindoga, W.; Rientjes TH, M.; Haile, A.T.; Reggiani, P.; Makurira, H. Propagation of CMORPH rainfall errors to REW streamflow simulation mismatch in the Upper Zambezi Basin. *J. Hydrol. Reg. Stud.* **2021**, *38*, 100966. [CrossRef]
13. Valdés-Pineda, R.; Demaría, E.M.C.; Valdés, J.B.; Wi, S.; Serrat-Capdevilla, A. Bias correction of daily satellite-based rainfall estimates for hydrologic forecasting in the Upper Zambezi, Africa. *Hydrol. Earth Syst. Sci.* **2016**, 1–28. [CrossRef]
14. Roy, T.; Serrat-Capdevila, A.; Gupta, H.; Valdes, J. A platform for probabilistic Multimodel and Multiproduct Streamflow Forecasting. *Water Resour. Res.* **2017**, *53*, 376–399. [CrossRef]
15. Beven, K.; Freer, J. A dynamic topmodel. *Hydrol. Processes* **2001**, *15*, 1993–2011. [CrossRef]
16. Clark, M.P.; Slater, A.G.; Rupp, D.E.; Woods, R.A.; Vrugt, J.A.; Gupta, H.V.; Hay, L.E. Framework for Understanding Structural Errors (FUSE): A modular framework to diagnose differences between hydrological models. *Water Resour. Res.* **2008**, *44*. [CrossRef]
17. Velázquez, J.A.; Anctil, F.; Ramos, M.H.; Perrin, C. Can a multi-model approach improve hydrological ensemble forecasting? A study on 29 French catchments using 16 hydrological model structures. *Adv. Geosci.* **2011**, *29*, 33–42. [CrossRef]
18. Seiller, G.; Anctil, F.; Perrin, C. Multimodel evaluation of twenty lumped hydrological models under contrasted climate conditions. *Hydrol. Earth Syst. Sci.* **2012**, *16*, 1171–1189. [CrossRef]
19. Wood, A.W.; Lettenmaier, D.P. An ensemble approach for attribution of hydrologic prediction uncertainty. *Geophys. Res. Lett.* **2008**, *35*. [CrossRef]
20. Wood, A.W.; Hopson, T.; Newman, A.; Brekke, L.; Arnold, J.; Clark, M. Quantifying streamflow forecast skill elasticity to initial condition and climate prediction skill. *J. Hydrometeorol.* **2016**, *17*, 651–668. [CrossRef]
21. Roy, T.; Valdés, J.B.; Serrat-Capdevila, A.; Durcik, M.; Demaria, E.M.; Valdés-Pineda, R.; Gupta, H.V. Detailed Overview of the multimodel multiproduct streamflow forecasting platform. *J. Appl. Water Eng. Res.* **2020**, *8*, 277–289. [CrossRef]
22. Valdés-Pineda, R.; Valdés, J.B.; Wi, S.; Serrat-Capdevila, A.; Tirthankar, R.; Demaria, E.; Durcik, M. Operational Daily Streamflow Forecasts by coupling Variational Ensemble Forecasting and Machine Learning (VEF-ML) approaches. In Proceedings of the EGU General Assembly Conference Abstracts, Virtual, 19–30 April 2021; p. EGU21-14087.
23. Tumbare, M.J.; Mukosa CF, G. A brief history of the creation of the Zambezi River Autority. In *Management of River Basins and Dams*; CRC Press: Boca Raton, FL, USA, 2021; pp. 43–47.
24. Hamududu, B.H.; Killingtveit, A. Hydropower production in future climate scenarios; the case for the Zambezi River. *Energies* **2016**, *9*, 502. [CrossRef]
25. Funk, C.C.; Peterson, P.J.; Landsfeld, M.F.; Pedreros, D.H.; Verdin, J.P.; Rowland, J.D.; Verdin, A.P. *A Quasi-Global Precipitation Time Series for Drought Monitoring*; US Geological Survey Data Series 832; Earth Resources Observation and Science (EROS) Center U.S. Geological Survey: Sioux Falls, SD, USA, 2014; pp. 1–12.
26. Funk, C.; Peterson, P.; Landsfeld, M.; Pedreros, D.; Verdin, J.; Shukla, S.; Michaelsen, J. The climate hazards infrared precipitation with stations—A new environmental record for monitoring extremes. *Sci. Data* **2015**, *2*, 150066. [CrossRef]
27. Sheffield, J.; Goteti, G.; Wood, E.F. Development of a 50-year high-resolution global dataset of meteorological forcings for land surface modeling. *J. Clim.* **2006**, *19*, 3088–3111. [CrossRef]
28. Huffman, G.J.; Bolvin, D.T.; Nelkin, E.J.; Wolff, D.B.; Adler, R.F.; Gu, G.; Stocker, E.F. The TRMM multisatellite precipitation analysis (TMPA): Quasi-global, multiyear, combined-sensor precipitation estimates at fine scales. *J. Hydrometeorol.* **2007**, *8*, 38–55. [CrossRef]
29. Joyce, R.J.; Janowiak, J.E.; Arkin, P.A.; Xie, P. CMORPH: A method that produces global precipitation estimates from passive microwave and infrared data at high spatial and temporal resolution. *J. Hydrometeorol.* **2004**, *5*, 487–503. [CrossRef]
30. Sorooshian, S.; Hsu, K.L.; Gao, X.; Gupta, H.V.; Imam, B.; Braithwaite, D. Evaluation of PERSIANN system satellite-based estimates of tropical rainfall. *Bull. Am. Meteorol. Soc.* **2000**, *81*, 2035–2046. [CrossRef]
31. Hong, Y.; Hsu, K.L.; Sorooshian, S.; Gao, X. Precipitation estimation from remotely sensed imagery using an artificial neural network cloud classification system. *J. Appl. Meteorol.* **2004**, *43*, 1834–1853. [CrossRef]
32. Saha, S.; Moorthi, S.; Pan, H.L.; Wu, X.; Wang, J.; Nadiga, S.; Liu, H. The NCEP climate forecast system reanalysis. *Bull. Am. Meteorol. Soc.* **2010**, *91*, 1015–1058. [CrossRef]
33. Bergström, S. *Development and Application of a Conceptual Runoff Model for Scandinavian Catchments No. 52*; Department of Water Resources Engineering, Lund Institute of Technology, University of Lund: Lund, Sweden, 1976.
34. Yang, Y.E.; Wi, S. Informing regional water-energy-food nexus with system analysis and interactive visualization—A case study in the Great Ruaha River of Tanzania. *Agric. Water Manag.* **2018**, *196*, 75–86. [CrossRef]

35. Boyle, D. Multicriteria Calibration of Hydrological Models. Ph.D. Thesis, University of Arizona, Tucson, AZ, USA, 2000.
36. Wi, S.; Yang, Y.C.E.; Steinschneider, S.; Khalil, A.; Brown, C.M. Calibration approaches for distributed hydrologic models in poorly gaged basins: Implication for streamflow projections under climate change. *Hydrol. Earth Syst. Sci.* **2015**, *19*, 857–876. [CrossRef]
37. Liang, X.; Lettenmaier, D.P.; Wood, E.F.; Burges, S.J. A simple hydrologically based model of land surface water and energy fluxes for general circulation models. *J. Geophys. Res.* **1994**, *99*, 14415–14428. [CrossRef]
38. Lohmann, D.; Raschke, E.; Nijssen, B.; Lettenmaier, D.P. Regional scale hydrology: I. Formulation of the VIC-2L model coupled to a routing model. *Hydrol. Sci. J.* **1998**, *43*, 131–141. [CrossRef]
39. Wang, Q.J. The genetic algorithm and its application to calibrating conceptual rainfall-runoff models. *Water Resour. Res.* **1991**, *27*, 2467–2471. [CrossRef]
40. Duan, Q.; Sorooshian, S.; Gupta, V. Effective and efficient global optimization for conceptual rainfall-runoff models. *Water Resour. Res.* **1992**, *28*, 1015–1031. [CrossRef]
41. Duan, Q.Y.; Gupta, V.K.; Sorooshian, S. Shuffled complex evolution approach for effective and efficient global minimization. *J. Optim. Theory Appl.* **1993**, *76*, 501–521. [CrossRef]
42. Beven, K.; Binley, A. The future of distributed models: Model calibration and uncertainty prediction. *Hydrol. Process.* **1992**, *6*, 279–298. [CrossRef]
43. Vrugt, J.A.; Gupta, H.V.; Bouten, W.; Sorooshian, S. A Shuffled Complex Evolution Metropolis algorithm for optimization and uncertainty assessment of hydrologic model parameters. *Water Resour. Res.* **2003**, *39*. [CrossRef]
44. Gupta, H.V.; Kling, H.; Yilmaz, K.K.; Martinez, G.F. Decomposition of the mean squared error and NSE performance criteria: Implications for improving hydrological modelling. *J. Hydrol.* **2009**, *377*, 80–91. [CrossRef]
45. Nash, J.E.; Sutcliffe, J.V. River flow forecasting through conceptual models part I—A discussion of principles. *J. Hydrol.* **1970**, *10*, 282–290. [CrossRef]
46. Liu, Y.; Gupta, H.V. Uncertainty in hydrologic modeling: Toward an integrated data assimilation framework. *Water Resour. research* **2007**, *43*, 43. [CrossRef]
47. Montanari, A.; Brath, A. A stochastic approach for assessing the uncertainty of rainfall-runoff simulations. *Water Resour. Res.* **2004**, *40*. [CrossRef]
48. Montanari, A. Large sample behaviors of the generalized likelihood uncertainty estimation (GLUE) in assessing the uncertainty of rainfall-runoff simulations. *Water Resour. Res.* **2005**, *41*. [CrossRef]
49. Montanari, A.; Grossi, G. Estimating the uncertainty of hydrologic forecasts: A statistical approach. *Water Resour. Res.* **2008**, *44*. [CrossRef]
50. Di Baldassarre, G.; Montanari, A. Uncertainty in river discharge observations: A quantitative analysis. *Hydrol. Earth Syst. Sci.* **2009**, *13*, 913–921. [CrossRef]
51. Sikorska, A.E.; Montanari, A.; Koutsoyiannis, D. Estimating the uncertainty of hydrological predictions through data-driven resampling techniques. *J. Hydrol. Eng.* **2015**, *20*, A4014009. [CrossRef]
52. Papacharalampous, G.; Koutsoyiannis, D.; Montanari, A. Quantification of predictive uncertainty in hydrological modelling by harnessing the wisdom of the crowd: Methodology development and investigation using toy models. *Adv. Water Resour.* **2020**, *136*, 103470. [CrossRef]
53. Clark, M.P.; Fan, Y.; Lawrence, D.M.; Adam, J.C.; Bolster, D.; Gochis, D.J.; Hooper, R.P.; Kumar, M.; Leung, L.R.; Mackay, D.S.; et al. Improving the representation of hydrologic processes in Earth System Models. *Water Resour. Res.* **2015**, *51*, 5929–5956. [CrossRef]
54. Wood, A.W.; Leung, L.R.; Sridhar, V.; Lettenmaier, D.P. Hydrologic implications of dynamical and statistical approaches to downscaling climate model outputs. *Clim. Chang.* **2004**, *62*, 189–216. [CrossRef]
55. Ines, A.V.; Hansen, J.W. Bias correction of daily GCM rainfall for crop simulation studies. *Agric. For. Meteorol.* **2006**, *138*, 44–53. [CrossRef]
56. Piani, C.; Haerter, J.O.; Coppola, E. Statistical bias correction for daily precipitation in regional climate models over Europe. *Theor. Appl. Climatol.* **2010**, *99*, 187–192. [CrossRef]
57. Crochemore, L.; Ramos, M.-H.; Pappenberger, F. Bias correcting precipitation forecasts to improve the skill of seasonal streamflow forecasts. *Hydrol. Earth Syst. Sci.* **2016**, *20*, 3601–3618. [CrossRef]
58. Cannon, A.J.; Sobie, S.R.; Murdock, T.Q. Bias correction of GCM precipitation by quantile mapping: How well do methods preserve changes in quantiles and extremes? *J. Clim.* **2015**, *28*, 6938–6959. [CrossRef]
59. Pastén-Zapata, E.; Jones, J.M.; Moggridge, H.; Widmann, M. Evaluation of the performance of Euro-CORDEX Regional Climate Models for assessing hydrological climate change impacts in Great Britain: A comparison of different spatial resolutions and quantile mapping bias correction methods. *J. Hydrol.* **2020**, *584*, 124653. [CrossRef]
60. Kim, B.K.; Kwon, H.H.; Han, D. Bias-correction schemes for calibrated flow in a conceptual hydrological model. *Hydrol. Res.* **2021**, *52*, 196–211. [CrossRef]
61. Nearing, G.S.; Gupta, H.V. Ensembles vs. information theory: Supporting science under uncertainty. *Front. Earth Sci.* **2018**, *12*, 653–660. [CrossRef]
62. DeChant, C.M.; Moradkhani, H. Improving the characterization of initial condition for ensemble streamflow prediction using data assimilation. *Hydrol. Earth Syst. Sci. Discuss.* **2011**, *15*, 3399–3410. [CrossRef]

63. Thiboult, A.; Anctil, F.; Boucher, M.A. Accounting for three sources of uncertainty in ensemble hydrological forecasting. *Hydrol. Earth Syst. Sci.* **2016**, *20*, 1809–1825. [CrossRef]
64. Arnal, L.; Wood, A.W.; Stephens, E.; Cloke, H.L.; Pappenberger, F. An efficient approach for estimating streamflow forecast skill elasticity. *J. Hydrometeorol.* **2017**, *18*, 1715–1729. [CrossRef]

hydrology MDPI

Article

A Hydraulic Analysis of Shock Wave Generation Mechanism on Flat Spillway Chutes through Physical Modeling

Muhammad Kaleem Sarwar [1], Muhammad Atiq Ur Rehman Tariq [2,3,*], Rashid Farooq [4], Hafiz Kamran Jaleel Abbasi [1], Faraz Ul Haq [1], Ijaz Ahmad [1], Muhammad Izhar Shah [5], Anne. W. M. Ng [6] and Nitin Muttil [2,3,*]

1 Centre of Excellence in Water Resources Engineering, University of Engineering and Technology, Lahore 54890, Pakistan; eng_Kaleem@yahoo.com (M.K.S.); kami.abbasi019@gmail.com (H.K.J.A.); engraraz@uet.edu.pk (F.U.H.); engr.ijaz786@gmail.com (I.A.)
2 College of Engineering and Science, Victoria University, Melbourne 8001, Australia
3 Institute for Sustainable Industries & Liveable Cities, Victoria University, P.O. Box 14428, Melbourne 8001, Australia
4 Department of Civil Engineering, International Islamic University, Islamabad 44000, Pakistan; rashidmeo50@gmail.com
5 Department of Civil Engineering, Abbottabad Campus, COMSATS University Islamabad, Abbottabad 22060, Pakistan; mizharshah.civ@uetpeshawar.edu.pk
6 College of Engineering, IT & Environment, Charles Darwin University, Darwin 0810, Australia; Anne.Ng@cdu.edu.au
* Correspondence: Atiq.Tariq@yahoo.com (M.A.U.R.T.); Nitin.Muttil@vu.edu.au (N.M.)

Citation: Sarwar, M.K.; Tariq, M.A.U.R.; Farooq, R.; Abbasi, H.K.J.; Haq, F.U.; Ahmad, I.; Shah, M.I.; Ng, A.W.M.; Muttil, N. A Hydraulic Analysis of Shock Wave Generation Mechanism on Flat Spillway Chutes through Physical Modeling. *Hydrology* **2021**, *8*, 186. https://doi.org/10.3390/hydrology8040186

Academic Editors: Carmelina Costanzo, Tommaso Caloiero and Roberta Padulano

Received: 29 November 2021
Accepted: 16 December 2021
Published: 17 December 2021

Abstract: Shock waves are generated downstream of spillways during flood operations, which have adverse effects on spillway operations. This paper presents the physical model study of shock waves at the Mohmand Dam Spillway project, Pakistan. In this study, hydraulic analysis of shock waves was carried out to investigate its generation mechanism. Different experiments were performed to analyze the rooster tail on a flat spillway chute and to examine the factors affecting the characteristics of the rooster tail. The study results show that shock wave height is influenced by spillway chute slope, pier shape, and flow depth. Moreover, the height of the shock wave can be minimized by installing a semi-elliptical pier on the tail part of the main pier. Further modifications in the geometry of the extended tail part of the pier are recommended for the elimination of the shock wave. Based on observed data collected from the model study, an empirical equation was developed to estimate the shock wave height generated on the flat slope spillway chutes (5° to 10°).

Keywords: shock wave; spillway; spillway pier; flat chute; physical modeling

1. Introduction

Concerns related to the effects of unpredictably high flows entering reservoirs, especially considering possible increased rainfall intensities, have led to a renewed general interest in reservoir spillway design. In the spillways, gates are mounted onto the crest of a free spillway that controls the head, discharge, reservoir volume, and reservoir level increase. The addition of these gates adds some new complex issues to the hydraulic subjects [1–3]. According to Ansar et al. [4], not all flow conditions can occur at most spillways; thus, flow conditions at a gated spillway tend to become controlled when the gate opening G_o is smaller than the critical depth y_c and submerged when the tailwater depth h is greater than y_c. Moreover, they developed generalized flow rating equations based on field flow measurements at the gated spillway. Al-Mansori et al. [5] found that, with increasing hydraulic head up to seven times that of the design head, the flow separation zone grows linearly.

Among gate discharge coefficients, the gate's location above the spillways, and separation of flow profile, the spillway transverse flows and waves are lesser-known issues. These waves are called by different names such as shock waves, lateral shock waves, and rooster tail waves [6,7]. The mechanism that induces the rooster tail is the gathering pressure generated by the joint flow. During that process, the kinetic energy of the diffused flow is converted into pressure energy and is transported to the bottom, subsequently. Thus, variations in the pressure of the spillway bottom can be used to reflect the intensity of the rooster tail [6]. Likewise, shock waves are frequently generated in dam spillways by uniform flow disturbances due to the presence of spillway piers, curves, or changes in the cross-section of spillway chutes [6,8,9]. Consequently, local maxima inflow depth are produced, and their magnitude can be much greater than the incoming uniform depth. This has important practical implications in the dimensioning of the chute walls, which must be designed with greater height to adequately convey the flow, with the obvious consequences in terms of magnitude and cost [10]. According to Jiang and Xiang [11], shock waves behind the pier may result in poor hydraulic performance such as reduced discharge capacity, erosion and swell, increased aeration, and asymmetry of water flow in the steep groove. The shock wave generated at the end of the pier causes the water surface downstream of the sluice to become a cross-shaped rhombic wave, which causes the water flow to be partially high and may lead to overflow. For spillways, overflow may result in erosion of the spillway foundation; thus, to prevent the overflow of water, it is necessary to increase the sidewall.

Moreover, the design of the spillway must tend to the constant cross-section and slope to approach uniform flow and thus avoid the shock wave phenomenon. However, these geometric particularities are frequently unavoidable due to the topographical and/or geological characteristics of the dam and spillway site [10]. Thus, shock waves can be generally grouped into different categories [6]. In slit-type energy dissipaters, a shock wave is generated due to the contraction of the channel section [5,12]. The second kind of shock wave is induced in stepped spillways due to their steep chute slope [13–16]. Rajaratnam [13] noted that the second kind of shock wave occurs during low flows. In another study, Chanson [17] observed the effect of step geometry and flow regime on the generation of a shock wave. Similarly, Carnacina et al. [16] conducted an experimental study to observe shock wave height on the stepped spillway and concluded that shock wave height is low in steep chutes as compared to flat ones. The third type of shock wave occurs due to the installation of an aerator on spillway chutes to avoid cavitation damages [18,19]. Studies indicate that the aerator ramp, the ratio of lateral jet length to bottom jet length, flow depth, and Froude number have a dominant effect on shock wave intensity. The fourth type of shock wave is generated directly downstream of the overflow spillway crest. In this case, diffusion of flows takes place due to spillway piers at their tail part [6]. A number of research studies have been carried out to study the hydraulic characteristic of this kind of shock wave. Behera et al. [20] concluded that a standing wave is created at the downstream end of the pier because the confluence of the flows from the two spillway bays is at a slight angle and higher discharge. The standing wave develops directly downstream of the pier, and the shock waves travel laterally, reaching the downstream of a bucket for higher discharge Q equals 810.54 m^3/s. These standing waves with shock waves are highly fluctuating and create additional scouring damage in the downstream spillway together with regular hydraulic jumps. Under these circumstances, such waves may be eliminated to save costs and avoid potential catastrophic hazards. Duan [21] developed a sloping tail pier to eliminate the shock wave generated on the spillway chute of a hydropower project. Later, an experimental study of [22–24] concluded that the height of shock wave depends on the ratio of approach flow depth to pier width. They performed several experiments in the horizontal channel and sloping chutes to investigate the hydraulic characteristics of a shock wave. In another study, Wu and Yan [25,26] investigated the formation of a shock wave in discharging tunnel due to pier. They developed a new type of pier to control shock wave in discharge tunnel. Recently, Xue et al. [6] also developed a composite tail for the spillway pier to reduce shock wave height or eliminate the shock wave on spillway chutes with slope variations of 12° to 20°. Xue et al. [6] noted that shock waves are influenced by spillway pier width and type, chute slope and flow depth in the spillway chute.

Despite many studies on the hydraulic characteristics of flow in a chute spillway, there is insufficient knowledge of the formation of shock waves. Investigations on the formation of the shock wave flow in a horizontal rectangular channel by Reinaur and Hager [22] showed that, for the state of constant flow depth H_0 and constant pier width b_p, the height of waves and their width increase with an increase in the Froude number. Recently, it was noticed that if the depth measurements were taken perpendicularly to the chute spillway, the results would be consistent with the studies performed in the horizontal channel [23,27]. Further, Wu and Yan [25] investigated the hydraulic characteristics of the shock wave's formation by the pier of the discharge tunnel of the Sanbanxi hydropower station. It was observed that the primary reason for the shock wave's inception was the concavity of the water surface.

The current literature on shock waves over spillways includes a large quantity of experimental work indicating that an intense shock wave can overtop the spillway chute walls and induce vibrations in spillway structures [25,28]. Ultimately, such a situation disturbs the flood operation of the spillway [29–31]. The literature survey also showed the lack of research on shock wave generation at flat spillway chutes (5^0 to 10^0). Keeping in view the adverse effects of a shock wave, a physical model of Mohmand Dam Spillway was constructed to investigate the hydraulic characteristics and generation mechanism of shock waves on flat spillway chutes. In the end, an empirical equation was developed based on observed data to estimate shock wave height for flat slope spillway chutes.

2. Study Area

Mohmand dam is proposed on Swat River, approximately 37 km north of Peshawar and 5 km upstream of existing Munda Headworks in Mohmand Agency of Pakistan's Federally Administrated Tribal Area (FATA), as shown in Figure 1. The gated type of spillway is provided to pass a design flood of 27,427 m^3/s. It is to be located on the left abutment of the dam. It comprises seven bays (07) and a long concrete chute. The crest level of the spillway is 539 m amsl, whereas the maximum reservoir operating level is 563 m amsl. To normalize the high flow velocity (23 to 45 m/s), a double stilling basin arrangement is provided to dissipate the energy.

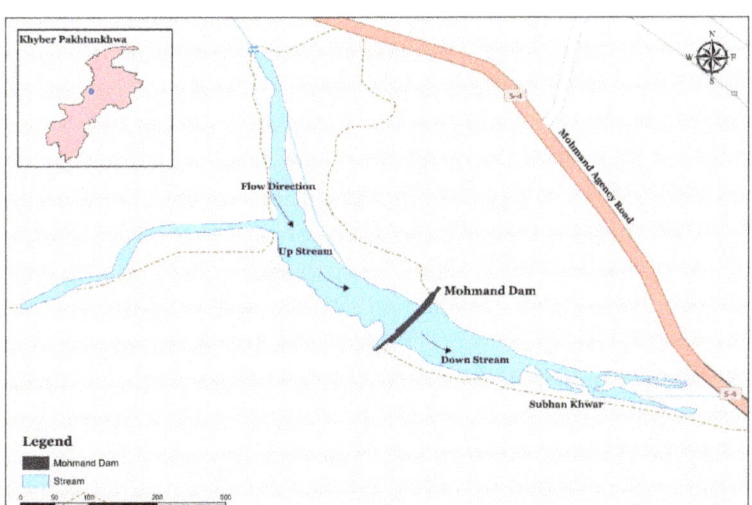

Figure 1. Location of Mohmand Dam project.

3. Experimental Setup

To study the hydraulic characteristics of shock waves on flat spillway chutes, a physical model of Mohmand Dam spillway was constructed in the model tray hall of Center of Excellence in Water Resources Engineering (CEWRE), Lahore, Pakistan. The spillway model was designed based on gravity similarity criteria. It was constructed with two bays at a scale of 1:100 considering discharge and space limitations.

Each bay was equipped with radial gates. As shown in Figure 2a, the model consists of: a water tank, V-notch, a small water tank, baffle walls, a spillway control section (provided with redial gates and piers), and a spillway chute. The width of each bay and pier was 15 and 5 mm, respectively. As depicted in Figure 2b, the tail part of the spillway pier has an elliptical or rectangular shape. The length of semi-major axis (a) and chute slope (θ) were considered for this study. Details regarding geometric parameters of the pier and spillway chute slope are discussed in Table 1 below (where $a = 0$ means tail part of the pier is rectangular, otherwise it is elliptical).

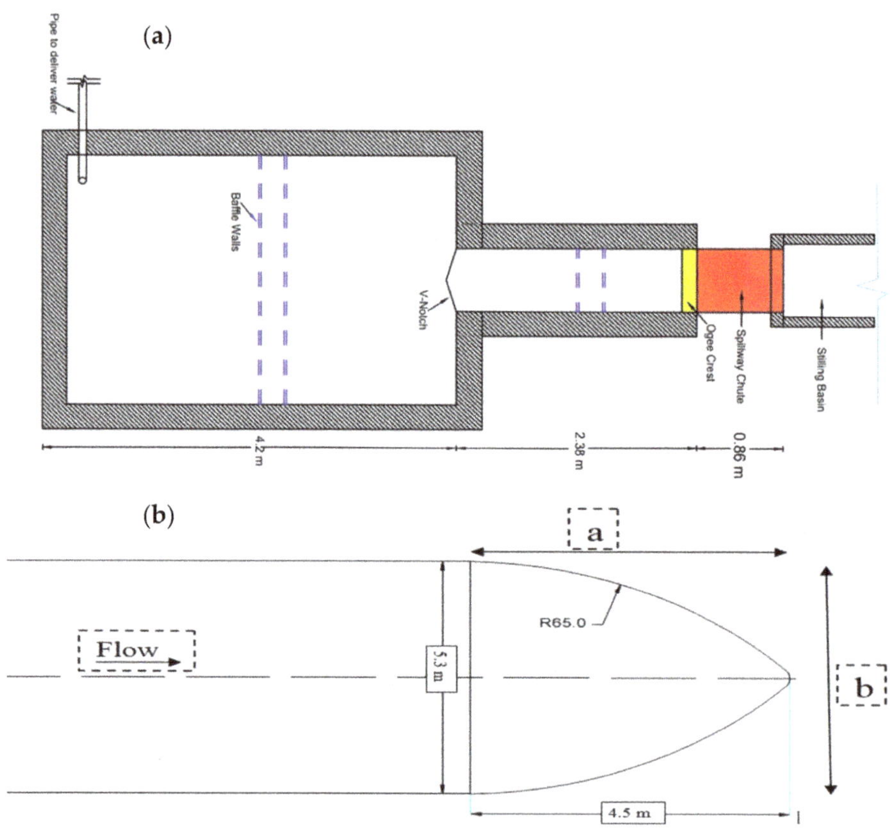

Figure 2. (**a**) Layout plane of the experimental setup. (**b**) Definition sketch for semi-major axis of spillway pier.

Table 1. Detail of geometric modifications.

Geometric Variables	M_1	M_2	M_3	M_4
Spillway slope (θ)	1:6.5	1:6.5	1:8.5	1:8.5
Length of semi major axis (m) (a)	0	4.5	0	4.5

Model operation was performed for each geometric variable shown in Table 1 at free-flow conditions by varying the reservoir levels 'H' from 541 to 558 m amsl (with an increment of 1 m). However, it was operated after its validation and dynamic similarity check, as discussed below. Hydraulic parameters observed during model operation include flow depth at upstream of the pier (h_o), shock wave height (h_m), depth of water surface cavity at confluence area of two defused flows (h_c), horizontal distance before the collision of the two diffused flows (I_c), and pressure head at the bottom of the chute (h). Flow depth and pressure heads were measured using point gauge and piezometer tubes, respectively. Figure 3 shows the definition sketch for observed parameters.

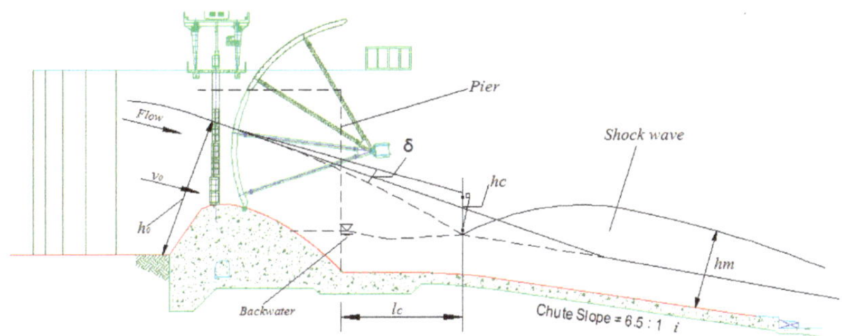

Figure 3. Definition sketch for observed hydraulic parameters.

4. Model Validation

Validation of the model was performed to obtain an idea about the accuracy of model results. For this purpose, observed discharge values were compared with computed ones. The comparison of observed discharge with computed discharge, as indicated in Table 2, at reservoir level of 541m average mean sea level (amsl) shows only a 1.19% error, whereas at maximum reservoir levels, this difference is 3.11%. Table 2 shows that the overall percentage of the difference between observed and computed values is up to 3%. It indicates that the physical model is a good representative of a prototype.

Table 2. Comparison between observed and computed discharge.

Sr. No.	Reservoir Level amsl (m)	Observed Discharge (m^3/s)	Computed Discharge (m^3/s)	Error (%)
1	539	0	0	0
2	541	156.99	155.14	1.19
3	543	453.08	446.8	1.41
4	545	848.05	837.4	1.27
5	547	1337.75	1314.28	1.79
6	549	1902.02	1870.5	1.69
7	551	2550.89	2498.5	2.10
8	558	5052.27	5214.2	3.11

5. Dynamic Similarity of Model

Dynamic similarity exists between the model and prototype if the ratio of inertial to gravity forces at some point in the model is the same at the corresponding point of the prototype. Froude similarity law is used in problems where gravity is important, i.e., in free-surface flows such as in case of flow over the spillway, weir, under sluice gates, open channel, etc. In the current study, Froude's model law was applied. The Froude number observed near the control section of the spillway at the reservoir level of 558 m amsl was 0.730, whereas the calculated value was 0.725. This indicates that the model is dynamically similar to the prototype.

6. Height of Shock Wave

The height of the shock wave provides an important reference for the computation of optimal height of spillway sidewalls. In the current study, shock wave height was computed by operating the Mohmand Dam Spillway model for geometric variables as shown in Table 1. The results show (Figure 4) that in the beginning, the height of the shock wave for all geometric variables (M_1 to M_4) was almost the same, but with the further increase in reservoir level (above 547 m amsl), a clear difference in shock wave height between M_1 and M_2 and M_3 and M_4 was noticed. For M_1, the shock wave height was increased with the increase in reservoir level up to 551m amsl, and after that, the shock wave height decreased with a further increase in reservoir level. For M_2, the shock wave height increased thoroughly with the increase in the reservoir level. The same trend was observed in the case of M_3 and M_4. It is also visible from Figure 4 that shock wave heights are also almost the same at the maximum reservoir level. It is noted that when discharge increases, the backwater flow also increases, and the distance between the two diffused flows and pier decreases, which affect the shock wave height at maximum discharge or reservoir level. A clear difference in shock wave height between M_1 and M_2 and M_3 and M_4 (from reservoir level 547 to 557 m amsl) indicates the impact of the extended part of the pier installed at its tail end (semi-major axis length).

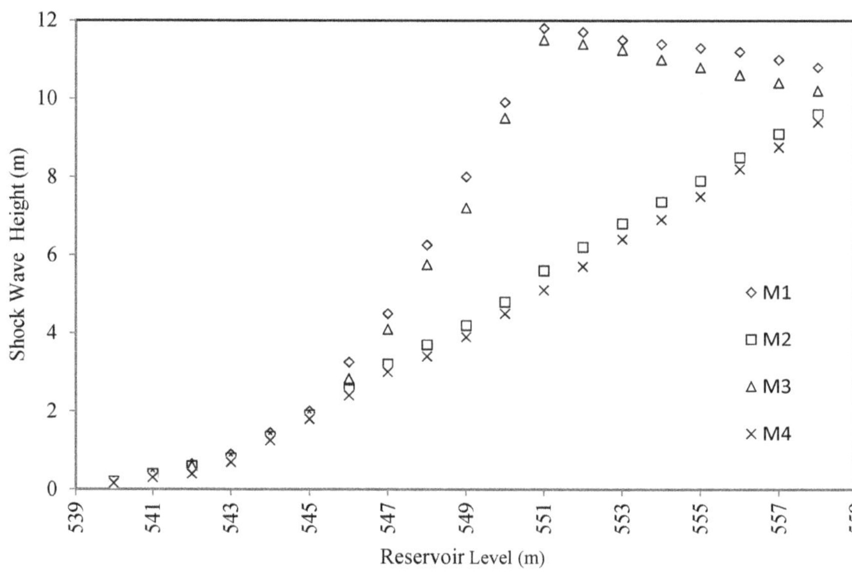

Figure 4. Observed shock wave heights for all geometric variables.

Figure 5a–c presents the profile view of shock wave formation at reservoir levels of 547, 549 and 551 m amsl, respectively, without geometric modification ($a = 0$), while Figure 5d–f shows the profile view of shock wave formation at above-mentioned reservoir levels with geometric modification ($a = 45$ mm). As shown in Figure 5a,d, for the same reservoir level, i.e., at 547 m amsl, the shock wave height is low in the case of 'd' as compared to case 'a' due to geometric modification. A similar trend can be noted at other reservoir levels (549 and 551 m amsl) from Figure 5.

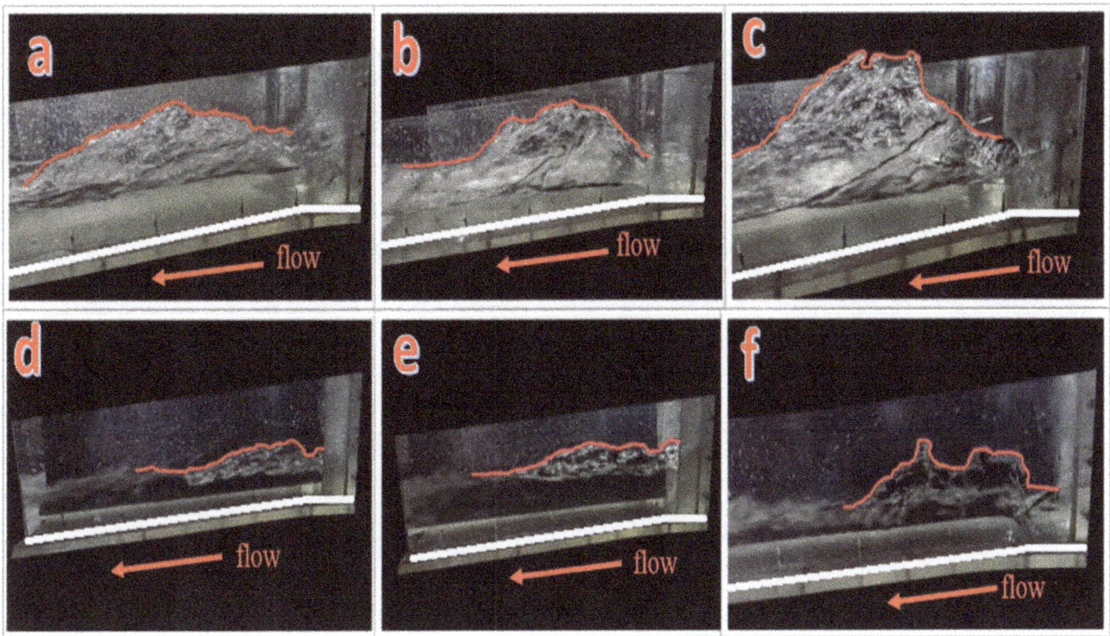

Figure 5. Profile views of shock wave formation for M_3 (**a–c**) and M_4 (**d–f**).

7. Evaluation of Cavity Angle and Diffused Flow Pressure

The flow was diffused laterally behind the pier because there was no constraint effect due to the pier sidewall. A cavity with backwater in the tail part of the pier was generated by the diffused flow, as shown in Figure 6a,b. Based on the experimental observations, it was noted that the size of the cavity can significantly affect the shock wave formation. The ratio of the horizontal diffusion distance to the depth of the water surface concave (I_c/h_c) can be used to describe the cavity form. Thus, the experimental data were used to investigate the relationship between the ratio of the shock wave height to the width of the pier (h_m/b) and I_c/h_c. The value of the coefficient of determination is 0.85, which indicates that there is a significant correlation between them (Figure 7), and an exponential relationship can be written as follows:

$$\frac{h_m}{b} = 3.0018\ \exp(-0.493\frac{I_c}{h_c}) \tag{1}$$

As per Figure 3, the equation for the dropping angle (δ) can be given as under:

$$cot\delta = \frac{I_c}{h_c}\frac{1}{cos^2\theta} + tan\theta \tag{2}$$

In Equation (2), θ represents the spillway chute angle. By rearranging Equation (2) and substituting it into Equation (1), we can obtain:

$$\frac{h_m}{b} = 3.0018\exp\left[-0.493(cot\delta - tan\theta)\cdot cos^2\theta\right] \tag{3}$$

It can be noted from Equation (3) that h_m/b increases with the increase in dropping angle (δ) when the spillway chute slope (θ) and the width of the pier (b) remain constant, which means that the rooster tail's height is significantly affected by the dropping angle.

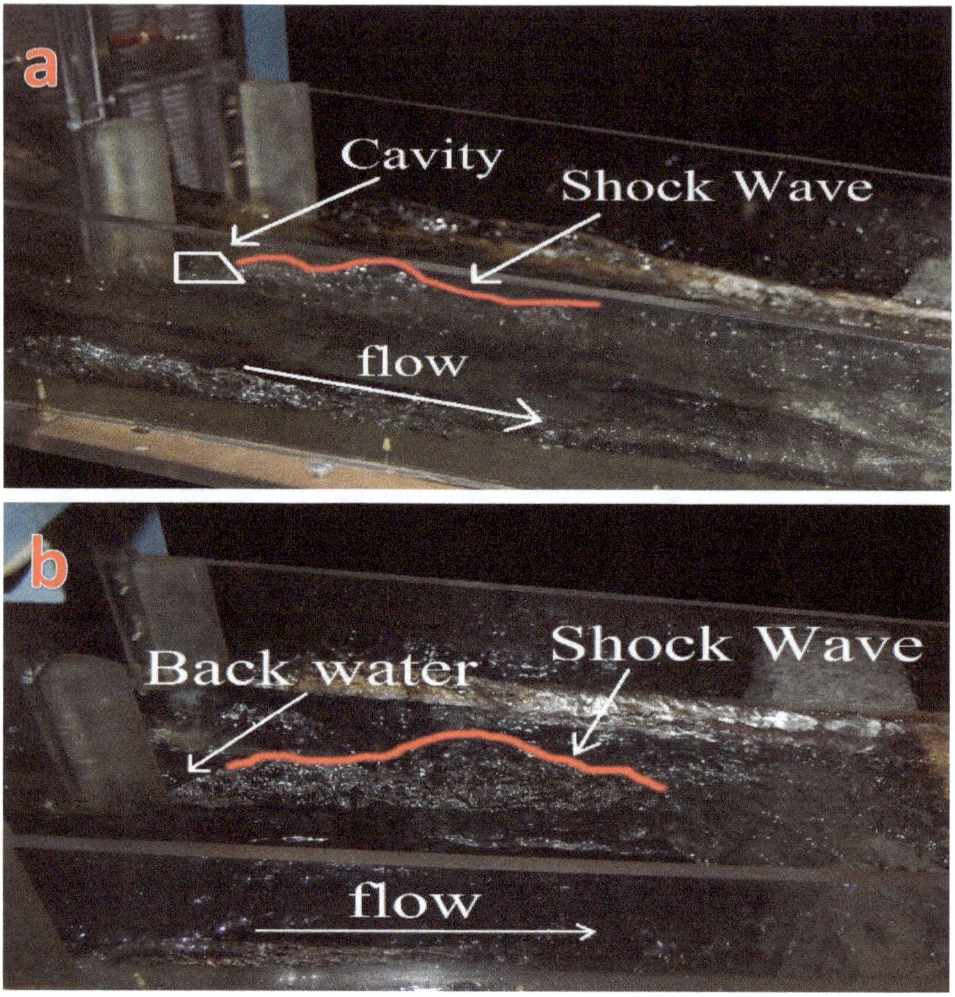

Figure 6. Cavity formation behind the pier. (**a**) Longitudinal view. (**b**) Plan view.

Pressure Energy

The other hydraulic parameter that induced the shock wave was the gathering pressure generated by the joint flow. During that process, the kinetic energy of the diffused flow is converted into pressure energy and is transported to the bottom, subsequently. Thus, variations in the pressure of the spillway bottom may be used to reflect the intensity of the shock wave.

The pressure distributions along the length of the chute for M_3 and M_4 are shown in Figures 8 and 9, respectively, in which the origin of the coordinate axis (x) is at the tail of the pier, 'p' is the spillway chute pressure, and 'γ' is the bulk density of water (γ = 9.8 kN/m^3). It can be concluded from Figures 8 and 9 that at different water heads (H), all the pressure peaks occur at the bottom near the tail part of the pier due to the formation of a shock wave. Moreover, the peak values for M_4 are slightly less as compared to M_3 due to extension in the tail part of the pier (length of semi-major axis).

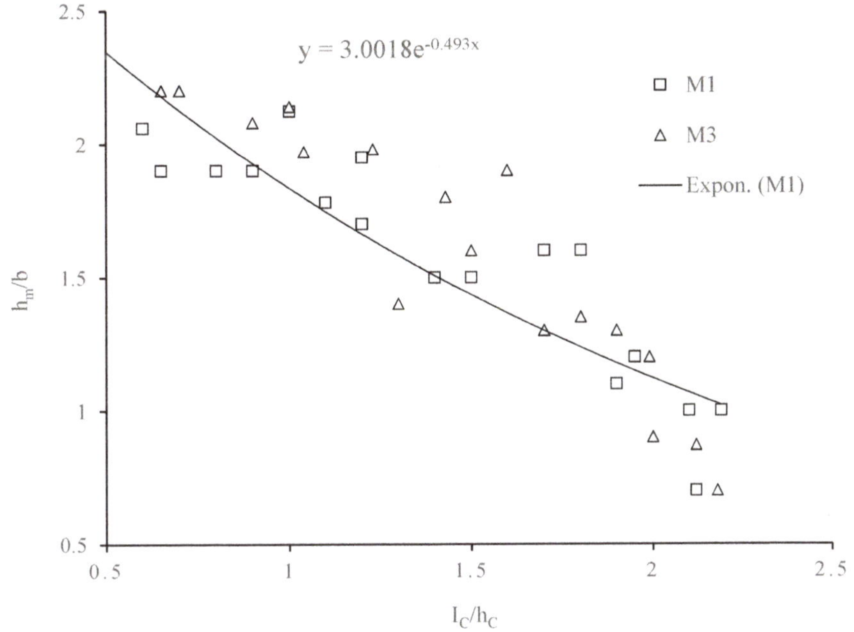

Figure 7. Relationship between the ratio of the rooster tail height to the width of the pier (h_m/b) and I_c/h_c.

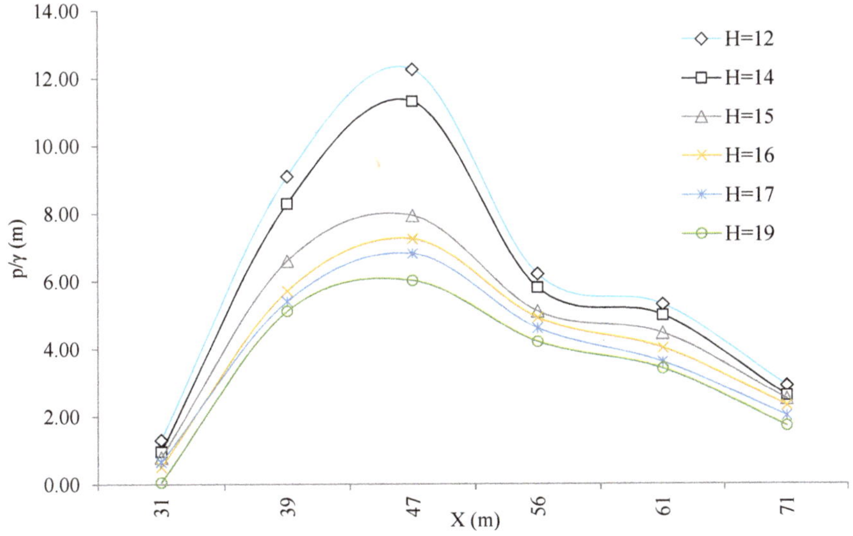

Figure 8. Spillway chute pressure distribution for M_3.

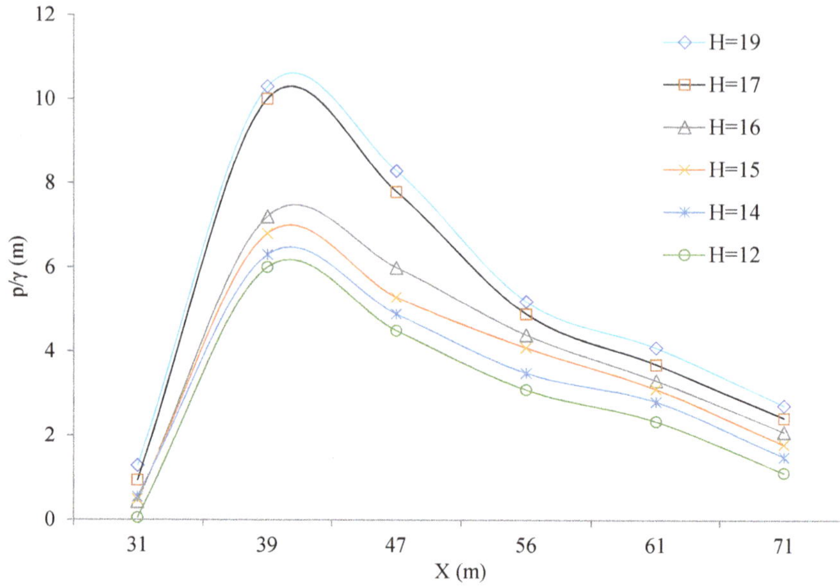

Figure 9. Spillway chute pressure distribution for M_4.

8. Evaluation of Cavity Angle and Diffused Flow Pressure

The equation presented by Xue et al. (2018) was modified for estimation of the shock wave height on flat spillway chutes (5° to 10°).

$$\frac{h_m}{h_o} = 2.5\left(\frac{h_o}{b}\right)^{0.5} \exp(-\frac{c_1 a}{h_o} - c_2 i) \tag{4}$$

In the above equation, 'c_1' and 'c_2' are experimental coefficients, whereas 'i' represents the slope ratio of the bottom. Analysis was performed on MiniTab software to examine the suitability of experimental coefficients for experimental data collected from the model study. Minitab is the statistical software that helps in taking out the complexities of statistical calculations. The experimental data were divided into training and testing values. Seventy percent (70%) of values were used as training values, while thirty percent (30%) values were considered as testing values. Training data were used to modify the Xue et al. (2018) equation. The modified form of Equation (4) is shown below.

$$\frac{h_m}{h_o} = 2.5\left(\frac{h_o}{b}\right)^{0.5} \exp(-\frac{0.78a}{h_o} - 13.67i) \tag{5}$$

Computed values of shock wave height using Equation (5) were correlated with the testing values (observed data) to find the coefficient of determination (R^2). Figure 10 shows that the coefficient of determination for computed and observed data of shock wave height is 0.91, which indicates that data is in good correlation. Hence, the equation developed (Equation (5)) through this study can be used for the estimation of shock wave height generated on flat spillway chutes (5° to 10°).

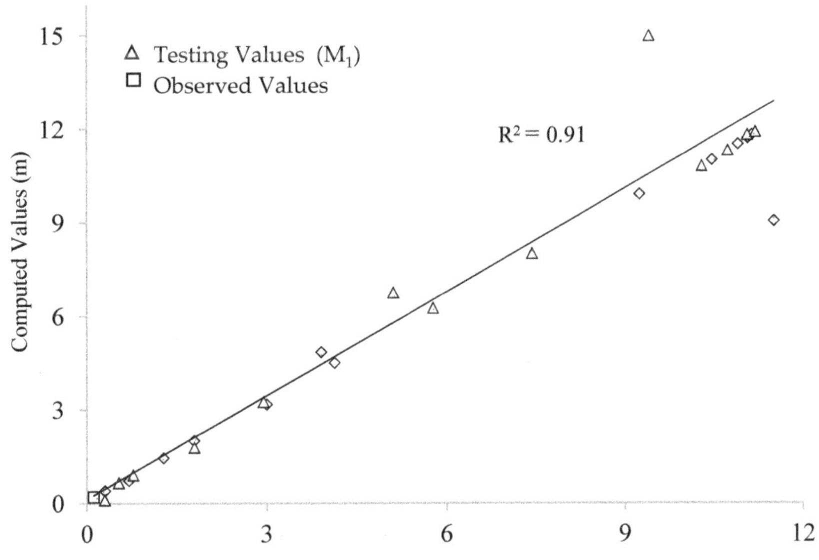

Figure 10. Plot between observed and computed shock wave height.

9. Conclusions

A physical study was performed on a flat spillway with varying geometric parameters of pier and spillway chute slope. While different chute configurations were tested, the focus of the study was on the hydraulic characteristics and generation mechanism of a shock wave on flat spillway chutes (5° to 10°). The spillway model was designed based on gravity similarity criteria. It was constructed with two bays at a scale of 1:100 considering discharge and space limitations.

The study results showed that shock wave height was slightly decreased with a decrease in spillway chute slope. However, extension in the tail part of the spillway pier (semi-major axis length) caused a further reduction in shock wave height. Overall, a 26% reduction in shock wave height was noted due to the extended part of the spillway pier. The dropping angle of diffused flow significantly affected the shock wave height. It was noted that shock wave height is directly proportional to the dropping angle of diffused flow. Variation in pressure energy along the spillway chute indicated the intensity of shock wave. Pressure peaks were observed near the tail part of the spillway pier due to the formation of a shock wave. In this study, an empirical equation was developed to estimate the shock wave height, which can be used to estimate the shock wave height at the flat spillway chute, varying from 5^0 to 10^0. Lastly, the main goal of the experiments was to support the planning of shock waves' adverse effects on spillway operations, therefore, the currently reduced-scale experiments may apply to prototype scale, considering the extension in the tail part of the spillway pier along with the reduction in chute slope for the elimination of shock waves.

Author Contributions: Conceptualization, M.K.S. and M.A.U.R.T.; Data curation, M.K.S., R.F. and H.K.J.A.; Formal analysis, M.K.S. and F.U.H.; Investigation, A.W.M.N.; Methodology, M.K.S., M.A.U.R.T., I.A. and N.M.; Validation, M.I.S.; Writing—original draft, M.K.S., M.A.U.R.T. and R.F.; Writing—review & editing, M.A.U.R.T. and R.F. All authors have read and agreed to the published version of the manuscript.

Funding: This research received no external funding.

Institutional Review Board Statement: No applicable.

Informed Consent Statement: No applicable.

Data Availability Statement: No applicable.

Acknowledgments: This study is supported by Centre of Excellence in Water Resources Engineering (CEWRE) UET, Lahore and WAPDA model study cell, Nandipur, Gujranwala Pakistan.

Conflicts of Interest: The authors declare no conflict of interest.

References

1. Lee, K.H. Simulation of Dam-Breach Outflow Hydrographs Using Water Level Variations. *Water Resour. Manag.* **2019**, *33*, 3781–3797. [CrossRef]
2. Eghlidi, E.; Barani, G.A.; Qaderi, K. Laboratory Investigation of Stilling Basin Slope Effect on Bed Scour at Downstream of Stepped Spillway: Physical Modeling of Javeh RCC Dam. *Water Resour. Manag.* **2020**, *34*, 87–100. [CrossRef]
3. Jie, Z.; Zhiming, C.; Matahel, A.; Zubayed, R.; Mark, W. Transitional Flow Analysis at Prototype Gated Spillways in South Florida. *J. Irrig. Drain. Eng.* **2021**, *147*, 4020042. [CrossRef]
4. Matahel, A.; Zhiming, C. Generalized Flow Rating Equations at Prototype Gated Spillways. *J. Hydraul. Eng.* **2009**, *135*, 602–608. [CrossRef]
5. Al-Mansori, N.J.H.; Alfatlawi, T.J.M.; Hashim, K.S.; Al-Zubaidi, L.S. The Effects of Different Shaped Baffle Blocks on the Energy Dissipation. *Civ. Eng. J.* **2020**, *6*, 961–973. [CrossRef]
6. Xue, H.; Diao, M.; Ma, Q.; Sun, H. Hydraulic Characteristics and Reduction Measure for Rooster Tails Behind Spillway Piers. *Arab. J. Sci. Eng.* **2018**, *43*, 5597–5604. [CrossRef]
7. Vayghan, V.H.; Mohammadi, M.; Ranjbar, A. Experimental Study of the Rooster Tail Jump and End Sill in Horseshoe Spillways. *Civ. Eng. J.* **2019**, *5*, 871–880. [CrossRef]
8. Vischer, D.L.; Hager, W.H.; Cischer, D. Intake structures. In *Dam Hydraulics*; John Wiley Sons: Chichester, UK, 1998; pp. 215–232.
9. Ghadimi, P.; Dashtimanesh, A.; Zamanian, R.; Feizi Chekab, M.A.; Mirhosseini, S.H.R. Rooster tail depression by originating a modified transom stern form using a Reynolds averaged Navier Stokes solver. *Sci. Iran.* **2015**, *22*, 765–777.
10. Salazar, F.; San-Mauro, J.; Celigueta, M.Á.; Oñate, E. Shockwaves in spillways with the particle finite element method. *Comput. Part. Mech.* **2020**, *7*, 87–99. [CrossRef]
11. JING, L.I.; XIANG, Z. Study on Formation Mechanism of Shock Wave after Gate Pier. In Proceedings of the E-Proceedings of the 38th IAHR World Congress, Panama City, Panama, 1–6 September 2019.
12. Huang, G.-b.; Hu, H.; Wang, C.-h.; Du, L. Shock waves and water wing in slit-type energy dissipaters. *J. Hydrodyn.* **2017**, *29*, 504–509. [CrossRef]
13. Rajaratnam, N. Skimming flow in stepped spillways. *J. Hydraul. Eng.* **1990**, *116*, 587–591. [CrossRef]
14. Tuna, M.C.; Emiroglu, M.E. Effect of step geometry on local scour downstream of stepped chutes. *Arab. J. Sci. Eng.* **2013**, *38*, 579–588. [CrossRef]
15. Toombes, L.; Wagner, C.; Chanson, H. Flow patterns in nappe flow regime down low gradient stepped chutes. *J. Hydraul. Res.* **2008**, *46*, 4–14. [CrossRef]
16. Carnacina, I.; Kurdistani, S.M.; Palermo, M.; Pagliara, S. El Chaparral dam model: Rooster tail formation on high sloped spillway. In Proceedings of the 3rd International Junior Researcher and Engineer Workshop on Hydraulic Structures, Brisbane, Australia, 2–3 May 2010; Volume 10, pp. 65–73.
17. Chanson, H. Prediction of the transition nappe/skimming flow on a stepped channel. *J. Hydraul. Res.* **1996**, *34*, 421–429. [CrossRef]
18. Najafi, M.R.; Zarrati, A.R. Numerical simulation of air–water flow in gated tunnels. *Proc. Inst. Civ. Eng. Water Manag.* **2010**, *163*, 289–295. [CrossRef]
19. Pagliara, S.; Kurdistani, S.M.; Roshni, T. Rooster tail wave hydraulics of chutes. *J. Hydraul. Eng.* **2011**, *137*, 1085–1088. [CrossRef]
20. Behera, J.; Kar, A.K.; Khamari, A. Scour at the downstream of Ghatakeswar spillway using non-cohesive hydraulic model. *ISH J. Hydraul. Eng.* **2021**, 1–11. [CrossRef]
21. Duan, W.J. The submerged sloping-tail pier—An effective measure to eliminate the crown of jumping flow. *J. Sichuan Univ. Eng. Sci. Ed.* **1982**, *1*, 63–67.
22. Reinauer, R.; Hager, W.H. Supercritical flow behind chute piers. *J. Hydraul. Eng.* **1994**, *120*, 1292–1308. [CrossRef]
23. Reinauer, R.; Hager, W.H. Pier waves in sloping chutes. *Int. J. Hydropower Dams* **1997**, *4*, 100–103.
24. Reinauer, R.; Hager, W.H. Shockwave in air-water flows. *Int. J. Multiph. flow* **1996**, *22*, 1255–1263. [CrossRef]
25. Wu, J.; Cai, C.; Wei, J.I.; Ruan, S. Hydraulic characteristics of water-wings for the middle-pier of a discharge tunnel. *J. Hydrodyn. Ser. B* **2006**, *18*, 567–571. [CrossRef]
26. Wu, J.; Yan, Z. Hydraulic characteristics of bottom underlay-type pier for water-wing control. *J. Hydrodyn.* **2008**, *20*, 735–740. [CrossRef]
27. Mooselu, M.G.; Nikoo, M.R.; Rayani, N.B.; Izady, A. Fuzzy multi-objective simulation-optimization of stepped spillways considering flood uncertainty. *Water Resour. Manag.* **2019**, *33*, 2261–2275. [CrossRef]

28. Sheng, C.; Jian, Z.; Ming, H.; Hazrati, A. Experimental study on water-wing characteristics induced by piers in flood drainage culverts. *Sci. Iran.* **2013**, *20*, 1320–1326.
29. Shao, Z.; Jahangir, Z.; Muhammad Yasir, Q.; Mahmood, S. Identification of potential sites for a multi-purpose dam using a dam suitability stream model. *Water* **2020**, *12*, 3249. [CrossRef]
30. Zhan, X.; Qin, H.; Liu, Y.; Yao, L.; Xie, W.; Liu, G.; Zhou, J. Variational Bayesian Neural Network for Ensemble Flood Forecasting. *Water* **2020**, *12*, 2740. [CrossRef]
31. Farooq, R.; Ahmad, W.; Hashmi, H.N.; Saeed, Z. Computation of Momentum Transfer Coefficient and Conveyance Capacity in Asymmetric Compound Channel. *Arab. J. Sci. Eng.* **2016**, *41*, 4225–4234. [CrossRef]

hydrology

MDPI

Article

Integrating Remote and In-Situ Data to Assess the Hydrological Response of a Post-Fire Watershed

Luca Folador [1], Alessio Cislaghi [1,2,*], Giorgio Vacchiano [1,2] and Daniele Masseroni [1]

1 Department of Agriculture and Environmental Sciences, University of Milan, Via Celoria 2, 20133 Milano, Italy; luca.folador@studenti.unimi.it (L.F.); giorgio.vacchiano@unimi.it (G.V.); daniele.masseroni@unimi.it (D.M.)

2 Centre of Applied Studies for the Sustainable Management and Protection of Mountain Areas (Ge.S.Di.Mont), University of Milan, Via Morino 8, Edolo, 25048 Brescia, Italy

* Correspondence: alessio.cislaghi@unimi.it

Abstract: Forest fire is a common concern in Mediterranean watersheds. Fire-induced canopy mortality may cause the degradation of chemical–physical properties in the soil and influence hydrological processes within and across watersheds. However, the prediction of the pedological and hydrological effect of forest fires with heterogenous severities across entire watersheds remains a difficult task. A large forest fire occurred in 2017 in northern Italy providing the opportunity to test an integrated approach that exploits remote and in-situ data for assessing the impact of forest fires on the hydrological response of semi-natural watersheds. The approach is based on a combination of remotely-sensed information on burned areas and in-situ measurements of soil infiltration in burned areas. Such collected data were used to adapt a rainfall–runoff model over an experimental watershed to produce a comparative evaluation of flood peak and volume of runoff in pre- and post-fire conditions. The model is based on a semi-distributed approach that exploits the Soil Conservation Service Curve Number (SCS-CN) and lag-time methods for the estimation of hydrological losses and runoff propagation, respectively, across the watershed. The effects of fire on hydrological losses were modeled by adjusting the CN values for different fire severities. Direct infiltration measurements were carried out to better understand the effect of fire on soil infiltration capacity. We simulated the hydrological response of the burned watershed following one of the most severe storm events that had hit the area in the last few years. Fire had serious repercussions in regard to the hydrological response, increasing the flood peak and the runoff volume up to 125% and 75%, respectively. Soil infiltration capacity was seriously compromised by fire as well, reducing unsaturated hydraulic conductivity up to 75% compared with pre-fire conditions. These findings can provide insights into the impact of forest fires on the hydrological response of a whole watershed and improve the assessment of surface runoff alterations suffered by a watershed in post-fire conditions.

Keywords: burned areas; hydrological modelling; infiltration capacity; SCS-CN; post-fire

Citation: Folador, L.; Cislaghi, A.; Vacchiano, G.; Masseroni, D. Integrating Remote and In-Situ Data to Assess the Hydrological Response of a Post-Fire Watershed. *Hydrology* **2021**, *8*, 169. https://doi.org/10.3390/hydrology8040169

Academic Editors: Carmelina Costanzo, Tommaso Caloiero and Roberta Padulano

Received: 12 October 2021
Accepted: 10 November 2021
Published: 12 November 2021

1. Introduction

Fires severely alter the hydrological response of watersheds to rainfall [1,2]. Fire-induced loss of canopy cover, litter consumption, and the formation of water repellent layers on the soil surface reduce canopy interception and soil infiltration, increasing flood peak and volume runoff [3–6].

Fire acts as a generalist herbivore that removes plant material above the ground surface. Depending on fire behavior and species-specific resistance, trees can suffer death or defoliation, survive, or re-sprout following fire. The ability to survive and re-sprout depends on tree height, scorch and char heights, tree species, age, stem and bark thickness, and fire intensity and residence time. In the hydrological cycle, this natural process leads to a reduction in evapotranspiration fluxes, leaf rainfall interception and tree suction capacity [7]. High temperatures can vaporize soil organic matter and generate a thin hydrophobic layer (from 1 to 7 cm), which strongly reduces soil infiltration capacity [8]. In addition, the burning process degrades soil structure and porosity, causing considerable nutrient loss through volatilization, leaching and erosion, and alters the quantity and specific composition of microbial and soil-dwelling invertebrate communities [9].

Such effects induce strong responses in watershed hydrology and flood hazard. In-situ measurements carried out in European and North American natural watersheds showed that the annual flood peak in post-fire conditions can increase between 1.2 and 10 times [10–17]. Such variability depends on watershed size, with stronger effects in smaller watersheds [18]. After individual flood events, post-fire hydrological responses are often sharper and quicker than in unburned sites, due to the simultaneous reduction in hydrological retention and time of concentration [19]. Flood magnitude may also increase after fire, e.g., from a 10-year to a 1-year return interval for the same discharge [20]. Youberg et al. [21] estimated that a 2-year return period rainfall in a burned watershed can produce a runoff similar to a 100-year event in pre-fire conditions. Such an exceptional increase in hazard has a catastrophic impact on the economy of a region, although assessing the costs of post-fire flooding is very difficult. The Emergency Events Database (EM-DAT), an international disaster database (*http://www.emdat.be/*, accessed on 8 October 2021), lists, for instance, damages of Euros 200 billion related to various disasters since 1900 in the Mediterranean countries, of which 85 billion are related to flooding.

Despite the knowledge accumulated on the effects of fire on soils, quantifying the impact of different burn severities on the soil's hydrological response and especially on infiltration losses across whole watersheds remains a difficult task [22]. In many cases, studies on post-fire hydrology have focused only on short time scales (from 1 to 3 years), thus limiting the assessment of the longevity of fire impacts [23]. The first years after fire are often considered the most critical in terms of hydrological instability [24,25]; however, in some cases a significant increase in runoff discharge has been observed up to a decade after burning [4,26]. Another reason for concern is the effect of climate change on both fire frequency and severity, and on rainfall frequency and intensity. Mediterranean watershed flood risk is expected to increase over time, with increasingly dangerous impacts on infrastructures, human and economic activities in the watersheds affected by fire [27].

In light of these concerns, the main purpose of this study is to develop a robust and rational methodology to assess the change in hydrological response of a post-fire watershed, especially where the scarcity or absence of hydrometric data does not allow the calibration of a more complex rainfall–runoff model. Thus, this study proposes an integrated approach that combines spatial information on burned areas and levels of fire severity, direct soil infiltration measurements, and rainfall–runoff modelling. Then it was developed, implemented, and tested on a burned forest catchment in order to better understand the repercussions of fire on the hydrological response of a natural watershed. Flood peak and volume were computed through the application of the Soil Conservation Service-Curve Number method (SCS-CN), whereas the flow propagation was simulated through a lag-time approach based on the time–area curve of the catchment. The curve number (CN) was adjusted according to the severity of burned areas, whereas direct soil infiltration measurements were carried out to corroborate information about the effect of fire on soil infiltration capacity. Finally, the proposed procedure could be a useful example for hydrologists and foresters engaged in designing post-disaster interventions and in planning silvicultural practices.

2. Study Area

The study domain is the Tinella watershed, a forested 2.6 km^2 area located in the province of Varese (Figure 1). It is an integral part of the Campo dei Fiori Regional Park, a natural protected area of about 6 km^2 where environment educational, cultural and sports activities are promoted for tourists and local communities. Inside the Tinella watershed, the elevation ranges between 470 m and 1200 m asl, while the average slope is about 46%. It is mostly south-facing, which increases the risk of forest fire [25,28]. The climate is mild, and generally warm and temperate, and is classified as humid subtropical climate (*Cfa*) according to the Köppen–Geiger classification. The average temperature is 12.0 °C with a total annual precipitation of ~1500 mm.

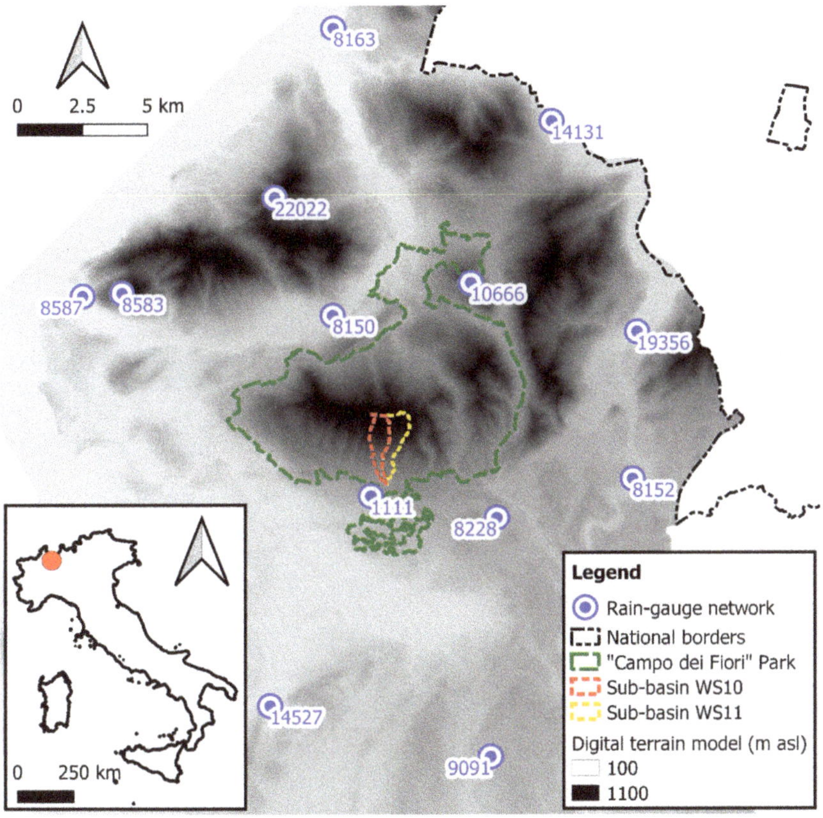

Figure 1. Position of the Tinella watershed, its two sub-basins (WS10 and WS11) and the rain gauge network present in the study area.

The Tinella watershed incorporates two nested sub-basins (WS10 and WS11 as shown in Figure 1) that are of similar size, ~1.3 km^2, and steepness ~45%, but they were affected differently by the fire. Table 1 summarizes several geomorphological features calculated by hydrological tools, implemented in QGis 3.10 and applied over the 5 m-resolution DEM, which is freely downloadable from the regional database at the link (*https://www.geoportale.regione.lombardia.it/web/geoportale/download-dati*, accessed on 8 October 2021).

Table 1. Morphometric features of the Tinella sub-basins WS10 and WS11.

Parameter	WS10 Western Sub-Basin	WS11 Eastern Sub-Basin
Area (km^2)	1.26	1.34
Perimeter (m)	8051	8751
Length of the main water stream (m)	2752.58	3426.99
Mean slope of the watershed (%)	45.07 ± 13.68	46.67 ± 15.84
Mean slope of the main water stream (%)	22.87	20.74
Minimum elevation (m)	475.72	475.00
Mean elevation (m)	841.56 ± 197.84	854.85 ± 198.77
Maximum elevation (m)	1221.02	1226.03

The Tinella watershed is mainly covered by mesophilous broad-leaved forests (about 47% of the sub-basin area in WS10 and 35% in WS11) dominated by chestnut (*Castanea sativa* Mill.) and maple-ash (*Acer psudoplatanus* L. and *Fraxinus excelsior* L.) on south-facing slopes (about 28% of the sub-basin area in WS10 and 30% in WS11), whereas the dominant species in the northern part is beech (*Fagus sylvatica* L.) (about 2% in WS10 of the sub-basin area and 24% in WS11). Concerning the soil, a 1:250.000 regional soil map shows that the watershed is mainly characterized by thin clay-loamy soil (about 80% of the sub-basin area, both in WS10 and WS11), with a minor presence of deep silty-loamy spoils in the southern part (10% of the sub-basin area).

The study area is located in the Lombardy region, one of the most forested regions in Italy (about 650,000 ha) with a high exposure to forest fire risk [29]. In October 2017, after a dry summer with little rain, a large fire burned 374 ha of the southern side of Campo dei Fiori, destroying 318 ha of forest. In-situ post-fire surveys revealed that the soil and the understory and overstory were severely compromised. The northern part of the Tinella watershed lost approximately 10 ha of forest and underwent significant hydrological changes (reduction in infiltration and water retention capacity). In fact, at each following severe rainfall event, some damage was registered at the watershed outlet in the municipality of Luvinate (in province of Varese), which was caused by a high-level of flow and sediment, mobilized by slope (soil instabilities) and along the channel network, bed and streambank erosion. The impacts of fire on post-rainfall runoff are still poorly understood since the watershed is not instrumented. However, it appears that hydraulic infrastructures are no longer able to channel the post-fire flood peaks and runoff volumes. To alleviate such threats, Lombardy region has invested about Euro 2 million for the restoration of burned areas and the building of new hydraulic infrastructures to reduce channel gradient, solid transport, and downstream runoff.

3. Materials and Methods

3.1. Storm Severity Analysis

Storm severity analysis was conducted on the rainfall data obtained by 13 rain gauges (Table 2). Twelve of them belong to the official network of meteorological stations managed by the Regional Environmental Protection Agency (ARPA) (*https://www.arpalombardia.it*, accessed on 8 October 2021), whereas the last one, very close to the outlet of the basin, belongs to the Centro Meteo Lombardo (CML) (*http://www.centrometeolombardo.com*, accessed on 8 October 2021). CML is a citizen's association with an interest in meteorology that manages a dense recreational network of rain gauges uniformly distributed over the regional territory. The distance between rain gauge positions and the outlet of the watershed varies from 1 to 16 km, whereas data acquisition at each station is every 10 min. Data from all rain gauges were weighted as a function of the inverse square distance from the center of the watershed to obtain a single vector of rainfall data to use as input for the rainfall–runoff model. In particular, the storm severity analysis was focused on an extreme rainfall event that occurred on the 24 September, 2020, which was the last day after a series of moderate rainfall events that occurred in 2018 and in 2019. Concerning the methodology, storm severity was represented through a graph showing the return periods of the storm for the different rainfall durations [30]. A moving-window procedure for rainfall accumulation was adopted to detect the maximum rainfall depth observed for each duration. Thus, the Depth–Duration–Frequency curve (DDF) was used to estimate the return period for each maximum rainfall depth. DDFs are described through the Generalized Extreme Value (GEV) probability distribution over the entire regional territory [31] and their parameters (for rainfalls 1–24 h) were available on a raster map with a spatial resolution of 2 km (*http://ita.arpalombardia.it/ITA/servizi/richiesta_dati/idro_pluvio_termo.asp*, accessed on 8 October 2021). The Bell [32] formula was applied for adjusting DDF curve parameters in case of rainfall duration of less than 1 h.

Table 2. Coordinates of the rain-gauge stations and their distance from the watershed outlet.

Sensor Code	Site	East *	North *	Distance from the Watershed Outlet (m)	Elevation (m asl)
8152	Arcisate	491,501	5,076,955	9451	334.66
9091	Castronno	486,058	5,066,146	12,658	365.76
8150	Cuveglio	480,047	5,083,338	5564	276.71
14,131	Lavena Ponte Tresa	488,398	5,090,913	14,193	273.53
8583	Laveno Mombello	471,939	5,084,222	11,848	950.00
8587	Laveno Mombello	470,425	5,084,109	13,121	194.37
8163	Luino	480,079	5,094,539	16,488	194.38
19,356	Porto Ceresio	491,668	5,082,687	10,550	279.99
22,022	Porto Valtravaglia	477,812	5,087,930	10,664	872.17
10,666	Valganna	485,317	5,084,612	7179	657.55
14,527	Varano Borghi	477,631	5,068,133	11,007	239.47
8228	Varese	486,300	5,075,452	4983	407.17
1111	Luvinate	481,511	5,076,293	1985	408.97

* WGS 84/UTM zone 32N—EPSG:32632.

3.2. Framework of the Integrated Approach

The integrated approach is characterized by a combination of (i) analysis of satellite images for detecting burned areas and level of fire severity, (ii) measurements of the soil infiltration capacity under different fire severities, and (iii) modelling rainfall–runoff processes of the watershed (Figure 2).

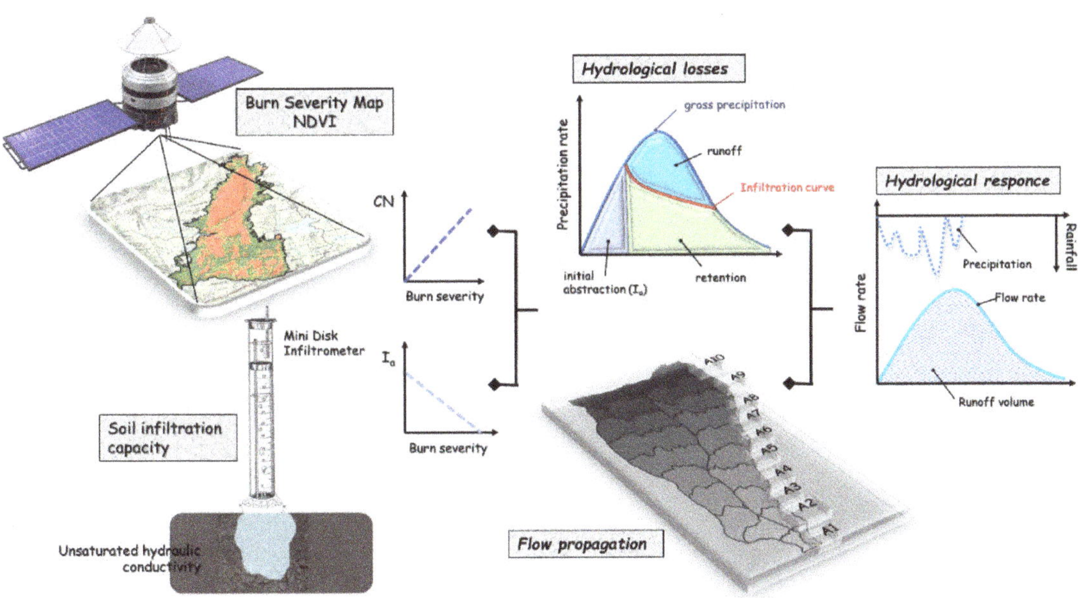

Figure 2. Steps of the integrated approach.

3.2.1. Detecting Burned Areas and Fire Severity

The spatial assessment of burned areas and fire severity was conducted by combining field-based measures of the Geometrically structured Composite Burn Index (GeoCBI) [33] and spectral signatures extracted from Sentinel-2 imagery. GeoCBI is a modification of the original CBI [34] that consists of a score based on a subjective assessment of fire impacts on five different vegetation layers. Canopy loss at the watershed scale was then assessed by the derivative differenced Normalized Difference Vegetation Index (NDVI), a traditional vegetation index commonly used to identify the photosynthetic activity [35].

3.2.2. Measurements of Soil Infiltration Capacity

Unsaturated hydraulic conductivity of soil (K) was considered as a proxy for soil infiltration capacity of the areas affected by different burn severity. K was measured directly in situ using a Mini-Disk Infiltrometer (MDI) (METER Group Inc., Pullman, WA, USA, 2020), a highly portable and inexpensive tension infiltrometer. These measurements were conducted following the standard protocols described in the MDI user manual [36]. In detail, MDI was used to measure cumulative infiltration with a pressure head (suction) of −2 cm, which is adequate for most soils—as reported in the MDI manual. Before placing the MDI, the soil surface must be cleaned and delicately levelled using a trowel and scissors to avoid the overturning of the device during measurements, as well as to allow perfect adherence between the sintered filter of the instrument and the soil surface. To ensure the contact between the sintered disk and soil surface, a thin layer of sand (about 0.5 cm thick) must be added. The water level inside the minidisk water reservoir is recorded at 30 s regular time intervals for no less than ~5 min for each test.

The unsaturated hydraulic conductivity is then derived by combining the obtained cumulative infiltration measures with knowledge of van Genuchten's soil parameters [37] using Zhang's method [38].

The cumulated infiltration rate measured with each test is expressed as a function of the square root of time and interpolated through the function proposed by Zhang (Equation (1)). C_2 is determined as the slope of the curve obtained with this method:

$$I = C_1 t^{0.5} + C_2 \tag{1}$$

where I is the cumulated infiltration rate (mm), t is time (s), C_1 (m s$^{-1/2}$) and C_2 (m s^{-1}) are two parameters related to soil sorptivity and unsaturated hydraulic conductivity.

The value of K (cm s^{-1}) was then calculated as shown in Equation (2):

$$K = \frac{C_2}{A_2} \tag{2}$$

where A_2 (−) is a parameter dependent on soil texture and suction and defined by Equation (3):

$$\begin{cases} A_2 = \dfrac{11.65\,(n^{0.1}-1)\exp[2.92(n-1.9)\alpha\, h_0]}{(\alpha\, r_0)^{0.91}} & n \geq 1.9 \\ A_2 = \dfrac{11.65\,(n^{0.1}-1)\exp[7.5(n-1.9)\alpha\, h_0]}{(\alpha\, r_0)^{0.91}} & n < 1.9 \end{cases} \tag{3}$$

where r_0 is the infiltrometer's radius (equal to 2.25 cm), n (−) and α (−) are retention parameters depending on soil texture and r_0, according to the values reported in the MDI manual, and h_0 is the pressure head of the infiltrometer (set to −2 cm).

3.2.3. Rainfall-Runoff Simulation Strategy

The assessment of hydrological losses and runoff propagation was performed through a simple approach widely described in the scientific literature and also easy to implement for ungauged basins or where details on physiographic characteristics of the watershed are scarce. The computation of hydrological losses was conducted using the SCS-CN method [39–41], a semi-spatial-distributed approach, implemented for providing direct runoff after the separation of initial abstraction and infiltration losses from total rainfall. This method is one of the commonly used heuristic approaches for estimating the surface runoff from watersheds. Direct runoff (P_n) (mm) is calculated through Equations (4) and (5):

$$P_n = \frac{(P - I_a)^2}{(P - I_a + S)} \quad (4)$$

$$S = \frac{25400}{CN} - 254 \quad (5)$$

where P (mm) is the gross rainfall depth, Ia (mm) the initial abstraction and S (mm) the maximum potential retention of the basin, which depends by CN. The CN value theoretically varies between 0 and 100. However, the practical values are typically in the range of 40 to 98 [42] and are usually determined by combining land use and hydrological group data through tables reported in technical manuals [43]. *Ia* includes surface storage, interception and infiltration aptitude before runoff begins and it is calculated as follows (Equation (6)):

$$I_\alpha = \alpha \cdot S \quad (6)$$

where α (−) is a constant, which in original formulation was set equal to 0.2, but recent works demonstrate that its value can range from 0.05 to 0.2 [42,44–46] as a function of the physiographic characteristics of the watershed, type of vegetation cover and severity of potential alterations suffered by the watershed such deforestation and fire [47,48].

Then, the propagation of direct runoff was simulated through a simple translation of water flow over the drainage watershed excluding natural storage. The runoff volumes originating from different parts of the watershed were cumulated assuming no interaction between them. The runoff travel time distribution was described by a time–area curve, obtained by the combination of a constant flow transfer velocity and the area–distance curve, which can be easily derived from the Digital Elevation Model (DEM) of the watershed using the 'Overland Flow Distance' tool of QGis. The corresponding time–area curves were then derived by scaling the x-axis with the average runoff propagation velocity obtained as the ratio of the length of the longest flow path and the time of concentration (t_c). In particular, t_c is another key parameter for estimating the hydrological response of a watershed. Despite the fact that the scientific literature describes a wide spectrum of empirical formulations [49–51], in this integrated approach, the Natural Resources Conservation Service-Soil Conservation Service (NRCS-SCS) method [41,52] was used. This method, described by Equation (7), directly depends on the length of the main water stream L (km), while it is inversely proportional to the average slope of the watershed (m m^{-1}) and CN:

$$t_c = 0.057\left(\frac{1000}{CN} - 9\right)^{0.7} L^{0.8}\, i_m^{-0.5} \quad (7)$$

where t_c in hours. It was considered suitable because it was tested on several natural watersheds. Moreover, the presence of the CN value as an input parameter allows us to include the potential shift in the flood peak before and after fire conditions.

3.2.4. Setup of the Integrated Approach in the Study Area

For determining fire severity, the GeoCBI was assessed in 73 georeferenced, 30 m-diameter plots randomly placed within the burned perimeter. The assessment of fire impacts was caried out on a scale of zero (unburned) to three (maximum severity) on five different vegetation layers (herbs, low shrubs <1 m tall, tall shrubs and trees 1–5 m tall, intermediate trees, and dominant or co-dominant trees), and scores were weighted on the fraction cover of each stratum [34]. Differenced NDVI (dNDVI) was calculated from pre-treated, cloud-free 10-m Sentinel-2 (level 2A) images taken in July one year before and one year after the fire, and bias-corrected using the average NDVI change across the whole Sentinel-2 image outside the fire perimeter, to account for phenological differences between years. Finally, dNDVI was then classified into three severity classes (low, moderate, and high) and unburned, based on GeoCBI measures in the field.

Concerning the determination of the soil infiltration capacity, 39 infiltration measurements at 13 points in the Tinella watershed were carried out over areas affected by different fire severity. The measurement points were selected randomly within the burned area as a function of the level of the fire severity. The van Genuchten's parameters (n and α) derived from soil texture were assessed by collecting ad hoc soil samples at each measurement point at a depth ranging from ~0 to 10 cm. This depth adequately represents both the soil layer that is mainly involved in the surface runoff [53] and that is mainly affected by burning. For each measurement point at least 3 repetitions were conducted.

Parametrizing the rainfall–runoff model is a fundamental step for assessing watershed runoff, especially determining the spatial distribution of CN and Ia values. Concerning CN, in the Tinella watershed, deciduous forests, evergreen coniferous and mix forests were considered as three macro-types of land use, whereas the hydrological groups were derived from soil characteristics, grouping sand, loamy sand, sandy loam in group A, silt loam in group B, silt, loam and sandy clay loam in group C, and lastly, silt clay loam, clay loam, sandy clay and clay in group D, according to the classification of USDA NRCS [54]. The results of these classifications are reported in Table 3.

Table 3. CN values (-) applied over Tinella watershed as a function of forest type (deciduous broadleaf, evergreen coniferous and mixed forests) and hydrological group (A, B, C and D).

Land Use	A	B	C	D
Deciduous broadleaf	42	66	79	85
Evergreen coniferous	34	60	73	79
Mixed forest	38	62	75	81

CN was assessed for each sub-area described by the area–distance curve and then adjusted according to the Antecedent Moisture Condition (AMC) first condition, since in the 5 days before 24 September, and more, no rainfall events occurred over the basin. Although there is consensus on the increase in flood peak after a fire, there is still no consistent methodology to estimate post-fire CN values [55]. In fact, the analysis of the hydrologic response of watersheds to fire is still a topic under investigation and only a few studies have been undertaken [56]. Most of them are mainly based on practical rules and/or experience and propose a modification of the CN value as a function of burn severity. In this study, two different approaches were adopted and implemented following two practical manuals provided by the U.S. National Forest Services: Uinta National Forest-Research Station [57] (hereafter called the HSR method) and Rocky Mountain-Research Station [58] (hereafter called the DS method). Table 4 summarizes the guidelines to choosing the post-fire CN. Concerning Ia, the α parameter was considered proportional to the level of fire severity, varying its value from 0.05 to 0.2 as a function of the soil infiltration capacity recorded in the different fire severity areas.

Table 4. Post-fire CN as a function of burn severity.

HSR Method	
High burn severity	$CN_{post} = CN_{pre} + 15$
Moderate burn severity	$CN_{post} = CN_{pre} + 10$
Low burn severity	$CN_{post} = CN_{pre} + 5$
DS Method	
High burn severity with water repellent soils	$CN_{post} = 95$
High burn severity without water repellent soils	$CN_{post} = 90$ to 91
Moderate burn severity with water repellent soils	$CN_{post} = 90$
Moderate burn severity without water repellent soils	$CN_{post} = 85$
Low burn severity	$CN_{post} = CN_{pre} + 5$

4. Results

4.1. Storm Severity

According to the rainfall data from the meteorological network, the storm event that occurred on 24 September hit the study area and the Tinella watershed at different times, intensity, and severity (Figure 3).

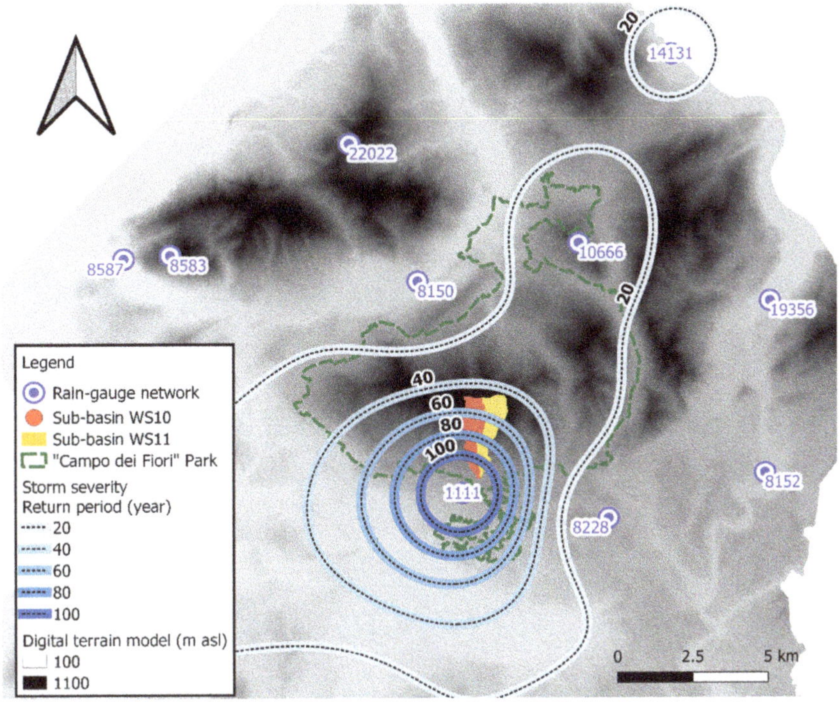

Figure 3. Spatial distribution of storm severity: the coloured contour lines and the labels written in in black describe the return period of storm (year), whereas the labels in blue describe the rain-gauge stations.

Rain gauges located in the northwestern part of the study area were the first to record the beginning of the storm event (at 5 p.m.). The storm was characterized by two rainfall peaks separated by ~1.5 h. Considering the rain gauges located in the north-western part of the study area, the first rainfall peak occurred between 5 and 6.15 p.m. and the intensity was between 80 and 100 mm h^{-1} (Figure 4a). The second peak was generally less intense (on average 55 mm h^{-1}) and occurred between 8 and 9 p.m. Considering data from the rain gauge close to the Tinella watershed (i.e., the rain gauge number 1111) the storm appeared to shift temporally, and the first rainfall peak occurred at about 7 p.m., with the second one at past 9 p.m. The rainfall intensity exceeded the 100 mm h^{-1} in the first peak, whereas in the second one it was close to 60 mm h^{-1}.

Severity graphs show that the storm was characterized by a return period of less than 10 years in large parts of the study area. However, in the northeastern part of the study area, the storm was characterized by a return period exceeding 20 years. In particular, severity calculated on data from rain gauge 10,666 and 14,131 showed a peak return period of 30 and 21 years, respectively, obtained for a critical rainfall duration of 150 and 180 min. Close to the Tinella watershed, however, the storm was extremely severe with two peak return periods exceeding 40 years. Specifically, the first one was about 120 years with a critical rainfall duration of about 40 min, whereas the second one was about 45 years with a critical rainfall duration of about 160 min. This contributed to the stress on the Tinella watershed during the storm event with significant repercussions on flood peak.

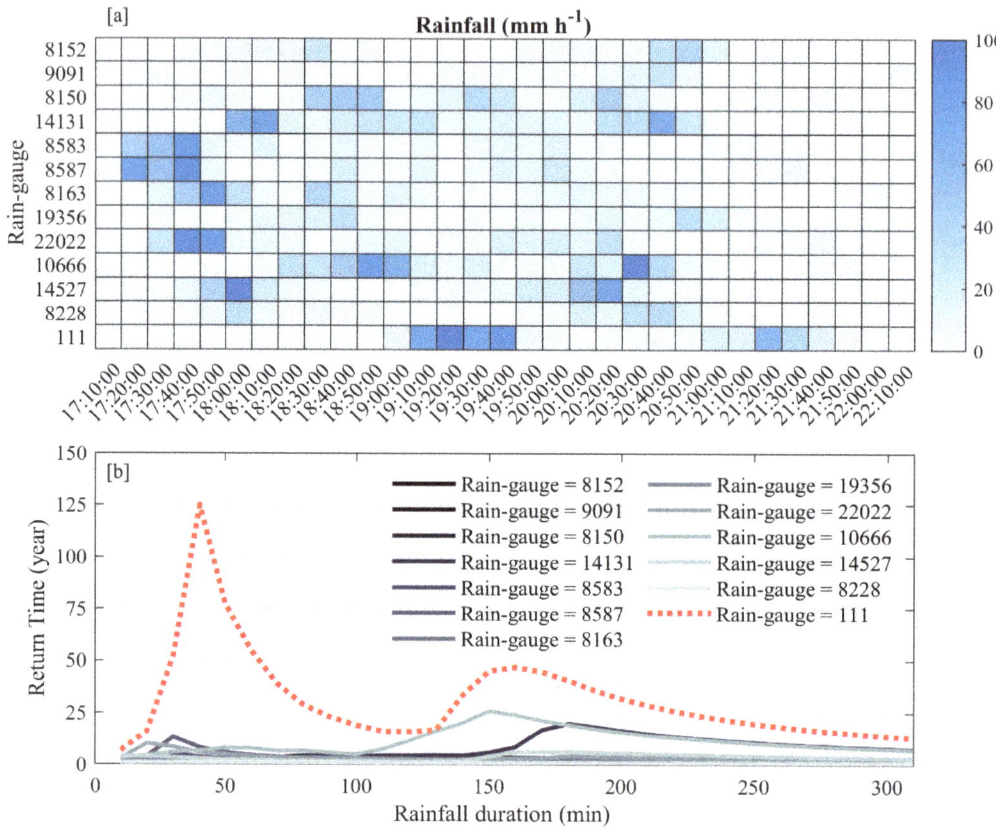

Figure 4. (**a**) Temporal distribution of rainfall intensity, and (**b**) severity graph obtained from each rain gauge located in the study area.

Concerning the spatial distribution of the return periods, the results of a linear interpolation of the maximum storm's return periods obtained from data from each rain gauge in the study area, show that the front direction of the storm was oriented from the north-east to the south-west of the study area, reaching the center of the Tinella watershed with its maximum severity (i.e., 120 years) as shown in Figure 4b.

4.2. Burned Area

The fire event of October 2017 damaged about 62% of the Tinella watershed area (Figure 5). Analyzing the WS10 and WS11 sub-basins separately showed they were affected by different levels of burn severity. In the WS10, 29% of the sub-basin area (i.e., 0.37 km^2) had a high level of burn severity, 42% (i.e., 0.53 km^2) had a moderate level of burn severity, and 4% (i.e., 0.05 km^2) had a low burn severity. Instead, in the WS11, 24% of the sub-basin area (i.e., 0.32 km^2) showed a high burn severity, 25% (i.e., 0.34 km^2) had a moderate level of burn severity, and 1% (i.e., 0.02 km^2) had a low burn severity.

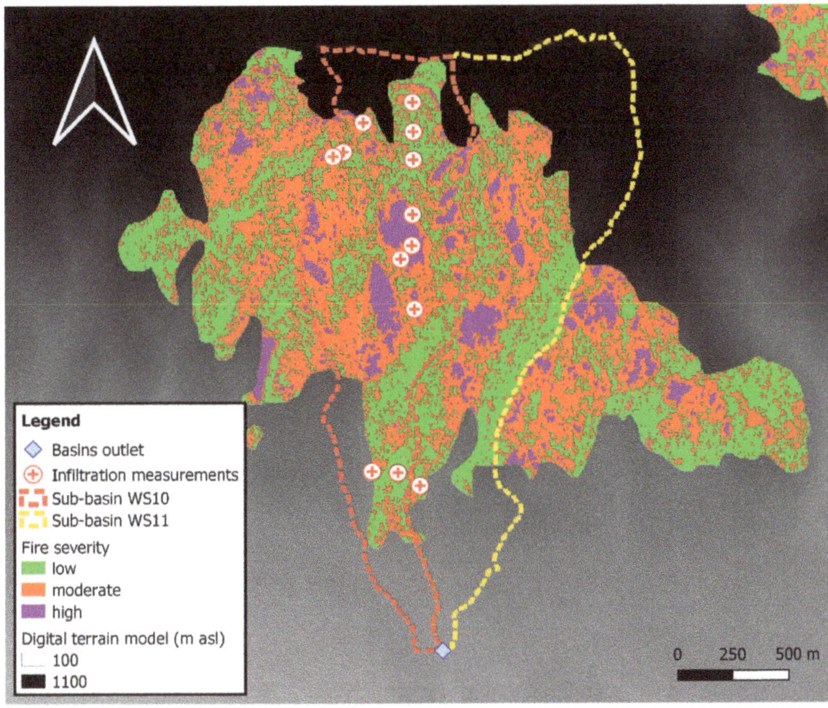

Figure 5. Areas affected by fire, burn severity and position of infiltration measurements.

4.3. Soil Infiltration Response to Burn Severity

Soil sample analysis showed that the texture of the first layer of the soil is homogenous across the entire study domain, with no significant spatial pattern or differences in areas affected by different burn severity levels (see Figure 5). Overall, the soils were sandy loam (for about 62% of samples), followed by loam (23%) and loam-sandy (15). A good level of skeleton ranging from 13 to 50% of sample weight was found in all detected points, whereas from a visual inspection, soil samples presented a clear component of ash that was observed in the soil–water mixture during the deposition step of the texture lab analysis.

Results of soil infiltration measurements performed over May–June 2021, revealed that unsaturated hydraulic conductivity was smaller in highly burned sites than in low and moderate burned areas (Figure 6). Specifically, K was on average 7.9, 17.7 and 32.7 mm h^{-1} in high, moderate and low burned areas, respectively. The reduction in infiltration capacity from low burned areas to moderate and high burned sites was on average of 45% and 75%, respectively. A slight increase in standard deviation was observed in measurements carried out in low burned areas (about 10 mm h^{-1}), whereas in moderate and high burned sites it was about 5 and 4 mm h^{-1}, respectively. No significant spatial pattern was observed across the watershed.

4.4. Hydrological Impact of Storm on Watershed Runoff in Pre- and Post-Fire Condition

One of the impacts of fire on watershed runoff is the timing of hydrological response. The presence of burned areas inside the Tinella watershed alters the time of concentration value. In particular, tc was estimated as ~50 min both in WS10 and WS11 sub-basins in pre-fire conditions, whereas its value decreases to ~40 min in post-fire condition using both the HSR and DS method. The reduction of 20% in time of concentration led to speeding up the hydrological response of both the watershed to the precipitation causing a shift of the flood peak in post-fire conditions close to the rainfall peak (Figure 7a,b).

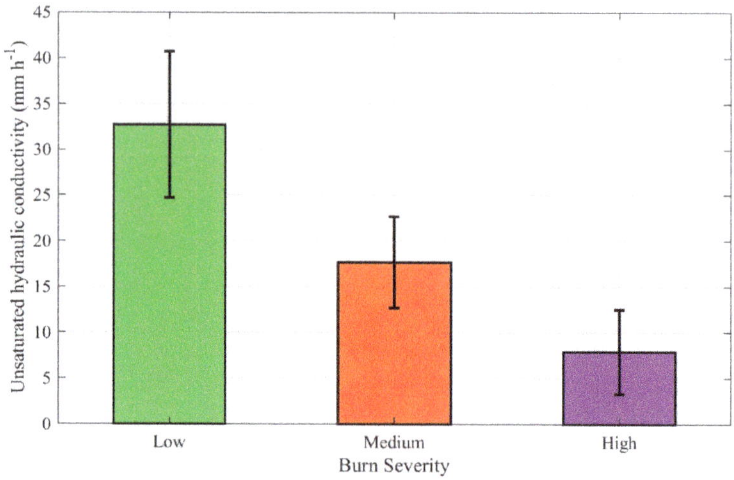

Figure 6. Unsaturated hydraulic conductivity measured in different forest areas affected by different burn severity levels.

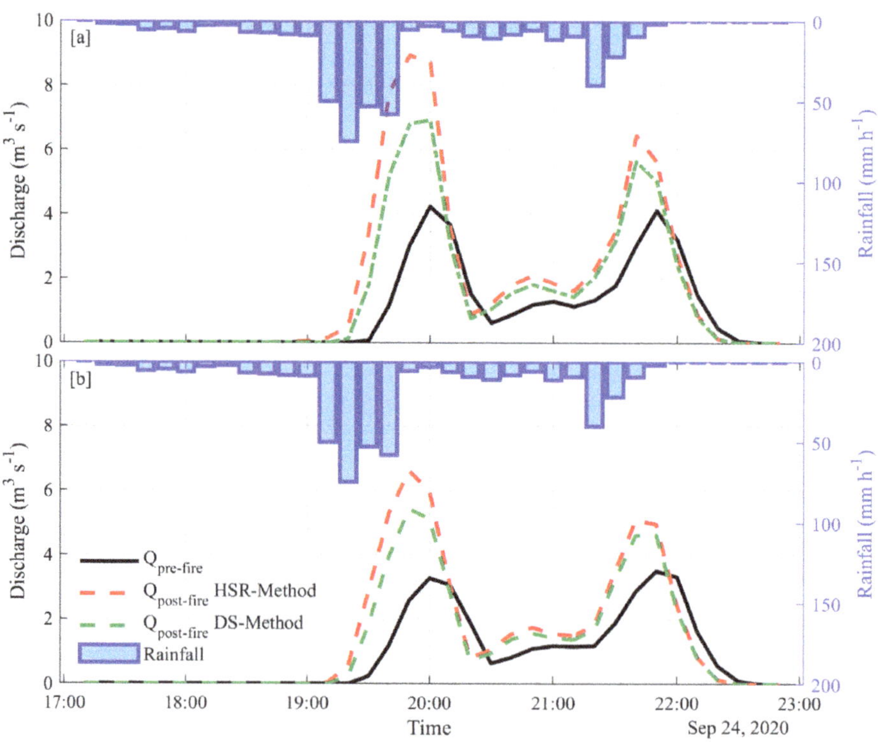

Figure 7. Tinella watershed runoff (Q) before and after the forest fire calculated with the HSR and DS methods. (**a**) Runoff formation from the WS10 sub-basin. (**b**) Runoff formation from the WS11 sub-basin. Rainfall intensity calculated as weighted average of the rainfall data recorded by the rain gauge network in the study area is also shown, indicating strikingly similar pluviographs.

In pre-fire conditions, the hydrological response of WS10 and WS11 sub-basins to the two consecutive rainfall events (the first characterized by rainfall peak of about 60 mm h^{-1} and the second characterized by rainfall peak of about 40 mm h^{-1}) was assessed to be very similar. This is also confirmed by the CN value found on average for the two portions of the Tinella watershed, which in pre-fire conditions was assessed to be about 66 in both WS10 and WS11.

In pre-fire conditions, the simulations produced flood peaks ranging between 3.5 and 4 $m^3 s^{-1}$ for both sub-basins (Figure 7a,b). The hydrological losses (considered through the CN value) in post-fire conditions were found to be 10% lower on average than in the pre-fire conditions. In post-fire conditions, the CN value was assessed to be 76 and 73 on average by applying the HSR method for the WS10 and WS11 sub-basin, respectively, whereas it was assessed to be 72 and 71 by applying the DS method. This led to a substantial increase in the flood peak in post-fire conditions in both the WS10 and WS11 sub-basins, however several differences were revealed. Concerning the WS10, the increase in the maximum flood peak from pre- to post-fire conditions was assessed to be 125%, more than doubling from about 4 to 9 $m^3 s^{-1}$ using the HSR method, and 75% using the DS method (from 4 to 7 $m^3 s^{-1}$). Slight differences in flood peaks were found during the second rainfall event using DS and HSR, with values ranging from 5.5 to 6.5 $m^3 s^{-1}$, respectively. In the WS11 sub-basin, the increase in the maximum flood peak from pre- to post-fire conditions was assessed to be 86% using the HSR method (i.e., from about 3.5 to 6.5 $m^3 s^{-1}$) and 57% using the DS method (i.e., from 3.5 to 5.5 $m^3 s^{-1}$). The flood peak difference between the WS10 and WS11 sub-basins in post-fire conditions was 2 $m^3 s^{-1}$ on average, whereas the maximum flood peak at the outlet of Tinella watershed was about 15 $m^3 s^{-1}$ (i.e., about two times the maximum flood peak in the pre-fire condition). The overall flood volume was estimated to be 39,700 m^3 in the pre-fire condition, whereas in post-fire conditions it was about 69,400 m^3 using the HSR method (i.e., 75% higher than the pre-fire condition) and 57,300 m^3 using the DS method (i.e., about 45% higher than the pre-fire condition).

5. Discussion

Forest fire removed the canopy cover, and reduced soil infiltration capacity, all of which then increased flood peaks after subsequent rainfall events. The reduction in the soil infiltration capacity (−45% and −75% after moderate and high severity fires relative to low severity areas) is consistent with those reported in previous studies [59,60]. Fire-induced water repellency contributes to reduced soil infiltration rates, with stronger effects in higher fire severities [59]. In particular, the clogging of soil pores, caused by small particles of ash, reduces infiltration across the top layer of the soil, thus facilitating the development of surface runoff [60]. Robichaud [61] found that hydraulic conductivity in natural watersheds after high severity fire was 50% lower than in low severity fire, as a consequence of soil surface crusting and sealing. Plaza-Álvarez et al. [62] found that soil hydraulic conductivity in burned Mediterranean pine (*Pinus pinaster* Aiton) forests ranged from 20 to 50 mm h^{-1} with a constant decrease over all seasons of about 10 mm h^{-1} relative to unburned sites.

Flood peaks calculated under pre-fire conditions showed small variations (below 0.2 $m^3 s^{-1}$) between the first and second flood peaks occurring during the storm event (as shown in Figure 7a,b). This can be explained considering the characteristics of the SCS-CN approach applied for calculating the hydrological losses in the watershed. In particular, the method allocated most of the hydrological losses in the first stage of precipitation, which produced a stronger abatement of the flood peak resulting from the first rainfall relative to the second one, although the intensity of the first rainfall event was about 50% higher. Flood peaks increased in both watersheds as a result of forest fire, especially after the first rainfall event (+75–125% and +57–86% in WS10 and WS11, respectively, according to the methods used). Fire effects on flood peaks were less evident in WS11, where fire affected only half of the watershed area. This evidence confirms that the difference in hydrological response in WS11 compared to the WS10 sub-basin can be mainly ascribed to the different size of the burned areas and local differences in fire severity.

The enhanced hydrological response clearly highlights the increased sensitivity of the burned basin to rainstorms. Additional reinforcing effects, although not analyzed in this study, could be attributed to the reduction in water losses by evapotranspiration after fire-induced tree mortality, and to the lower surface roughness produced by the consumption of ground vegetation and litter [63]. Besides the impact on peak discharge, forest fire can also strongly increase sediment yields, with repercussions on soil erosion rates, water quality, and debris flow hazard [64].

The workflow illustrated in this study may contribute to the analysis of fire effects in ungauged watersheds, integrating analysis of canopy mortality, soil hydrological properties, and hydraulic responses in catchments with spatially heterogenous disturbance severities. These findings provide additional information that may help to improve the understanding of the effect of forest fires on hydrological response, which is especially relevant in the context of human-induced climate change. Global warming may in fact exacerbate hydrological hazards as a consequence of forest fire, as a consequence of increased fire frequency and severity [65], altered rainfall patterns [66], and loss of forest resilience due to disturbances [67]. In this context, more research is required to improve the model parametrization and calibration, while explicitly accounting for the spatial and temporal variability of hydrological responses. Lastly, an integrated approach for the analysis of hydrological consequences of forest fire in ungauged catchments could also support fire prevention and recovery activities by highlighting spatial priorities for silvicultural actions aimed at improving forest resistance to fire and accelerating post-fire recovery by afforestation measures, especially where natural vegetation response is not fast enough to ensure hydrogeologic protection against future extreme rainfall [68].

6. Conclusions

In this study, an integrated approach that exploits a combination of remote and in-situ data for a comprehensive evaluation of the impact of forest fire on hydrological response of natural watersheds was tested on a real case study located in northern Italy. Specifically, the NDVI obtained by Sentinel-2 images was used to identify burned areas in the watershed and their level of fire severity, whereas direct infiltration measurements performed in those areas through a Mini Disk Infiltrometer were used to better understand the infiltration capacity of the soils in post-fire conditions. These data were included in a semi-distributed rainfall–runoff model based on the SCS-CN method and lag-time for the calculation of hydrological losses and flow propagation over the watershed, respectively. Specifically, the impact of forest fire on the hydrological response was investigated through a modelling experiment on the Tinella watershed, which was affected by a forest fire about four years ago. On 24 September 2020, the Tinella watershed was hit by a storm event with a maximum recorded return period of 120 years, which resulted in several damaged sites within the watershed and downstream to the outlet. The characteristics of this storm event were used as meteorological input to the modelling approach. The CN value was adjusted as a function of the level of fire severity, whereas direct infiltration measurements carried out in watershed areas affected by different burn severity were adopted to obtain the proportional behavior of the initial abstraction as a function of burn severity level.

The results of this study show that the infiltration capacity of burned soil was deeply affected by the forest fire with an unsaturated hydraulic conductivity that decreased by 45% and 75% from low to moderate, and from low to high burned sites, respectively. The reduction in soil hydraulic conductivity and tree canopy cover actively altered the surface runoff. Runoff hydrographs were found to be much sharper with considerably higher flood peaks and runoff volumes. In particular, the flood peak and runoff volume in post-fire conditions increased by up to 2.2 and 1.7 times compared to the pre-fire conditions.

Author Contributions: Conceptualization, D.M. and G.V.; methodology, A.C. and D.M.; software, L.F. and D.M., field campaign: L.F.; investigation, L.F., A.C., G.V. and D.M.; resources, D.M.; data curation, L.F. and A.C.; writing—original draft preparation, L.F.; writing—review and editing, A.C., G.V. and D.M. All authors have read and agreed to the published version of the manuscript.

Funding: This research was conducted in the context of the traineeship at the *Campo dei Fiori* Regional Park and of the bachelor's degree Thesis in Agrotechnology for the Environment and the Territory by Luca Folador at the University of Milan, and was part of the project TREE:HERO, acronym of "TREE distribution patterns: HillslopE failuRe prevention through forest management", entirely funded by Fondazione Cariplo (Italy; Ref. 2017-0714) in the framework of "Research dedicated to hydrogeological instability: a contribution to the prevision, prevention and risk mitigation".

Institutional Review Board Statement: Not applicable.

Informed Consent Statement: Not applicable.

Acknowledgments: The authors would like to thank Antonia Moreno Carrera for her support in the soil laboratory analysis, Marco Pistocchini and Alessandro Nicoloso for their support during the field campaign.

Conflicts of Interest: The authors declare no conflict of interest.

References

1. Stoof, C.R.; Vervoort, R.W.; Iwema, J.; van den Elsen, E.; Ferreira, A.J.D.; Ritsema, C.J. Hydrological Response of a Small Catchment Burned by Experimental Fire. *Hydrol. Earth Syst. Sci.* **2012**, *16*, 267–285. [CrossRef]
2. Lavabre, J.; Torres, D.S.; Cernesson, F. Changes in the Hydrological Response of a Small Mediterranean Basin a Year after a Wildfire. *J. Hydrol.* **1993**, *142*, 273–299. [CrossRef]
3. DeBano, L.F.; Neary, D.G.; Ffolliott, P.F. *Fire Effects on Ecosystems*; John Wiley & Sons: Hoboken, NJ, USA, 1998; ISBN 0-471-16356-2.
4. Inbar, M.; Tamir, M.; Wittenberg, L. Runoff and Erosion Processes After a Forest Fire in Mount Carmel, a Mediterranean Area. *Geomorphology* **1998**, *24*, 17–33. [CrossRef]
5. Shakesby, R.; Doerr, S. Wildfire as a Hydrological and Geomorphological Agent. *Earth-Sci. Rev.* **2006**, *74*, 269–307. [CrossRef]
6. Moody, J.A.; Shakesby, R.A.; Robichaud, P.R.; Cannon, S.H.; Martin, D.A. Current Research Issues Related to Post-Wildfire Runoff and Erosion Processes. *Earth-Sci. Rev.* **2013**, *122*, 10–37. [CrossRef]
7. Soulis, K.X.; Generali, K.A.; Papadaki, C.; Theodoropoulos, C.; Psomiadis, E. Hydrological Response of Natural Mediterranean Watersheds to Forest Fires. *Hydrology* **2021**, *8*, 15. [CrossRef]
8. DeBano, L.F. Water Repellency in Soils: A Historical Overview. *J. Hydrol.* **2000**, *231–232*, 4–32. [CrossRef]
9. Certini, G. Effects of Fire on Properties of Forest Soils: A Review. *Oecologia* **2005**, *143*, 1–10. [CrossRef] [PubMed]
10. Hoyt, W.G.; Troxell, H.C. Forests and Stream Flow. *Trans. Am. Soc. Civ. Eng.* **1934**, *99*, 1–30. [CrossRef]
11. Anderson, H.W. Fire Effects on Water Supply, Floods, and Sedimentation. *Proc. Tall Timbers Fire Ecol. Conf.* **1976**, *15*, 249–260.
12. Hessling, M. Hydrological Modelling and a Pair Basin Study of Mediterranean Catchments. *Phys. Chem. Earth Part B Hydrol. Ocean. Atmos.* **1999**, *24*, 59–63. [CrossRef]
13. Loáiciga, H.A.; Pedreros, D.; Roberts, D. Wildfire-Streamflow Interactions in a Chaparral Watershed. *Adv. Environ. Res.* **2001**, *5*, 295–305. [CrossRef]
14. Conedera, M.; Peter, L.; Marxer, P.; Forster, F.; Rickenmann, D.; Re, L. Consequences of Forest Fires on the Hydrogeological Response of Mountain Catchments: A Case Study of the Riale Buffaga, Ticino, Switzerland. *Earth Surf. Process. Landf.* **2003**, *28*, 117–129. [CrossRef]
15. Rulli, M.C.; Rosso, R. Hydrologic Response of Upland Catchments to Wildfires. *Adv. Water Resour.* **2007**, *30*, 2072–2086. [CrossRef]
16. Pierson, F.B.; Robichaud, P.R.; Moffet, C.A.; Spaeth, K.E.; Hardegree, S.P.; Clark, P.E.; Williams, C.J. Fire Effects on Rangeland Hydrology and Erosion in a Steep Sagebrush-dominated Landscape. *Hydrol. Process.* **2008**, *22*, 2916–2929. [CrossRef]
17. Seibert, J.; McDonnell, J.J.; Woodsmith, R.D. Effects of Wildfire on Catchment Runoff Response: A Modelling Approach to Detect Changes in Snow-Dominated Forested Catchments. *Hydrol. Res.* **2010**, *41*, 378–390. [CrossRef]
18. Versini, P.A.; Velasco, M.; Cabello, A.; Sempere-Torres, D. Hydrological Impact of Forest Fires and Climate Change in a Mediterranean Basin. *Nat. Hazards* **2013**, *66*, 609–628. [CrossRef]
19. Rulli, M.C.; Bozzi, S.; Spada, M.; Bocchiola, D.; Rosso, R. Rainfall Simulations on a Fire Disturbed Mediterranean Area. *J. Hydrol.* **2006**, *327*, 323–338. [CrossRef]
20. Rosso, R.; Rulli, M.C.; Bocchiola, D. Transient Catchment Hydrology after Wildfires in a Mediterranean Basin: Runoff, Sediment and Woody Debris. *Hydrol. Earth Syst. Sci. Discuss.* **2007**, *11*, 125–140. [CrossRef]
21. Youberg, A.M.; Loverich, J.B.; Kellogg, M.J.; Fuller, J.E. Before the Fire: Assessing Post-Wildfire Flooding and Debris-Flowhazards for Pre-Disaster Mitigation. *Nat. Hazards Earth Syst. Sci. Discuss.* **2019**. [CrossRef]
22. Wieting, C.; Ebel, B.A.; Singha, K. Quantifying the Effects of Wildfire on Changes in Soil Properties by Surface Burning of Soils from the Boulder Creek Critical Zone Observatory. *J. Hydrol. Reg. Stud.* **2017**, *13*, 43–57. [CrossRef]
23. Moody, J.A.; Martin, D.A. Post-Fire, Rainfall Intensity-Peak Discharge Relations for Three Mountainous Watersheds in the Western USA. *Hydrol. Process.* **2001**, *15*, 2981–2993. [CrossRef]
24. Marqués, M.A.; Mora, E. The Influence of Aspect on Runoff and Soil Loss in a Mediterranean Burnt Forest (Spain). *Catena* **1992**, *19*, 333–344. [CrossRef]

25. Cerda, A.; Imeson, A.C.; Calvo, A. Fire and Aspect Induced Differences on the Erodibility and Hydrology of Soils at La Costera, Valencia, Southeast Spain. *Catena* **1995**, *24*, 289–304. [CrossRef]
26. Mayor, A.G.; Bautista, S.; Llovet, J.; Bellot, J. Post-Fire Hydrological and Erosional Responses of a Mediterranean Landscpe: Seven Years of Catchment-Scale Dynamics. *Catena* **2007**, *71*, 68–75. [CrossRef]
27. Fowler, H.J.; Ali, H.; Allan, R.P.; Ban, N.; Barbero, R.; Berg, P.; Blenkinsop, S.; Cabi, N.S.; Chan, S.; Dale, M.; et al. Towards Advancing Scientific Knowledge of Climate Change Impacts on Short-Duration Rainfall Extremes. *Philos. Trans. R. Soc. A* **2021**, *379*, 20190542. [CrossRef]
28. Konstantinidis, P.; Tsiourlis, G.; Xofis, P. Effect of Fire Season, Aspect and Pre-Fire Plant Size on the Growth of Arbutus Unedo L. (Strawberry Tree) Resprouts. *For. Ecol. Manag.* **2006**, *225*, 359–367. [CrossRef]
29. Lari, S.; Frattini, P.; Crosta, G.B. Integration of Natural and Technological Risks in Lombardy, Italy. *Nat. Hazards Earth Syst. Sci.* **2009**, *9*, 2085–2106. [CrossRef]
30. Norbiato, D.; Borga, M.; Sangati, M.; Zanon, F. Regional Frequency Analysis of Extreme Precipitation in the Eastern Italian Alps and the August 29, 2003 Flash Flood. *J. Hydrol.* **2007**, *345*, 149–166. [CrossRef]
31. Braca, G.; Bussettini, M.; Lastoria, B.; Mariani, S. *Linee Guida Per L'analisi e L'elaborazione Statistica Di Base Delle Serie Storiche Di Dati Idrologici*; Istituto Superiore per la Protezione e la Ricerca Ambientale: Roma, Italy, 2013; p. 166.
32. Bell, F.C. Generalized Rainfall-Duration-Frequency Relationships. *J. Hydraul. Div.* **1969**, *95*, 311–328. [CrossRef]
33. De Santis, A.; Chuvieco, E. GeoCBI: A Modified Version of the Composite Burn Index for the Initial Assessment of the Short-Term Burn Severity from Remotely Sensed Data. *Remote Sens. Environ.* **2009**, *113*, 554–562. [CrossRef]
34. Key, C.H.; Benson, N.C. Landscape Assessment: Ground Measure of Severity, the Composite Burn Index, and Remote Sensing of Severity, the Normalized Burn Index. In *FIREMON: Fire Effects Monitoring and Inventory System*; Lutes, D.C., Keane, R.E., Caratti, J.F., Key, C.H., Benson, N.C., Sutherland, S., Gangi, L.J., Eds.; General Technical Report RMRS GTR-164-CD; USDA Forest Service, Rocky Mountain Research Station: Ogden, UT, USA, 2006; pp. 1–51.
35. Veraverbeke, S.; Gitas, I.; Katagis, T.; Polychronaki, A.; Somers, B.; Goossens, R. Assessing Post-Fire Vegetation Recovery Using Red–near Infrared Vegetation Indices: Accounting for Background and Vegetation Variability. *ISPRS J. Photogramm. Remote Sens.* **2012**, *68*, 28–39. [CrossRef]
36. METER Group, Inc. *Mini Disk Infiltrometer*; METER Group, Inc.: Pullman, WA, USA, 2020.
37. van Genuchten, M.T. A Closed-Form Equation for Predicting the Hydraulic Conductivity of Unsaturated Soils. *Soil Sci. Soc. Am. J.* **1980**, *44*, 892–898. [CrossRef]
38. Zhang, R. Determination of Soil Sorptivity and Hydraulic Conductivity from the Disk Infiltrometer. *Soil Sci. Soc. Am. J.* **1997**, *61*, 1024–1030. [CrossRef]
39. McCuen, R.H. *Hydrologic Analysis and Design*, 2nd ed.; Pearson Education/Prentice Hall: Upper Saddle River, NJ, USA, 1998; ISBN 978-0-13-134958-2.
40. Mishra, S.K.; Singh, V.P. Validity and Extension of the SCS-CN Method for Computing Infiltration and Rainfall-Excess Rates. *Hydrol. Process.* **2004**, *18*, 3323–3345. [CrossRef]
41. Mockus, V. Estimation of Direct Runoff from Storm Rainfall. In *National Engineering Handbook*; U.S. Department of Agriculture, Soil Conservation Service: Washington, DC, USA, 1964; Volume 4, p. 28.
42. Ponce, V.M.; Hawkins, R.H. Runoff Curve Number: Has It Reached Maturity? *J. Hydrol. Eng.* **1996**, *1*, 11–19. [CrossRef]
43. USDA National Resources Conservation Service. *National Engineering Handbook: Part 630-Hydrology*; U.S. Department of Agriculture: Washington, DC, USA, 2004.
44. Yuan, Y.; Mitchell, J.K.; Hirschi, M.C.; Cooke, R.A.C. Modified SCS Curve Number Method for Predicting Subsurface Drainage Flow. *Trans. ASAE* **2001**, *44*, 1673–1682. [CrossRef]
45. Jain, M.K.; Mishra, S.K.; Suresh Babu, P.; Venugopal, K.; Singh, V.P. Enhanced Runoff Curve Number Model Incorporating Storm Duration and a Nonlinear Ia-S Relation. *J. Hydrol. Eng.* **2006**, *11*, 631–635. [CrossRef]
46. Jain, M.K.; Mishra, S.K.; Suresh Babu, P.; Venugopal, K. On the Ia–S Relation of the SCS-CN Method. *Hydrol. Res.* **2006**, *37*, 261–275. [CrossRef]
47. Caletka, M.; Šulc Michalková, M.; Karásek, P.; Fučík, P. Improvement of SCS-CN Initial Abstraction Coefficient in the Czech Republic: A Study of Five Catchments. *Water* **2020**, *12*, 1964. [CrossRef]
48. Krajewski, A.; Sikorska-Senoner, A.E.; Hejduk, A.; Hejduk, L. Variability of the Initial Abstraction Ratio in an Urban and an Agroforested Catchment. *Water* **2020**, *12*, 415. [CrossRef]
49. Gericke, O.J.; Smithers, J.C. Review of Methods Used to Estimate Catchment Response Time for the Purpose of Peak Discharge Estimation. *Hydrol. Sci. J.* **2014**, *59*, 1935–1971. [CrossRef]
50. Ravazzani, G.; Boscarello, L.; Cislaghi, A.; Mancini, M. Review of Time-of-Concentration Equations and a New Proposal in Italy. *J. Hydrol. Eng.* **2019**, *24*, 04019039. [CrossRef]
51. de Almeida, I.K.; Almeida, A.K.; Anache, J.A.A.; Steffen, J.L.; Alves Sobrinho, T.A. Estimation on Time of Concentration of Overland Flow in Watersheds: A Review. *Geociências* **2014**, *33*, 661–671.
52. Woodward, D.E. Chapter 15. Time of Concentration. In *Hydrology National Engineering Handbook*; Hoeft, C.C., Humpal, A., Cerrelli, G., Eds.; USDA: Washington, DC, USA, 2010; p. 29.
53. Galli, A.; Peruzzi, C.; Beltrame, L.; Cislaghi, A.; Masseroni, D. Evaluating the Infiltration Capacity of Degraded vs. Rehabilitated Urban Greenspaces: Lessons Learnt from a Real-World Italian Case Study. *Sci. Total Environ.* **2021**, *787*, 147612. [CrossRef]

54. USDA National Resources Conservation Service. *Urban Hydrology for Small Watersheds TR55*; U.S. Department of Agriculture: Washington, DC, USA, 1986; p. 164.
55. Leopardi, M.; Scorzini, A.R. Effects of Wildfires on Peak Discharges in Watersheds. *iForest* **2015**, *8*, 302–307. [CrossRef]
56. Lucas-Borja, M.E.; Bombino, G.; Carrà, B.G.; D'Agostino, D.; Denisi, P.; Labate, A.; Plaza-Alvarez, P.A.; Zema, D.A. Modeling the Soil Response to Rainstorms after Wildfire and Prescribed Fire in Mediterranean Forests. *Climate* **2020**, *8*, 150. [CrossRef]
57. Higginson, B.; Jarnecke, J. *Salt Creek BAER-2007 Burned Area Emergency Response*; Hydrology Specialist Report; Uinta National Forest: Provo, UT, USA, 2007; p. 11.
58. Foltz, R.B.; Robichaud, P.R.; Rhee, H. *A Synthesis of Postfire Road Treatments for BAER Teams: Methods, Treatment Effectiveness, and Decisionmaking Tools for Rehabilitation*; U.S. Department of Agriculture, Forest Service, Rocky Mountain Research Station: Fort Collins, CO, USA, 2008; p. RMRS-GTR-228.
59. Doerr, S.H.; Shakesby, R.A.; Walsh, R.P.D. Soil Hydrophobicity Variations with Depth and Particle Size Fraction in Burned and Unburned Eucalyptus Globulus and Pinus Pinaster Forest Terrain in the Águeda Basin, Portugal. *Catena* **1996**, *27*, 25–47. [CrossRef]
60. Hendrayanto; Kosugi, K.; Takahisa, M. Scaling Hydraulic Properties of Forest Soils. *Hydrol. Process.* **2000**, *14*, 521–538. [CrossRef]
61. Robichaud, P.R. Fire Effects on Infiltration Rates after Prescribed Fire in Northern Rocky Mountain Forests, USA. *J. Hydrol.* **2000**, *231–232*, 220–229. [CrossRef]
62. Plaza-Álvarez, P.A.; Lucas-Borja, M.E.; Sagra, J.; Zema, D.A.; González-Romero, J.; Moya, D.; De las Heras, J. Changes in Soil Hydraulic Conductivity after Prescribed Fires in Mediterranean Pine Forests. *J. Environ. Manag.* **2019**, *232*, 1021–1027. [CrossRef] [PubMed]
63. Stoof, C.R.; Ferreira, A.J.D.; Mol, W.; Van den Berg, J.; De Kort, A.; Drooger, S.; Slingerland, E.C.; Mansholt, A.U.; Ferreira, C.S.S.; Ritsema, C.J. Soil Surface Changes Increase Runoff and Erosion Risk after a Low–Moderate Severity Fire. *Geoderma* **2015**, *239–240*, 58–67. [CrossRef]
64. Grangeon, T.; Vandromme, R.; Cerdan, O.; Girolamo, A.M.D.; Lo Porto, A. Modelling Forest Fire and Firebreak Scenarios in a Mediterranean Mountainous Catchment: Impacts on Sediment Loads. *J. Environ. Manag.* **2021**, *289*, 112497. [CrossRef]
65. Turco, M.; Rosa-Cánovas, J.J.; Bedia, J.; Jerez, S.; Montávez, J.P.; Llasat, M.C.; Provenzale, A. Exacerbated Fires in Mediterranean Europe Due to Anthropogenic Warming Projected with Non-Stationary Climate-Fire Models. *Nat. Commun.* **2018**, *9*, 3821. [CrossRef]
66. Rajczak, J.; Schär, C. Projections of Future Precipitation Extremes over Europe: A Multimodel Assessment of Climate Simulations. *J. Geophys. Res. Atmos.* **2017**, *122*, 10–773. [CrossRef]
67. Stevens-Rumann, C.S.; Kemp, K.B.; Higuera, P.E.; Harvey, B.J.; Rother, M.T.; Donato, D.C.; Morgan, P.; Veblen, T.T. Evidence for Declining Forest Resilience to Wildfires under Climate Change. *Ecol. Lett.* **2018**, *21*, 243–252. [CrossRef] [PubMed]
68. Soulis, K.X. Estimation of SCS Curve Number Variation Following Forest Fires. *Hydrol. Sci. J.* **2018**, *63*, 1332–1346. [CrossRef]

hydrology

MDPI

Article

Numerical Simulation of Flow in Parshall Flume Using Selected Nonlinear Turbulence Models

Mehdi Heyrani *, Abdolmajid Mohammadian and Ioan Nistor

Department of Civil Engineering, University of Ottawa, Ottawa, ON K1N 6N5, Canada; amohamma@uottawa.ca (A.M.); inistor@uottawa.ca (I.N.)
* Correspondence: mheyr054@uottawa.ca

Abstract: This study uses a computational fluid dynamics (CFD) approach to simulate flows in Parshall flumes, which are used to measure flowrates in channels. The numerical results are compared with the experimental data, which show that choosing the right turbulence model, e.g., $v^2 - f$ and LC, is the key element in accurately simulating Parshall flumes. The Standard Error of Estimate (SEE) values were very low, i.e., 0.76% and 1.00%, respectively, for the two models mentioned above. The Parshall flume used for this experiment is a good example of a hydraulic structure for which the design can be more improved by implementing a CFD approach compared with a laboratory (physical) modeling approach, which is often costly and time-consuming.

Keywords: Venturi flume; CFD; OpenFOAM; RANS; nonlinear model; turbulence model; numerical simulation; Parshall flume

Citation: Heyrani, M.; Mohammadian, A.; Nistor, I. Numerical Simulation of Flow in Parshall Flume Using Selected Nonlinear Turbulence Models. *Hydrology* **2021**, *8*, 151. https://doi.org/10.3390/hydrology8040151

Academic Editors: Tommaso Caloiero, Carmelina Costanzo and Roberta Padulano

Received: 11 September 2021
Accepted: 8 October 2021
Published: 10 October 2021

1. Introduction

Data from downscaled physical models of different hydraulic structures, such as dams, weirs, etc., were, in the past, the main resource for predicting the consequences of extreme damage. In recent years, with advancements in computing facilities and numerical modeling methods, numerical simulations have become a powerful and popular approach in the design of hydraulic structures [1].

There are various reasons, such as irrigation and quality control, for the importance of measuring the flowrate in an open channel, and this has led different individuals to come up with various ideas and designs for discharge measurement devices. One of the most popular devices is the Parshall flume, a modification of the Venturi flume, developed by Ralph L. Parshall in the 1930s. The major difference between this flume and the Venturi flume is the drop that was introduced in the throat's bed elevation. This design, with a negative bed slope starting at the beginning of the throat section, helps fluid gain speed and, shortly before exiting the throat, a relatively gentle positive slope reduces the speed of the fluid at the exit of the throat section. The relationship between the head at two locations within the flume, i.e., the throat and upstream, provides a value for the flowrate in the open channel [2].

The available sizes for Parshall flumes are limited, and in addition, within this limited range, manufacturers tend to contravene the original specifications provided by Parshall as the inventor. To create a custom-size Parshall flume, many experiments have to be undertaken by the manufacturers to ensure the accuracy of the flowrate within the device. It is costly and time-consuming to run the necessary laboratory experiments for a new size, and using a Computational Fluid Dynamics (CFD) model can significantly accelerate the process [3].

Computer simulations are an essential tool in the design and optimization of hydraulic structures at present, and recent advancements in computing hardware now also allow researchers and engineers to solve previously impossible equations. Fluid motion is one of the most complicated engineering phenomena, and a particular approach to solving a fluid's governing equations depends on the hardware limitations and available time. Various turbulence models are available within different computational fluid dynamics simulation software, and obtaining the best possible hydraulic structural design is possible through the use of CFD simulations. It is important to choose the best model with respect to the cost of calculations and accuracy. Therefore, in this paper, three nonlinear turbulence models from the RANS family were chosen to simulate the flow of water in a 3-inch Parshall flume, and the data from the simulations are compared with the experimental results from a study conducted by Dursun [4].

The study by Wright et al. [5] on the Parshall flume rating curve revealed that calibration for low-discharge flows for the Parshall flume had not been carried out; therefore, there was a bias in the results provided by Parshall himself for the proposed relationship. In their paper, they tried to provide a solution to this flaw, and so a numerical model was established to address the effect of the viscosity of the fluid on the depth discharge relationship. Experiments on a variety of flumes that carried only 15% of the recommended discharge revealed that the flowrate was overpredicted by 25%. The proposed numerical model for the low discharges provided a good match with the experimental data obtained in the laboratory.

Khosronejad et al. [6] implemented a Large Eddy Simulation (LES) model to determine the accuracy of Parshall flume discharge results in comparison with the experimental data. Their study was conducted on two Parshall flumes that were placed in a parallel arrangement, and the results were taken either from the flow passing through an individual flume when the other flume was closed or with the flow passing through both flumes at the same time. In addition to the flow measurement device used in this experiment, a dye dilution approach was also implemented to determine flow rates in the field. The difference between the standard rating curve value and the modeled value according to their study was, at maximum, 10%, while the discharge was at the lower flow rate for all three different scenarios, i.e., flows passing through flumes individually or in parallel, and was a minimum of 1.3% when the discharge was between 1.13 and 1.7 m^3/s in the parallel flow case. It was concluded that a Parshall flume could provide more accurate results when operated at higher flowrates.

Davis and Deutsch [7] conducted studies on Parshall flumes with nonstandard positioning: the slope of the stream, the upstream velocity profile, and alterations in Parshall flume geometry were investigated in this research. Due to the implementation of SOFA-LUMP, a 3D finite-difference code, the simulated flowrates were accurate enough and the computational cost was under the expected budget. A downside of this study was the neglect of the viscosity effects in the numerical model; however, the numerical results were close to the experimental findings. The authors concluded that the proposed numerical model could be used as a guideline to determine the results for nonstandard Parshall flumes, and that the numerical model was the best substitute for laboratory experiments or field installations for accurate results.

Sun et al. [8] investigated the flow in a flume with symmetrical curve obstructions on the flume's sides, and the results revealed that there was an incremental velocity increase within the throat section and a sudden flowrate decrease due to the introduction of a submerged flow condition at the end of that section. A comparison of water levels between the laboratory experiments and the numerical simulations showed a 4.7% error value, which was described as a good agreement. Due to its high accuracy and lower head loss, the proposed curved flume was believed to be an ideal choice for implementation from mild sloped flows to flat ones, e.g., for agriculture and irrigation systems.

Savage et al. [9] tackled the common problem of nonstandard entrance wingwalls in Parshall flumes, which is often neglected. To obtain proper results, it is important to know the best upstream location to measure the head for the flume. It was shown that CFD is a better tool, providing more accurate data compared to the costly physical "build and test" method. This paper introduced a correction factor for a range of different sizes (2–8 ft) of Parshall flumes, to adjust their results, and the implementation of this study for a nonstandard Parshall flume with a free-flow condition increased the accuracy of the discharge results from a 60% error to just +/−5%.

In a study by Heyrani et al. [10], the data from seven different turbulence models were compared with the experimental results from Dursun [4]. In the paper, it was concluded that, among the Reynolds-average Navier–Stokes (RANS), Large Eddy Simulation (LES), and Detached Eddy Simulation (DES) models, the best performance was achieved by the $k-\varepsilon$ model from the RANS family, while the Dynamic K LES model was in second place. The water level results from the CFD simulations provided an error of less than 1.93%–2.08% compared to the experimental findings and were reasonably acceptable for further implementation. Although several turbulence models were examined in the study, some important ones remained unused, which are the subject of the present paper.

The objective of this paper is to extend the study by Heyrani et al. (2021) with more sophisticated, and potentially more accurate, turbulence models in order to develop highly accurate yet efficient modeling approaches for Parshall flumes. Two nonlinear $k-\varepsilon$ models, which have proved to be highly accurate in certain fluid problems, are considered. In addition, the v^2-f model, which is a compromise between the computational efficiency of two-equation models and the accuracy of the Reynolds Stress Models (RSM), is also considered in this study.

This paper is organized as follows. Governing equations and description of turbulence models are provided in Section 2, and numerical details such as mesh, boundary, and initial conditions are then described in Section 3. Next, results and discussions are presented in Section 4, and some concluding results complete the study.

2. Methodology

2.1. Description of CFD Model

As one of the most reliable tools to analyze the behavior of fluids, Computational Fluid Dynamics (CFD) is capable of calculating a wide range of related parameters by taking advantage of the recent development of computer processors. The parameters that are accurately calculated by CFD include flow velocity, temperature and pressure.

CFD models are capable of providing solutions for the flow equations or describing the behavior of a fluid when it interacts with rigid boundaries or obstacles along its path. With respect to the conservation of mass, energy, and momentum, the Navier–Stokes system of equations is derived for viscous fluids [11].

As a reliable open-source solver for computational fluid dynamics models, OpenFOAM is implemented in this study to perform reliable simulations. This computing platform uses the Finite Volume Method (FVM) and includes many specific libraries developed in C++. It is an object-oriented toolbox that can simulate a wide range of flow problems, such as two-phase flows and free-surface flows, with a wide range of turbulence models. It also has the ability to numerically solve continuum mechanics problems [12].

Three nonlinear turbulence models are used to simulate the flow motion in this paper, i.e., the LC low-Reynold, SQ low-Reynold, and v^2-f models, which are briefly discussed in the following section.

2.2. Governing Equations

A viscous incompressible fluid flow is governed by a set of general three-dimensional systems of equations called the Navier–Stokes system, which consists of momentum and continuity equations. The system is described as follows [13,14]:

$$\frac{\partial u}{\partial x}+\frac{\partial v}{\partial y}+\frac{\partial w}{\partial z}=0 \quad (1)$$

Equation (1) is the continuity equation, followed by the three momentum equations (Equations (2)–(4)) for different directions, i.e., x, y, and z directions.

$$\frac{\partial u}{\partial t} + u\frac{\partial u}{\partial x} + v\frac{\partial u}{\partial y} + w\frac{\partial u}{\partial z} = -\frac{1}{\rho}\frac{\partial p}{\partial x} + v^2 u \quad (2)$$

$$\frac{\partial v}{\partial t} + u\frac{\partial v}{\partial x} + v\frac{\partial v}{\partial y} + w\frac{\partial v}{\partial z} = -\frac{1}{\rho}\frac{\partial p}{\partial y} + v^2 v \quad (3)$$

$$\frac{\partial w}{\partial t} + u\frac{\partial w}{\partial x} + v\frac{\partial w}{\partial y} + w\frac{\partial w}{\partial z} = -\frac{1}{\rho}\frac{\partial p}{\partial z} + v^2 w - g \quad (4)$$

In the above equations, fluid density is denoted by ρ, the three spatial directions of velocity are represented by u, v, and w, p denotes the total pressure, time is symbolized by t, and g is the gravitational acceleration.

2.2.1. RANS

The Reynolds-Averaged Navier–Stokes Model is the dominant practical method for simulating the motion of a fluid. Other methods, such as Large Eddy Simulation (LES), are computationally more expensive and still impractical for engineering applications. The viscosity-related properties of the fluid are essentially used to estimate the impact of turbulence. The variation in the turbulent kinetic energy (k) is described by an equation for k.

A variety of turbulence models are available under this category, and the application of three non-linear approaches to form a comparison with the experimental data forms the main objective of this study.

$v^2 - f$ Model

A modified version of the $k - \varepsilon$ model, where k represents the turbulent kinetic energy and ε denotes the energy dissipation, is called the $v^2 - f$ turbulence model, which consists of four equations to simulate the effect of turbulence to find a solution for fluid flow motion. It has two extra equations for velocity and relaxation factors, as well as the two general kinetic energy and dissipation equations. This model falls between the categories of the Reynolds Stress Models (RSM) and the original $k - \varepsilon$ model. In order to evaluate eddy viscosity with this model, the new term v^2, which represents the velocity, is implemented instead of the term for kinetic energy. The governing equations of this model are as follows:

$$\frac{\partial k}{\partial t} + \frac{\partial k u_i}{\partial x_i} = P - \varepsilon + \frac{\partial}{\partial x_j}(Dk_{eff}\frac{\partial k}{\partial x_j}) + S_k \quad (5)$$

$$\frac{\partial \varepsilon}{\partial t} + \frac{\partial \varepsilon u_i}{\partial x_i} = \frac{C'_{\varepsilon 1}P - C_{\varepsilon 2}\varepsilon}{T}\frac{\partial}{\partial x_j}(D\varepsilon_{eff}\frac{\partial \varepsilon}{\partial x_j}) + S_\varepsilon \quad (6)$$

$$\frac{\partial \overline{V^2}}{\partial t} + \frac{\partial \overline{V^2}u_i}{\partial x_i} = kf - 6\overline{V^2}\frac{\varepsilon}{k}\frac{\partial}{\partial x_j}(Dk_{eff}\frac{\partial \overline{V^2}}{\partial x_j}) + S_{\overline{V^2}} \quad (7)$$

$$f - L^2\frac{\partial^2 f}{\partial {x_i}^2} = (C_1 - 1)\frac{\frac{2}{3} - \frac{\overline{V^2}}{k}}{T} + C_2\frac{P}{k} + (\frac{5\overline{V^2}/k}{T}) + S_f \quad (8)$$

In the above equations, the length and time-scales for turbulence are denoted by L and T, while f represents the solution to the last equation. The elliptic operator is used by the $v^2 - f$ model to calculate a similar term to the strain–pressure correlation term of the RSM. There are four different constant C terms, i.e., C_1, C_2, $C'_{\varepsilon 1}$, and $C_{\varepsilon 2}$, which are considered the constants, and the four S terms, i.e., S_k, S_ε, $S_{\overline{V^2}}$, and S_f, which are expected to be defined by the user as source terms. The reader is referred to [15] for further details and values of the coefficients.

Shih et al. (1998) Quadratic $k-\varepsilon$ Model (SQ)

This model is derived from the direct implementation of a basic turbulence relationship. To propose a novel algebraic equation for the Reynolds stress, an essential turbulent relationship has been applied. Two limitations are defined based on their realizability and the theory of rapid distortion and, within the inertia sublayer, the coefficients of the model are regulated using simple flows, such as surface flow and homogenous shear flow.

Quadratic and cubic terms of average velocity in the model were proposed for Reynolds stresses. This is a short version of the general formula for mean velocity gradients and turbulent stresses [16]. The rapid distortion theory was used to determine the coefficients of the model's constraints [17] of realizability [18].

The equations used to model the general turbulent shear flow are:

$$\rho_{,t} + \left(\rho U_j\right)_{,j} = 0 \tag{9}$$

$$\left(\rho U_i\right)_{,t} + \left(\rho U_i U_j\right)_{,j} = -P_{,i} + \left[\mu\left(U_{i,j} + U_{j,i} - \frac{2}{3}U_{k,k}\delta_{ij}\right) - \rho\overline{u_i u_j}\right]_{,j} \tag{10}$$

$$\left(\rho k\right)_{,t} + \left(\rho U_i k\right)_{,i} = \left[\left(\mu + \frac{\mu_T}{\sigma_k}\right)k_{,j}\right]_{,j} - \rho\overline{u_i u_j}U_{i,j} - \rho\varepsilon \tag{11}$$

$$\left(\rho\varepsilon\right)_{,t} + \left(\rho U_i\varepsilon\right)_{,i} = \left[\left(\mu + \frac{\mu_T}{\sigma_\varepsilon}\right)\varepsilon_{,j}\right]_{,j} + C_1 f_1 \rho S\,\varepsilon - C_2 f_2 \rho\frac{\varepsilon^2}{k + \sqrt{v\varepsilon}} + C_3\frac{\mu\mu_T}{\rho}S_{,j}S_{,j} \tag{12}$$

where S in the above equation is defined as:

$$S = \sqrt{2S_{ij}S_{ij}}\,,\;\; S_{ij} = \frac{1}{2}\left(U_{i,j} + U_{j,i}\right)$$

The nonlinear model for the Reynolds stresses is:

$$\begin{aligned} -\rho\overline{u_i u_j} = &-\tfrac{2}{3}\rho k\,\delta_{ij} + \mu_T\left(U_{i,j} + U_{j,i} - \tfrac{2}{3}U_{k,k}\delta_{ij}\right) + A_3\tfrac{\rho k^3}{2\varepsilon^2}\left(U_{k,i}U_{k,j} - U_{i,k}U_{j,k}\right) \\ &+ A_5\tfrac{\rho k^4}{\varepsilon^3}\Big[U_{k,i}U_{k,p}U_{p,j} + U_{k,j}U_{k,p}U_{p,i} - \tfrac{2}{3}\Pi_3\delta_{ij} - \tfrac{1}{2}U_{l,l}\left(U_{i,k}U_{k,j} + U_{j,k}U_{k,i} - \tfrac{2}{3}\Pi_1\delta_{ij}\right) \\ &- \tfrac{1}{2}U_{l,l}\left(U_{k,i}U_{k,j} + U_{i,k}U_{j,k} - \tfrac{2}{3}\Pi_2\delta_{ij}\right)\Big] \end{aligned} \tag{13}$$

$$\Pi_1 = U_{i,j}U_{j,i}\;\;,\;\; \Pi_2 = U_{i,j}U_{i,j}\;,\;\; \Pi_3 = U_{i,k}U_{i,p}U_{p,k} \tag{14}$$

The equations used to obtain the coefficients value of μ_T and A_3 to A_5 are provided below:

$$\mu_T = C_\mu f_\mu \rho\frac{k\left(k + \sqrt{v\varepsilon}\right)}{\varepsilon},\; A_3 = \frac{\sqrt{1 - \frac{9}{2}C_\mu^2\left(\frac{kS^*}{\varepsilon}\right)^2}}{0.5 + \frac{3}{2}\frac{k^2}{\varepsilon^2}\Omega^*\,S^*},\; A_5 = \frac{1.6\,\mu_T}{\frac{pk^4}{\varepsilon^3}\frac{7(S^*)^2 + (\Omega^*)^2}{4}} \tag{15}$$

$$\begin{aligned} &C_\mu = \tfrac{1}{4 + A_s U^*\frac{k}{\varepsilon}}\,,\;\; C_1 = \max\left\{0.43, \tfrac{\eta}{5+\eta}\right\},\; C_2 = 1.9,\; C_3 = 1.0, \\ &\sigma_k = 1.0,\; \sigma_\varepsilon = 1.2,\; U^* = \sqrt{\left(S^*\right)^2 + \left(\Omega^*\right)^2}\,,\; S^* = \sqrt{S_{ij}^*S_{ij}^*}, \\ &\Omega^* = \sqrt{\Omega_{ij}\Omega_{ij}},\; \Omega_{ij} = \tfrac{1}{2}\left(U_{i,j} - U_{j,i}\right),\; S_{ij}^* = S_{ij} - \tfrac{1}{3}S_{kk}\delta_{ij}, \\ &\eta = \tfrac{S\,k}{\varepsilon},\; A_s = \sqrt{6}\cos\phi,\; \phi = \tfrac{1}{3}\arccos\left(\sqrt{6}W^*\right),\quad W^* = \tfrac{S_{ij}^*S_{ji}^*S_{ki}^*}{\left(S^*\right)^3} \end{aligned} \tag{16}$$

The reader is referred to [19] for further details and values of the coefficients.

Lien (1996) Cubic Turbulence Model (LC)

A new version of an eddy-viscosity model for turbulent flows with high Reynolds numbers was derived in [15] by implementing a nonlinear association between the parameters of strain and Reynolds stresses. For low-Reynolds conditions, vorticity tensors were also included in this relationship to identify all the variations in the turbulence length scale close to the wall in an asymptotic manner.

Using series-expansion, a general and coordinate invariant formula for strains and stresses is possible, as follows:

$$\frac{\overline{u_i'u_j'}}{k} = \frac{2}{3}\delta_{ij} - \frac{\nu_T}{k}S_{ij} + C_1\frac{\nu_T}{\epsilon}\left[S_{ik}S_{kj} - \frac{1}{3}\delta_{ij}S_{kl}S_{kl}\right] + C_2\frac{\nu_T}{\epsilon}\left[\Omega_{ik}S_{kj} + \Omega_{jk}S_{ki}\right] + C_3\frac{\nu_T}{\epsilon}\left[\Omega_{ik}\Omega_{jk} - \frac{1}{3}\delta_{ij}\Omega_{kl}\Omega_{kl}\right] + HOT \tag{17}$$

where C_μ and C_1 to C_3, proposed by [20] and only applicable to high-Reynolds areas, are:

$$C_\mu = \frac{0.667}{A_1 + S + 0.9\Omega}|_{A_1=1.25}\,, \tag{18}$$

$$C_1 = \frac{3/4}{(1000+S^3)}, C_2 = \frac{15/4}{(1000+S^3)}, C_3 = \frac{19/4}{(1000+S^3)} \tag{19}$$

$$S_{ij} = \frac{\partial u_i}{\partial x_j} + \frac{\partial u_j}{\partial x_i}\,,\ \Omega_{ij} = \frac{\partial u_i}{\partial x_j} + \frac{\partial u_j}{\partial x_i}\,, S = \frac{k}{\epsilon}\sqrt{\frac{1}{2}S_{ij}S_{ij}}\,,\ \Omega = \frac{k}{\epsilon}\sqrt{\frac{1}{2}\Omega_{ij}\Omega_{ij}}\,. \tag{20}$$

To inspect the consequence of streamline curvature, a cubic correction proposed by [21] is also used:

$$HOT = C_4\frac{\nu_T k}{\epsilon^2}\left(S_{ki}\Omega_{lj} + S_{kj}\Omega_{li}\right)S_{kl} + C_5\frac{\nu_T k}{\epsilon^2}(S_{kl}S_{kl} - \Omega_{kl}\Omega_{kl})S_{ij} \tag{21}$$

where:

$$C_4 = -10C_\mu^2\ C_5 = -2C_\mu^2 \tag{22}$$

The turbulent viscosity ν_T obtained from the $k-\varepsilon$ modelling framework is described as:

$$\nu_T = C_\mu\frac{k^2}{\epsilon} \tag{23}$$

The reader is referred to [22] for further details and values of the coefficients.

2.3. Numerical Setup

The interFoam solver from the OpenFOAM family was chosen as the solver in this study, as it provides a blend of applications of the VoF method and the finite-volume method [10]. The Euler and Crank–Nicolson schemes were implemented as first- and second-order time schemes, respectively, to discretize the temporal term, while the Gauss linear method was applied for the gradient terms. The results from the two different temporal discretization schemes used in this study showed no significant differences, i.e., no significant improvement was observed when the second-order scheme was used. Therefore, using either method has no effect on the reduction in error. In other words, the time scheme has a negligible impact as the source of error. Within this solver, different schemes were used for different purposes, such as the corrected Gauss linear scheme for the Laplacian scheme and a linear scheme for the purpose of discretization of the interpolation terms.

As the initial condition, the inflows of the flume for different scenarios were constant, i.e., 10 l/s, 20 l/s, and 30 l/s. Similar to Heyrani et al. (2021), the flow passing through the walls was considered to be zero, and no dissipation or acceleration was initially defined in the model.

2.3.1. Boundary Conditions

Figure 1 provides a schematic side and top view of the boundary condition considered in this simulation, where the flow enters and exits from one end to another while passing above the bed, which was defined as a wall. Over the flow is the atmosphere boundary, and the condition at the outlet is zero gradient. The volume of fluid method was implemented for the surface of the flow with regard to the zero-pressure state where the two fluids, i.e., liquid and air, meet.

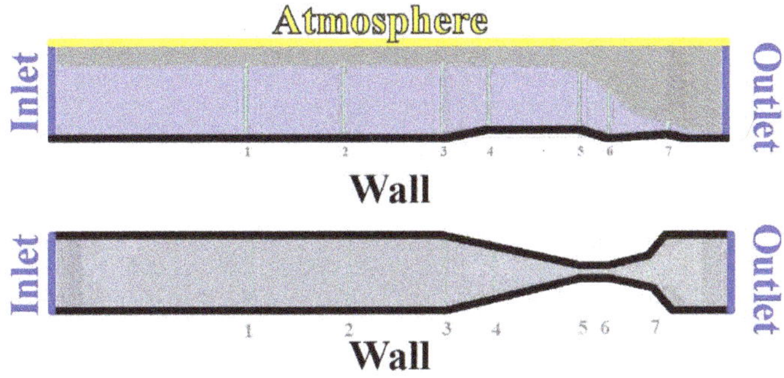

Figure 1. Side and top view of boundary condition of the modeled Parshall flume.

2.3.2. Mesh Sensitivity Analysis

Implementing the right mesh size, i.e., the mesh closest to the optimum grid size, allows for the simulation to produce the results that are the closest to the actual data, i.e., experimental results, with an optimal computational cost. For the simulations in this study, a mesh sensitivity analysis was performed to determine the best grid size for the structured mesh that was used.

In this procedure, the refined mesh resolution was progressively increased until no further changes were obtained in the results. Figure 2 describe the four steps taken to find the optimum grid size in this study. This was started with 52,000 cells in total, progressing to 270,000 over three steps. The data quality resulting from the progression to the second step, i.e., from 52,000 to 75,000 cells, had significant changes, but on proceeding to 270,000, there were no significant changes recorded in the quality of the simulated data. Therefore, no further increase in the number of cells is recommended after 75,000.

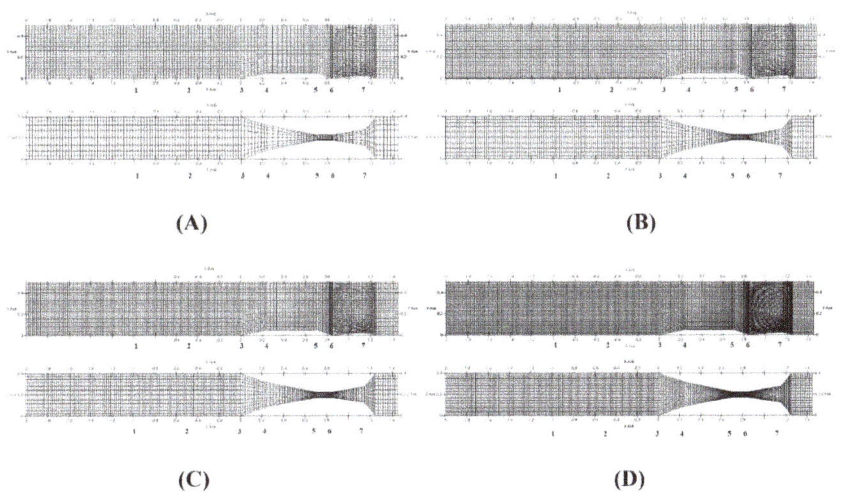

Figure 2. Mesh sensitivity analysis: side and top views of the Parshall flume mesh with different cell numbers: (**A**) with 27,000 cells; (**B**) with 52,000 cells; (**C**) with 75,000 cells; (**D**) with 270,000 cells.

2.4. Data

The trials that resulted in the experimental data were conducted by [4] at the hydraulic laboratory of Firat University in Elazig, Turkey. All experiments were completed in a flume with a rectangular shape and fixed dimensions of 0.4 m width, 5 m length, and 0.6m depth. Although the scope of Dursun's study was the measurement of dissolved oxygen in the fluid before and after entering and exiting the flume, in the present study, only data for water levels were used. Flowrates of 10, 20, 30, and 40 l/s were chosen, which were measured with the help of an electromagnetic flow meter within a modifiable Parshall flume to obtain results that were sufficient to draw conclusions.

The time taken for the simulation to reach steady state was 50 seconds. With respect to the existing hardware that performed the simulation, i.e., Intel Xeon Processor E5-2683 v3 (35M Cache, 2.00 GHz) the total time taken to achieve steady-state, i.e., 50 seconds, was approximately 4 hours. Considering the total number of the cells used in all three simulations, i.e., 75000, the nonlinear model is not a costly model and could be considered in the future by other researchers.

The optimum number of time-steps to achieve a steady water level was found to be 150. The maximum height fluctuations were found to be less than 2% of the steady level height. ParaView was used as post-processing software to demonstrate the water levels and other properties of the flow passing through the flumes. To determine the height of a column of water as a representative segment in each selected cross-section, the line of intersection between two perpendicular planes passing through the column point was found. Then, using the value of the Y coordinate of each datapoint, the water level was determined.

3. Results

The water levels at different sections of the flume were measured for comparison with the experimental results obtained by Dursun (2016). Figure 3 illustrates the geometry and dimensions of the Parshall flume used in the simulation. Water enters the main channel, which has a width of 40 cm, i.e., cross-sections 1 and 2, and, with the help of wing walls, it gradually enters the throat section, which has a 5-cm wall-to-wall distance. The length of the throat, i.e., the distance between cross-sections 5 and 6, is 15 cm. Finally, the flow passes the divergence section, where the slope of the bed gradually becomes positive after cross-section 6.

As shown in Figure 3, seven locations were chosen along the x-axis to assess the water levels for this experiment. As, in the previous study by Heyrani et al. [10], the adjustment of the first sampling location was suggested to obtain more accurate results, cross-section number one was shifted forward, to where fewer fluctuations occur. The locations of the remaining cross-sections were selected as at the beginning of each transition in the flume, i.e., cross-sections 3 and 4 were where the convergence section starts and cross-section 5 was at the start of the throat section. The remaining cross-sections followed the same pattern.

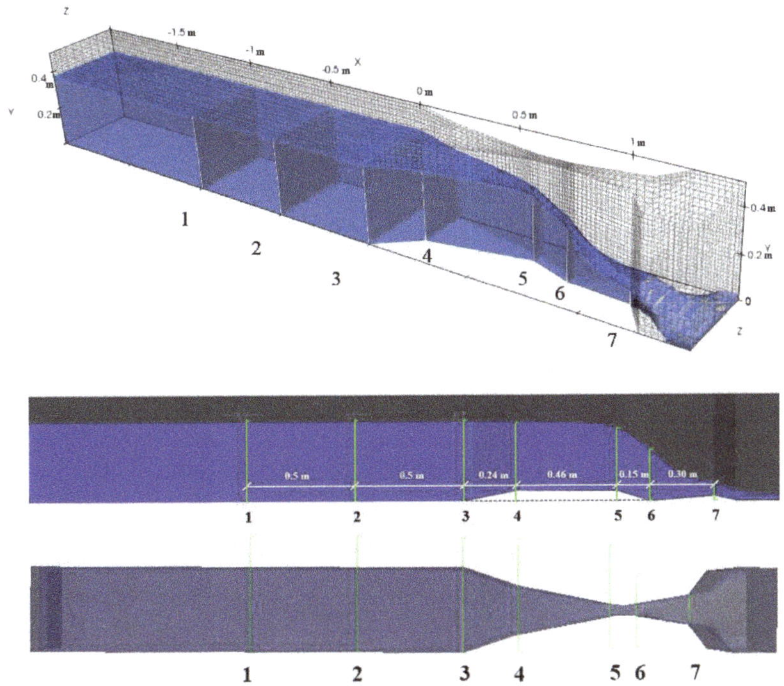

Figure 3. Location of cross-sections: 3D view (**top**), side view (**middle**), and top view (**bottom**).

The model was run with four different grid sizes to find the most suitable one, to obtain better-quality data. Among the different cell quantities tried in this study, i.e., 27,000, 52,000, 75,000, and 275,000, the results tended to remain the same with cell numbers of 52,000 and above.

The models were also run with three different flowrates, i.e., 10 l/s, 20 l/s and 30 l/s, and the smallest error was achieved for the 20 l/s discharge.

Figure 4 illustrates the water levels obtained using the three different turbulence models versus the experimental results from Dursun (2016). The performance of the nonlinear models was found to be more precise compared to the other turbulence models used by Heyrani et al. (2021). The error value derived with Equation (25) for the $v^2 - f$ model, which was the lowest among the three, was 0.76%.

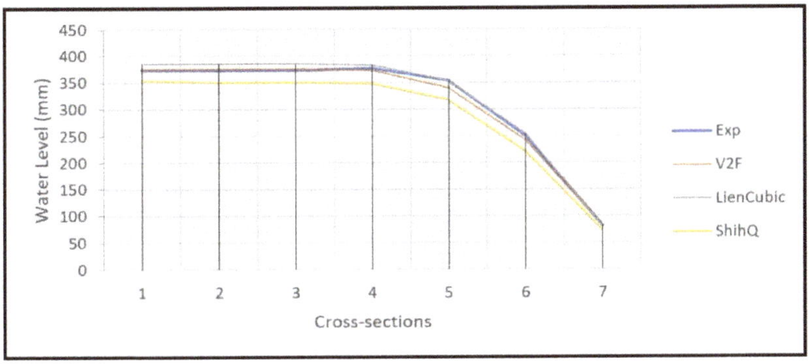

Figure 4. Comparison of water levels over the flume for the $v^2 - f$, LC, and SQ k-ε models versus the experimental results.

Figure 5 shows the velocity gradient of the flow passing through the flume. The contraction at the beginning of the throat, i.e., cross-section 5, forces the flow to gain velocity until it reaches its highest point at cross-section 7, where it experiences the maximum velocity downstream at the second diverging section. Parshall flume's design leads to an increase in flow velocity at certain sections, while the flowrate remains constant.

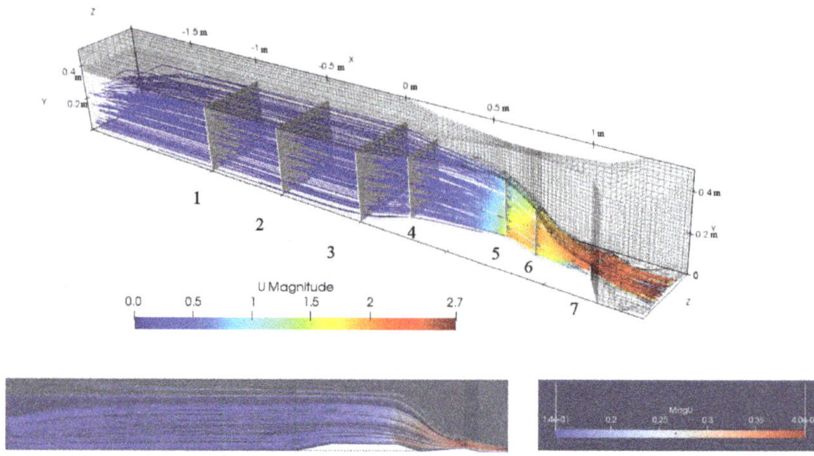

Figure 5. Velocity distribution gradient map.

The velocity profiles at cross-sections 5, 6, and 7 are presented in Figure 6. As shown in Figure 5, the flow speed variation gradually increases along the flow path. At different cross-sections in Figure 6, the maximum speeds were 0.97, 1.21, and 2.45 m/s, respectively, from cross-sections 5 to 7. Due to the shape of the flume, the distribution of the velocity profiles was varied in shape, e.g., at cross-section 5 it was distributed evenly, but as the flow moves forward, the velocity concentration shifted toward the center of the cross-section. The diverging shape of section 7 is the reason that the velocity distribution was concentrated at the sides and not the center.

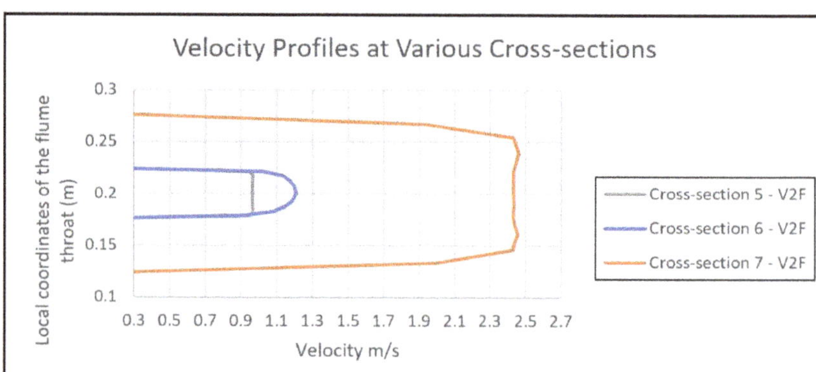

Figure 6. Velocity profiles for different cross-sections.

As illustrated in Figure 7, the flow's pressure field gradient was the lowest when the flow achieved a higher speed from cross-sections 5 to 7. Due to the presence of the throat contraction, a higher pressure was present downstream over the entire flow up to the start of the throat section.

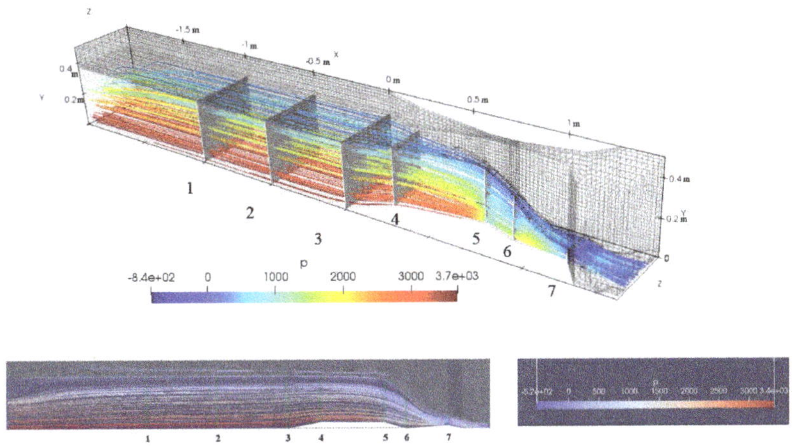

Figure 7. Pressure distribution gradient map.

4. Discussion

Different methods of comparing of the water levels estimated from OpenFOAM versus the experimental results from the case study are illustrated in Tables 1 and 2. The relationships used to calculate the error values are as follows:

$$Error = \frac{|x_{exp} - x_{sim}|}{x_{exp}} \tag{24}$$

$$Standard\ Error\ of\ Estimate\ (SEE) = \sqrt{\frac{\sum_{i=1}^{n}\left(x_{exp_i} - x_{sim_i}\right)^2}{n-2}} \tag{25}$$

$$R^2 = \frac{\sum_{i=1}^{n}\left(x_{exp_i} - \overline{x_{exp}}\right)^2}{\sum_{i=1}^{n}\left(x_{sim_i} - \overline{x_{exp}}\right)^2} \tag{26}$$

Table 1. Error percentage calculated by Equation (24) of the estimated values for the three turbulence models across 7 cross-sections.

	Error Percentage						
Cross-Sections	**1**	**2**	**3**	**4**	**5**	**6**	**7**
V2-F	0.29%	0.55%	0.79%	1.05%	3.92%	3.13%	2.83%
LienCubic(LC)	2.97%	3.23%	3.48%	1.60%	0.25%	2.10%	1.60%
ShihQ(SQ)	5.59%	6.42%	5.93%	7.95%	10.70%	11.57%	12.67%

Table 2. Average error, Standard Error of Estimate (SEE), Square of Correlation coefficient R^2 vs. experimental data.

Turbulence Model	Average Error %	SEE %	R^2	$(1-R^2)$%
V2F	1.79%	0.76%	0.9971	0.29 %
LienCubic(LC)	2.17%	1.00%	0.9985	0.15 %
ShihQ(SQ)	8.69%	3.09%	0.9959	0.41 %

The Standard Error of Estimate (SEE) and the correlation coefficient (R) were calculated to estimate the errors of the simulation data. Tables 1 and 2 shows the calculated error values.

Equation (24) was applied to each individual cross-section and returned a separate error value for each of them, while Equations (25) and (26) provided a single overall value for each dataset, i.e., data from the LC simulation or the $v^2 - f$ model. To represent a single value for an error obtained by Equation (24), an average value was considered for all seven cross-sections.

In an overall analysis of the error percentages, for the first four cross-sections, the $v^2 - f$ turbulence model provided the least amount of error, while the SQ model returned the highest amount of error for the same sections. Moving to the subsequent sections, $v^2 - f$ lost its superiority over the LC turbulence model, where, for all three remaining cross-sections, the LC model provided the least amount of error. The SQ model in this study delivered unacceptable results compared to the other two, and as the average error percentages in Table 2 show, the $v^2 - f$ model was higher than the rest.

As described above, different methods were used to determine how far the simulation results were from the experimental ones. The standard square of estimate is one useful method for estimating the exactness of any prediction. The values produced by this method were 0.76% for the $v^2 - f$ model, 1.00% for the LC model, and 3.09% for the SQ model. This is another proof of the accuracy of the $v^2 - f$ model for this scenario. The correlation coefficient and the root mean square value are also counted as two other proofs of the superiority of the $v^2 - f$ turbulence model.

4.1. SQ Model

The error values from this turbulence model's estimates show higher error percentages than the different methods, and the overall performance of this model was the poorest among the three. As shown in Figures 8 and 9, the average error values for the different cross-sections increased rapidly compared to the other two models. The maximum error value is recorded when the mean value of the error percentages for the seven different cross-sections is calculated.

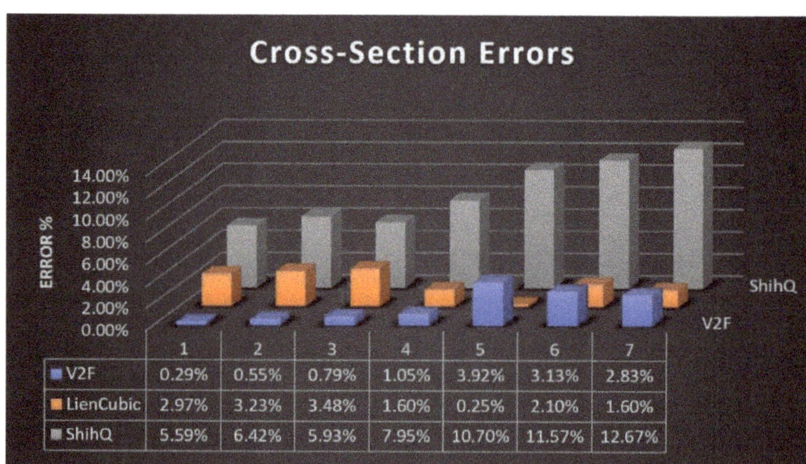

Figure 8. Calculated error magnitude for different cross-sections.

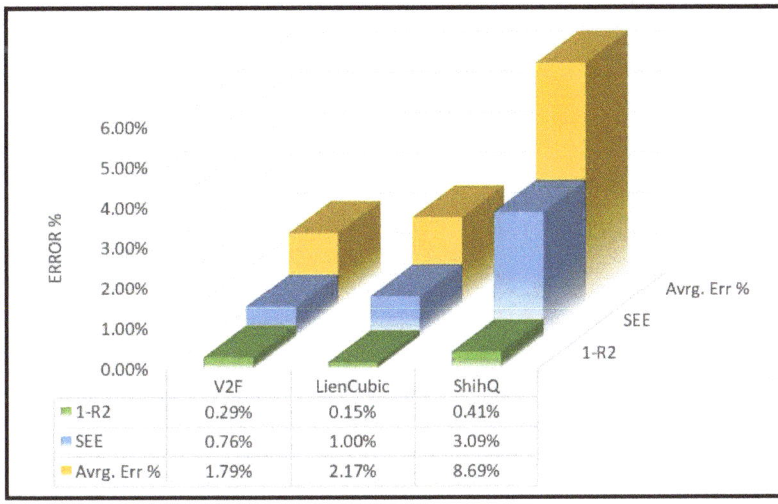

Figure 9. SEE, Average error, and $1 - R^2$ values.

4.2. LC Model

The correlation coefficient values that were closest to the experimental data were obtained from this model, i.e., $1 - 0.15\% = 0.9985$ (99.85%), as shown in Figure 9. Additionally, the last three cross-sections tended to have the lowest error percentage values under the Lien Cubic model. This model produces error values that are almost identical to the $v^2 - f$ model, but slightly higher. Compared to the SQ model, after the $v^2 - f$ model, the data from the LC model are considered reliable.

4.3. $v^2 - f$ Model

Of all the methods used to determine the error percentages of the estimated water levels, compared to the experimental one, this model provided the least error and was the most accurate. Aside from the lack of accuracy regarding the last three cross-sections, i.e., 5, 6, and 7, this model is the most recommended based on the results of this study.

5. Conclusions

The objective of this research was to study the reliability of numerical simulations of a Parshall flume using various nonlinear turbulence models. This is the first time that these state-of-the-art turbulence models have been employed to investigate the hydrodynamic performance of the Parshall flumes.

This study was specifically performed in alignment with the previous research conducted by Heyrani et al. [10]; however, the methods and turbulence models described here have been enhanced to provide more accurate results. Additionally, the recommendations offered in the other paper are implemented in this study where applicable, i.e., in the selection of locations for cross-section number 1. The following points are the main highlights of this research:

- The comparison of three nonlinear turbulence models, i.e., the LC, SQ, and $v^2 - f$ models, reveals that the results obtained from these nonlinear models, except for the SQ model, lead to higher accuracy when compared with the experimental data, as shown in Tables 1 and 2, where the mean error values over all seven cross-sections are 1.79% and 2.17% for the $v^2 - f$ and LC models, respectively.
- The use of $v^2 - f$ and LC models in this study is considered a significant improvement since, in the previous study by Heyrani et al. [10], with the similar initial criteria of the model parameters, none of the seven turbulence models from the three different

families of RANS, LES and DES were able to provide the accuracy of the $v^2 - f$ and LC results when compared to the experimental data, especially the $v^2 - f$ model. However, these two non-linear turbulence models could be considered when the highest accuracy is demanded under similar conditions.

- The performance of the quadratic model in this simulation was not adequate and led to a high error percentage, well beyond the desired boundary. Hence, it is strongly recommended that, if the nonlinear model is chosen, the quadratic model is not used when dealing with Parshall flume modeling with similar specifications and flowrate value.
- The results strongly support the possibility of using CFD simulation as a reliable and cost-effective solution for a variety of different hydraulic projects. It was not only proven to be cost-effective compared to laboratory-scale simulations, but is also less time-consuming, depending on how powerful the computer system is.
- With respect to the enhancement of the Parshall flume design, implementing CFD software is a key element to improving common designs that are used and approved by many different authorities. More studies and laboratory experiments are needed to determine the optimum design, but the results from this study support the possibility of skipping or supplementing the use of experimental data and substituting them with bias-corrected simulation data.

Although the findings presented in this study show acceptable level of error, which are less than 1% with the Standard Error of Estimate method, more research is needed to determine the best combination of different turbulence models for use under various hydraulic conditions, to design Parshall flumes. This will be undertaken with the aim of determining the best numerical approach.

Author Contributions: Conceptualization, A.M. and I.N.; Data curation, M.H.; Formal analysis, M.H.; Funding acquisition, A.M. and I.N.; Investigation, M.H.; Methodology, A.M. and I.N.; Project administration, A.M. and I.N.; Resources, M.H.; Software, M.H., A.M. and I.N.; Supervision, A.M. and I.N.; Validation, M.H.; Visualization, M.H.; Writing—original draft, M.H.; Writing—review & editing, A.M. and I.N. All authors have read and agreed to the published version of the manuscript.

Funding: This research received no external funding.

Data Availability Statement: Part or all of the data is available by the corresponding author upon request.

Conflicts of Interest: The authors declare no conflict of interest.

References

1. Akan, A.O.; Houghtalen, R.J. *Urban Hydrology, Hydraulics, and Stormwater Quality: Engineering Applications and Computer Modeling*; John Wiley & Sons: Hoboken, NJ, USA, 2003.
2. Figuérez, J.A.; González, J.; Galán, Á. Accurate Open Channel Flowrate Estimation Using 2D RANS Modelization and ADCP Measurements. *Water* **2021**, *13*, 1772. [CrossRef]
3. Vermeyen, T. *Laboratory Calibration of a Nonstandard Parshall Flume*; Hydraulics Branch Research and Laboratory Services Division: Denver, CO, USA, 1991.
4. Dursun, O.F. An Experimental Investigation of the Aeration Performance of Parshall Flume and Venturi Flumes. *KSCE J. Civ. Eng.* **2016**, *20*, 943–950. [CrossRef]
5. Wright, S.J.; Tullis, B.P.; Long, T.M. Recalibration of Parshall Flumes at Low Discharges. *J. Irrig. Drain. Eng.* **1994**, *120*, 348–362. [CrossRef]
6. Khosronejad, A.; Herb, W.; Sotiropoulos, F.; Kang, S.; Yang, X. Assessment of Parshall Flumes for Discharge Measurement of Open-Channel Flows: A Comparative Numerical and Field Case Study. *Measurement* **2020**, *167*, 108292. [CrossRef]
7. Davis, R.W.; Deutsch, S. A Numerical-Experimental Study of Parshall Flumes. *J. Hydraul. Res.* **1980**, *18*, 135–152. [CrossRef]
8. Sun, B.; Zhu, S.; Yang, L.; Liu, Q.; Zhang, C.; Zhang, J. ping Experimental and Numerical Investigation of Flow Measurement Mechanism and Hydraulic Performance on Curved Flume in Rectangular Channel. *Arab. J. Sci. Eng.* **2020**. [CrossRef]
9. Savage, B.M.; Heiner, B.; Barfuss, S.L. Parshall Flume Discharge Correction Coefficients through Modelling. *Proc. Inst. Civ. Eng. Water Manag.* **2014**, *167*, 279–287. [CrossRef]
10. Heyrani, M.; Mohammadian, A.; Nistor, I.; Dursun, O.F. Numerical Modeling of Venturi Flume. *Hydrology* **2021**, *8*, 27. [CrossRef]
11. Jasak, H. OpenFOAM: Open Source CFD in Research and Industry. *Int. J. Nav. Arch. Ocean Eng.* **2009**, *1*, 89–94. [CrossRef]

12. Weller, H.G.; Tabor, G.; Jasak, H.; Fureby, C. A Tensorial Approach to Computational Continuum Mechanics Using Object-Oriented Techniques. *Comput. Phys.* **1998**, *12*, 620. [CrossRef]
13. Imanian, H.; Mohammadian, A. Numerical Simulation of Flow over Ogee Crested Spillways under High Hydraulic Head Ratio. *Eng. Appl. Comput. Fluid Mech.* **2019**, *13*, 983–1000. [CrossRef]
14. Alfonsi, G. Reynolds-Averaged Navier–Stokes Equations for Turbulence Modeling. *Appl. Mech. Rev.* **2009**, *62*, 040802. [CrossRef]
15. Durbin, P.A. Separated Flow Computations with the K-Epsilon-v-Squared Model. *AIAA J.* **1995**, *33*, 659–664. [CrossRef]
16. Shih, T.-H.; Lumley, J.L. Kolmogorov Behavior of Near-Wall Turbulence and Its Application in Turbulence Modeling. *Int. J. Comput. Fluid Dyn.* **1993**, *1*, 43–56. [CrossRef]
17. Reynolds, W.C. *Fundamentals of Turbulence for Turbulence Modeling and Simulation*; Stanford University, Department of Mechanical Engineering: Stanford, CA, USA, 1987.
18. Shih, T.-H.; Zhu, J.; Lumley, J.L. A New Reynolds Stress Algebraic Equation Model. *Comput. Methods Appl. Mech. Eng.* **1995**, *125*, 287–302. [CrossRef]
19. Shih, T.-H.; Liu, N.-S.; Chen, K.-H. A Non-Linear k-Epsilon Model for Turbulent Shear Flows. In Proceedings of the 34th AIAA/ASME/SAE/ASEE Joint Propulsion Conference and Exhibit, Cleveland, OH, USA, 13–15 July 1998; p. 3983.
20. Shih, T.-H. *A Realizable Reynolds Stress Algebraic Equation Model*; National Aeronautics and Space Administration, Lewis Research Center, Institute for Computational Mechanics in Propulsion: Cleveland, OH, USA, 1993; Volume 105993.
21. Suga, K. Eddy-Viscosity Modelling with Deformation Invariants and Non-Linear Elements. Ph.D. Thesis, UMIST, Department of Mechanical Engineering, Manchester, UK, 1993.
22. Lien, F.S. Low-Reynolds-Number Eddy-Viscosity Modelling Based on Non-Linear Stress-Strain/Vorticity Relations. In Proceedings of the 3rd Symposium on Engineering Turbulence Modelling and Measurement, Heraklion-Crete, Greece, 27–29 May 1996.

hydrology

MDPI

Article

Hydrological Modeling in Agricultural Intensive Watershed: The Case of Upper East Fork White River, USA

George Bariamis * and **Evangelos Baltas**

Department of Water Resources & Environmental Engineering, School of Civil Engineering, National Technical University of Athens, Str. Iroon Politexniou 9, 157 80 Zografou, Greece; baltas@chi.civil.ntua.gr
* Correspondence: bariamis@mail.ntua.gr

Abstract: Identifying the core hydrological processes of catchments is a critical step for operative hydrological modeling. This study attempts to assess the long-term alterations in streamflow in three adjacent catchments of Upper East Fork White River, Indiana USA, by employing the SWAT hydrological model. The model simulations are spanning from 1980 up to 2015 and distributed in three configurations periods to identify monthly alterations in streamflow. For this purpose, water abstraction, land use, tillage, and agricultural field drainage practices have been incorporated in the model to provide accurate data input. The model setup also integrates spatially disaggregated sectorial water use data from surface and groundwater resources integrating the significant increases of water abstractions mainly for agricultural and public water supply purposes. The land cover of the study area is governed by rotating crops, while agricultural practices and tile drainage are crucial model parameters affecting the regional hydrological balance. Streamflow prediction is based on the SUFI-2 algorithm and the SWAT-CUP interface has been used for the monthly calibration and validation phases of the model. The evaluation of model simulations indicate a progressively sufficient hydrological model setup for all configuration periods with NSE (0.87, 0.88, and 0.88) and PBIAS (14%, −7%, and −2.8%) model evaluation values at the Seymour outlet. Surface runoff/precipitation as well as percolation/precipitation ratios have been used as indicators to identify trends to wetter conditions. Model outputs for the upstream areas, are successful predictions for streamflow assessment studies to test future implications of land cover and climate change.

Keywords: hydrological modeling; streamflow; water balance; SWAT

Citation: Bariamis, G.; Baltas, E. Hydrological Modeling in Agricultural Intensive Watershed: The Case of Upper East Fork White River, USA. *Hydrology* **2021**, *8*, 137. https://doi.org/10.3390/hydrology8030137

Academic Editor: Giorgio Baiamonte

Received: 20 July 2021
Accepted: 7 September 2021
Published: 10 September 2021

1. Introduction

Growing population is expected to reach 10.9 billion by 2100 [1]. As a result, living standards and dietary lifestyles are progressively change in many regions around the world exerting even more pressures to the global food production system. Agricultural products are also used in the livestock and aquaculture industries increasing even more the global competition on water resources. Currently, agriculture uses 70–80% of global water resources to produce the necessary quantities ensuring food security in the supply chain [2]. In USA, agriculture is a key economic sector consuming 40% of the total water use in the country [3]. Key agricultural products are cultivated in several farming belts where the climate conditions are favorable for improved crop yields [4]. Corn and soybean are two of the most cultivated products in upper Mississippi River where intensive agriculture and crop rotation schemes are being practiced for more than a century. An area as large as the Corn Belt is subjected to changes in crop patterns, areal coverage, harvested lands and crop yields affected by the agricultural practices as well as by the climate conditions. A recent study estimated the crop rotation corn-soybean pattern is extending over the 70% of the Corn Belt area [5].

The state of Indiana is one of the key producer states of agricultural and livestock products in the Corn Belt region, centering a critical part of its economy and employment [6].

During the last decades corn and soybean have been significantly increased by 133% (80 bu/acre to 187 bu/acre) and by 93% (30 bu/acre to 58 bu/acre) respectively [7].

The intensive agricultural development activities in the catchments' areas require consistent monitoring in key hydrologic parameters which affect the overall quality and quantity of the crop yields in the region like precipitation, temperature, soil properties, plantation, harvesting period, etc. Furthermore, the effective cultivation of highly valuable crops requires soil preparation during the early weeks of spring, as well as the application of pesticides and additional nutrients to ensure high yields [8,9].

Considering the soil preparation in the region, tile drainage is one of the most common practice which ensures the proper soil nutrients concentrations as well as moisture levels for a proper growing season. The installation of such subsurface tile drainage systems was a significant infrastructure investment in the region, which consequently enriched the soil with air by removing excess water and transforming wetland regions into highly nutrient content valleys for agricultural development [10].

Hydromorphological pressures [11], over-fertilization, short-term land use management [12,13] are some serious problems in intensive agricultural areas, resulting in the collapse of surface and groundwater resources and consequently in the deterioration of ecosystem and their services [14,15]. It is well known that many of the agricultural practices applied in the upstream regions of the Mississippi River are the key drivers for serious impacts in the downstream riparian, coastal, and sea ecosystems in the Gulf of Mexico (eutrophication, chronic, and seasonal hypoxia) [16–19].

As a result of the extensive plowing and overall land use change dynamics over the years, with urbanization and agricultural areas expansion rates at high levels [20], soil structure is greatly unsettled, resulting in increased erosion risk and sediment transport phenomena. More specifically, sediment (suspended and wash load) is the primary mean of pollutant transport in the downstream areas which pose not only geomorphological degradation [21] but also risks for ecosystems status, issues which have to be considered by the current and future management practices applied by river basins committees and authorities [22]. The future of agriculture has to face considerable rise in food demand while trying to decrease its global footprint on natural resources [23].

Model-based methods for hydrological modeling are usually time consuming and require extensive time series of several water related parameters, therefore observation-based methods have been developed to provide quite accurate and early estimates of the human or natural influence on hydrological deficits/droughts [24]. Pair catchment analysis by using unsteady water balance equation and double mass curve techniques, can effectively separate climate change effects from the watershed disturbance (e.g., seasonal effects of forest coverage in hydrological drought). However, some of the limitations of such approaches are to find suitable catchment pairs with long-time series of available data on the pre-disturbance period, and relevant climate, land use characteristics along with detailed physical properties of the watersheds [25].

Hydrological modeling and computational techniques in hydrology have been offered very important advancements the last years due to the constant integration of more accurate algorithmic routines, predicting several hydrological cycle components with remarkable accuracy, as well as in the significant increase of computational power [26]. However, hydrological models are heavily dependent on rainfall observations which must capture accurate precipitation patterns and trends (in case of climate change impacts studies) to effectively simulate the water cycle, while climate/landscape models require further development to better describe spatial scale, magnitude, accuracy, and complexity issues [27]. Data inputs are the primary source of information which is used in the calibration and validation phases which cover a wide range of typology (from ground-based monitoring stations to satellite collected data) [28,29].

As the hydrological models principally attempt to provide the best available estimates of water–land–soil dynamics and regime in study areas simulated, there are a lot of intermediate preparatory steps, decisions and actions made by their operators to provide accurate simulations and targeted outputs. Model results are providing valuable feedback in assessing the surface/groundwater links [30], improvement of management practices under different scenarios [31–33], future projections for coupled land use—climate change impact assessments [34–36] and contribution to large scale modeling [37–41].

Models' applications have been extensively used in either local, regional or even continental scale simulations, where the quantitative and qualitative assessment can provide important conclusions for effective water management [42–46]. Hydrological models outputs have evolved to the level that they can provide to decision makers and policy planners the necessary inputs to protect the environment and ensure water security by the application of best management practices varying from focused sectorial measures of sustainable water use to emissions regulations in order to protect water quality and all dependent ecosystems' chain [47–50]. There is a need for continuous streamlined and monitoring programs for the anthropogenic and natural pressures on water resources based on the principles of collaborative adaptive management. Such approaches integrate experiences and the collective perspectives of managers, stakeholders and scientists in a way which minimizes the sources of uncertainties while supporting informed management decisions in complex and competitive watersheds [51].

Responding to the need for transparent and accessible data inputs and interpretation of model outcomes, the UN Statistics Division has been working in the last few decades with international organizations as well as along with environmental and economic institutes in environmental accounting approaches [52]. Environmental accounting provides the standardization of environmental and economic information in a way to identify their interactions with the anthropogenic socioeconomic environment, which as a process can be defined as the starting point of future long-term planning of the utilization of natural resources. Environmental accounting methodologies have been applied in the domains of water, land, forestry, ecosystems, and energy [53–55].

2. Materials and Methods

2.1. Watershed Description

The study area of headwaters of Upper East Fork White River (UEFWR) consists of three adjacent catchments in the headwaters of the Patoka White River, a tributary of the Wabash River in the state of Indiana, USA. They cover an area of approximately 5700 km^2. The three catchments are the cataloging units of Driftwood (HUC8—05120204), Flatrock-Haw (HUC8—05120205) and the Upper East Fork White (HUC8—05120206) based on the USGS Watershed Boundary Dataset [56]. The study area is drained in the USGS monitoring location 03365500, at East Fork White River at Seymour, Indiana, 95 km southeast from the State's capital City of Indianapolis. The study area is delineated within the boundaries of Bartholomew, Marion, Hancock, Henry, Johnson, Shelby, Rush, Decatur, Jackson, and Jennings counties of Indiana where a population of more than 440,000 people reside (Figure 1a).

The elevation of the watershed ranges from 158 to 358 m with an average value of 258 m as shown in Figure 1b. Elevation data have been acquired from the NASA SRTM program [57]. The soil characteristics have been integrated based on the STATSGO soil dataset, integrating 19 different soil types [58,59]. The relief is majorly formed in light slopes (average slope 2%) forming flatlands ideal for extensive agricultural development. The study area was grouped into three slope classes: (a) <2%, (b) 2–5%, and (c) >5% as shown in Figure 1c.

Within the study area, there were used nine weather and four streamflow stations. The average elevation of the weather stations is 238 m, while climate normal in the Shelbyville station for the 1981–2010 period are for total average precipitation 1106 mm/year (min/max: 60/134 mm) and for the annual average temperature 6.3 °C (min/max: −4.5/17.6 °C) [60]. The weather stations (for precipitation and min/max temperatures) included in the databases were acquired from the NOAA database [61] and are in Columbus (USC00121747), Greenfield (USC00123527), Greensburg (USC00123547), New Castle (USC00126164), Rushville (USC00127646), Seymour

(USC00127935), Shelbyville Sewage plant (USC00127999), Indianapolis International Airport (USW00093819), and Martinsville (USC00125407).

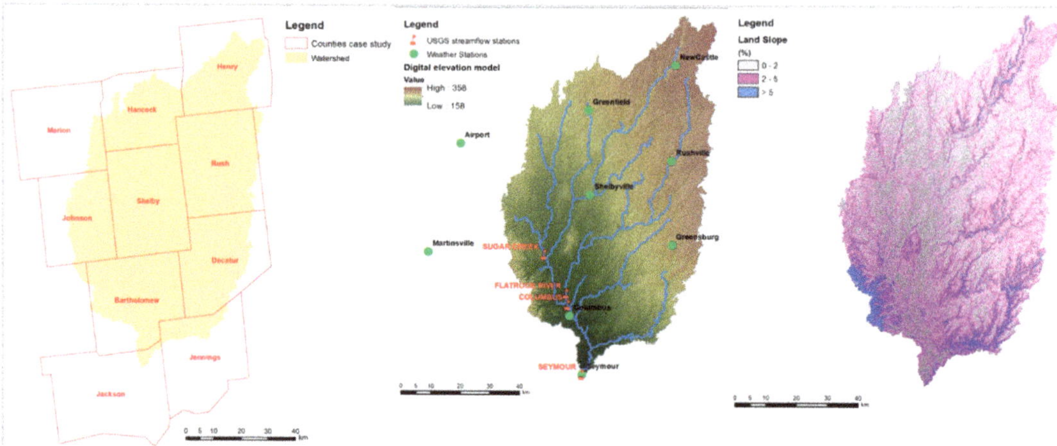

(**a**) (**b**) (**c**)

Figure 1. Study area with counties delineation (**a**), digital elevation model with weather and streamflow stations (**b**), slope distribution (**c**).

2.2. Land Use

In order to successfully perform long term hydrological evaluation via modeling, necessary data inputs had to be collected and curated prior to their integration in the respective SWAT working database. One of the major components affecting the overall rainfall-runoff regime in hydrological models is the land cover status. In our case study, we organized the modeling period into three configuration periods in an attempt to adequately capture the dynamic change of each of the water balance components without adding unnecessary computational burden. The above periods refer to the following years; C_1 (1980–1992), C_2 (1993–2002), and C_3 (2003–2015). For this purpose, three land cover and use instances (NLCD version of 1992, USDA CropScape versions of 2001 and 2011) have been acquired and integrated in each of the model configuration as shown in Figure 2b [62–64].

The distribution of land use types describes a relative stable land use conditions over the last 30 years with the agriculture to be the dominant one [65–67]. In detail, corn and soyabean cultivated in crop rotation pattern, occupy 65% of the total area. The land cover types balance for the latest year of the assessment in 2011 is: 65% agriculture, 14% forest, 11% pastures, 10% urban development, and 1% water surfaces [67]. During the period 1980–2015 in the basin there was a substantial increase in the developed areas (from 3% to 10%) and in forested areas (from 9% to 14%), while pasture areas cover was decreased (from 20% down to 11%) as shown in Figure 2a.

2.3. Streamflow Data

The streamflow data have been acquired from the USGS National Water Information System (NWIS) at four gauging sites; Starting from the headwaters Sugar Creek (USGS STATID: 03362500) draining 18% of the watershed subbasins, Flatrock River at Columbus (USGS STATID: 03363900) draining 23%, East Fork White River at Columbus (USGS STATID: 03364000) draining 78% of the area and Seymour (USGS STATID: 03365500) which is the watershed's outlet. The monthly average summaries of the streamflow stations for each of the configuration periods are presented in Figure 3.

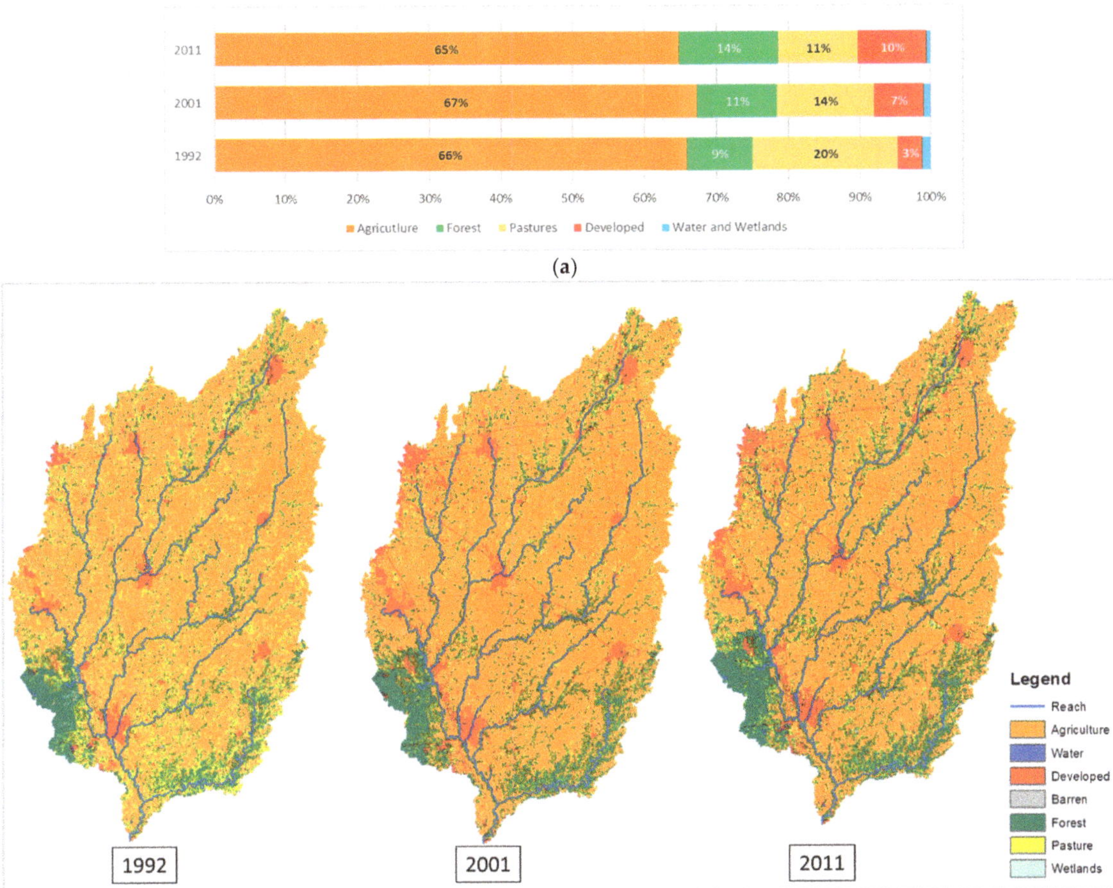

(b)

Figure 2. Land use distribution in the UEFWR basins (**a**) and land use maps integrated in the databases for each of the configuration periods: C_1—1992, C_2—2001, and C_3—2011 (**b**).

2.4. SWAT Model

The SWAT model is a river basin scale model developed by USDA Agricultural Research Service [68]. Its main characteristics are that it is physically based, of continuous simulation, semi distributed, utilizing already available data inputs in a computationally efficient way for simulations which can span for several decades and large study areas [69].

The land phase of the hydrological cycle in SWAT is described by the Equation (1) below:

$$SW_t = SW_0 + \sum_{i=1}^{t} \left(R_{day} - Q_{surf} - E_a - W_{seep} - Q_{gw} \right) \quad (1)$$

where, SW_t, is the final soil water content, SW_0 is the initial soil water content, R_{day} is the amount of daily precipitation, Q_{surf} is the amount of daily surface runoff, E_a the daily actual evapotranspiration, W_{seep} the groundwater recharge to the unsaturated (vadose) zone, and Q_{gw} the daily groundwater return flow to the stream, in mm [69].

It is a model which is continually upgraded and expanded in critical water related processes throughout the years [68,70–72]. The SWAT model is organizing the watershed in smaller partitions with unique land cover, soil and management arrangements. These areas are the hydrologic response units (HRUs) where the land and routing phases of the hydrologic cycle are modelled to provide accurate predictions. The land surface of the hydrologic cycle includes climate parameters, natural hydrologic processes, land cover/plant growth, erosion, nutrients, pesticides, and other management practices occurring in each of the HRUs of the model. A recent release of the model (SWAT2012 rev. 681) was used and in combination with the ArcGIS (version 10.3) interface (ArcSWAT 2012.10.3.19) was used in this paper [73].

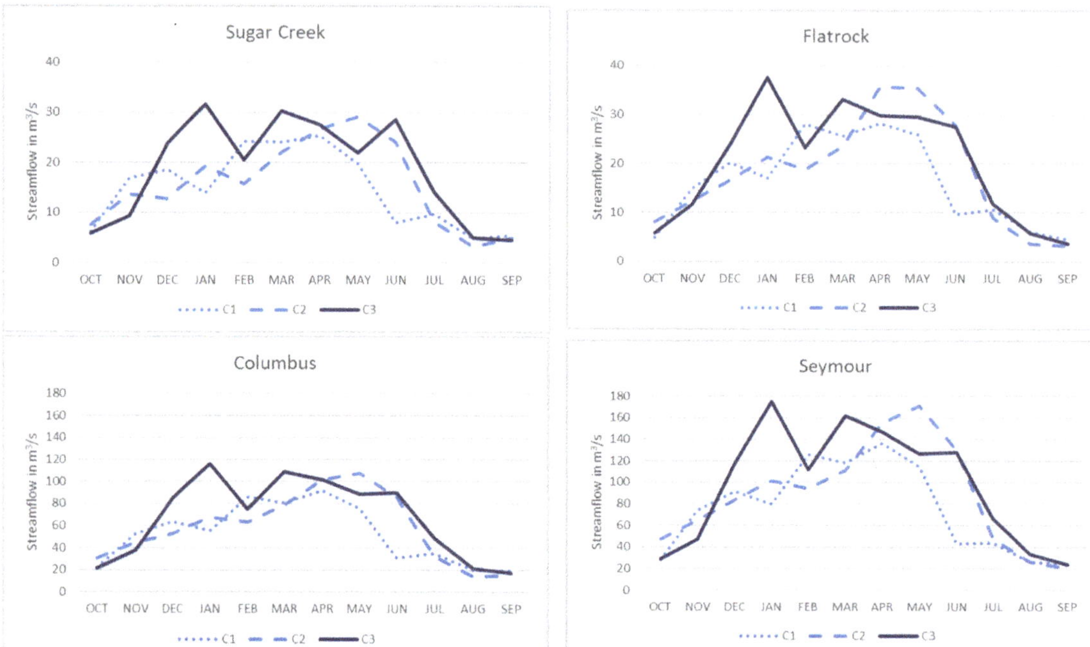

Figure 3. Monthly average streamflow (in m^3/s) in water years for the gauging stations distributed in the configuration periods.

2.5. Agricultural Practices Data

The case study area is one of the most productive in the state of Indiana in terms of agricultural yield. In the study area the crop rotating plantation of corns and soyabeans is being majorly applied [74,75]. Since this agricultural practice is being applied in two or three-year cycles, it has been integrated via the management operations module for better model functioning and water–land relations parameterization. This includes an April to October entire agricultural season with tillage operations, plantations and five-month growing seasons, fertilizer application, and final harvesting [76]. Based on USDA the UEFWR has somewhat poorly and moderately well natural soils drainage characteristics. Therefore, the tile drains were simulated with at 1200 mm depth, 24 h the time to drain a soil to field capacity, and 72 h the amount of time required to release water from a drain tile to a stream reach [76,77]. Tile drainage has been applied in agricultural corn and soyabean fields where slope is in the 0–2% range class.

2.6. Water Abstraction

Water removal from the environment plays a key role while elaborating with hydrological modeling, as it affects the overall water availability as well as from which water resource type (surface or groundwater), the respective volumes have been withdrawn. The case where detailed data collection on water abstractions can be available, it can serve the overall modeling purpose by affecting the respective components as well as supporting more detailed model outputs via improved calibration approach. In the case study area, the Indiana Department of Natural Resources (DNR) operates the Significant Water Withdrawal Facility (SWWF) database where facilities with a capability of abstracting more than 100,000 gallons of water a day (groundwater, surface water, or their combination) are being registered [78]. The UEFWR watershed is intersecting with seven Indiana counties where relative data were available; Bartholomew, Hancock, Henry, Johnson, Shelby, Rush, and Decatur. A spatial analysis has been conducted in order to select the number of facilities within each of the counties, and afterwards to integrate the water abstraction volumes in the SWAT database in the respective subbasins. This analysis also allowed to assess the type of surface or groundwater resource from where it was abstracted (Figure 4a), as well as for which servicing sector this water volumes have been used for (Figure 4b). As the SWWF volumes dataset was not available for the whole period due to technical reasons (size and complexity) the water abstraction data from 1980–2015 by USGS Water Use Data for the Nation program have been used [3,79–85]. This option was selected in order to support the modeling period in the three configuration periods. Based on our analysis the total annual water abstraction for each of the configuration periods are $C_1 = 45.2$ hm^3, $C_2 = 62.5$ hm^3, and for $C_3 = 57.5$ hm^3. Starting from 1984 there was a significant increase of water withdrawal facilities from 158 up to 487, while 90% of the facilities were pumping stations from groundwater (Figure 3). The pumping capacity has been climbed to an average 51 hm^3/month in 2015 starting from 18 hm^3/month in 1984. As for the sectorial distribution of water abstracted 74% is being allocated for public water supply, 13% for industrial uses, and 12% for agricultural production. The water abstraction data have been integrated in the SWAT database (.wus table) on monthly average basis.

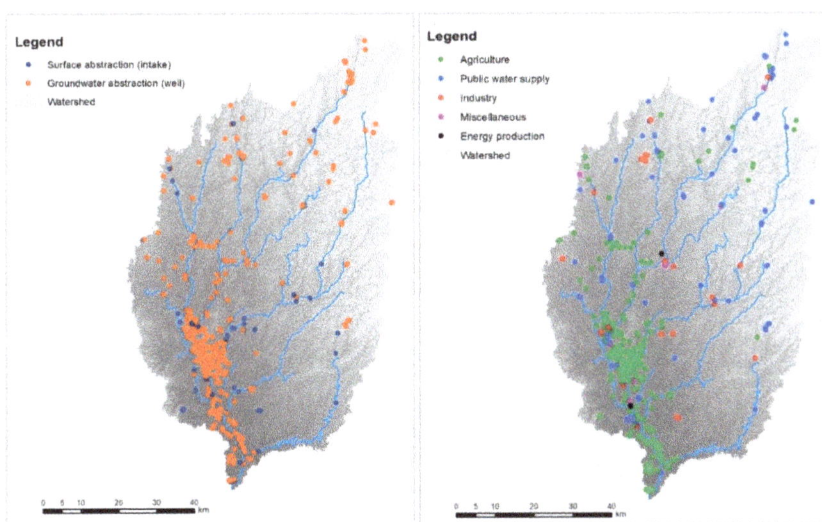

(a) (b)

Figure 4. Water abstraction facilities (as of 2015) in the study area that significantly abstract water by source type (**a**) and by sector type (**b**).

2.7. Calibration and Validation Approach

The purpose of the calibration process is to optimize the performance of the model given the inherent modelling uncertainties already in place while trying to predict the function of each of the hydrological processes. The calibration approach followed in this study was focused on the routing phase of the hydrological cycle by mainly adjusting surface runoff and baseflow model components.

The calibration/validation processes have been conducted in the locations of the four USGS gauging stations at Sugar Creek, Flatrock, Columbus, and the Seymour outlet. Sugar Creek and Flatrock upstream areas are considered hydrologically independent as no other evidence and data are known in case of significant water transfers between them or among other locations. Columbus is the station directly downstream of Flatrock, while Seymour in the outlet is affected by the upstream Columbus calibration process and from the remaining 12% of the ungauged area southeast of the UEFWR basin.

Each of the upstream drainage areas of the respective stations, have been parameterized and calibrated with the SWAT-CUP software [86], an interface which delivers significantly improved results by utilizing the Sequential Uncertainty Fitting algorithm (SUFI-2). Based on similar studies review utilizing SUFI-2 [49,87–90], eight sensitive streamflow related parameters have been selected to be optimized in order to provide improved streamflow calibration results. The selection of the list of calibrated parameters is based on a selection of the most sensitive streamflow related parameters [76,91–93].

The initial SCS runoff curve number (CN2.mgt) and available water capacity of the first soil layer (SOL_AWC(1).sol) have been optimized by relevant variation in the range of [−20%, +20%], while baseflow alpha factor (ALPHA.BF.gw), groundwater delay (GW_DELAY.gw), threshold depth of water in the shallow aquifer for return flow (GWQMN.gw), groundwater re-evaporation coefficient (GW_REVAP.gw), deep aquifer percolation (RCHRG_DP.gw), and soil evaporation compensation factor (ESCO.hru) have been set to vary within a range of absolute values as shown in Table 1. For the snow parameters in the basins, a soft calibration approach has been selected by assigning minimum and maximum snow melt rate (SMFMX.bsn and SMFMN.bsn) to 1 mm of H2O/°C-day and for the snow pack temperature lag factor to 0.4 (TIMP.bsn) [76]. At first, the Sugar Creek (seven upstream basins) and Flatrock (seven upstream basins) upstream regions have been calibrated, then Columbus (14 upstream basins) as the right after in the downstream route of the river, and in the end the Seymour outlet (14 upstream basins) as shown in Figure 5. In total, there were 42 basins that the respective parameters have been calibrated for a set of 500 simulations for each one of them. A set of consecutive years within each of the configuration periods have been used for warm-up (3 y), calibration (≥8 y) and for validation the last 3–4 y of each of the configuration periods. Regarding the suspended sediment load data, there were not available for any of the configuration periods since their data collection in Seymour outlet ended in 1981.

Table 1. Optimized parameters with the SUFI-2 algorithm implemented in SWAT-CUP.

SWAT Parameter	Physical Explanation	Range
CN2 (.mgt)	Initial SCS runoff curve number for moisture condition II	[−20%, +20%]
SOL_AWC(1) (.sol)	Available water capacity of first soil layer (mm/mm)	[−20%, +20%]
ALPHA_BF (.gw)	Baseflow alpha factor (days)	0–1
GW_DELAY (.gw)	Groundwater delay (days)	0–300
GWQMN (.gw)	Threshold depth of water in shallow aquifer for return flow (mm H_2O)	0–300
GW_REVAP (.gw)	Groundwater "revap" coefficient	0–0.2
RCHRG_DP (.gw)	Deep aquifer percolation fraction	0–0.5
ESCO (.hru)	Soil evaporation compensation factor	0.7–1

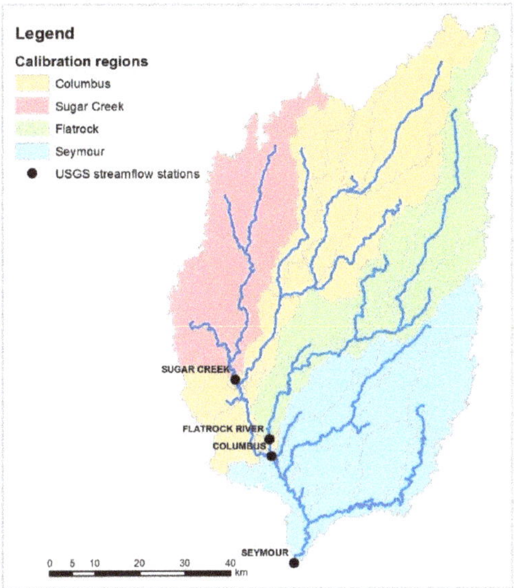

Figure 5. Delineation of the upstream areas, defined by the USGS gauging stations, where calibration phase has been implemented.

3. Results

3.1. Streamflow Calibration Results

As a result of the sequential parametric fitting, the new values of the parameters have been acquired, evaluating the calibration performance against measured streamflow in the four USGS monitoring sites, integrated in the SWAT databases, and run for the years of validation's evaluation. The performance of the model in each of the calibration and validation phases has been evaluated with the Nash–Sutcliffe (NSE) model efficiency coefficient [94], PBIAS, and coefficient of determination (R^2) based on the calibration and validation evaluation approach [95,96] and results [91].

Based on [95,97] the recommended evaluation criteria for recommended statistical performance for modelled flow as model outputs response on monthly scale and 'very good' indication are NSE > 0.8, PBIAS (%) < ±5% and R^2 > 0.85. The new parameters' ranges or absolute values are presented in the Table 2 below for the best simulation fit of SUFI-2 algorithm which was also was the basis to update SWAT model parameters used for the validation phase of this study.

The implementation of auto calibration with SUFI-2 algorithm has been successfully completed based on our approach to calibrate in four locations, in absence of other spatial datasets that could potentially support more detailed parameterization of the SWAT parameters. The evaluation of the calibration and validation results is presented in the Table 3 below for each of the stations and for every configuration period assessed.

Table 2. Calibration results of the SUFI-2 algorithm for each of the three configuration periods and each of the stations' upstream drainage areas.

	C_1				C_2				C_3			
	Sugar Crk	**Flatrock**	**Columbus**	**Seymour**	**Sugar Crk**	**Flatrock**	**Columbus**	**Seymour**	**Sugar Crk**	**Flatrock**	**Columbus**	**Seymour**
CN2.mgt	−19.2%	−6.5%	1.6%	−16.4%	2.0%	−2.0%	−13.3%	−16.4%	2.3%	1.6%	−18.8%	4.6%
ALPHA_BF.gw	0.06	0.06	0.06	0.65	0.81	0.16	0.06	0.65	0.97	0.06	0.71	0.93
RCHRC_DP.gw	0.43	0.21	0.05	0.09	0.83	0.10	0.03	0.09	0.38	0.05	0.10	0.09
GW_DELAY	12.74	20.70	20.70	22.50	14.25	56.10	32.70	22.50	9.90	20.70	38.70	18.90
GWQMN.gw	58.00	128.70	164.70	191.70	130.35	249.30	255.90	191.70	62.70	164.70	3.30	112.50
GW_REVAP.gw	0.07	0.06	0.13	0.03	0.15	0.19	0.17	0.03	0.17	0.13	0.18	0.11
ESCO.hru	0.99	0.83	0.90	0.98	0.98	0.70	0.93	0.98	0.98	0.90	0.74	0.99
SOL_AWC.sol	−3.4%	−16.4%	8.4%	−19.7%	−13%	−17%	19.6%	−19.7%	−14.9%	8.4%	−7.3%	−11.4%

Table 3. Streamflow calibration and validation results for each of the gauging stations and the respective configuration periods.

	Calibration						Validation			
	Station	p	**R^2**	**NSE**	**PBIAS (%)**		**Station**	**R^2**	**NSE**	**PBIAS (%)**
C_1 (1983–1989)	Sugar Crk	0.39	0.66	0.63	13.5	C_1 (1990–1992)	Sugar Crk	0.84	0.66	33.00
	Flatrock	0.86	0.88	0.87	6.00		Flatrock	0.84	0.82	14.00
	Columbus	0.69	0.88	0.88	5.20		Columbus	0.88	0.83	16.50
	Seymour	0.46	0.89	0.88	9.20		Seymour	0.90	0.87	14.50
C_2 (1993–1999)	Sugar Crk	0.73	0.92	0.89	15.30	C_2 (2000–2002)	Sugar Crk	0.83	0.83	−4.30
	Flatrock	0.89	0.90	0.90	−2.30		Flatrock	0.82	0.54	−40.20
	Columbus	0.76	0.94	0.94	5.20		Columbus	0.91	0.87	−11.60
	Seymour	0.69	0.94	0.92	7.70		Seymour	0.89	0.88	−6.90
C_3 (2003–2011)	Sugar Crk	0.71	0.88	0.88	1.00	C_3 (2012–2015)	Sugar Crk	0.83	0.83	3.20
	Flatrock	0.77	0.86	0.86	0.20		Flatrock	0.85	0.65	−13.40
	Columbus	0.62	0.90	0.90	3.70		Columbus	0.85	0.84	5.40
	Seymour	0.53	0.90	0.88	6.50		Seymour	0.89	0.88	−2.80
			Proposed ranges of model evaluation criteria							
			R^2	NSE	PBIAS (%)					
	Satisfactory		0.7–0.8	0.55–0.7	±(10–15)					
	Good		0.8–0.85	0.7–0.85	±(3–10)					
	Very good		0.85–1	0.85–1	±(0–3)					

Additionally, the model performance for wet (October–March) and dry (April–September) seasons are also presented in Table 4. We used the recommended model performance measures for both calibration and validation phases for a SWAT set-up on monthly temporal resolution in an effort to minimize the uncertainties in ungauged parts of the UEFWR basin, especially for Flatrock and Columbus upstream basins [91,95]. For the first and second configuration periods (C_1 and C_2), both coefficient of determination (R^2) and Nash–Sutcliffe efficient results are satisfactory in Flatrock, Columbus, and Seymour outlet, while for the Sugar Creek, lower values of the evaluators were observed with a negligible trend of the model to underestimate the streamflow. In the calibration results of the third configuration period (C_3), in all four stations the results are very satisfactory and withing the acceptable limits.

Table 4. Model performance for low (dry) and high (wet) flow seasons.

		Sugar Crk	**Flatrock**	**Columbus**	**Seymour**
Low flow season	NSE	0.77	0.75	0.84	0.93
	PBIAS (%)	14%	−5%	4%	4%
	R^2	0.69	0.87	0.88	0.88
High flow season	NSE	0.87	0.92	0.94	0.94
	PBIAS (%)	6%	2%	5%	8%
	R^2	0.73	0.93	0.91	0.94

For the evaluation of the validation period, we observed satisfactory performance for R^2 and NSE model evaluators in all periods for the outlet, while for C1 in Sugar Creek, in C2 and C3 in Flatrock we noted that model underestimated the streamflow presenting lower performance values. For the basin's outlet at Seymour, the validation results were R^2 = 0.89, NSE = 0.88, and PBIAS = −2.80% which were considered very satisfactory. The full time series of the modeling period is presented for each of the stations and covering both calibration and validation phases in Figure 6 below.

Figure 6. Monthly discharge (in m^3/s) timeseries of observed (**red**) and modeled (**blue**) outputs for Sugar Creek, Flatrock, Columbus, and Seymour (outlet) monitoring sites. With light green are highlighted the intermediate validation time range among the different configuration periods.

3.2. Precipitation—Snowmelt

Based on the predominant climate conditions in the State of Indiana and in the study area, snowfall is characteristic element of the weather in the region greatly affecting the crop production (by defining level of soil conditions, proper for crop plantation) in each of the growing seasons, while providing increased spring/early summer seasonal streamflow from snowmelt.

Analyzing the model outputs for the precipitation/snowmelt balance, we observed that annual average snow melt accounted for an average 9% of precipitation for all configuration periods. In detail, the maximum average monthly snowmelt to precipitation ratio was 39% in C_1, 47% in C_2, and 38% in C_3 period, which were all observed in February as shown in Figure 7.

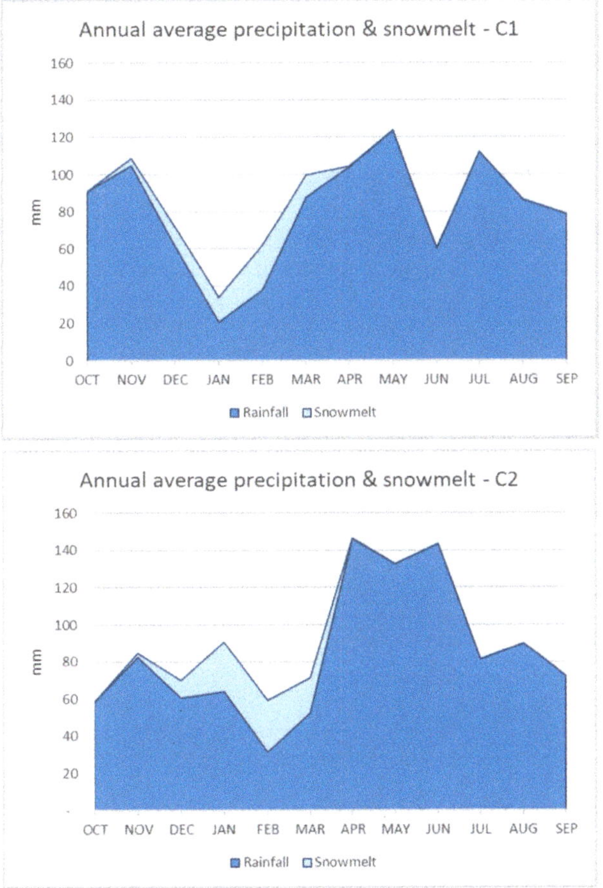

Figure 7. *Cont.*

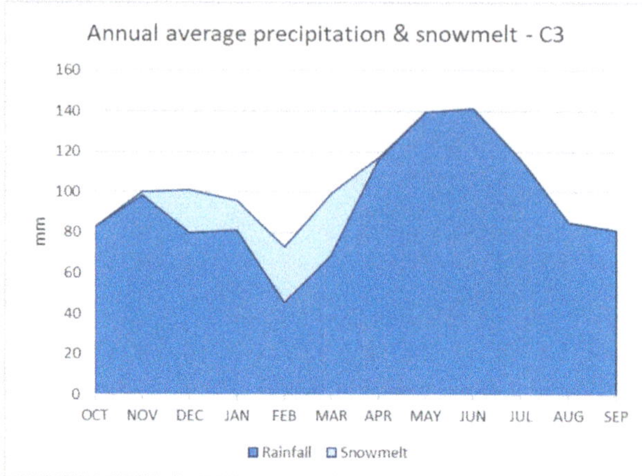

Figure 7. Annual average precipitation and snowmelt (in mm) for the three configuration periods.

3.3. *Water Balance*

Analyzing the water balance components on monthly basis, we observed for the precipitation more humid conditions occurring after January, while in actual evapotranspiration marginal increase is presented mainly during the summer months as shown in Figure 8b. Consequently, surface runoff and water yield shown a relevant increase throughout the year from 274 to 429 mm and from 369 to 556 mm, respectively. In more detail the spatial patterns of key water balance components are presented in Figure 8.

By the completion of the calibration phase, water balance components ratios were extracted from the SWAT results' databases, in annual average terms as shown in the following Table 5. Precipitation is used as the main hydrological cycle component when elaborating these ratios since it is the main water input in the SWAT model. We observed that annual average precipitation has been increased by 20% between C_1 and C_3 periods while actual evapotranspiration remained marginally stable for the same period. This resulted that between the C1–C3 periods we had an increase in effective precipitation from 365 mm up to 589 mm on average. Moreover, the increase in the average monthly precipitation had as a result an increase in surface runoff and total flow (summary of surface runoff, lateral flow, and groundwater return to the streams).

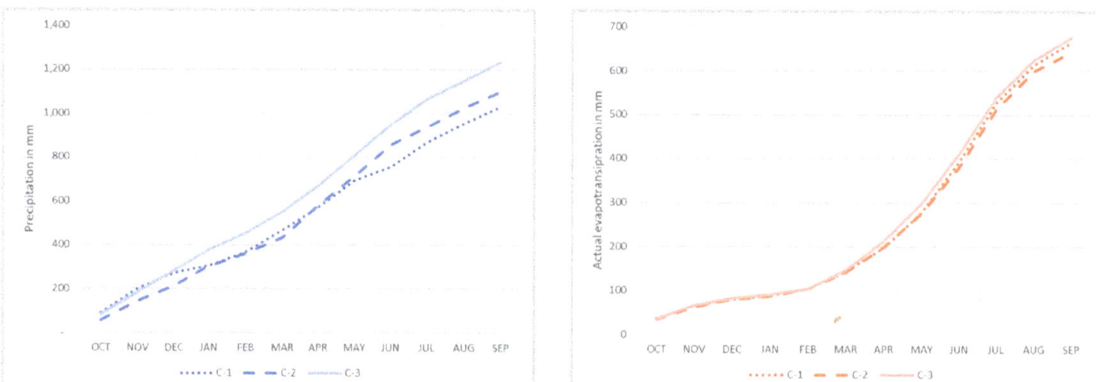

(a) (b)

Figure 8. *Cont.*

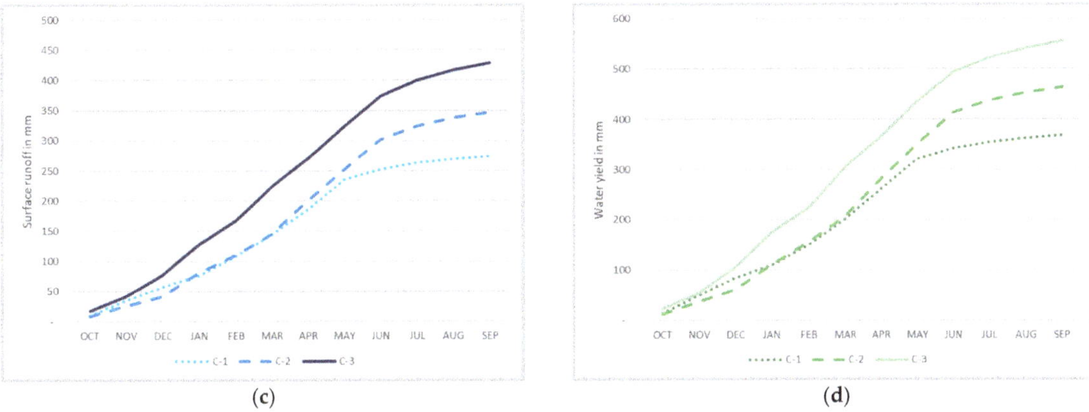

Figure 8. Cumulative water balance monthly components, on water year basis, per configuration period for (**a**) precipitation, (**b**) actual evapotranspiration, (**c**) surface runoff, and (**d**) produced water yield. All numbers are in mm.

Table 5. Annual average water balance ratios among different hydrological cycle parameters based on SWAT outputs for all configuration periods.

Water Balance Components Ratios	C_1	C_2	C_3
Streamflow/Precipitation	27%	31%	35%
Baseflow/Total flow	39%	41%	28%
Surface runoff/Total flow	61%	59%	72%
Percolation/Precipitation	11%	13%	11%
Deep recharge/Precipitation	2%	1%	1%
Evapotranspiration/Precipitation	64%	58%	54%

4. Discussion

4.1. Model Performance

The purpose of this hydrological modeling study was to assess the streamflow status and indirect related streamflow parameters for a long period of time in an agricultural intensive watershed. The extensive data collection system existing in the study area for weather, streamflow, water use, and agricultural practices covering the period from 1980 until 2015 and provided the necessary data inputs for a SWAT model setup. The integration of all the model inputs, have spatially explicit characteristics which were important on the design of the model evaluation. To succeed this, the period of the assessment has been split into three configuration periods where the weather, land cover, and water use data where adequately representative to conduct the calibration and validation phases of the model. Moreover, the three calibration and validation places have been selected in monitoring sites where streamflow data were available for all the period and represented in a balanced way the characteristics of their upstream draining areas. The calibration periods had a range from seven (for C1 and C2) up to nine years (for C3) where a three-year warm up period has been used for each one of them. The streamflow sensitive parameters have been optimized via the use the sensitivity analysis algorithm SUFI-2 of SWAT-CUP software.

The calibration results for all the monitoring sites have been improved for all the three evaluation criteria selected: R^2, NSE, and PBIAS. In the first and second configuration periods the PBIAS for Sugar Creek are 13.5% and 15.3% respectively indicated underestimation of streamflow in the site. This might be a result of marginally increased losses mechanisms not releasing water in the main reach downstream to the calibration site. Other than this, the calibration performed on good and very good basis, regarding the PBIAS indicator for all monitoring sites since it has been calculated within the range of ± 3–10%. For NSE only the Sugar Creek station, for the C1 period presented satisfactory calibration results with a value of 0.60 (satisfactory range 0.55–0.70) which was expected as PBIAS already identified streamflow underestimation as already described. For the rest of the stations and configuration periods calibration's scores were very high and above 0.85 which is the threshold for very good hydrological simulations.

The validation phase of the model is the necessary process which verifies that the calibrated model can adequately perform for a period which was not initially used during the calibration of its parameters. The values of PBIAS in the two basins upstream of Sugar Creek (33%) and Flatrock (14%), indicate underestimation of streamflow in C1, while for the C2 and C3 periods, Flatrock indicated overestimation (−40% and −13.4% respectively). This implies that despite the fact detailed tile drainage data and tillage practices have been integrated in the model setup, more detailed exploration is required to improve the model output within satisfactory levels. This might be a result of a change in agricultural practices followed, construction or change of operation of small water regulation infrastructure. An additional calibration point upstream of Flatrock, might support better model performance for the whole upstream subbasin. In the C3 period though, PBIAS calculated within the satisfactory values range. Considering NSE indicator, the model performed at least satisfactory at all monitoring sites.

As an additional measure of evaluation of the long period simulation in our study, we assessed the model outputs for low and high flow seasons as presented in Table 4. For the low flow seasons and according to NSE indicator, the model performs good for the upstream basins of Sugar Creek and Flatrock, while for the downstream Columbus basin and the Seymour outlet the model performs very well at 0.84 and 0.93 NSE values respectively. The model presents satisfactory and good results for PBIAS values where Flatrock indicates marginal underestimation of the streamflow, while the rest of the three stations show higher (Sugar Creek 14%) or within the good evaluation overestimation values (Columbus and Seymour at 4%). For the high flow season, the model simulation presents very good performance according to NSE indicator as at all monitoring sites the values area above 0.87 and very good level of streamflow magnitude overestimation from 2% at Flatrock up to 8% in the Seymour outlet.

The Pearson correlation coefficient R^2, even though it is a sensitive indicator in cases where high/low values are met, presented high level of collinearity between modeled and observed values for both the dry and wet flow seasons. The overall model validation results, indicate a progressively improving model predictability for streamflow from C1 to C3 configuration periods, for high flow season and from upstream to downstream locations indicating that all intermediate calibration steps at the upstream regions play a significant accumulating role for this very good model performance score at the Seymour outlet.

4.2. Land Cover Dynamics in the Sub-Basins

SWAT provides model outcomes in analytical database format where detailed data exports can support to further assess key outcomes of the simulations conducted. After exploring the results, we selected the four most critical water balance parameters to explore the spatial patterns of their development during the configuration periods. These parameters are the precipitation, actual evapotranspiration, surface runoff, and percolation (groundwater recharge to shallow aquifers). The spatial assessment has been conducted in terms of the areas upstream of the four calibration points as shown in Figures 9 and 10. The assessment of the water balance components is cross studied with the overall land use status for agricultural use (AGRI), pastures/hay (PAST), urbanized areas (URBAN), and forest (FOREST) as shown below. In all the four upstream areas a decreasing trend is observed in pasture/hay areas in favor of urbanization and forest expansion. Urbanization is a key parameter in altering the hydrological regime despite the fact that peri urban and independent forests within the areas presented a respective increase [98].

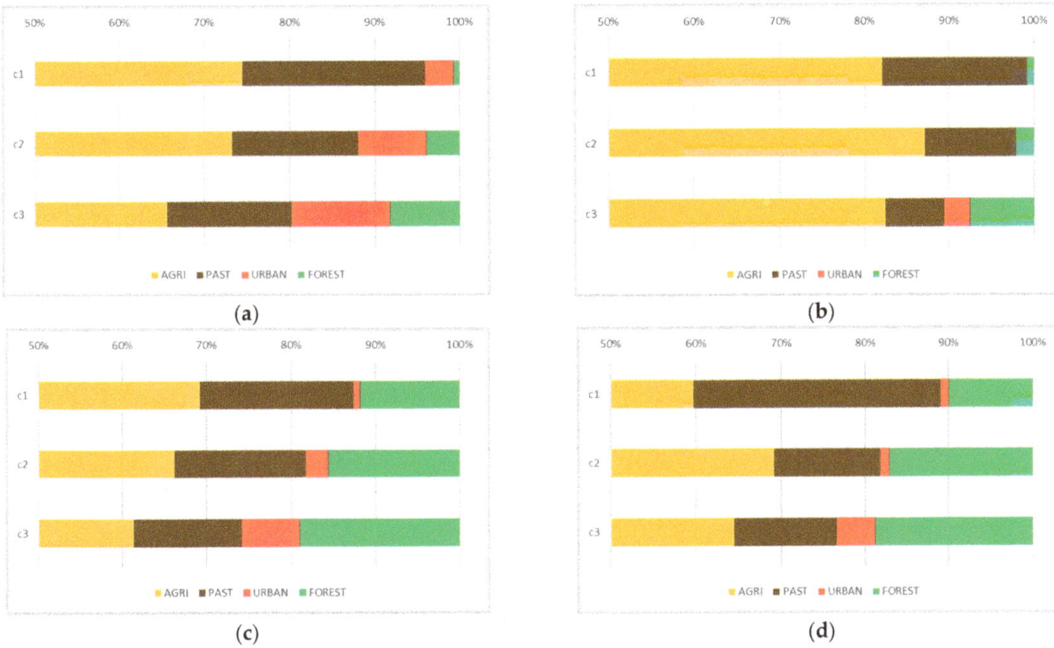

Figure 9. Land cover dynamics in the four upstream areas for (**a**) Sugar Creek, (**b**) Flatrock, (**c**) Columbus, and (**d**) Seymour outlet.

4.3. Precipitation—Actual Evapotranspiration

The amount of precipitation has been uniformly increased in all the four upstream regions where less increase has been observed in the headwaters of Columbus and in the intermediate zones of the river in both Columbus and Flatrock regions. While precipitation increased throughout the state, some places experienced larger increases than others with the southern and west-central regions of the state observed the largest increases [99–101].

Actual evapotranspiration did not have any major change in the UEFWR, except in upstream of Columbus and within the major two out of three tributaries. The significant reduction of pastures areas in both Flatrock and Seymour is one of the reasons affecting the regional marginal reduction of the evapotranspiration from natural vegetated areas.

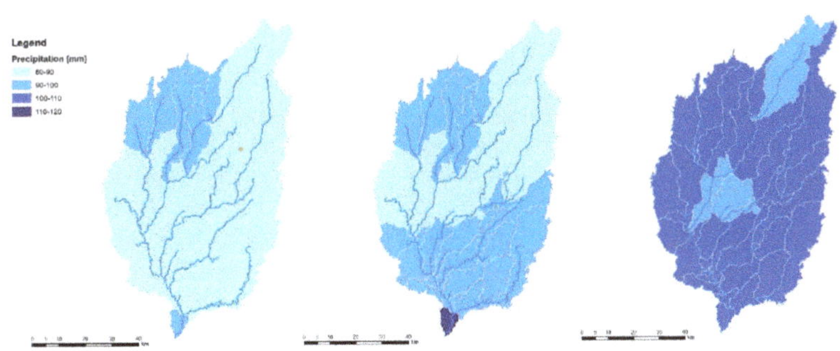

Figure 10. *Cont.*

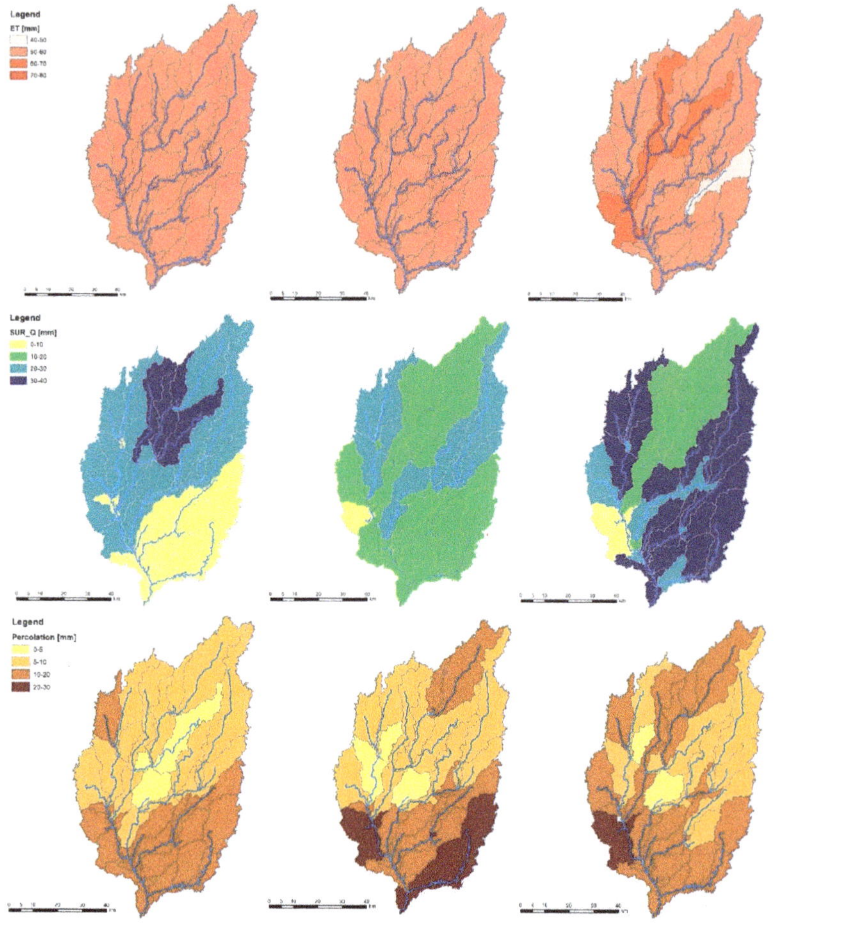

Figure 10. Precipitation, actual evapotranspiration, surface runoff, and percolation maps at subbasin level for each of the configuration periods (C_1, C_2, and C_3) from left to right. All the values presented are in mm.

4.4. Surface Runoff—Percolation—Tile Drainage

As a result of the overall increase in precipitation, this consequently affected the surface runoff throughout the basins and more specifically the upstream areas of Flatrock and Seymour. As the agricultural areas in Seymour have been increased, more subsurface drainage networks have been installed since the latest available data indicate at least a 2773 km^2 increase in land drained by tile, rising to a total of 27% at state level (26,000 km^2) [102,103]. The dipole between surface runoff and percolation seem to interact in opposite directions due to the existence of extended tile drainage in the agricultural parts of the study area. In detail, percolation affected by the existence of crops cultivation rather than surface runoff which is controlled by forest and urban expansion. In Sugar Creek land cover change dynamics are affected more by the extensive urban expansion, with the headwater regions present a slight increase in percolation rates. For Flatrock, the high level of croplands (~80%) govern the water balance especially in the downstream regions close to Columbus keeping percolation/precipitation rates within the range of 6–8%, while less than 10% increase of forests positively affect the surface runoff.

Upstream of Columbus, the reduction of agricultural and pasture areas effectively increased the percolation rates, which resulted in reducing the surface runoff/precipitation rates. More detailed results for all the stations and configuration periods are presented in Figure 11 below.

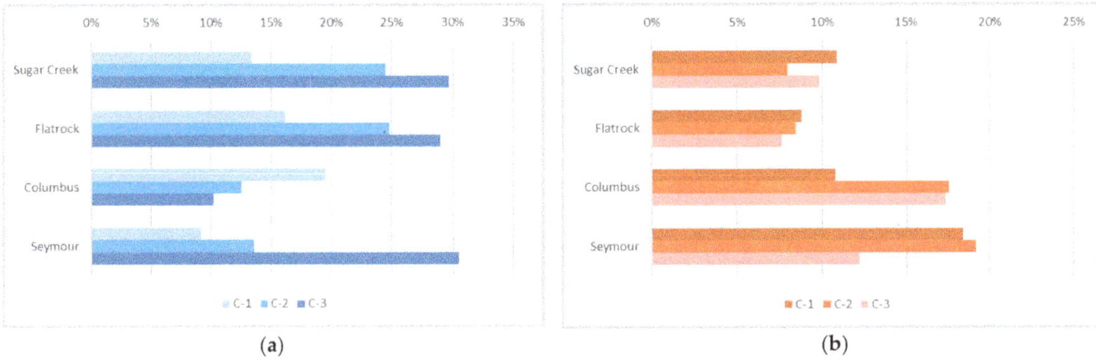

Figure 11. Water balance ratios for each of the upstream drainage areas and per configuration period for (**a**) surface runoff/precipitation and (**b**) percolation/precipitation.

5. Conclusions

In this study, we examined the spatial variability of the main water balance components in an intensively agricultural area in the headwaters of Upper East Fork White River in Indiana, USA. The assessment time range has been divided in three configuration periods to adequately capture potential weather (precipitation and temperature) and land cover changes during the years from 1980 to 2015; following urbanization trends against pastures and peri-urban agricultural areas. Extensive data collection was necessary to provide the best possible data input for a SWAT model set up for the simulation. In order to optimize the data outputs, a spatial calibration approach has been implemented in four gauging sites, by using the auto calibration SUFI-2 algorithm for eight sensitive land cover, soil, and groundwater parameters, where the results were very satisfactory for all the configuration periods and calibration points.

The model evaluation criteria for both calibration (R^2 = 0.90, NSE = 0.88, PBIAS = 6.50%) and validation periods (R^2 = 0.89, NSE = 0.88, PBIAS = −2.8%) at Seymour outlet, indicate a strong correlation and goodness of fit for the most recent configuration period as well as for the two initial ones, while this very good performance in the outlet is being sustained when evaluating both dry (R^2 = 0.88, NSE = 0.93, PBIAS = 4%) and wet (R^2 = 0.94, NSE = 0.94, PBIAS = 8%) seasons.

It was confirmed that in areas with intensified agricultural development, an activity which heavily disturbs the land phase of the hydrological cycle, it is important and critical for hydrological models to integrate data input such water use and relevant agricultural management practices. Those proved to considerably affect the regional hydrological balance as indicated while using precipitation ratios with surface runoff and percolation. The practice of easy-to-use model outputs expressed as ratios can provide a brief description of the water balance dynamics in the study area, allowing proper adjustment of the hydrological model parameters.

The results of this study can be generalized to other watersheds with extensive agricultural land cover and crop synthesis to the extent that supportive information is available on the agricultural practices applied in them such as tile drainage, tillage as well as the soil characteristics of plainlands. They can ideally provide insights for further long-term assessments and improvements in the hydrological cycle representation via the SWAT model, in a way that will promote environmental sustainability (scenario building) while ensuring socioeconomic stability and production of critical crops.

Author Contributions: Conceptualization, G.B. and E.B.; Methodology, G.B.; Software, G.B.; Validation, G.B.; Formal analysis, G.B.; Investigation, G.B.; Resources, G.B. and E.B.; Data curation, G.B.; Writing—original draft preparation, G.B. and E.B.; Writing—review and editing, G.B.; Visualization, G.B.; Supervision, E.B. Both authors have read and agreed to the published version of the manuscript.

Funding: This research received no external funding.

Conflicts of Interest: The authors declare no conflict of interest.

References

1. United Nations. *World Population Prospects 2019*; Nations United: San Francisco, CA, USA, 2019; ISBN 9789211483161.
2. FAO Aquastat—Water Use Thematic Assessment. Available online: http://www.fao.org/aquastat/en/overview/methodology/water-use (accessed on 1 July 2021).
3. Dieter, C.A.; Maupin, M.A.; Caldwell, R.R.; Harris, M.A.; Ivahnenko, T.I.; Lovelace, J.K.; Barber, N.L.; Linsey, K. *Estimated Use of Water in the United States in 2015*; U.S. Geological Survey: Reston, VA, USA, 2018.
4. NCEI U.S. Agricultural Belts. Available online: https://www.ncdc.noaa.gov/monitoring-references/maps/us-ag-belts.php#winter-wheat (accessed on 1 July 2021).
5. Socolar, Y.; Goldstein, B.R.; De Valpine, P.; Bowles, T.M. Biophysical and policy factors predict simplified crop rotations in the US Midwest. *Environ. Res. Lett.* **2021**, *16*, 054045. [CrossRef]
6. ISDA: About Indiana Agriculture. Available online: https://www.in.gov/isda/about/about-indiana-agriculture/ (accessed on 1 July 2021).
7. USDA. Census of Agriculture. Available online: https://www.nass.usda.gov/AgCensus/index.php (accessed on 1 July 2021).
8. David, M.B.; Drinkwater, L.E.; McIsaac, G.F. Sources of Nitrate Yields in the Mississippi River Basin. *J. Environ. Qual.* **2010**, *39*, 1657–1667. [CrossRef]
9. Kladivko, E.J.; Van Scoyoc, G.E.; Monke, E.J.; Oates, K.M.; Pask, W. Pesticide and Nutrient Movement into Subsurface Tile Drains on a Silt Loam Soil in Indiana. *J. Environ. Qual.* **1991**, *20*, 264–270. [CrossRef]
10. Valayamkunnath, P.; Barlage, M.; Chen, F.; Gochis, D.J.; Franz, K.J. Mapping of 30-meter resolution tile-drained croplands using a geospatial modeling approach. *Sci. Data* **2020**, *7*, 1–10. [CrossRef]
11. Elosegi, A.; Díez, J.; Mutz, M. Effects of hydromorphological integrity on biodiversity and functioning of river ecosystems. *Hydrobiologia* **2010**, *657*, 199–215. [CrossRef]
12. Valle Junior, R.F.; Varandas, S.G.P.; Pacheco, F.A.L.; Pereira, V.R.; Santos, C.F.; Cortes, R.M.V.; Sanches Fernandes, L.F. Impacts of land use conflicts on riverine ecosystems. *Land Use Policy* **2015**, *43*, 48–62. [CrossRef]
13. Quintas-Soriano, C.; Castro, A.J.; Castro, H.; García-Llorente, M. Impacts of land use change on ecosystem services and implications for human well-being in Spanish drylands. *Land Use Policy* **2016**, *54*, 534–548. [CrossRef]
14. Ju, X.T.; Xing, G.X.; Chen, X.P.; Zhang, S.L.; Zhang, L.J.; Liu, X.J.; Cui, Z.L.; Yin, B.; Christie, P.; Zhu, Z.L.; et al. Reducing environmental risk by improving N management in intensive Chinese agricultural systems. *Proc. Natl. Acad. Sci. USA* **2009**, *106*, 8077. [CrossRef] [PubMed]
15. Theodoropoulos, C.; Iliopoulou-Georgudaki, J. Response of biota to land use changes and water quality degradation in two medium-sized river basins in southwestern Greece. *Ecol. Indic.* **2010**, *10*, 1231–1238. [CrossRef]
16. Alexander, R.B.; Smith, R.A.; Schwarz, G.E. Effect of stream channel size on the delivery of nitrogen to the Gulf of Mexico. *Nature* **2000**, *403*, 758–761. [CrossRef] [PubMed]
17. Faust, D.R.; Kröger, R.; Moore, M.T.; Rush, S.A. Management Practices Used in Agricultural Drainage Ditches to Reduce Gulf of Mexico Hypoxia. *Bull. Environ. Contam. Toxicol.* **2018**, *100*, 32–40. [CrossRef] [PubMed]
18. Turner, R.E.; Rabalais, N.N.; Justic, D. Gulf of Mexico hypoxia: Alternate states and a legacy. *Environ. Sci. Technol.* **2008**, *42*, 2323–2327. [CrossRef] [PubMed]
19. Whittaker, G.; Barnhart, B.L.; Srinivasan, R.; Arnold, J.G. Cost of areal reduction of gulf hypoxia through agricultural practice. *Sci. Total Environ.* **2015**, *505*, 149–153. [CrossRef] [PubMed]
20. Worku, T.; Khare, D.; Tripathi, S.K. Modeling runoff–sediment response to land use/land cover changes using integrated GIS and SWAT model in the Beressa watershed. *Environ. Earth Sci.* **2017**, *76*, 1–14. [CrossRef]
21. Myers, D.N.; Metzker, K.D.; Davis, S. Status and trends in suspended-sediment discharges, soil erosion, and conservation tillage in the Maumee River basin-Ohio, Michigan, and Indiana. In *Water-Resources Investigation Report*; U.S. Geological Survey: Reston, VA, USA, 2000.
22. Meshesha, T.W.; Wang, J.; Melaku, N.D. Modelling spatiotemporal patterns of water quality and its impacts on aquatic ecosystem in the cold climate region of Alberta, Canada. *J. Hydrol.* **2020**, *587*, 124952. [CrossRef]
23. Mueller, N.D.; Gerber, J.S.; Johnston, M.; Ray, D.K.; Ramankutty, N.; Foley, J.A. Closing yield gaps through nutrient and water management. *Nature* **2012**, *490*, 254–257. [CrossRef] [PubMed]
24. Van Loon, A.F.; Rangecroft, S.; Coxon, G.; Naranjo, J.A.B.; Van Ogtrop, F.; Van Lanen, H.A.J. Using paired catchments to quantify the human influence on hydrological droughts. *Hydrol. Earth Syst. Sci.* **2019**, *23*, 1725–1739. [CrossRef]
25. Aryal, Y.; Zhu, J. Effect of watershed disturbance on seasonal hydrological drought: An improved double mass curve (IDMC) technique. *J. Hydrol.* **2020**, *585*, 124746. [CrossRef]

26. Devia, G.K.; Ganasri, B.P.; Dwarakish, G.S. A Review on Hydrological Models. *Aquat. Procedia* **2015**, *4*, 1001–1007. [CrossRef]
27. Kiprotich, P.; Wei, X.; Zhang, Z.; Ngigi, T.; Qiu, F.; Wang, L. Assessing the impact of land use and climate change on surface runoff response using gridded observations and swat+. *Hydrology* **2021**, *8*, 48. [CrossRef]
28. Rajib, A.; Kim, I.L.; Golden, H.E.; Lane, C.R.; Kumar, S.V.; Yu, Z.; Jeyalakshmi, S. Watershed modeling with remotely sensed big data: MODIS leaf area index improves hydrology and water quality predictions. *Remote Sens.* **2020**, *12*, 2148. [CrossRef]
29. Rajib, M.A.; Merwade, V.; Yu, Z. Multi-objective calibration of a hydrologic model using spatially distributed remotely sensed/in-situ soil moisture. *J. Hydrol.* **2016**, *536*, 192–207. [CrossRef]
30. Yifru, B.A.; Mitiku, D.B.; Tolera, M.B.; Chang, S.W.; Chung, I.M. Groundwater Potential Mapping Using SWAT and GIS-Based Multi-Criteria Decision Analysis. *KSCE J. Civ. Eng.* **2020**, *24*, 2546–2559. [CrossRef]
31. Liu, Y.; Wang, R.; Guo, T.; Engel, B.A.; Flanagan, D.C.; Lee, J.G.; Li, S.; Pijanowski, B.C.; Collingsworth, P.D.; Wallace, C.W. Evaluating efficiencies and cost-effectiveness of best management practices in improving agricultural water quality using integrated SWAT and cost evaluation tool. *J. Hydrol.* **2019**, *577*, 123965. [CrossRef]
32. Oh, H.C.; Shimizu, Y.; Kim, S.K. Effect Analysis Regarding Different Scenarios to Improve Water Quality of the Lake Paldang Basin in Korea. *KSCE J. Civ. Eng.* **2018**, *22*, 3246–3253. [CrossRef]
33. Moriasi, D.N.; Starks, P.J.; Steiner, J.L.; Zhang, X.C.; Garbrecht, J.D.; Glasgow, S. An overview of research into conservation practice effects on soil and water resources in the Upper Washita Basin, Oklahoma, United States. *J. Soil Water Conserv.* **2020**, *75*, 330–339. [CrossRef]
34. Carvalho-Santos, C.; Marcos, B.; Nunes, J.P.; Regos, A.; Palazzi, E.; Terzago, S.; Monteiro, A.T.; Honrado, J.P. Hydrological impacts of large fires and future climate: Modeling approach supported by satellite data. *Remote Sens.* **2019**, *11*, 2832. [CrossRef]
35. Pokhrel, Y.; Burbano, M.; Roush, J.; Kang, H.; Sridhar, V.; Hyndman, D.W. A review of the integrated effects of changing climate, land use, and dams on Mekong river hydrology. *Water* **2018**, *10*, 266. [CrossRef]
36. Wang, H.; Stephenson, S.R.; Qu, S. Quantifying the relationship between streamflow and climate change in a small basin under future scenarios. *Ecol. Indic.* **2020**, *113*, 106251. [CrossRef]
37. White, M.J.; Beiger, K.; Gambone, M.; Haney, E.; Arnold, J.; Gao, J. Development of a hydrologic connectivity dataset for SWAT assessments in the US. *Water* **2017**, *9*, 892. [CrossRef]
38. Meng, X.; Wang, H.; Shi, C.; Wu, Y.; Ji, X. Establishment and evaluation of the China meteorological assimilation driving datasets for the SWAT model (CMADS). *Water* **2018**, *10*, 1555. [CrossRef]
39. Abbaspour, K.C.; Vaghefi, S.A.; Yang, H.; Srinivasan, R. Global soil, landuse, evapotranspiration, historical and future weather databases for SWAT Applications. *Sci. Data* **2019**, *6*, 1–11. [CrossRef]
40. Rajib, A.; Merwade, V. Hydrologic response to future land use change in the Upper Mississippi River Basin by the end of 21st century. *Hydrol. Process.* **2017**, *31*, 3645–3661. [CrossRef]
41. Parshotam, A. An evaluation of Climate Forecast System Reanalysis (CFSR) data for use in models that require meteorological weather station data in New Zealand. *Agron. N. Z.* **2019**, *49*, 51–62.
42. Ali, Y.S.A.; Crosato, A.; Mohamed, Y.A.; Abdalla, S.H.; Wright, N.G. Sediment balances in the Blue Nile River Basin. *Int. J. Sediment Res.* **2014**, *29*, 316–328. [CrossRef]
43. Haimann, M.; Liedermann, M.; Lalk, P.; Habersack, H. An integrated suspended sediment transport monitoring and analysis concept. *Int. J. Sediment Res.* **2014**, *29*, 135–148. [CrossRef]
44. Yen, H.; White, M.J.; Jeong, J.; Arabi, M.; Arnold, J.G. Evaluation of alternative surface runoff accounting procedures using the SWAT model. *Int. J. Agric. Biol. Eng.* **2015**, *8*, 54–68. [CrossRef]
45. Freund, E.R.; Abbaspour, K.C.; Lehmann, A. Water resources of the Black Sea Catchment under future climate and landuse change projections. *Water* **2017**, *9*, 598. [CrossRef]
46. Farjad, B.; Gupta, A.; Razavi, S.; Faramarzi, M.; Marceau, D.J. An integrated modelling system to predict hydrological processes under climate and land-use/cover change scenarios. *Water* **2017**, *9*, 767. [CrossRef]
47. Arnold, J.G.; White, M.J.; Allen, P.M.; Gassman, P.W.; Bieger, K. Conceptual Framework of Connectivity for a National Agroecosystem Model Based on Transport Processes and Management Practices. *J. Am. Water Resour. Assoc.* **2021**, *57*, 154–169. [CrossRef]
48. Singh, J.; Knapp, H.V.; Arnold, J.G.; Demissie, M. Hydrological modeling of the Iroquois River watershed using HSPF and SWAT. *J. Am. Water Resour. Assoc.* **2005**, *41*, 343–360. [CrossRef]
49. Tan, M.L.; Gassman, P.W.; Yang, X.; Haywood, J. A review of SWAT applications, performance and future needs for simulation of hydro-climatic extremes. *Adv. Water Resour.* **2020**, *143*, 103662. [CrossRef]
50. Fidelis, T.; Roebeling, P. Water resources and land use planning systems in Portugal—Exploring better synergies through Ria de Aveiro. *Land Use Policy* **2014**, *39*, 84–95. [CrossRef]
51. Hubbart, J.A.; Kellner, E.; Zeiger, S.J. A case-study application of the experimental watershed study design to advance adaptive management of contemporary watersheds. *Water* **2019**, *11*, 2355. [CrossRef]
52. UNSD. *System of Environmental-Economic Accounting, Central Framework*; ST/ESA/STAT/Ser.F/109; UNSD: New York, NY, USA, 2012.
53. Zal, N.; Bariamis, G.; Zachos, A.; Baltas, E.; Mimikou, M. *Use of Freshwater Resources in Europe 2002–2014*; EEA: København, Copenhagen, 2017.
54. Bariamis, G.; Paschos, G.; Baltas, E. Land Accounts in the River Basin Districts of Greece. *Environ. Process.* **2018**, *5*. [CrossRef]

55. Bagstad, K.J.; Ancona, Z.H.; Hass, J.; Glynn, P.D.; Wentland, S.; Vardon, M.; Fay, J. Integrating physical and economic data into experimental water accounts for the United States: Lessons and opportunities. *Ecosyst. Serv.* **2020**, *45*, 101182. [CrossRef]
56. USGS Watershed Boundary Dataset. Available online: https://www.usgs.gov/core-science-systems/ngp/national-hydrography/watershed-boundary-dataset?qt-science_support_page_related_con=4#qt-science_support_page_related_con (accessed on 1 July 2021).
57. NASA Shuttle Radar Topography Mission. Available online: https://www2.jpl.nasa.gov/srtm/cbanddataproducts.html (accessed on 1 July 2021).
58. USDA United States General Soil Map (STATSGO2). Available online: https://data.nal.usda.gov/dataset/united-states-general-soil-map-statsgo2 (accessed on 1 July 2021).
59. USDA Description of STATSGO2 Database | NRCS Soils. Available online: https://www.nrcs.usda.gov/wps/portal/nrcs/detail/soils/survey/geo/?cid=nrcs142p2_053629 (accessed on 1 July 2021).
60. NOAA. *1991–2020 Climate Normals Indiana*; NOAA: Washington, DC, USA, 2021.
61. USGS National Climatic Data Center (NCDC). Available online: https://www.ncdc.noaa.gov/cdo-web/datasets (accessed on 1 July 2021).
62. Boryan, C.; Yang, Z.; Mueller, R.; Craig, M. Monitoring US agriculture: The US department of agriculture, national agricultural statistics service, cropland data layer program. *Geocarto Int.* **2011**, *26*, 341–358. [CrossRef]
63. USGS Indiana National Land Cover Dataset 1992. Available online: https://maps.indiana.edu/previewMaps/Environment/Land_Cover_1992.html (accessed on 18 July 2021).
64. Han, W.; Yang, Z.; Di, L.; Mueller, R. CropScape: A Web service based application for exploring and disseminating US conterminous geospatial cropland data products for decision support. *Comput. Electron. Agric.* **2012**, *84*, 111–123. [CrossRef]
65. Wickham, J.D.; Stehman, S.V.; Smith, J.H.; Yang, L. Thematic accuracy of the 1992 National Land-Cover Data for the western United States. *Remote Sens. Environ.* **2004**, *91*, 452–468. [CrossRef]
66. Wickham, J.D.; Stehman, S.V.; Fry, J.A.; Smith, J.H.; Homer, C.G. Thematic accuracy of the NLCD 2001 land cover for the conterminous United States. *Remote Sens. Environ.* **2010**, *114*, 1286–1296. [CrossRef]
67. USDA CropScape-NASS CDL Program. Available online: https://nassgeodata.gmu.edu/CropScape/ (accessed on 1 July 2021).
68. Arnold, J.G.; Srinivasan, R.; Muttiah, R.S.; Williams, J.R. Large area hydrologic modelling and assessment Part I: Model development. *J. Am. Water Resour. Assoc.* **1998**, *34*, 73–89. [CrossRef]
69. Neitsch, S.; Arnold, J.; Kiniry, J.; Williams, J. Soil & Water Assessment Tool Theoretical Documentation Version 2009. *Texas Water Resour. Inst.* **2011**, 1–647. [CrossRef]
70. Arnold, J.G.; Allen, P.M.; Bernhardt, G. A comprehensive surface-groundwater flow model. *J. Hydrol.* **1993**, *142*, 47–69. [CrossRef]
71. Arnold, J.G.; Williams, J.R.; Maidment, D.R. Continuous-Time Water and Sediment-Routing Model for Large Basins. *J. Hydraul. Eng.* **1995**, *121*, 171–183. [CrossRef]
72. Srinivasan, R.; Ramanarayanan, T.; Arnold, J.G.; Bednarz, S. Large area hydrologicl modelling and assessment Part II: Model application. *J. Am. Water Resour. Assoc.* **1998**, *34*, 91–101. [CrossRef]
73. SWAT ArcSWAT Interface Version 24 October 2012. Available online: https://swat.tamu.edu/software/arcswat/ (accessed on 1 July 2021).
74. USDA NASS. *Indiana Corn County Estimates 2019*; USDA NASS: Washington, DC, USA, 2020; Volume 48823.
75. USDA NASS. *Indiana Soybeans County Estimates 2019*; NASS: Washington, DC, USA, 2020; Volume 48823.
76. Panagopoulos, Y.; Gassman, P.W.; Jha, M.K.; Kling, C.L.; Campbell, T.; Srinivasan, R.; White, M.; Arnold, J.G. A refined regional modeling approach for the Corn Belt-Experiences and recommendations for large-scale integrated modeling. *J. Hydrol.* **2015**, *524*, 348–366. [CrossRef]
77. Schilling, K.E.; Jha, M.K.; Zhang, Y.-K.; Gassman, P.W.; Wolter, C.F. Impact of land use and land cover change on the water balance of a large agricultural watershed: Historical effects and future directions. *Water Resour. Res.* **2008**, *44*, 1–12. [CrossRef]
78. DNR Significant Water Withdrawal Facility Data. Available online: https://www.in.gov/dnr/water/water-availability-use-rights/significant-water-withdrawal-facility-data/ (accessed on 1 July 2021).
79. Solley, W.B.; Chase, E.B.; Mann, W.B.I. *Estimated Use of Water in the United States in 1980*; U.S. Geological Survey Circular: Reston, VA, USA, 1983.
80. Solley, W.B.; Merk, C.F.; Pierce, R.R. *Estimated Use of Water in the United States in 1985*; U.S. Geological Survey Circular: Reston, VA, USA, 1988.
81. Solley, W.B.; Pierce, R.R.; Perlman, H.A. *Estimated Use of Water in the United States in 1990*; U.S. Geological Survey Circular: Reston, VA, USA, 1993.
82. Hutson, S.S.; Barber, N.L.; Kenny, J.F.; Linsey, K.S.; Lumia, D.S.; Maupin, M.A. *Estimated use of water in the United States in 2000*; U.S. Geological Survey Circular: Reston, VA, USA, 2004; ISBN 060797818X.
83. Kenny, J.F.; Barber, N.L.; Hutson, S.S.; Linsey, K.S.; Lovelace, J.K.; Maupin, M.A. *Estimated Use of Water in the United States in 2005*; U.S. Geological Survey Circular: Reston, VA, USA, 2005; Volume 1344, ISBN 9781411326002.
84. Maupin, M.A.; Kenny, J.F.; Hutson, S.S.; Lovelace, J.K.; Barber, N.L.; Linsey, K.S. *Estimated Use of Water in the United States in 2010*; U.S. Geological Survey Circular: Reston, VA, USA, 2010; ISBN 9781411338623.
85. Coast, G.; Catahoula, T.; Counties, G.; Counties, J.; County, W.; County, J.; County, H.; Water, T.; Board, D.; District, H.C.S. *Estimated Use of Water in the United States in 1995*; U.S. Geological Survey Circular: Reston, VA, USA, 1998; ISBN 060790075X.

86. Abbaspour, K.C.; Johnson, C.A.; van Genuchten, M.T. Estimating Uncertain Flow and Transport Parameters Using a Sequential Uncertainty Fitting Procedure. *Vadose Zo. J.* **2004**, *3*, 1340–1352. [CrossRef]
87. Du, L.; Rajib, A.; Merwade, V. Large scale spatially explicit modeling of blue and green water dynamics in a temperate mid-latitude basin. *J. Hydrol.* **2018**, *562*, 84–102. [CrossRef]
88. Jang, W.S.; Engel, B.; Ryu, J. Efficient flow calibration method for accurate estimation of baseflow using a watershed scale hydrological model (SWAT). *Ecol. Eng.* **2018**, *125*, 50–67. [CrossRef]
89. Woldesenbet, T.A.; Elagib, N.A.; Ribbe, L.; Heinrich, J. Hydrological responses to land use/cover changes in the source region of the Upper Blue Nile Basin, Ethiopia. *Sci. Total Environ.* **2017**, *575*, 724–741. [CrossRef] [PubMed]
90. White, J.; Stengel, V.; Rendon, S.; Banta, J. The importance of parameterization when simulating the hydrologic response of vegetative land-cover change. *Hydrol. Earth Syst. Sci.* **2017**, *21*, 3975–3989. [CrossRef]
91. Arnold, J.G.; Moriasi, D.N.; Gassman, P.W.; Abbaspour, K.C.; White, M.J.; Srinivasan, R.; Santhi, C.; Harmel, R.D.; Van Griensven, A.; Van Liew, M.W.; et al. SWAT: Model use, calibration, and validation. *Trans. ASABE* **2012**, *55*, 1491–1508. [CrossRef]
92. Panagopoulos, Y.; Gassman, P.W.; Kling, C.L.; Cibin, R.; Chaubey, I. Water Quality Assessment of Large-scale Bioenergy Cropping Scenarios for the Upper Mississippi and Ohio-Tennessee River Basins. *J. Am. Water Resour. Assoc.* **2017**, *53*, 1355–1367. [CrossRef]
93. Abbaspour, K.C.; Vaghefi, S.A.; Srinivasan, R. A guideline for successful calibration and uncertainty analysis for soil and water assessment: A review of papers from the 2016 international SWAT conference. *Water* **2017**, *10*, 6. [CrossRef]
94. Nash, J.E.; Sutcliffe, J.V. River flow forecasting through conceptual models PART I-A discussion of principles. *J. Hydrol.* **1970**, *10*, 282–290. [CrossRef]
95. Moriasi, D.N.; Gitau, M.W.; Pai, N.; Daggupati, P. Hydrologic and water quality models: Performance measures and evaluation criteria. *Trans. ASABE* **2015**, *58*, 1763–1785. [CrossRef]
96. Moriasi, D.N.; Arnold, J.G.; Van Liew, M.W.; Bingner, R.L.; Harmel, R.D.; Veith, T.L. Model evaluation guidelines for systematic quantification of accuracy in watershed simulations. *Trans. ASABE* **2007**, *50*, 885–900. [CrossRef]
97. ASABE. Guidelines for Calibrating, Validating, and Evaluating Hydrologic and Water Quality (H/WQ) Models. *Trans. ASABE* **2017**, *61*, 1–15.
98. Aboelnour, M.; Gitau, M.W.; Engel, B.A. Hydrologic response in an urban watershed as affected by climate and land-use change. *Water* **2019**, *11*, 1603. [CrossRef]
99. Widhalm, M.; Hamlet, A.; Byun, K.; Robeson, S.; Baldwin, M.; Staten, P.; Huber, M. *Indiana's Past & Future Climate: A Report from the Indiana Climate Change Impacts Assessment*; Purdue Climate Change Research Center, Purdue University: West Lafayette, IN, USA, 2018; pp. 1–13. [CrossRef]
100. Gowda, P.; Steiner, J.L.; Olson, C.; Boggess, M.; Farrigan, T.; Grusak, M.A. *Impacts, Risks, and Adaptation in the United Stated: Fourth National Climate Assessment, Volume II: Agriculture and Rural Communities*; U.S. Global Change Research Program: Washington, DC, USA, 2018; Volume II.
101. Angel, J.R.; Swanson, C.; Boustead, B.M.; Conlon, K.; Hall, K.R.; Jorns, J.L.; Kunkel, K.E.; Lemos, M.C.; Lofgren, B.M.; Ontl, T.; et al. *Midwest. Impacts, Risks, and Adaptation in the United States: The Fourth National Climate Assessment*; U.S. Global Change Research Program: Washington, DC, USA, 2018; Volume II.
102. USDA. *Agricultural Census*; USDA: Washington, DC, USA, 2012; Volume 1, pp. 196–209.
103. USDA. *Agricultural Census*; USDA: Washington, DC, USA, 2017; pp. 192–205.

hydrology

MDPI

Article

Reinvestigating the Parabolic-Shaped Eddy Viscosity Profile for Free Surface Flows

Rafik Absi

ECAM-EPMI, LR2E-Lab, Quartz-Lab (EA 7393), 13 Boulevard de l'Hautil, 95092 Cergy-Pontoise, France; rafik.absi@yahoo.fr or r.absi@ecam-epmi.com; Tel.: +33-1-30756921

Abstract: The flow in rivers is turbulent. The main parameter related to turbulence in rivers is the eddy viscosity, which is used to model a turbulent flow and is involved in the determination of both velocities and sediment concentrations. A well-known and largely used vertical distribution of eddy viscosity in free surface flows (open channels and rivers) is given by the parabolic profile that is based on the logarithmic velocity profile assumption and is valid therefore only in the log-law layer. It was improved thanks to the log-wake law velocity profile. These two eddy viscosities are obtained from velocity profiles, and the main shortcoming of the log-wake profile is the empirical Coles' parameter. A more rigorous and reliable analytical eddy viscosity model is needed. In this study, we present two analytical eddy viscosity models based on the concepts of velocity and length scales, which are related to the exponentially decreasing turbulent kinetic energy (TKE) function and mixing length, namely, (1) the exponential-type profile of eddy viscosity and (2) an eddy viscosity based on an extension of von Karman's similarity hypothesis. The eddy viscosity from the second model is Re_*-independent, while the eddy viscosity from the first model is Re_*-dependent (where Re_* is the friction Reynolds number). The proposed analytical models were validated through computation of velocity profiles, obtained from the resolution of the momentum equation and comparisons to experimental data. With an additional correction function related to the damping effect of turbulence near the free surface, both models are similar to the log-wake-modified eddy viscosity profile but with different values of the Coles' parameter, i.e., $\Pi = 0.2$ for the first model and $\Pi = 0.15$ for the second model. These values are similar to those found in open-channel flow experiments. This provides an explanation about the accuracy of these two analytical models in the outer part of free surface flows. For large values of Re_* ($Re_* > 2000$), the first model becomes Re_* independent, and the two coefficients reach asymptotic values. Finally, the two proposed eddy viscosity models are validated by experimental data of eddy viscosity.

Keywords: river flow; open channels; eddy viscosity; parabolic profile; streamwise velocity distribution; turbulent kinetic energy (TKE); mixing length; log law; log wake

Citation: Absi, R. Reinvestigating the Parabolic-Shaped Eddy Viscosity Profile for Free Surface Flows. *Hydrology* **2021**, *8*, 126. https://doi.org/10.3390/hydrology8030126

Academic Editors: Tommaso Caloiero, Carmelina Costanzo and Roberta Padulano

Received: 19 July 2021
Accepted: 17 August 2021
Published: 20 August 2021

1. Introduction

Determination of velocity distribution in open-channel flows and rivers is a topic of high interest and is involved in different practical applications [1–7]. However, the hydrodynamic in rivers and open-channel flows is strongly influenced by turbulence [8–11]. Different experimental studies were conducted to better understand the effect of turbulence on the streamwise velocity distribution [12–16].

Open-channel flow studies are considered an important preliminary step to investigate more complex river flows. The main interest in laboratory investigations of free surface turbulent flows is related to the experimental conditions that are chosen to be in agreement with the assumptions related to the models. Therefore, experiments in laboratory flumes allowed analytical models of turbulence to be developed that are in the form of analytical solutions, semi-theoretical or empirical relationships [17–20]. These models were proposed for mean velocities, turbulent kinetic energy (TKE), mixing length, eddy viscosity with their link to flow parameters (flow depth, friction velocity) [11,18]. The developed analytical models are mostly two-dimensional and for uniform flows, while free surface turbulent flows in rivers are generally strongly three-dimensional and non-uniform. Even if the assumptions are far from the real-life conditions in rivers, and the developed analytical models cannot account for the full complexity of turbulent river flows, these models present high practical interest. For example, measured data in the central part of rivers are well described by the analytical expressions obtained from laboratory open-channel flows [21].

In open-channel flows and rivers, in the classical two-layer approach, the flow is divided into two regions: an inner ($\xi < 0.2$) and an outer ($\xi > 0.2$) region (where $\xi = y/h$ is the ratio of the distance from the bed y to flow depth h). For smooth open-channel flows, log law allows accurate description of mean streamwise velocities $U^+(y^+)$ in the logarithmic layer ($30 < y^+ < 0.2Re_*$ or $30/Re_* < \xi < 0.2$, where in wall unit $y^+ = y\,u_*/\nu$, u_* is the friction velocity, ν the kinetic viscosity, and $Re_* = h\,u_*/\nu$ is the friction Reynolds number) [11]. For river flows, Franca and Lemmin [22] found from analysis of experimental data from a field study of extremely rough, three-dimensional river flows, that in more than 65% of the profiles, the log law can be applied up to $\xi = 0.4$, while above this value, mean velocities *show* deviations from the logarithmic profile.

In the outer region, in addition to the simple power law [23,24], the log-wake law is largely used. It is an extension of the log law by adding the Coles' wake function, which contains the Coles' parameter Π [25,26]. However, this method is empirical and no physical-based approach is available to determine the Coles' parameter Π [27]. The value of Π is therefore not universal. From experiments in zero-pressure-gradient boundary layers, Cebeci and Smith [28] found that Π increases with the Reynolds number and becomes $\Pi = 0.55$ at high Reynolds numbers. For open-channel flows over smooth beds, the following values were found: Π increases from zero with Re_* and becomes $\Pi \approx 0.2$ for $Re_* = 2000$ [18], $\Pi \approx 0.08$ [29], $\Pi = 0.1$ [30], $\Pi = 0.3$ [31], $\Pi = 0.45$ for velocity distribution with dip-phenomenon in narrow open channels [32].

However, the more rigorous method for the prediction of velocity profiles is based on the resolution of the momentum equation. This method needs a reliable model for the eddy viscosity, which is the main parameter related to turbulence used in free surface flows since it is involved in the determination of both velocities and sediment concentrations (through the sediment diffusivity, i.e., the product of the eddy viscosity by the inverse of the turbulent Schmidt number) [33,34]. A well-known and largely used vertical distribution of eddy viscosity in free surface flows (open channels and rivers) is given by the parabolic profile [11]. This profile is based on the logarithmic velocity profile assumption and is valid therefore only in the log-law layer. It was improved thanks to the log-wake law velocity profile. These two eddy viscosities are obtained from velocity profiles. As for velocity profiles in the outer region, the main shortcoming of the log-wake eddy viscosity profile is the non-universal Coles' parameter Π.

An analytical eddy viscosity model is therefore needed to predict velocity profiles. In this study, we present two eddy viscosity models based on the concepts of velocity and length scales, which are related, respectively, to the exponentially decreasing turbulent kinetic energy (TKE) function [11] and mixing length, namely, (1) the exponential-type profile of eddy viscosity [35,36] and (2) an eddy viscosity based on an extension of von Karman's similarity hypothesis [37–39]. An additional correction is used in order to account for the damping effect of turbulence near the free surface. The proposed analytical models are validated through computation of velocity profiles, obtained from the resolution of the momentum equation, and comparisons to experimental data. This study aims to provide an explanation and a theoretical foundation to the empirical well-known eddy viscosity profiles.

2. Literature for Eddy Viscosity Models for Open-Channel Flows

2.1. Parabolic Eddy Viscosity

The widely used eddy viscosity (ν_t) formulation is the parabolic profile given by [40,41]

$$\nu_t(y) = \kappa u_* y\left(1 - \frac{y}{h}\right), \tag{1}$$

where y is the vertical distance from the bed, κ is the von Karman constant, h the flow depth, and u_* the friction or shear velocity. Equation (1) is based on shear stress, which decreases linearly with distance from channel bed y and a logarithmic velocity profile [33].

2.2. Log-Wake-Modified Eddy Viscosity Profile

When used for the resolution of the momentum equation, the parabolic eddy viscosity profile (1) is unable to predict accurately velocities outside the log layer [36]. In order to improve the parabolic eddy viscosity (Equation (1)), it is corrected in accordance with Coles' log-wake law for velocities as [18]

$$\nu_t(y) = \frac{\kappa u_* y\left(1 - \frac{y}{h}\right)}{1 + \pi\,\Pi\,\frac{y}{h}\sin\left(\frac{\pi y}{h}\right)} \tag{2}$$

where Π is the Coles' parameter. In Equation (2), the eddy viscosity (1) is corrected by dividing the parabolic eddy viscosity profile by the term $1 + \pi\Pi(y/h)\sin(\pi y/h)$ of the log-wake velocity profile.

2.3. Mixing Length and Mixing Velocity

In order to predict the velocity profile over the entire flow depth, it is more suitable to define the eddy viscosity from the concepts of velocity and length scales, which are here given by mixing velocity and mixing length as

$$\nu_t = w_m l_m \tag{3}$$

From the parabolic profile given by Equation (1), mixing length and mixing velocity are given, respectively, by $l_m = \kappa y\sqrt{1 - y/h}$ and $w_m = u_*\sqrt{1 - y/h}$. The eddy viscosity of Equation (2) allows for the following expression:

$$l_m(y) = \frac{\kappa y\sqrt{1 - y/h}}{1 + \pi\Pi(y/h)\sin(\pi y/h)} \tag{4}$$

Equation (4) consists of a correction of the "parabolic" mixing length $\kappa y\sqrt{1 - y/h}$ by dividing it by the term $1 + \pi\Pi(y/h)\sin(\pi y/h)$ as in Equation (2). The related mixing velocity is given by $w_m = u_*\sqrt{1 - y/h}$ [42].

3. Proposed Eddy Viscosity Models for Free Surface Flows

The eddy viscosity is related to turbulent kinetic energy (TKE) as

$$\nu_t = C_\mu^{\frac{1}{4}}\sqrt{k}\, l_m \tag{5}$$

where k is the turbulent kinetic energy (TKE), and $C_\mu = 0.09$.

3.1. Mixing Velocity from TKE Profile

A semi-theoretical function for TKE is given by [11]:

$$k(\xi) = D_k {u_*}^2 e^{-2C_k\xi} \tag{6}$$

where $\xi = y/h$, C_k and D_k are empirical constants, $D_k = 4.78$, and $C_k = 1$ [11]. Equation (6) was validated by direct numerical simulation DNS data [43]. Instead of the mixing velocity given in the above section by $w_m = u_*\sqrt{1-y/h}$, the shape of mixing velocity should be supported by turbulence intensity measurements [38] and in agreement with the TKE formulation of Equation (6).

From Equations (3) and (5), the mixing velocity is related to TKE as follows:

$$w_m = C_\mu{}^{\frac{1}{4}}\sqrt{k} = C_\mu{}^{\frac{1}{4}}\sqrt{D_k}u_* e^{-C_k\xi} \tag{7}$$

With the assumption $\sqrt{D_k} = 1/C_\mu{}^{\frac{1}{4}}$ (based on log-law and local equilibrium assumption [39]) and with $C_k = 1$, the mixing velocity reverts to

$$w_m(\xi) = u_* e^{-\xi}, \tag{8}$$

which shows that the mixing velocity decreases exponentially with distance from the bed and is in agreement with observations of turbulence intensity and TKE.

3.2. Damping Function for Free Surface

An additional correction is required in order to account for the damping effect of turbulence near the free surface [44–46]. In order to decrease turbulent viscosity near the free surface, Hosoda [47] proposed a damping function as

$$f(\xi) = 1 - e^{-B_f(1-\xi)} \tag{9}$$

where B_f is a damping coefficient.

If we include the free surface damping function in Equation (3), the eddy viscosity reverts to

$$\nu_t = w_m l_m f(\xi) \tag{10}$$

In the following sections, two eddy viscosity formulations will be presented.

3.3. First Formulation: Exponential-Type Profile of Eddy Viscosity

In the equilibrium region, where TKE production is balanced by dissipation, the velocity gradient is given by $\frac{dU}{dy} = \frac{C_\mu{}^{1/4}\sqrt{k}}{l_m}$. In the log-law layer, $\frac{dU}{dy} = \frac{u_*}{\kappa\, y}$, and with a TKE given by $\sqrt{k} = C_\mu{}^{-1/4}u_* f(y)$, the mixing length should read as $l_m = \kappa y f(y)$ [38]. Since in the equilibrium region ($y^+ > 50$), TKE is given by Equation (6), and the velocity profile is given by the log law for ($y^+ > 30$). These two conditions that are given by $\frac{dU}{dy} \approx \frac{\sqrt{k}}{l_m} = \frac{u_*}{\kappa\, y}$ show that the mixing length should be as $l_m = \kappa\; y\; e^{-C_k(y/h)}$. The 1st eddy viscosity is therefore given by

$$\nu_t(y) = \alpha_1 \kappa u_* y e^{-C_1\xi} \tag{11}$$

where α_1 and C_1 are two coefficients, α_1 is related to C_μ and D_k, while C_1 is related to C_k. Equation (11), i.e., the exponential-type profile of eddy viscosity is consistent with the exponentially decreasing mixing velocity (8). Equation (11) was proposed empirically and was used in the planetary boundary layer [48] and coastal engineering [49–53]. However, in order to allow accurate description for different flow conditions, Equation (11) was written in a Re_*-dependent form as [35,36]

$$\nu_t(y) = u_* y e^{-\frac{y^+ + 0.34Re_* - 11.5}{0.46Re_* - 5.98}} \tag{12}$$

where in the wall units, $y^+ = yu_*/\nu$, and the friction Reynolds number $Re_* = hu_*/\nu$. In other words,

$$\nu_t{}^+(y^+) = y^+ e^{-\frac{y^+ + 0.34Re_* - 11.5}{0.46Re_* - 5.98}}$$

where $\nu_t{}^+ = \nu_t/\nu$. The link between Equations (11) and (12) is given by

$$C_\alpha = \alpha_1\kappa = e^{-\frac{0.34Re_*-11.5}{0.46Re_*-5.98}} \text{ and } C_1 = \frac{Re_*}{0.46Re_* - 5.98}$$

3.4. Second Formulation: Eddy Viscosity Formulation Based on Mixing Length Equation from Similarity Hypothesis

This formulation uses a mixing length, which was derived from an extension of von Karman's similarity hypothesis, energy equilibrium assumption, and Nezu and Nakagawa's (1993) TKE function as [37–39,54] (Appendix A)

$$l_m(\xi) = \kappa h\left(1 - e^{-\xi}\right) \tag{13}$$

Using Equations (8)–(10) and (13), the second proposed eddy viscosity is given by

$$\nu_t(\xi) = \kappa h u_* e^{-\xi}\left(1 - e^{-\xi}\right)\left(1 - e^{-B_f(1-\xi)}\right) \tag{14}$$

4. Results

The following ordinary differential equation for velocity distribution U in open-channel flows was obtained from analysis of the Reynolds-averaged Navier–Stokes equations [32,55].

$$\frac{dU}{dy} = \frac{u_\tau^2}{\nu + \nu_t}\left[\left(1 - \frac{y}{h}\right) - \alpha\frac{y}{h}\right] \tag{15}$$

where α is a parameter related to dip-phenomenon. For wide-open channels (ratio of channel width to flow depth >5) $\alpha = 0$, and Equation (15) reverts in the outer region (in the wall units) to the following:

$$\frac{dU^+}{dy^+} = \frac{1}{\nu_t{}^+}\left(1 - \frac{y^+}{Re_*}\right) \tag{16}$$

Mean streamwise velocities are obtained from the numerical resolution of Equation (16). To solve Equation (16), the eddy viscosity $\nu_t{}^+$ is calculated using the two proposed analytical eddy viscosity models given above, i.e., the first is given by Equation (12), while the second by Equation (14).

Both proposed models are validated by experimental data of velocities in open-channel flows for $923 < Re_* < 6139$ [18]. The measurements were carried out in a rectangular cross section and a hydraulically smooth wall open channel. The total length of the channel is 20 m with a cross-sectional size (60 cm wide × 65 cm deep). The width of the channel is sufficiently large to neglect the wall effect. A flow depth of 10 cm was kept constant with varying discharge to examine at various Froude and Reynold numbers. Moreover, measurement was carried out at a length of 18 m from the inlet position. Laser Doppler Anemometer (LDA) was used to carry out velocity measurements under different flow conditions. Table 1 summarizes experimental hydraulic conditions. The flow conditions are listed in Table 1 [18].

Table 1. Flow conditions [18].

Case	Depth of Flow, h (cm)	Width to Depth Ratio	Reynolds Number [1], Re= $\frac{4RU_m}{\nu}$	Froude Number, Fr= $\frac{U_m}{\sqrt{gh}}$	Friction Reynolds Number, $Re_*=\frac{hu_*}{\nu}$
P2	10.3	5.9	5.5×10^4	0.189	923
P3	10.0	6.0	14.3×10^4	0.488	2156
P4	10.0	6.0	21.0×10^4	0.704	3001
P5	10.5	5.7	44.0×10^4	1.170	6139

[1] U_m = mean bulk velocity, R = hydraulic radius.

4.1. Velocity Profiles from the First Eddy Viscosity Formulation: Exponential-Type Profile

Computed mean streamwise velocity profiles are obtained from (16) with the first eddy viscosity given by Equation (12) and are validated by experimental data [18]. The following boundary condition for the velocity is applied at $\xi = 0.2$ (or $y^+ = 0.2\ Re_*$): $U^+(y^+ = 0.2Re_*) = (1/\kappa)\ln(0.2Re_*) + B$; where $\kappa = 0.41$ and $B = 5.29$ [11].

Figures 1 and 2 show comparisons between computed velocity profiles (red solid lines) and experimental data (symbols). The model allows the prediction of log law (black thin dashed lines) to be improved in the outer region.

In order to improve the results, the first eddy viscosity given by (12) is used with the condition of an eddy viscosity equal to zero at the free surface, which requires the use of the damping function given by Equation (9), Equation (12) reverts to

$$\nu_t{}^+ = y^+ e^{-\frac{y^+ + 0.34Re_* - 11.5}{0.46Re_* - 5.98}} \left(1 - e^{-B_f\left(1 - \frac{y^+}{Re_*}\right)}\right) \quad (17)$$

Results obtained with Equation (17) (magenta thick dashed curves) allow the prediction to be improved, particularly for high Reynolds numbers.

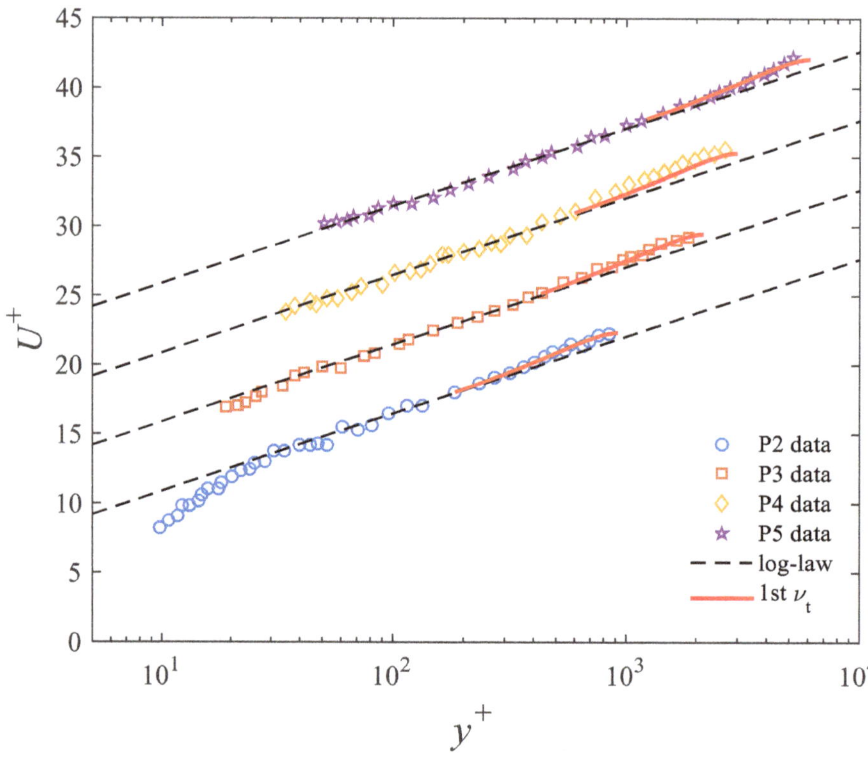

Figure 1. Velocity profiles (red solid lines) obtained from (16) with the first eddy viscosity (12); symbols: experimental data [18]; dashed lines: log law (profiles shifted by 5 units).

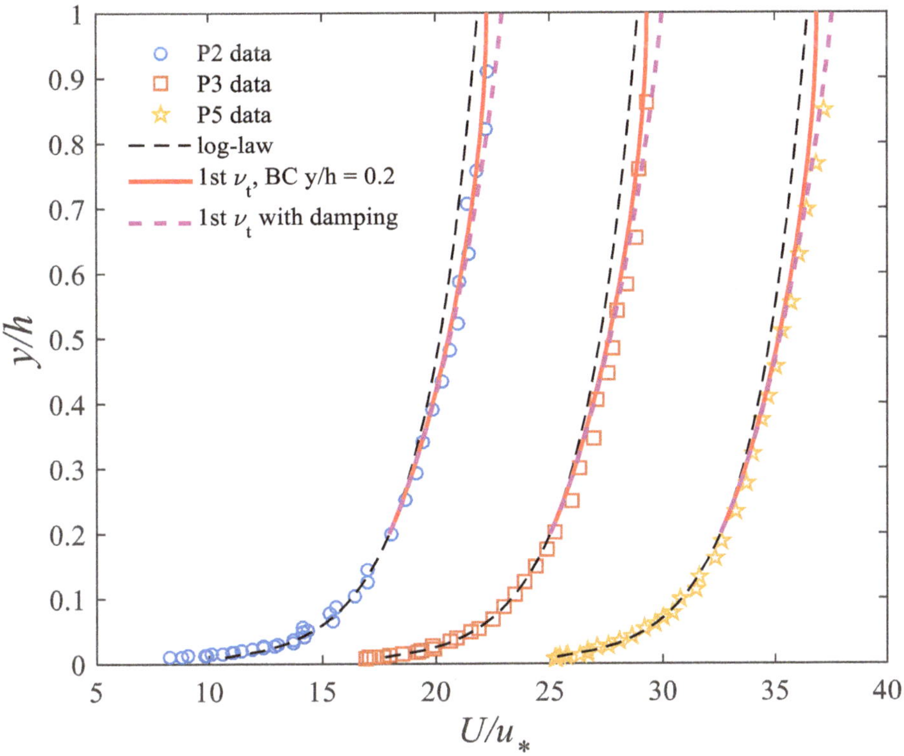

Figure 2. Velocity profiles obtained from (16) with the first eddy viscosity (12) (red solid lines) and from (16) with (17) ($B_f = 6$) magenta thick dashed lines; symbols: experimental data [18]; dashed lines: log law (profiles shifted by 5 units).

4.2. Velocity Profiles from the Second Eddy Viscosity Formulation Based on Von Karman's Similarity Hypothesis

Computed mean streamwise velocity profiles are obtained from (16), and the second eddy viscosity given by (14) and are validated by experimental data [18] (Table 1). The same boundary condition is applied (velocity equal to the log-law value at $y^+ = 0.2\ Re_*$). Figure 3 shows comparisons between computed velocity profiles (red solid lines) and experimental data (symbols). The model allows log-law profiles (black thin dashed lines) to be improved in the outer region. Figure 3 shows that computed velocity profiles (red solid lines) show good agreement and improve the prediction of log law (black dashed thin lines). In order to improve the results, the momentum equation is resolved with a second boundary condition at the lower limit of the logarithmic layer, i.e., $y^+ = 30$, where the velocity is given by the log law as $U^+(30) = (1/\kappa)\ln(30) + B$. Results (magenta thick dashed curves) allow the prediction to be improved, particularly for high Reynolds numbers.

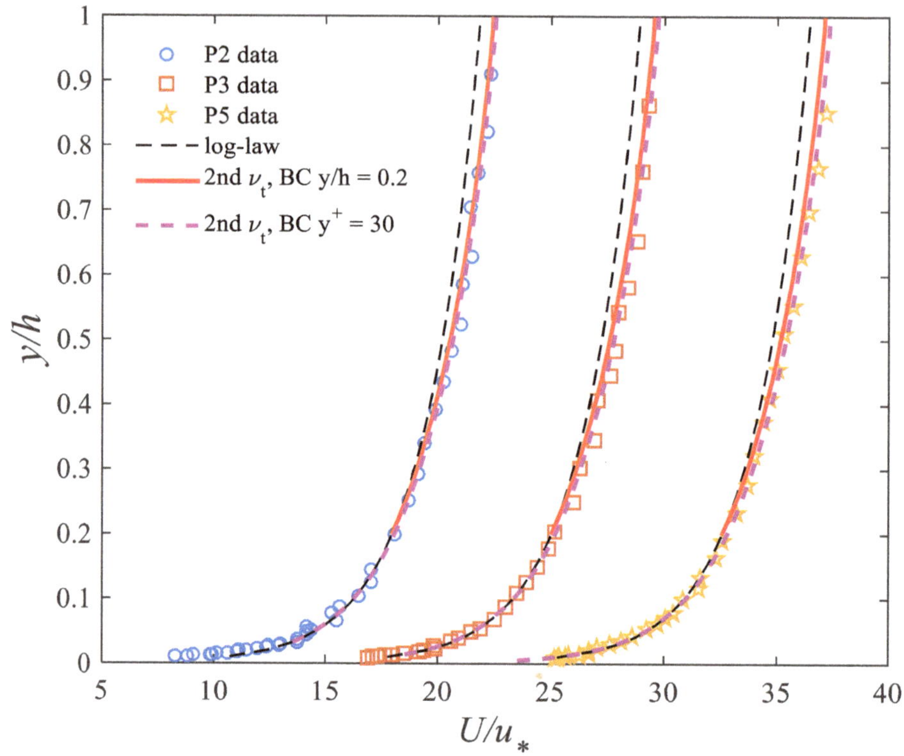

Figure 3. Velocity profiles obtained from (16) with the second eddy viscosity (14) ($B_f = 4$) with the first boundary condition at $\xi = 0.2$ (red solid lines), and the second boundary condition at $y^+ = 30$ (magenta thick dashed lines); symbols: experimental data [18]; dashed lines: log law (profiles shifted by 5 units).

4.3. Eddy Viscosity Profiles

In order to compare the two proposed analytical eddy viscosity models (Equations (12), (14) and (17)) with the existing parabolic and log-wake-modified eddy viscosity profiles (Equations (1) and (2)), all equations are written in the same dimensionless form as

$$\frac{\nu_t}{h\,u_*} = \kappa\frac{y}{h}\left(1-\frac{y}{h}\right) \tag{18}$$

$$\frac{\nu_t}{h\,u_*} = \frac{\kappa\frac{y}{h}\left(1-\frac{y}{h}\right)}{1+\pi\Pi\frac{y}{h}\sin\left(\frac{\pi y}{h}\right)} \tag{19}$$

$$\frac{\nu_t}{h\,u_*} = \xi e^{-\frac{(\xi+0.34)Re_*-11.5}{0.46Re_*-5.98}} \tag{20}$$

$$\frac{\nu_t}{h\,u_*} = \xi e^{-\frac{(\xi+0.34)Re_*-11.5}{0.46Re_*-5.98}}\left(1-e^{-B_f(1-\xi)}\right) \tag{21}$$

$$\frac{\nu_t}{h\,u_*} = \kappa e^{-\xi}\left(1-e^{-\xi}\right)\left(1-e^{-B_f(1-\xi)}\right) \tag{22}$$

We notice that eddy viscosities from Equations (18), (19), and (22) are Re_*-independent, while Equations (20) and (21) are Re_* dependent.

Figures 4 and 5 show comparisons of the vertical distribution of the different eddy viscosity models. The three Re_*-independent Equations (18), (19), and (22) are first compared.

Figure 4 shows that the shape of the eddy viscosity given by the second model (Equation (22)) (red solid line) is similar to the parabolic profile (green dashed line), where the maximum value is located at the same position, i.e., half water depth ($\xi = 0.5$). Even though the second model exhibits a similar shape, it predicts smaller values than the parabolic profile. The profile obtained from the second model is compared to log-wake-modified (dash-dotted lines) profiles. The magenta dash-dotted curve is from the log-wake-modified eddy viscosity given by Equation (19) with a Coles' parameter $\Pi = 0.2$. With a smaller value of Π, the blue dash-dotted curve (for $\Pi = 0.15$) is closer to the eddy viscosity given by the second model. This value ($\Pi = 0.15$) is close to values found for open-channel flow experiments [18,30]. This provides an explanation about the accuracy of the computed velocity profiles obtained by the second model in the outer part of free surface flows.

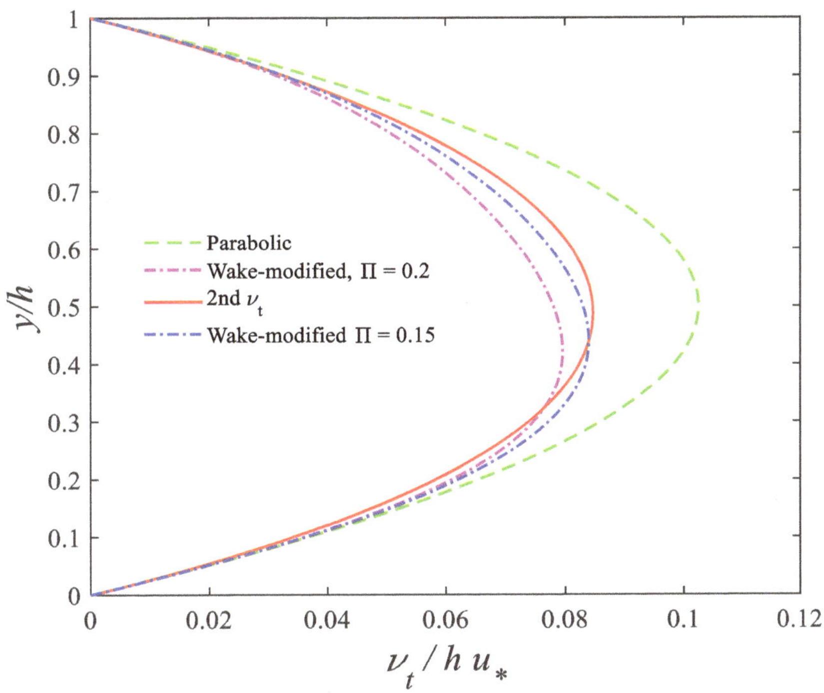

Figure 4. Vertical distribution of eddy viscosity; green dashed line: parabolic eddy viscosity (Equation (18)); dash-dotted lines: log-wake-modified eddy viscosity (Equation (19)); solid line: second eddy viscosity (Equation (22)) with $B_f = 4$.

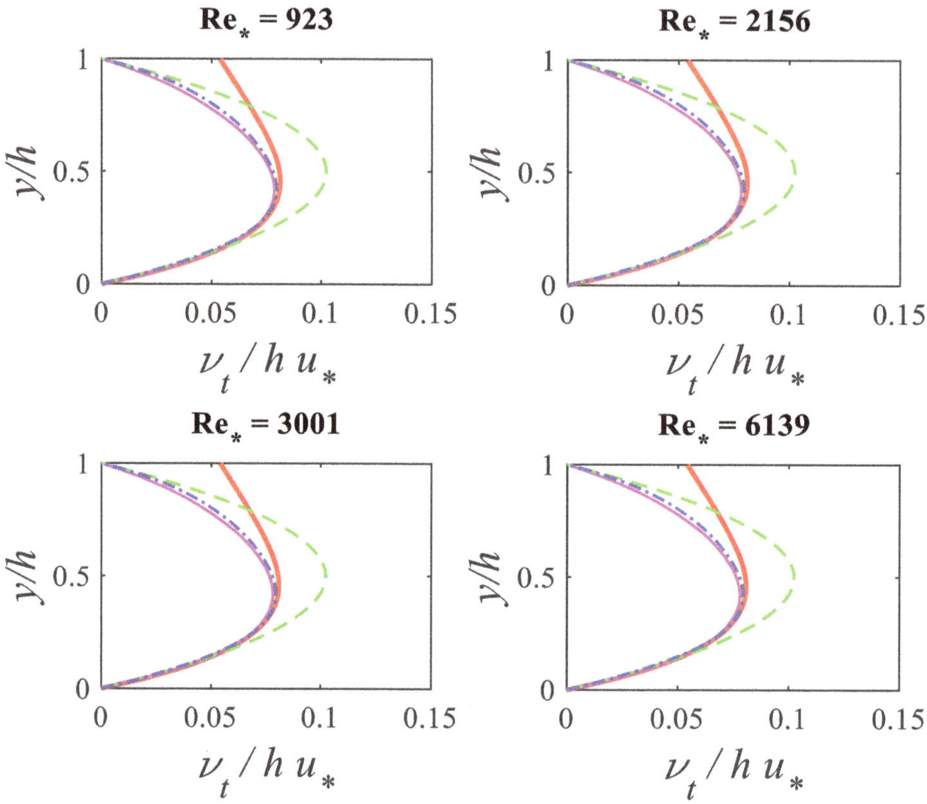

Figure 5. Eddy viscosity profiles; thick red solid lines: first eddy viscosity (Equation (20)); thin magenta solid lines: first eddy viscosity with free surface damping function (Equation (21)); blue dash-dotted lines: log-wake-modified eddy viscosity (Equation (19)) with $\Pi = 0.2$; green dashed lines: parabolic eddy viscosity (Equation (18)).

For the Re_*-dependent eddy viscosity given by the first model (Equations (20) and (21)), Figure 5 shows comparisons for the vertical distribution of the eddy viscosity for the four friction Reynolds number. The eddy viscosity curves (thick red solid lines) obtained from the first model (Equation (20)) show that the eddy viscosity increases from the bed—the maximum value is located at around the half water depth and then decreases until the free surface. However, the eddy viscosity does not vanish at the free surface. With the condition of an eddy viscosity equal to zero at the free surface, the profiles with the damping function (Equation (21)) predict everywhere smaller values than the parabolic eddy viscosity (thin magenta solid lines) and have a shape similar to the log-wake-modified eddy viscosity profile (blue dash-dotted lines). Interestingly, with a value of Coles' parameter $\Pi = 0.2$, eddy viscosities from both the first model (Equation (21)) and log-wake-modified profile (Equation (19)) are almost superimposed. The value $\Pi = 0.2$ of Coles' parameter is the same as that proposed from open-channel flow experiments [18]. This provides an explanation about the accuracy of the computed velocity profiles obtained by the first model (Equation (21)) in the outer part of free surface flows.

The profiles obtained for the four Re_*. numbers seem similar. For large values of Re_*, the two coefficients $C_\alpha = \alpha_1 \kappa$ and C_1. of the first model reach asymptotic values equal, respectively, to $C_\alpha = \alpha_1\kappa = e^{-(0.34/0.46)} = 0.477$ and $C_1 = 1/0.46 = 2.17$ (Figure 6). Equations (20) and (21) reverts to the following Re_*-independent forms:

$$\frac{\nu_t}{h\,u_*} = C_\alpha \xi e^{-C_1\xi} \tag{23}$$

$$\frac{\nu_t}{h\,u_*} = C_\alpha \xi \left(1 - e^{-B_f(1-\xi)}\right) e^{-C_1\xi} \tag{24}$$

where $C_\alpha = 0.477$, $C_1 = 2.17$, and $B_f = 6$.

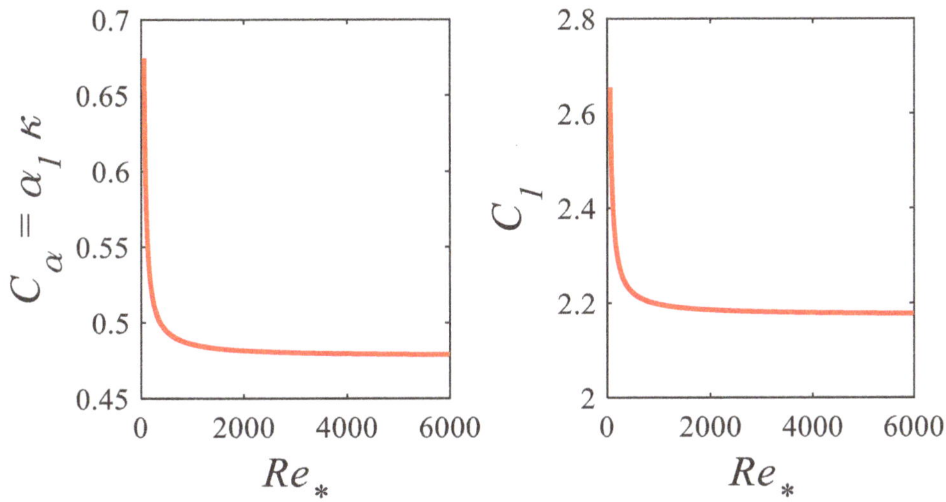

Figure 6. Asymptotic behavior of the two coefficients of Equations (23) and (24): $C_\alpha = \alpha_1\kappa = 0.477$ and $C_1 = 2.17$.

Figure 7 shows the comparison between the two proposed eddy viscosity models (24, 22), both with free surface damping function and parabolic and log-wake-modified eddy viscosity profiles. Figure 7 shows that with the free surface damping function both exhibit smaller values than the parabolic profile. Both models predict profiles similar to log-wake-modified eddy viscosity (Equation (19)). The first model (Equation (24)) (red solid line) is similar to (Equation (19))with a Coles' parameter $\Pi = 0.2$ (red dash-dotted line), while the second model (Equation (22)) (blue solid line) is similar to (Equation (19))with a Coles' parameter $\Pi = 0.15$ (blue dash-dotted line). Interestingly, the two coefficients of the first model are found to be equal, respectively, to $C_\alpha = \alpha_1\kappa = 0.477$ and $C_1 = 2.17$. The value of the first coefficient C_α results in $\alpha_1 = 1.16$ (with $\kappa = 0.41$), which is close to $\alpha_1 = 1$ (related to the assumption $\sqrt{D_k} = 1/C_\mu{}^{\frac{1}{4}}$) (see also [49,50]). The value of C_α is also between the two values $0.41 < 0.478 < 0.49$ obtained, respectively, with the assumption $\sqrt{D_k} = 1/C_\mu{}^{\frac{1}{4}}$, $\sqrt{D_k}C_\mu{}^{\frac{1}{4}}\kappa = \kappa = 0.41$ and the empirical value $D_k = 4.78$ [56], $\sqrt{D_k}C_\mu{}^{\frac{1}{4}}\kappa = 0.49$. The value of the second coefficient allows the coefficient in TKE to be defined as $C_k = C_1/2 = 1.088$, which is close to the empirical value $C_k = 1$ [55].

Figure 8 shows a comparison between the two proposed eddy viscosity models (24, 22) and experimental data of eddy viscosity [18,57]. In addition to the experimental data of Nezu and Rodi [18], data from experiments of Ueda et al. [57] are used, which seem to confirm the same behavior. Figure 8 shows that profiles obtained from both models show good agreement with experimental data of eddy viscosity.

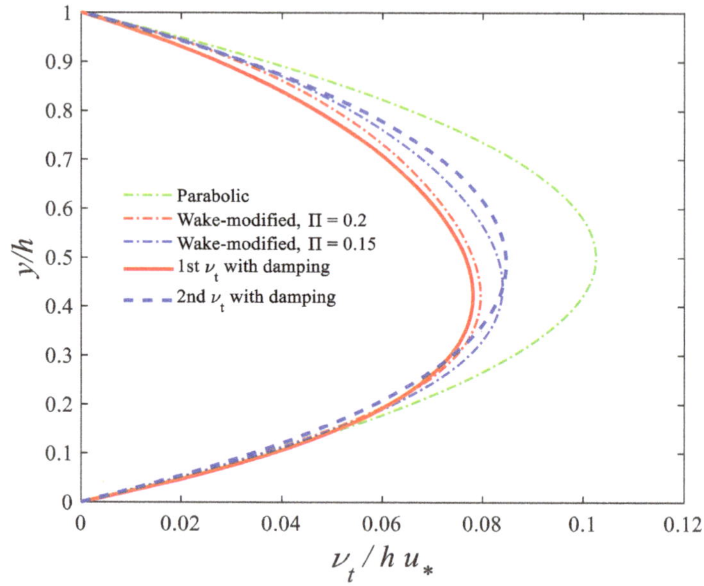

Figure 7. Eddy viscosity profiles; red solid line: first eddy viscosity (Equation (24)); blue dashed lines: second eddy viscosity (Equation (22)) both with free surface damping function; red dash-dotted line: log-wake-modified (Equation (2b)) with $\Pi = 0.2$; blue dash-dotted line: Equation (2) with $\Pi = 0.15$; green dash-dotted line: parabolic eddy viscosity (Equation (18)).

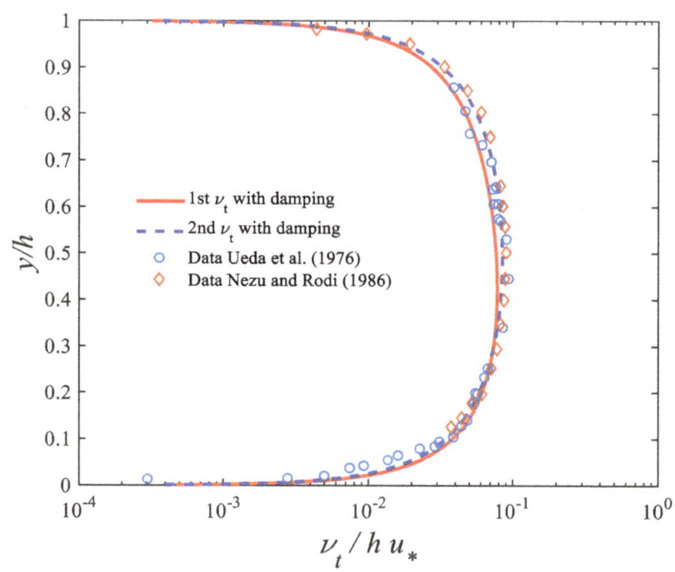

Figure 8. Validation by experimental data; eddy viscosity profiles; red solid line: first eddy viscosity (Equation (24)); blue dashed lines: second eddy viscosity (Equation (22)) both with free surface damping function; symbols: experimental data of Nezu and Rodi (1986) and Ueda et al. (1976).

5. Conclusions

The parabolic eddy viscosity is based on the log-law velocity profile and is valid only in the log layer. The improved log-wake-modified eddy viscosity was obtained from the log-wake law velocity profile. However, both were obtained from velocity profiles. The main shortcoming of the log-wake profile is the uncertainty in the value of the empirical Coles' parameter.

In this study, the eddy viscosity is defined as a product between a velocity scale (related to the root-square of TKE, which is given by a semi-theoretical exponentially decreasing function) and a length scale (related to mixing length). From this definition, two analytical eddy viscosity models are proposed, namely, (1) the exponential-type profile of eddy viscosity and (2) an eddy viscosity based on an extension of von Karman's similarity hypothesis with an additional correction in order to account for the damping effect of turbulence near the free surface. As for the parabolic and log-wake-modified profiles, the eddy viscosity from the second model is Re_* independent, while the eddy viscosity from the first model is Re_* dependent. The proposed analytical models are validated through computation of velocity profiles, obtained from the resolution of the momentum equation and comparisons to experimental data.

Mean streamwise velocity profiles were obtained by solving the momentum equation. For both the first and second proposed analytical eddy viscosity models, a boundary condition is applied at the lower limit of the outer region, i.e., $y^+ = 0.2\ Re_*$ or $\xi = 0.2$, where the velocity is given by the logarithmic law. Computed mean velocities are compared to experimental data of open-channel flows for $923 < Re_* < 6139$. Computed velocity profiles show good agreement in the outer region.

In order to improve the results, the first eddy viscosity was used with the free surface damping function. Results allow the prediction to be improved, particularly for high Reynolds numbers. For the second analytical eddy viscosity model, a second boundary condition was used at the lower limit of the logarithmic layer, i.e., $y^+ = 30$ where the velocity is given by the log law. Results allow predicted velocity profiles to be improved, particularly for high Reynolds numbers. The results for velocity profiles show the ability of these analytical eddy viscosity models to predict accurately the velocities in the outer region from the momentum equation.

Finally, the vertical distribution of eddy viscosity from both proposed analytical models was analyzed. Both profiles from the first and second analytical eddy viscosity models with the free surface damping function are similar to the log-wake-modified profiles but with different values of Coles' parameter. $\Pi = 0.2$ for the first model and $\Pi = 0.15$ for the second model. These values are close to values found from open-channel flow experiments. This provides an explanation of the accuracy of the computed velocity profiles in the outer part of free surface flows.

For large values of Re_* ($Re_* > 2000$), the first model becomes Re_* independent, and the two coefficients reach asymptotic values equal to $C_\alpha = \alpha_1 \kappa = 0.477$ and $C_1 = 2.17$. Interestingly, with a value of Coles' parameter $\Pi = 0.2$, eddy viscosities from both the first model and log-wake-modified profile are almost superimposed. The value $\Pi = 0.2$ of Coles' parameter is the same as that proposed from open-channel flow experiments. The analysis of these two coefficients allowed the models' assumptions to be verified and former empirical values to be found. The two proposed eddy viscosity models are validated by two experimental data. The comparison shows that profiles from both models show good agreement with experimental data of eddy viscosity.

Funding: This research received no external funding.

Institutional Review Board Statement: Not applicable.

Informed Consent Statement: Not applicable.

Data Availability Statement: Not applicable.

Acknowledgments: The author would like to thank Xiaofeng Liu from Penn State University for providing the experimental data of eddy viscosity of references [18,57] used in Figure 8.

Conflicts of Interest: The author declares no conflict of interest.

Appendix A

The von Karman's similarity hypothesis allows writing the mixing length as [58]

$$l_m{}^+ = -\kappa \frac{dU^+/dy^+}{d^2U^+/dy^{+2}} \tag{A1}$$

where U^+ is the streamwise mean velocity. With $dU^+/dy^+ \approx \sqrt{k^+}/l_m{}^+$, Equation (A1) becomes

$$l_m{}^+ = -\kappa \frac{\sqrt{k^+}/l_m{}^+}{d\left(\sqrt{k^+}/l_m{}^+\right)/dy^+} \tag{A2}$$

Introducing the function $f^+ = \sqrt{k^+}/l_m{}^+$, (A2) becomes

$$\sqrt{k^+} = -\kappa \frac{f^{+2}}{df^+/dy^+} \tag{A3}$$

We write (A3) in the following form:

$$-\frac{df^+/dy^+}{f^{+2}} = \frac{\kappa}{\sqrt{k^+}} \tag{A4}$$

The integration of the LHS term of Equation (A4) from A_0 to y^+, is given by

$$\int_{A_0}^{y^+} -\frac{df^+/dy^+}{f^{+2}} dy^+ = \frac{1}{f^+(y^+)} - \frac{1}{f^+(A_0)} \tag{A5}$$

Integrating (A4) from A_0 to y^+, provides, therefore, the mixing length as

$$l_m{}^+(y^+) = \sqrt{k^+}\left(\kappa \int_{A_0}^{y^+} \frac{1}{\sqrt{k^+}} dy^+ + \frac{l_m{}^+(A_0)}{\sqrt{k^+(A_0)}}\right) \tag{A6}$$

Using Equation (6) for TKE and taking the boundary condition $l_m{}^+(A_0) = \kappa A_0$ yields

$$l_m{}^+(y^+) = \kappa e^{-C_k y^+/Re_*}\left(\frac{Re_*}{C_k}\left(e^{C_k y^+/Re_*} - e^{C_k A_0/Re_*}\right) + A_0 e^{C_k A_0/Re_*}\right) \tag{A7}$$

Rearranging the terms of (A7) allows writing the mixing length as [37–39,54]

$$l_m{}^+(y^+) = \alpha\left(Re_* - (Re_* - C_k A_0)e^{-C_k(y^+ - A_0)/Re_*}\right) \tag{A8}$$

where $\alpha = \kappa/C_k$. Since $A_0 \ll Re_*$ we write (18) in a simplified form that does not depend on A_0, as

$$l_m{}^+(y^+) = \alpha\, Re_*\left(1 - e^{-C_k y^+/Re_*}\right) \tag{A9}$$

By taking $C_k = 1$, Equation (A9) reverts to Equation (13).

References

1. Figuérez, J.; Galán, Á.; González, J. An Enhanced Treatment of Boundary Conditions for 2D RANS Streamwise Velocity Models in Open Channel Flow. *Water* **2021**, *13*, 1001. [CrossRef]
2. Nazari-Sharabian, M.; Karakouzian, M.; Hayes, D. Flow Topology in the Confluence of an Open Channel with Lateral Drainage Pipe. *Hydrology* **2020**, *7*, 57. [CrossRef]

3. Zhu, Z.; Yu, J.; Dou, J.; Peng, D. An Expression for Velocity Lag in Sediment-Laden Open-Channel Flows Based on Tsallis Entropy Together with the Principle of Maximum Entropy. *Entropy* **2019**, *21*, 522. [CrossRef]
4. MacVicar, B.J.; Roy, A.G. Hydrodynamics of a forced riffle pool in a gravel bed river: 1. Mean velocity and turbulence intensity. *Water Resour. Res.* **2007**, *43*, W12401. [CrossRef]
5. Liu, X.; Nayamatullah, M. Semianalytical Solutions for One-Dimensional Unsteady Nonequilibrium Suspended Sediment Transport in Channels with Arbitrary Eddy Viscosity Distributions and Realistic Boundary Conditions. *J. Hydraul. Eng.* **2014**, *140*, 04014011. [CrossRef]
6. Johnson, E.D.; Cowen, E.A. Estimating bed shear stress from remotely measured surface turbulent dissipation fields in open channel flows. *Water Resour. Res.* **2017**, *53*, 1982–1996. [CrossRef]
7. Macvicar, B.J.; Sukhodolov, A.N. Sampling strategies to improve scaling parameter estimates in rivers. *J. Hydraul. Res.* **2019**, *57*, 1–10. [CrossRef]
8. Sukhodolov, A.N.; Krick, J.; Sukhodolova, T.A.; Cheng, Z.; Rhoads, B.L.; Constantinescu, G.S. Turbulent flow structure at a discordant river confluence: Asymmetric jet dynamics with implications for channel morphology. *J. Geophys. Res. Earth Surf.* **2017**, *122*, 1278–1293. [CrossRef]
9. Sukhodolov, A.N. Field-based research in fluvial hydraulics: Potential, paradigms and challenges. *J. Hydraul. Res.* **2014**, *53*, 1–19. [CrossRef]
10. Gualtieri, C.; Shao, D.; Angeloudis, A. Advances in Environmental Hydraulics. *Water* **2021**, *13*, 1192. [CrossRef]
11. Nezu, I.; Nakagawa, H. *Turbulence in Open Channel Flows*; Balkema, A.A., Ed.; Balkema: Rotterdam, The Netherlands, 1993.
12. Yokosi, S. The structure of river turbulence. *Bull. Disaster Prev. Res. Inst. Kyoto Univ.* **1967**, *17*, 1–29.
13. McQuivey, R.S. Summary of turbulence data from rivers, conveyance channels and laboratory flumes. *US Geol. Surv. Prof.* **1973**, *N802B*, B1–B66.
14. Grinvald, D.I. *Turbulence of Natural Flows (in Russian)*; Hydrometeoizdat: Leningrad, Russia, 1974.
15. Iwasa, Y.; Asano, T. Characteristics of turbulence in rivers and conveyance channels. In Proceedings of the Third International Symposium on Stochastic Hydraulics, Delft, The Netherlands, 5–8 August 1980; pp. 565–576.
16. Nikora, V.I.; Smart, G.M. Turbulence characteristics of New Zealand gravel-bed rivers. *J. Hydraul. Eng.* **1997**, *123*, 764–773. [CrossRef]
17. Grass, A.J. Structural features of turbulent flow over smooth and rough boundaries. *J. Fluid Mech.* **1971**, *50*, 233–255. [CrossRef]
18. Nezu, I.; Rodi, W. Open channel measurements with a laser Doppler anemometer. *J. Hyd. Eng. ASCE* **1986**, *112*, 335–355. [CrossRef]
19. Absi, R. A simple eddy viscosity formulation for turbulent boundary layers near smooth walls. *C. R. Mec.* **2009**, *337*, 158–165. [CrossRef]
20. Guo, J. Modified log-wake-law for smooth rectangular open channel flow. *J. Hydraul. Res.* **2014**, *52*, 121–128. [CrossRef]
21. Sukhodolov, A.; Thiele, M.; Bungartz, H. Turbulence Structure in a River Reach with Sand Bed. *Water Resour. Res.* **1998**, *34*, 1317–1334. [CrossRef]
22. Franca, M.J.; Lemmin, U. A field study of extremely rough, three-dimensional river flow. In Proceedings of the 4th IAHR International Symposium on Environmental Hydraulics, Hong Kong, China, 15–18 December 2004.
23. Afzal, N.; Seena, A.; Bushra, A. Power Law Velocity Profile in Fully Developed Turbulent Pipe and Channel Flows. *J. Hydraul. Eng. ASCE* **2007**, *133*, 1080–1086. [CrossRef]
24. Cheng, N.S. Power-law index for velocity profiles in open channel flows. *Adv. Water Res.* **2007**, *30*, 1775–1784. [CrossRef]
25. Coles, D.E. The law of the wake in the turbulent boundary layer. *J. Fluid Mech.* **1956**, *1*, 191–226. [CrossRef]
26. Hinze, J.O. *Turbulence*; McGraw-Hill: New York, NY, USA, 1975.
27. Krug, D.; Philip, J.; Marusic, I. Revisiting the law of the wake in wall turbulence. *J. Fluid Mech.* **2017**, *811*, 421–435. [CrossRef]
28. Cebeci, T.; Smith, M.O. *Analysis of Turbulent Boundary Layers*; Academic: San Diego, CA, USA, 1974.
29. Cardoso, A.H.; Graf, W.H.; Gust, G. Uniform flow in a smooth open channel. *J. Hydraul. Res.* **1989**, *27*, 603–616. [CrossRef]
30. Kirkgoz, M.S. The turbulent velocity profiles for smooth and rough channel flow. *J. Hydraul. Eng.* **1989**, *115*, 1543–1561. [CrossRef]
31. Li, X.; Dong, Z.; Chen, C. Turbulent flows in smooth-wall open channels with different slope. *J. Hydraul. Res.* **1995**, *33*, 333–347.
32. Absi, R. An ordinary differential equation for velocity distribution and dip-phenomenon in open channel flows. *J. Hydraul. Res.* **2011**, *49*, 82–89. [CrossRef]
33. Absi, R. Rebuttal on A mathematical model on depth-averaged β-factor in open-channel turbulent flow. *Environ. Earth Sci.* **2020**, *79*, 113. [CrossRef]
34. Gualtieri, C.; Angeloudis, A.; Bombardelli, F.; Jha, S.; Stoesser, T. On the Values for the Turbulent Schmidt Number in Environmental Flows. *Fluids* **2017**, *2*, 17. [CrossRef]
35. Absi, R. Eddy viscosity and velocity profiles in fully-developed turbulent channel flows. *Fluid Dyn.* **2019**, *54*, 137–147. [CrossRef]
36. Absi, R. Analytical eddy viscosity model for velocity profiles in the outer part of closed- and open-channel flows. *Fluid Dyn.* **2021**, *56*, 577–586. [CrossRef]
37. Absi, R. Wave boundary layer instability near flow reversal. In *Coastal Engineering 2002, Proceedings of the 28th International Conference Coastal Engineering, ASCE, Wales, UK, 19 November 2002*; Smith, J.M., Ed.; World Scientific Publishing: Singapore, 2002; Volume 1, pp. 532–544.

38. Absi, R. Comment on Turbulent diffusion of momentum and suspended particles: A finite-mixing-length theory. *Phys. Fluids* **2005**, *17*, 079101. [CrossRef]
39. Absi, R. A roughness and time dependent mixing length equation. *JSCE Doboku Gakkai Ronbunshuu B Jpn. Soc. Civ. Eng.* **2006**, *62*, 437–446. [CrossRef]
40. Yalin, M.S. *Mechanics of Sediment Transport*; Pergamon Press: Oxford, UK, 1977.
41. Absi, R. Comments on Turbulent velocity profile in fully-developed open channel flows. *Environ. Fluid Mech.* **2008**, *8*, 389–394. [CrossRef]
42. Absi, R. Comments on Modeling stratified suspension concentration distribution in turbulent flow using fractional advection–diffusion equation. *Environ. Fluid Mech.* **2021**. Submitted.
43. Absi, R. Analytical solutions for the modeled k-equation. *ASME J. Appl. Mech.* **2008**, *75*, 044501. [CrossRef]
44. Nakagawa, H.; Nezu, I.; Ueda, H. Turbulence of open channel flow over smooth and rough beds. *Jpn. Soc. Civ. Eng.* **1975**, *241*, 155–168. [CrossRef]
45. Langhi, M.; Hosoda, T.; Dey, S. Analytical Solution of k-ϵ Model for Nonuniform Flows. *J. Hydraul. Eng. ASCE* **2018**, *144*, 04018033. [CrossRef]
46. Welderufael, M.; Absi, R.; Mélinge, Y. Assessment of velocity profile models for turbulent smooth wall open channel flows. *ISH J. Hyd. Eng.* **2021**, in press.
47. Hosoda, T. Turbulent Diffusion Mechanism in Open Channel Flows. Ph.D. Thesis, Kyoto University, Kyoto, Japan, 1990.
48. Businger, J.A.; Arya, S.P.S. Heights of the mixed layer in the stable stratified planetary boundary layer. *Adv. Geophys.* **1974**, *18A*, 73–92.
49. Hsu, T.W.; Jan, C.D. Calibration of Businger-Arya type of eddy viscosity model's parameters. *J. Waterw. Port Coastal. Ocean. Eng. ASCE* **1998**, *124*, 281–284. [CrossRef]
50. Absi, R. Discussion of Calibration of Businger-Arya type of eddy viscosity model's parameters. *J. Waterw. Port Coastal. Ocean. Eng. ASCE* **2000**, *126*, 108–109. [CrossRef]
51. Absi, R. Time-Dependent Eddy Viscosity Models for Wave Boundary Layers. In *Coastal Engineering 2000, Proceedings of the 27th International Conference Coastal Engineering, Sydney, Australia, 16–21 July 2000*; Edge, B.L., Ed.; ASCE: Reston, VA, USA; pp. 1268–1281.
52. Absi, R. Concentration profiles for fine and coarse sediments suspended by waves over ripples: An analytical study with the 1-DV gradient diffusion model. *Adv. Water Res.* **2010**, *33*, 411–418. [CrossRef]
53. Absi, R.; Tanaka, H.; Kerlidou, L.; André, A. Eddy viscosity profiles for wave boundary layers: Validation and calibration by a k-ω model. In Proceedings of the 33th International Conference Coastal Engineering, Santander, Spain, 1–6 July 2012.
54. Absi, R. Stress length and momentum cascade in turbulent channel flows. *Phys. Rev. Fluids* **2021**. Submitted.
55. Yang, S.Q.; Tan, S.K.; Lim, S.Y. Velocity distribution and dip-phenomenon in smooth uniform open channel flows. *J. Hydraul. Eng.* **2004**, *130*, 1179–1186. [CrossRef]
56. Nezu, I. Open-Channel Flow Turbulence and Its Research Prospect in the 21st Century. *J. Hydraul. Eng. ASCE* **2005**, *131*, 229–246. [CrossRef]
57. Ueda, H.; Moller, R.; Komori, S.; Mizushina, T. Eddy diffusivity near the free surface of open channel flow. *Int. J. Heat Mass Transf.* **1977**, *20*, 1127–1136. [CrossRef]
58. von Kármán, T. Mechanische ahnlichkeit und turbulenz. *Math. -Phys. Kl.* **1930**, *5*, 58–76.

hydrology

MDPI

Article

Simulation of Dam Breaks on Dry Bed Using Finite Volume Roe-TVD Method

Ebrahim Alamatian [1], Sara Dadar [1] and Bojan Đurin [2,*]

[1] Khavaran Institute of Higher Education, Mashhad 9184168619, Iran; e.alamatian@profs.khi.ac.ir (E.A.); sara.dadar@mail.um.ac.ir (S.D.)

[2] Department of Civil Engineering, University North, 42000 Varaždin, Croatia

* Correspondence: bojan.durin@unin.hr

Abstract: Dams are one of the most important hydraulic structures. In view of unrecoverable damages occurring after a dam failure, analyzing a dams' break is necessary. In this study, a dam located in Iran is considered. According to adjacent tourist and entertainment zones, the breaking of the dam could lead to severe problems for the area and bridges downstream of the river. To investigate the issue, a numerical FORTRAN code based on the 2D finite volume Roe-TVD method on a fixed bed is provided to assess the effects of the dam break. Turbulence terms and dry bed conditions were considered in the code. A numerical wave tank (NWT) with a triangular barrier in the bed was numerically modeled and compared with analytical models to verify the capability of the code. Comparing numerical, experimental and analytical results showed that estimated water level and mass conservation in the numerical model is in good agreement with the experimental data and analytical solutions. The 2D approach used has reduced the cost of computing compared to a 3D approach while obtaining accurate results. The code is finally applied to a full-scale dam-break flood. Six KM of the natural river downstream of the dam, including two bridges, B1 and B2, is considered. Flood flow hydrographs and water level variations at bridges B1 and B2 are presented. The results denoted that bridges B1 and B2 will be flooded after 12 and 21 min, respectively, and are at risk of the potential break. Thus, it is necessary to announce and possibly evacuate the resort area alongside the dam in order to decrease losses.

Keywords: numerical model; shallow water equations; dam break; dry and wet beds modeling

Citation: Alamatian, E.; Dadar, S.; Đurin, B. Simulation of Dam Breaks on Dry Bed Using Finite Volume Roe-TVD Method. *Hydrology* **2021**, *8*, 88. https://doi.org/10.3390/hydrology8020088

Academic Editors: Tommaso Caloiero, Ali Fares, Carmelina Costanzo and Roberta Padulano

Received: 19 April 2021
Accepted: 29 May 2021
Published: 3 June 2021

1. Introduction

In recent years, the statistics of the dam break phenomenon have been collected to improve the knowledge for the construction dams. The primary goals of the studies are considering and implementing the new findings in designs to have safe dams adjacent to urban areas and keeping the projects time and cost-efficient. On the other hand, the consequences of a dam break need to be considered in the designing stage in order to prevent probable damage to nearby infrastructure and resort areas.

Modeling and analyzing dam breaks have been carried out by several researchers. In some one-dimensional models, the non-conservative form of the shallow water equations has been solved [1,2]. Zoppou et al., in 2000 [3], described a two-dimensional numerical model for dam-break problems. The model was able to consider shocks, complex geometries, including steep bed slopes, and was capable of simulating the wetting and drying process. In this research, the performance of 20 numerical schemes used to solve the SWE for simulating the dam-break problem was examined, and some form of flux or slope limiter was used to eliminate oscillations.

Bradford et al., in 2002, developed a model based on the finite-volume method for 2D unsteady, shallow-water flow over arbitrary topography with moving lateral boundaries. They also introduced a new technique to prevent numerical truncation errors caused by the pressure and bed slope [4]. In the same year, Valiani et al. validated their code with the Malpasset dam-break [5]. In 2003 also, Ying et al. [6] developed a numerical model to simulate flood inundation due to a dam break. Sanders et al., in 2008, by applying Reynolds transport theorem to a finite control volume, derived shallow-water equations appropriate for urban flood modeling [7]. Ni et al., in 2018, presented an approximate solution to dam breaks [8]. The proposed method only worked on uniform slope channels and cannot be applied to real dam-break problems.

Fent et al. [9] investigated the vertical velocity distribution of a dam-break wave using experimental equipment. Using the OpenFOAM, a new code for modeling the two-phase flow in the dam break problem that was created by Park et al. [10], the accuracy of the new method was confirmed by comparing the results with laboratory data. The solution domain was wet in the proposed model. The effect of an obstacle in the river on the hydraulic characteristics of the dam-break flow was investigated by Issakhov et al. [11].

Different forms of obstacles were investigated numerically and experimentally. Different numerical methods for dam-break flow modeling were reviewed by Yang et al. in 2017 [12]. This study showed that Navier–Stokes equations with turbulence modeling have good accuracy. The finite volume method has been applied to a wide range of numerical methods [13]. In this method, volume integrals in a partial differential equation are converted into surface integrals, using the divergence theorem. In this regard, determining the passing flux through the surface is the primary goal of this method. However, estimation of the flux on the dry surface is more complicated. Haltas et al., for the failure problem, reviewed the calculation of the flux at the dry boundary [14]. An individual system is conducted to specify the dry, semi-dry and wet elements in any models. The wet cells remain in the simulation, and dry cells remove from the calculated boundary. The wet-dry method was applied in a three-dimensional finite-difference model by Medeiros and Hagen in 2013 [15]. In their model, different conditions have been used to ensure that the cell is wet and remains in the computational interval during each time step. The drought control was performed based on a length scale, defining the bed roughness. The points that must be removed from the calculated domain are specified after estimating the water depth in each cell side and cell center. In order to reconsider the cells into the calculated domain, the length scale is used, and it is compared with the water depth of the surrounding points. If it guarantees the conditions of the flow, then the cell returns to the next calculations time step. Casulli and Walters combined a three-dimensional difference-finite volume model with a dry-wet method [16]. In their method, the water depth was estimated, and the distance of vertical points was also updated based on each time step. By measuring the water depth as zero, the surface velocity and height were set to zero. Hou et al., in 2013 [17], experimentally investigated dam-break flow on a dry bed. The effect of topography was investigated in this study. Ji et al. used the Environmental Fluid Dynamics Code (EFDC) [18] and applied the dry-wet method presented by Hamrick and William to analyze flood and drought behaviors [19]. In this method, determining dry cells are the key, and if the cells are specified as dry, then the transmitted flux would be zero. In Brufau's finite volume model, the dry side is similar to the boundary, and water flow is controlled by the depth [20,21]. In this light, to preserve the conservation of mass, bed elevation deference is redefined locally. The finite volume Roe-TVD scheme was applied to the 2D shallow water equation by Cea [22]. He considers three depth average turbulence models by considering the behavior of wet-dry fronts for river flow. Song et al. used the Godunov-type finite volume model in order to simulate two-dimensional dam-break floods over complex topography with wetting and drying [23]. Vichiantong et al., in 2019, used a well-balanced finite volume method for flood simulation [24]. In their work, a well-balanced scheme with bottom slope approximation was developed. The accurate simulation of dam-break problems was investigated numerically by Antunes et al. in 2019 [25]. They presented an efficient technique for switching from Serre [26] to Saint-Venant in the breaking zone. The majority of these models are only applicable to dam-break flows over fixed beds [27]. Spinewine and Zech [28] used two-layer models to simulate dam-break flows over mobile beds. These models were applied to the morphological changes caused predominately by the non-equilibrium transporting of bedload. The impact of a dam-break wave on an erodible embankment with a steep slope was studied by Di Cristo et al. [29]. The simulation was carried out using a two-phase depth-integrated model. Recently, smoothed particle hydrodynamics (SPH) modeling represents a valuable numerical method of CFD particularly suitable for the simulation of dam-break floods [30]. The internal boundary condition was used to apply the bridge piers boundary condition of 2D shallow water equations by Dazzi et al. [31]. They confirmed the capability of this boundary condition in different flow regimes. Different methods of considering the bridge piers on 2D shallow water equations were investigated by Ratia et al. [32]. They applied the headloss method (HL) and solid wall boundary (MD) to consider the effects of the bridge piers, and it was concluded that the MD approach achieves results close to reality even when regimes other than free surface flow are involved.

When a dam breaks, it can cause casualties. These casualties are dependent on the extent of the inundation area, the population in the flood zone and the amount of alert time available. An investigation of the history of dam failures in the world shows that a large number of dams have broken and caused great financial and human losses. For example, the failure of dams Vajont (1963) in Italy, Johnstown (1889) in the USA and Macho (1979) in India caused 2600, 2200 and 2000 deaths, respectively [33]. According to Coasta [34], the death toll when a dam breaks without a flood warning or weak warning is 19 times higher than if there is a proper flood warning system in the area. Due to the critical situation at the time of dam failure, field data of the flood flow characteristics, such as flow depth, flow velocity, etc., are very limited. Therefore, the use of numerical modeling can be useful to determine the flood zone or the duration of response to the flood wave due to dam failure [33]. In this regard, a dam located in East Asia is considered to investigate the consequences after breaching. Breaking of the dam can initiate a detrimental risk to downstream structures, highways, and infrastructure and flow could cause the failure of two bridges, B1 and B2, and also an important highway. In this paper, the 2D finite volume method (Roe-TVD) is used on a fixed bed in order to model the dam break and assess its potential risks. Therefore, a numerical code is developed to model the dry bed. The maximum time that is necessary to apply precautionary behaviors to prevent potential risks is also presented. Due to the limited availability of real-scale flood data caused by dam break, code validation has only been performed with laboratory-scale data [35].

2. Shallow Water Equations

Shallow water equations are used when the length of the flume is much longer than depth, i.e., rivers. These equations can be obtained by integrating the 3D Navier–Stokes equations over the flow depth, considering the incompressibility of fluid and the hydrostatic pressure distribution. The equations are applied to study a wide range of physical phenomena such as dam break, flow in an open channel, flood waves, forces on offshore and nearshore structures [36,37] and pollution transfer. The two-dimensional form is [22]:

$$\frac{\partial W}{\partial t}+\frac{\partial F_x}{\partial x}+\frac{\partial F_y}{\partial y}=\sum_{k=1}^{3}G_k, \tag{1}$$

where

$$W=\begin{pmatrix} h \\ q_x \\ q_y \end{pmatrix}; F_y=\begin{pmatrix} q_y \\ \frac{q_x q_y}{h} \\ \frac{q_y^2}{h}+\frac{gh^2}{2} \end{pmatrix} F_x=\begin{pmatrix} q_x \\ \frac{q_x^2}{h}+\frac{gh^2}{2} \\ \frac{q_x q_y}{h} \end{pmatrix}; \tag{2}$$

where W is the vector of the conserved variables, including the water depth h, and the unit discharges in each direction are q_x and q_y as well as the vectors F_x and F_y account for the convective fluxes in the x and y directions, and g is the acceleration due to gravity, respectively. The vector G_k is a source term composed of the bed slope G_1, bed friction G_2, and turbulence terms G_3:

$$G_1 = \begin{pmatrix} 0 \\ -gh\frac{\partial Z_b}{\partial x} \\ -gh\frac{\partial Z_b}{\partial y} \end{pmatrix}; G_2 = \begin{pmatrix} 0 \\ -\frac{\tau_{b,x}}{\rho} \\ -\frac{\tau_{b,y}}{\rho} \end{pmatrix}; G_3 = \begin{pmatrix} 0 \\ -\frac{\partial h \ \overline{u_i' u_j'}}{\partial x} \\ -\frac{\partial h\overline{u_i' u_j'}}{\partial y} \end{pmatrix} \quad (i,j = 1,2) \tag{3}$$

where Z_b is the bed elevation, $\tau_{b,x}$ and $\tau_{b,y}$ are the bed shear stresses due to the friction in the x and y directions, ρ is the fluid density and $\overline{u_i' u_j'}$ is Reynolds stress.

3. Numerical Solution for Equations by Finite Volume Method

Different methods have been used for generating a mesh in finite volume methods. In the unstructured method, triangular cells are used, and the center of the triangular cell is the basis of numerical computation. Bermudez et al., in 1998, introduced unstructured quadrilateral cells [38]. In this method, first, the domain is divided into some triangles (Figure 1a). Then the middle of each side of the triangle is considered as the center of a quadrilateral cell. The two corners of this latter cell are the start and the end of the triangular cell side, and the other two corners are the triangular cell centers around the line (Figure 1b).

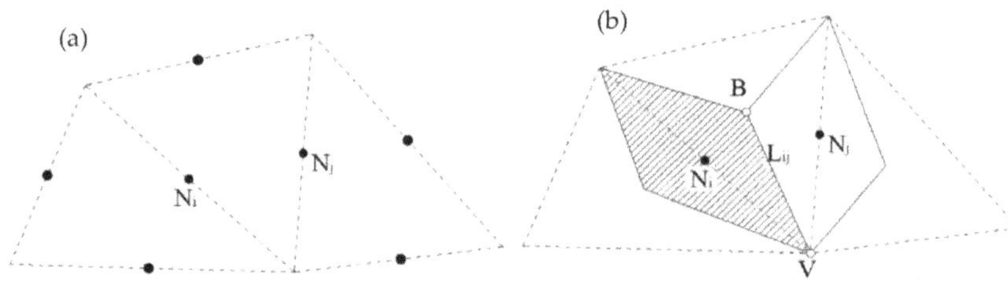

Figure 1. Finite volume cells [22]: (**a**) Initial triangle cell; (**b**) control volume.

A numerical code in the FORTRAN by using the finite volume Roe-TVD method is written to solve shallow water equations. A second-order method is applied by discretizing equations in time and using a semi-step. A multi-dimensional slope limiter is proposed to achieve second-order accuracy. Two depth-averaged turbulence models, including the $k - \varepsilon$ model and algebraic stress model (ASM), are used to calculate turbulence terms (G_3). The flowchart of the code is presented in Figure 2.

By time discretization of the system (1) and simplification, the following equations are obtained with a second-order of accuracy in time, [39]:

$$\begin{aligned} W^{n+\frac{1}{2}} &= W^n - \frac{\Delta t}{2}\left(\frac{\partial F_x}{\partial x}(W^n) + \frac{\partial F_y}{\partial y}(W^n)\right) + \frac{\Delta t}{2}\sum_{k=1}^{3} G_k^n \\ W^{n+1} &= W^n - \Delta t\left(\frac{\partial F_x}{\partial x}\left(W^{n+\frac{1}{2}}\right) + \frac{\partial F_y}{\partial y}\left(W^{n+\frac{1}{2}}\right)\right) + \Delta t \sum_{k=1}^{3} G_k^{n+\frac{1}{2}} \end{aligned} \tag{4}$$

where W^n is the vector of conserved variables at time t_n, and Δt is the time step. For spatial discretization, an upwind model may be implemented. In the upwind method, the numerical flux is defined as:

$$
\begin{gathered}
\varphi_{ij} = \frac{Z(W_i,n_{ij})+Z(W_j,n_{ij})}{2} - \frac{1}{2}|\phi(W_i, W_j, n_{ij})\ |(W_j - W_i) \\
|\phi| = X|D|X^{-1};\ Z = F_x n_x + F_y n_y \\
|D| = \begin{bmatrix} |\overline{\lambda}_1| & 0 & 0 \\ 0 & |\overline{\lambda}_2| & 0 \\ 0 & 0 & |\overline{\lambda}_3| \end{bmatrix}; \\
X = \begin{bmatrix} 0 & 1 & 1 \\ -\tilde{n}_y & \overline{U}_x + \overline{C}\,\tilde{n}_x & \overline{U}_x - \overline{C}\,\tilde{n}_x \\ \tilde{n}_x & \overline{U}_y + \overline{C}\,\tilde{n}_y & \overline{U}_y - \overline{C}\,\tilde{n}_y \end{bmatrix} \\
\overline{\lambda}_1 = n_x\overline{U}_x + n_y\overline{U}_y\,;\ \overline{\lambda}_2 = \overline{\lambda}_1 + \overline{C}\,L_{ij}\,; \\
\overline{\lambda}_3 = \overline{\lambda}_1 - \overline{C}L_{ij}
\end{gathered} \tag{5}
$$

The averaged values in the first-order scheme of Roe at each cell are defined as:

$$
\overline{U}_x = \frac{\sqrt{h_i}U_{x,i} + \sqrt{h_j}U_{x,j}}{\sqrt{h_i} + \sqrt{h_j}} \overline{U}_y = \frac{\sqrt{h_i}U_{y,i} + \sqrt{h_j}U_{y,j}}{\sqrt{h_i} + \sqrt{h_j}} \overline{C} = \sqrt{g\frac{h_i + h_j}{2}} \tag{6}
$$

where $U_{x,i}$ and $U_{y,i}$ are the velocities of the flow in cell i in the x- and y- directions, respectively, φ_{ij} is the numerical flux at the cell face ij, hi is the water depth in cell i, L_i is the boundary of the cell i, and $\tilde{n} = (\tilde{n}_x, \tilde{n}_y)$ is the unit vector normal to the cell face, Figure 1a. To achieve a second order accuracy in the method of Roe, the conserved variables at the triangular cell faces are reconstructed using a spatial limiting technique [38], Figure 3b:

$$
W_{I\,j} = W_i + r\nabla_i^1 W \tag{7}
$$

in which $W_{I\,j}$ is the value of W_i at the boundary with cell j, r is the distance vector between the cell area center i and the middle of L_{ij}, and ∇_i^1 is the limited gradient of variables at cell i, defined by:

$$
\nabla_i^1 W = w_a \nabla W_a + w_b \nabla W_b + w_c \nabla W_c \tag{8}
$$

where w_a, w_b and w_c are weighting factors, and ∇W_a, ∇W_b and ∇W_c are unlimited gradients of the three surrounding cells a, b and c. The unlimited gradient for cell i is computed using the area-weighted average gradients at the three faces:

$$
\nabla W_i = \frac{A_{i1a2}(\nabla W)_1 + A_{i2b3}(\nabla W)_2 + A_{i3c1}(\nabla W)_3}{A_{i1a2} + A_{i2b3} + A_{i3c1}} \tag{9}
$$

where A_{i1a2} is the area of quadrilateral $i1a2$ (Figure 3a), and $(\nabla W)_m$ is the gradient of the variable W at the face m of cell i. This gradient may be computed from the divergence theorem and an area-weighted average of two triangles around each face. For example, for face 1 in Figure 1a, it can be written:

$$
\begin{gathered}
(\nabla W)_1 = \frac{A_{1a2}\,\nabla W|_{1a2} + A_{1i2}\,\nabla W|_{1i2}}{A_{1a2} + A_{1i2}}; \\
\nabla W|_{1a2} = \frac{1}{A_{1a2}} \oint_\Gamma W n d\Gamma
\end{gathered} \tag{10}
$$

in which Γ is the integral path along the circumference of each sub-triangle (e.g., $1a2$). The weighting functions w_a, w_b and w_c are defined as:

$$
\begin{gathered}
w_a = \frac{(g_b g_c + \varepsilon^2)}{(g_a^2 + g_b^2 + g_c^2 + 3\varepsilon^2)};\ w_b = \frac{(g_a g_c + \varepsilon^2)}{(g_a^2 + g_b^2 + g_c^2 + 3\varepsilon^2)}; \\
w_c = \frac{(g_a g_b + \varepsilon^2)}{(g_a^2 + g_b^2 + g_c^2 + 3\varepsilon^2)}
\end{gathered} \tag{11}
$$

where ε is a small value (on the order of 10^{-4} and less), and g_a, g_b and g_c are the functions of the gradients of variables in the cells surrounding cell i (i.e., cells a, b and c in Figure 1a) defined by:

$$g_a = \|\nabla W_a\|_2^2; g_b = \|\nabla W_b\|_2^2; g_c = \|\nabla W_c\|_2^2 \quad (12)$$

In Equation (12), $\|\nabla W\|_2^2$ is the second norm of the unlimited gradient of a specified variable. By definition, the norm L_2 of a vector is the sum of the squares of its elements. Using the above equations, the limited gradient of a variable may be determined, and the data may be reconstructed at the cell boundaries.

The slope and friction source terms, G_1 and G_2, in Equation (3) may be defined as, [39]:

$$G_1 = \begin{pmatrix} 0 \\ -ghS_{0x} \\ -ghS_{0y} \end{pmatrix}; G_2 = \begin{pmatrix} 0 \\ -c_f U_x |U_x| \\ -c_f U_y |U_y| \end{pmatrix}; c_f = \frac{gn^2}{h^{\frac{1}{3}}} \quad (13)$$

where S_{0x} and S_{0y} are the bed slopes in the x- and y- directions, respectively, c_f is the bed coefficient of friction, and n is the Manning roughness. During the numerical computations, G_1 and G_2 are calculated based on the data obtained for the cell centers.

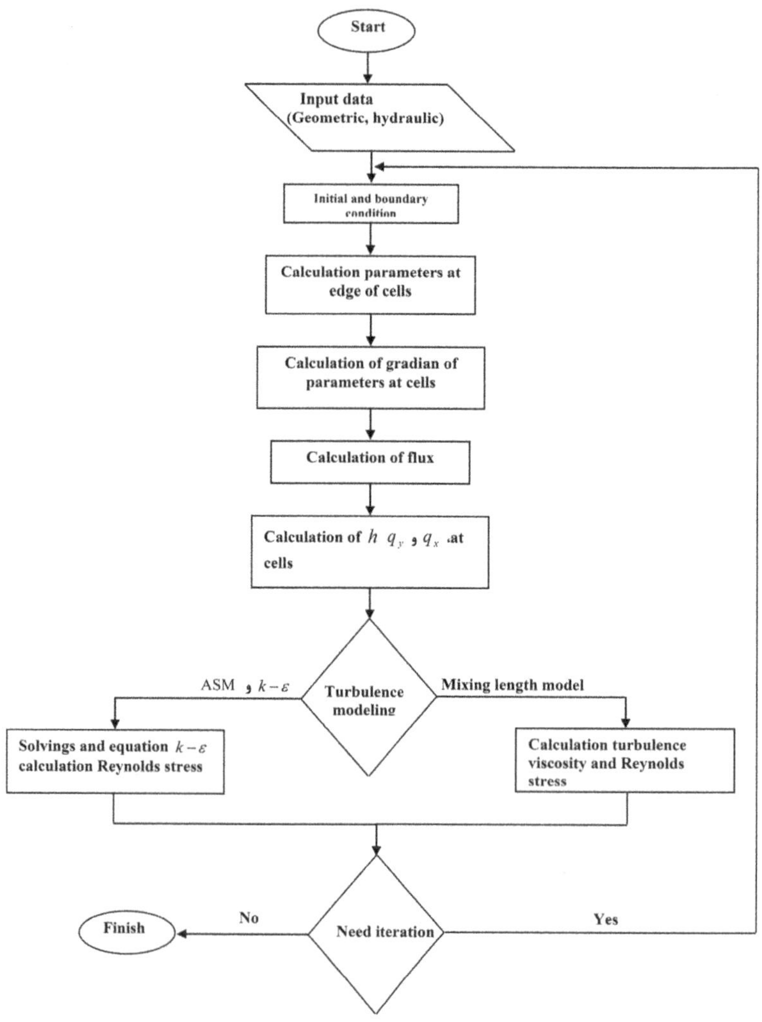

Figure 2. Flowchart of the written code.

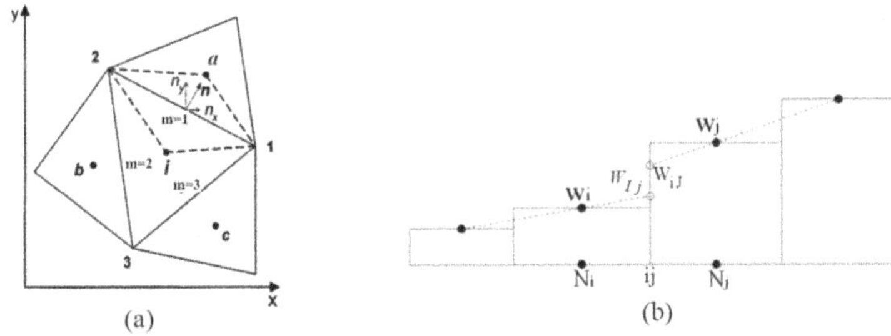

Figure 3. (**a**) A typical initial control volume cell. (**b**) Reconstruction of the conservative variables from the cell centers to the cell faces.

The Boussinesq assumption is the basis of all of the turbulence eddy viscosity models. It relates the Reynolds stresses to the mean velocity gradients via the eddy viscosity. Using this assumption in the averaged Reynolds stress models, the effects of the Reynolds stress in the shallow flows (G_3 in Equation (3)) may be written as:

$$G_3 = \begin{pmatrix} 0 \\ \frac{\partial}{\partial x}\left(v_t h \frac{\partial U_x}{\partial x}\right) + \frac{\partial}{\partial y}\left(v_t h \frac{\partial U_x}{\partial y}\right) \\ \frac{\partial}{\partial x}\left(v_t h \frac{\partial U_y}{\partial x}\right) + \frac{\partial}{\partial y}\left(v_t h \frac{\partial U_y}{\partial y}\right) \end{pmatrix} \tag{14}$$

where v_t is the turbulence eddy viscosity. To discretize this term, a semi-implicit method may be applied. When the viscosity is large, it is required to discretize the diffusive term implicitly. This term can be divided in two parts:

$$G_3 = G_{3,\perp} + G_{3,\parallel} \tag{15}$$

where $G_{3,\perp}$ is the orthogonal viscosity, and $G_{3,\parallel}$ is the non-orthogonal viscosity. For the momentum component in the x-direction, the two components of viscosity may be calculated from [39]:

$$G_{3,\perp,x} = \sum_{j \in K_i} \Gamma_{D_\perp} U_{x,j} - \frac{\Gamma_{D_\perp}}{h_i} q_{x,i} \tag{16}$$

$$G_{3,\parallel} = \sum_{j \in K_i} v_{t,ij} h_{ij} \frac{d_{ij}}{d_{\perp,ij}} (U_{x,B} - U_{x,V}) (\widetilde{\alpha}_{x,ij} \widetilde{n}_{x,ij} + \widetilde{\alpha}_{y,ij} \widetilde{n}_{y,ij}) \tag{17}$$

in which $\Gamma_{D_\perp} = v_{t,ij} h_{ij} \frac{|n_{ij}|}{d_{\perp,ij}}$ is the orthogonal diffusion, $q_{x,i}$ is the unit discharge at a cell i in the x-direction, h_{ij} and $v_{t,ij}$ are the averages of the depth and turbulent eddy viscosity in cells i and j, $U_{x,B}$ and $U_{x,V}$ are velocities in the x-direction at points B and V in Figure 4, $\widetilde{\alpha}_{ij} = (\widetilde{\alpha}_{x,ij}, \widetilde{\alpha}_{y,ij})$ is a unit vector perpendicular to the line that connects the centers of the cells i and j, and $d_{\perp,ij}$ is the projection of the distance between the two cell centers i and j over a line perpendicular to the common face of the two cells. All of the variables in Equation (16) are evaluated at time t_n except the unit discharge $q_{x,i}$, which is calculated at t_{n+1}. Therefore, no additional system of equations must be solved to increase the computational cost. The turbulence terms in the y-direction may be calculated similarly.

In the above equations, the eddy viscosity term may be computed using any of the turbulence modeling theories. The details of methods are presented by Alamatian and Jafarzadeh [39].

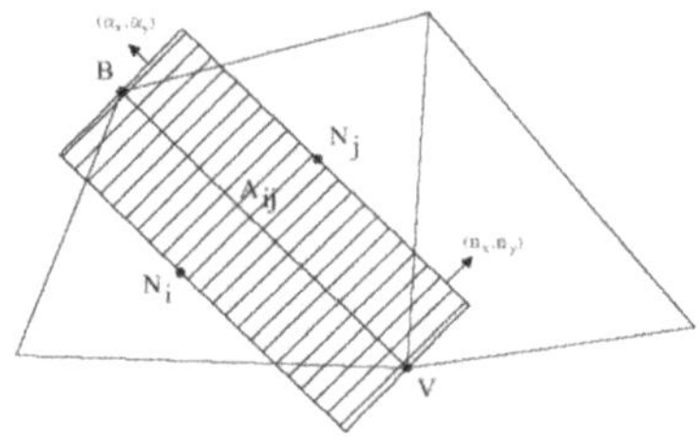

Figure 4. Discretization of the turbulent diffusion term.

4. Modeling of Dry Bed

Dam break simulation is possible in both dry and wet bed conditions. In 2012, Alamatian and Jafarzadeh [39] considered the initial computational domain as wet. In the dry bed simulation domain, the riverbed of the dam is initially considered as dry. In the wet bed modeling, by approaching the water depth as zero in every single cell ($h \approx 0$), the flux increases sharply and creates instability in numerical modeling. It is possible to consider a minimum initial water depth in computational domain cells to avoid a dry bed condition; however, the accuracy of this method for dam-break simulation is limited. Various methods for modeling dry and wet bed conditions have been introduced in different studies [22]. In some methods, the computational domain cells can consider both the dry and wet domain of the solution [6].

In this paper, the active method is used for wet-dry behavior [22]. This method of work is based on activating and deactivating each computational cell for wet and dry bed conditions, respectively. These zones may be added or removed from the computational domain, and computational cells would be directly considered as active (wet) and inactive (dry) in the solution matrix. In this paper, the cell is assumed dry when its water depth is less than 1.0 mm. When the cell is inactive (dry bed), the computational flux in the boundary of that cell is ignored. In this method, the bed elevation is updating for obtaining an exact balance at the wet-dry front between the bed slope and the hydrostatic pressure term for hydrostatic conditions. If the wet-dry front occurs between the cells *Ci* and *Cj*, the modified bed elevation at the front is defined as [22]:

$$\Delta_{zb,ij} = \begin{cases} h_i - h_j & if \\ zb,j - zb,i & otherwise \end{cases} \tag{18}$$

This method is time-efficient since the computational domain decreases by drying the bed. When the computational domain contains shocks, this method is more accurate [15].

4.1. Numerical Model Verification

The numerical model results are compared with experimental data to verify the performance of the code. The experiments were carried out by Soares-Frazao in 2007 [40]. The length and width of the considered flume are 5.6 and 0.5 m, respectively. A sluice gate is placed at $x = 2.39$ with a reservoir depth of 0.111 m upstream. A symmetrical triangular barrier with 0.065 m height and ± 0.14 slope on each side, at $x = 4.45$, is considered. The water depth of 0.02 m at the downstream side of the barrier is considered, and the other parts are assumed as dry (Figure 5).

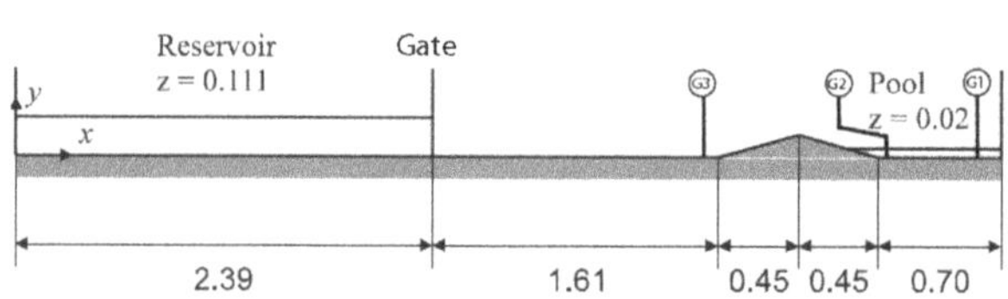

Figure 5. Geometrical properties of experimental model [40].

By immediately opening the sluice gate, which simulates the breaking of the dam, three probes start measuring the water depth in different locations. In the numerical simulation, the no-slip condition is applied for all wall boundaries [22]. A total of 14,618 unstructured triangular cells are considered with a Manning roughness coefficient and Courant–Friedrichs–Lewy (CFL) number of 0.011 and 0.9, respectively. The water level fluctuation of G_1 and G_2 during the 10 and 30 s after the dam break is investigated to verify the numerical simulation. The results of the numerical modeling and the experimental results are shown in Figures 6 and 7.

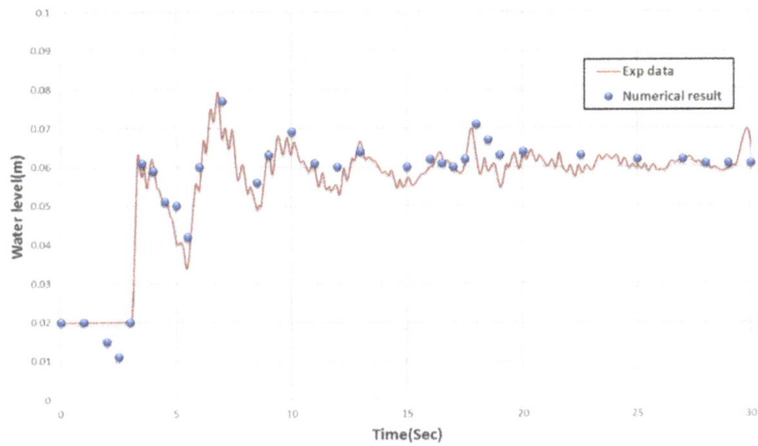

Figure 6. Water level variations at the station of G_1 during the 30 s after dam break.

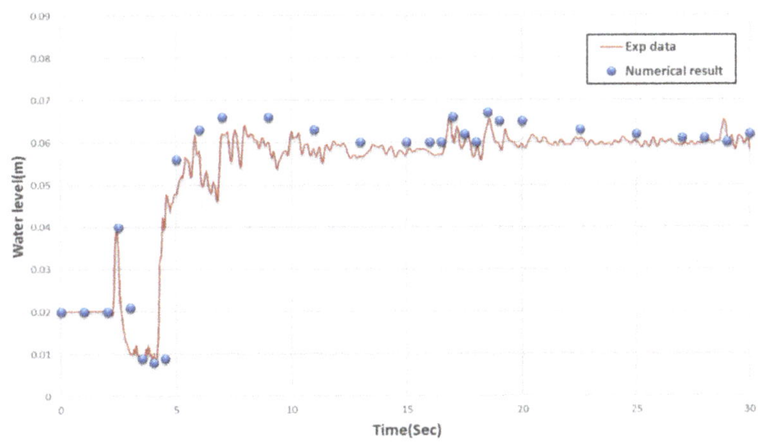

Figure 7. Water level variations at the station of G_2 during the 30 s after dam break.

Conservation of Mass Verification

One of the crucial issues in every CFD numerical code is to check the continuity or conservation of mass. To verify the ability of the code to conserve the mass in a dry canal, a numerical wave tank (NWT) with a barrier is considered. The NWT is defined with a length and width of 4.5 and 0.5 m, respectively. A triangular barrier is modeled in the center of the canal with a base of 1.5 m, a height of 0.5 m and 0.5 m in width (Figure 8).

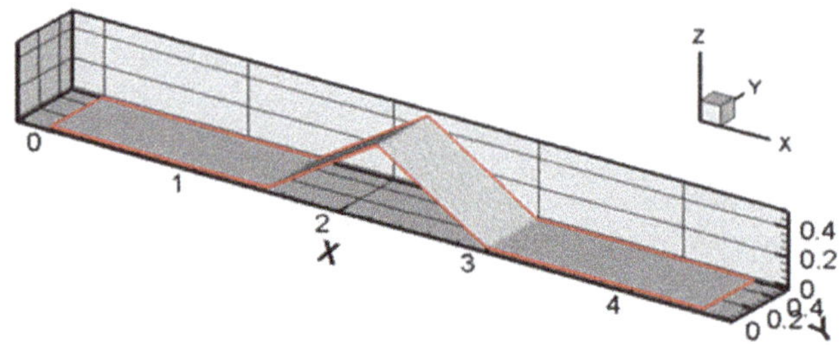

Figure 8. The geometry of the NWT in the test of flow passing barrier and stability above the barrier level.

The computational domain is considered as 12120 unstructured triangular cells. The no-slip condition is applied for all wall boundaries [22]. The initial conditions are composed of a water column with a height of 2 m and a length of 1 m in which its width is equal to the canal's width. The acceptable error of 10^{-8} is considered for the numerical code. Figure 9 shows the initial conditions of the test.

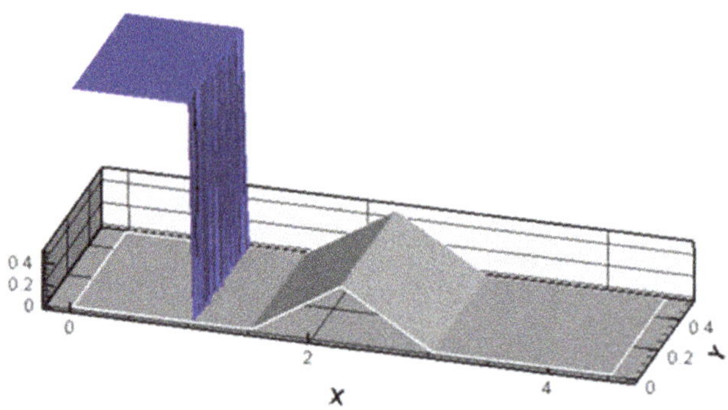

Figure 9. Initial conditions in the test of flow passing barrier and stability above the barrier level.

Figure 10 shows the flow passing through the barrier and the stability of water above the barrier at different time steps. The numerical test result is compared with the analytical method. In this regard, the volume of water and barrier volume are added together, and by dividing this volume with the solution domain, the water level after stability can be obtained. Table 1 presents the analytical solution and water level after stability.

Accordingly, the water level will be equal to 0.53 m along the canal after reaching the stable stage. Figure 11 shows numerical and analytical solution results in comparison. It is observed that numerical and analytical solution results are in good agreement, and the conservation of mass is preserved favorably.

Table 1. Analytical solution and water level after stability.

Properties	Water Volume (m^3)	Barrier Volume (m^3)	Total Volume of Barrier and Water (m^3)	Domain Area (m^2)	Water Level after Stability (m)
Values	$2 \times 1 \times 0.5 = 1$	$[(0.5 \times 1.5)/2] \times 0.5 = 0.188$	$1 + 0.188 = 1.188$	4.5×0.5	$1.188/(4.5 \times 0.5) = 0.53$

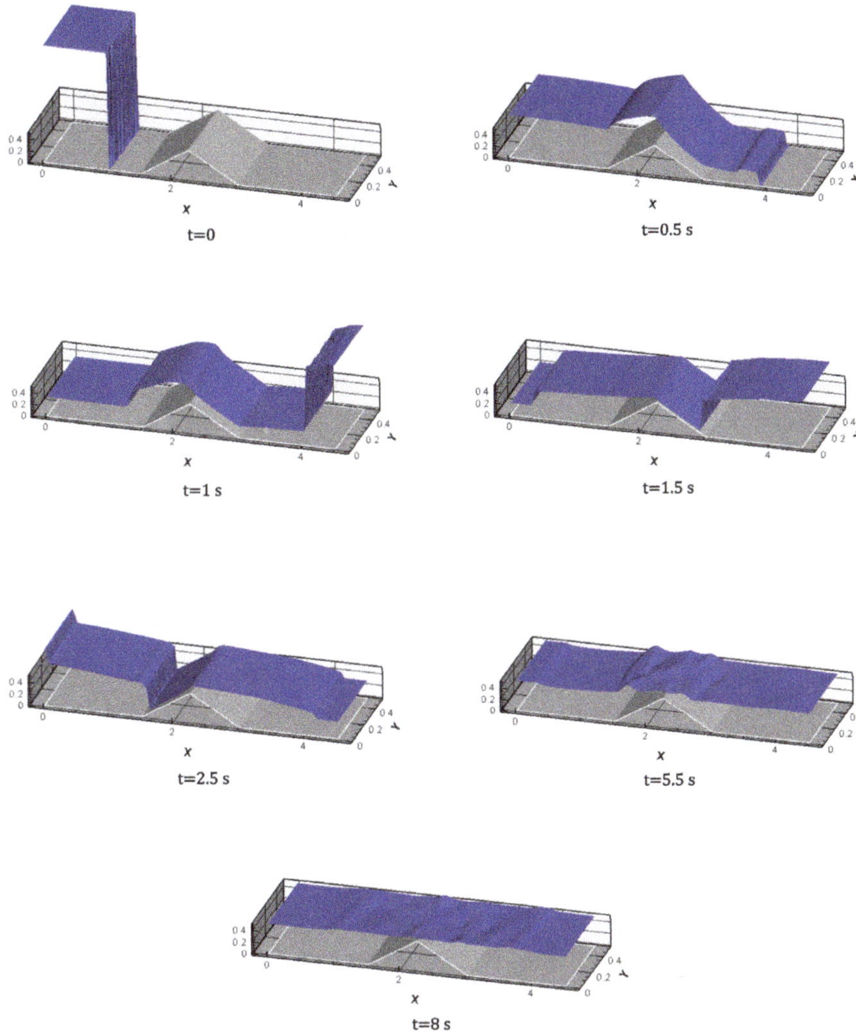

Figure 10. Progress of a wave in the test of flow passing the barrier and stability above the barrier level.

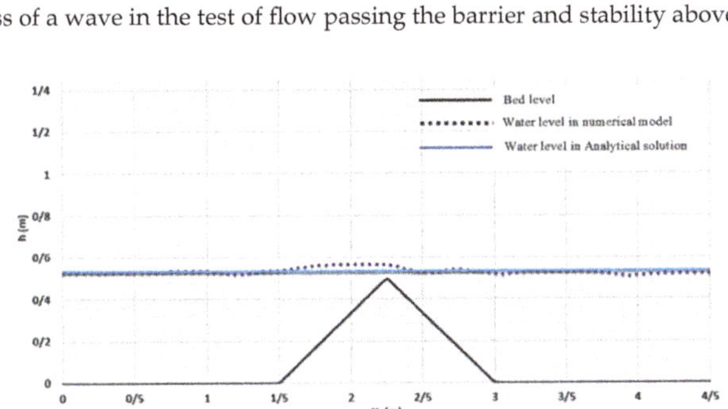

Figure 11. Numerical and analytical results in the test of flow passing barrier and stability above barrier level.

5. Modeling of the Dam Break

The Golestan dam has been constructed on a river in the north of Iran. It has a length of 30.6 m at the bottom with a crest length of 130 m, it is 20 m in height, and its reservoir volume is 2.7 million cubic meters. The area of the watershed dam is 1,010,000 hectares. The Golestan dam is a historic barrier dating back 150 years that was built using traditional materials. This dam has been exposed to various earthquakes. The changing of the land use upstream of the dam has caused increasing water to enter it and increased the likelihood of failure. In this paper, the length of 5400 m of the river downstream of the dam has been modeled. The topography of the area, dam reservoir and pathway plan have been obtained. Figure 12 shows the river plan and land risks.

Two bridges, B1 and B2, are located at distances of 3900 m and 5320 m downstream of the dam, respectively. These two bridges are critical for the area, and the dam break could cause a serious problem for these bridges. Their properties are presented in Table 2.

Table 2. Bridges Properties.

Bridge Details	B1 Bridge	B2 Bridge
Pier radius	0.6 m	0.6 m
Number of piers	48	16
Bridge height	6 m	7 m

Figure 12. River plan and land risks.

Only bridge piers have been applied in the modeling, and due to the size of the computational domain, a slip condition has been used at the wall boundaries of the bridge piers [22]. At the downstream end, the dependent variables (h, qx and qy) were interpolated from the solution domain, which is reasonable due to the supercritical nature of the flow [39]. The Courant–Friedrichs–Lewy (CFL) number is considered as 0.8. Figure 13 shows a change of maximum flow depth with different numbers of computational cells at B1 and B2 bridges. According to this figure, 35,628 unstructured triangular cells were chosen.

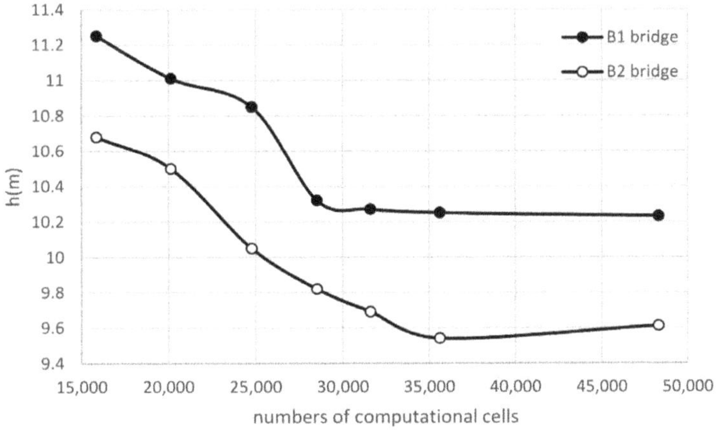

Figure 13. Maximum flow depth versus different numbers of computational cells.

Due to the sand-bed river without vegetation downstream of the dam, the value of the Manning roughness coefficient varies between 0.025 and 0.033 [41]. The maximum depths of flood flow versus Manning roughness are shown in Figure 14 for the points attached to bridge B1 and B2. It is observed that with increasing Manning roughness from 0.025 to 0.033, the maximum depth of the flood wave increases by about 11%. For reliability, Manning roughness is considered as 0.033.

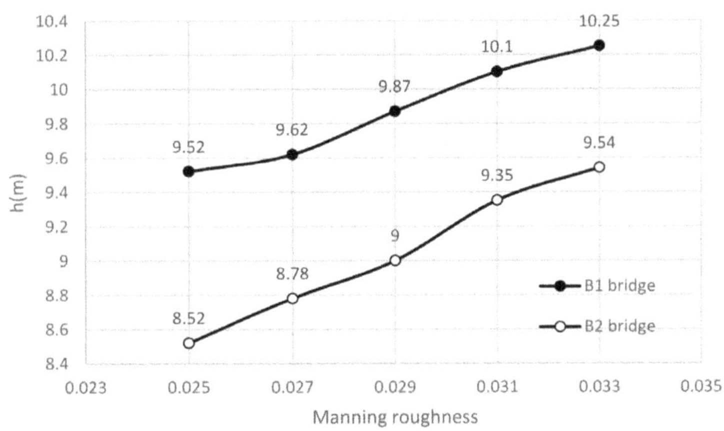

Figure 14. Maximum depths of flood flow versus Manning roughness.

6. Results and Discussion

The dam break is simulated using the above information. The analysis continued up to the point where water level approaches the elevation of the bridge. The travel time duration of the first wave that hits the bridges is calculated. Figure 15 shows flood zoning in different periods. In this figure, the river border and computational domain are distinguished by red lines. Flood zoning indicates that water depth decreases gradually in the reservoir after the dam breaks, and it takes 22 min and 30 s.

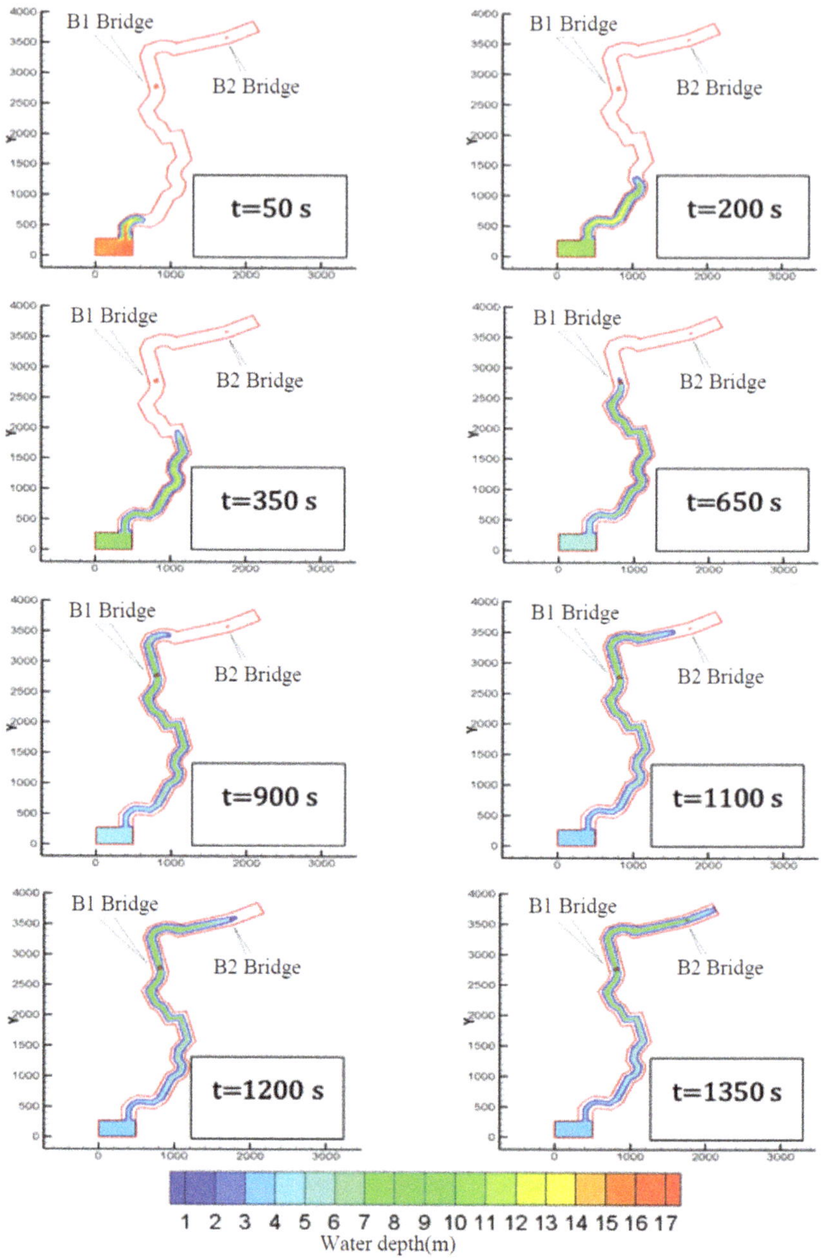

Figure 15. Flooded zoning due to the dam break.

Figures 14 and 15 show a closer look at the flood wave when it arrives at the B1 bridge at a time of 620 and 660 s, respectively. The flow depth distribution across the river and the bridge's piers are seen in Figures 16 and 17.

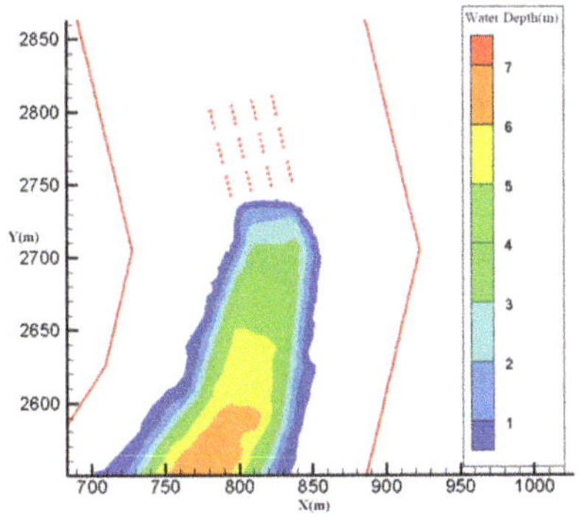

Figure 16. Water level at B1 bridge at time 620 s.

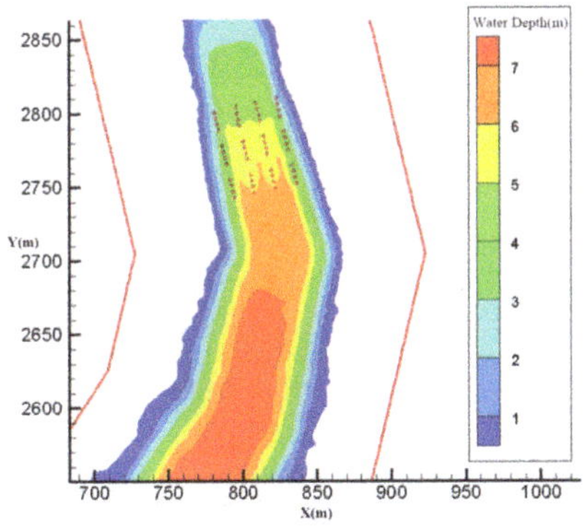

Figure 17. Water level at B1 bridge at time 660 s.

The velocity vectors and the contours of streamlines at the B1 bridge at time 1100 s are shown in Figure 18. According to this figure, the proper performance of the wall boundary condition is confirmed.

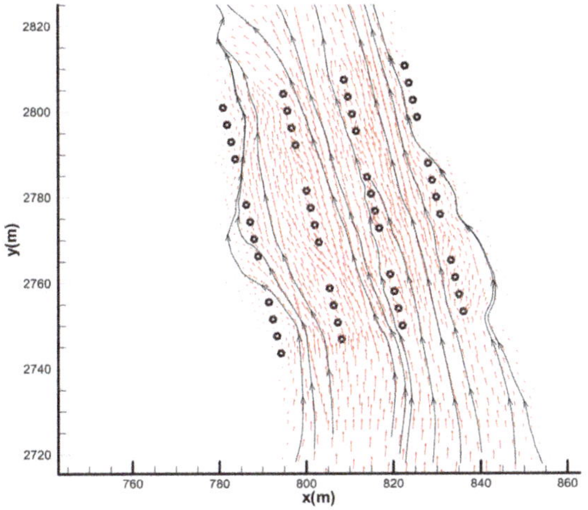

Figure 18. The velocity vectors and the contours of streamlines at the B1 bridge (t = 1100 s).

Flood flow hydrographs at bridges B1 and B2 are presented in Figure 19. It can be seen that the maximum flood discharge at the locations of bridges B1 and B2 is 120 (at time 16.5 min) and 110.2 (at time 28.5 min), respectively.

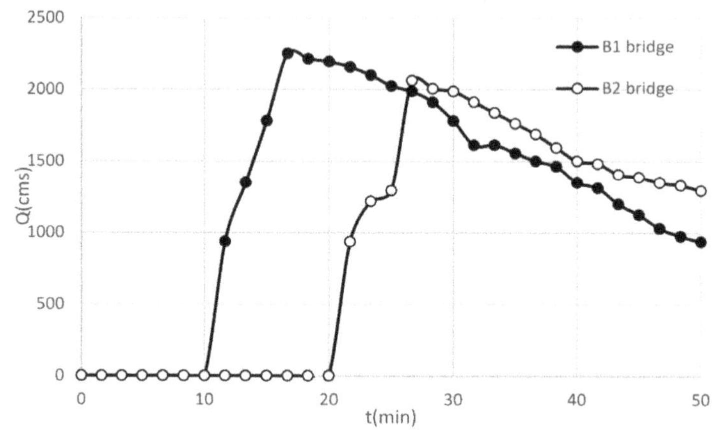

Figure 19. Flood flow hydrographs at bridges B1 and B2.

Figures 20 and 21 present the water level at B1, B2 pier locations. The B1 deck bridge would suffer from flooding at 11 to 40 min after the break and encounter dangerous waves with greater heights than the deck bridge elevation. It is observed that the height of the waves reaches up to 4 m above deck bridge level, which could cause a disaster. According to Figure 20, it is observed that 11 min after the dam break, the height of the flood wave reaches the level of the bridge deck. Consequently, for this bridge, there are only 11 min after the breaking of the dam to apply necessary proceedings and evacuate the adjacent roads. Precautionary measures should be taken during the safe periods.

The results based on Figure 21 showed that the B2 bridge would be flooded from 21 to 50 min. The maximum wave height of 9.54 m will occur about 27 min after the breaking of the dam, which is 2.5 m greater than the bridge elevation. Therefore, for bridge B2, there is only 21 min available before the bridge is completely flooded, and road evacuation should be considered to prevent future risks.

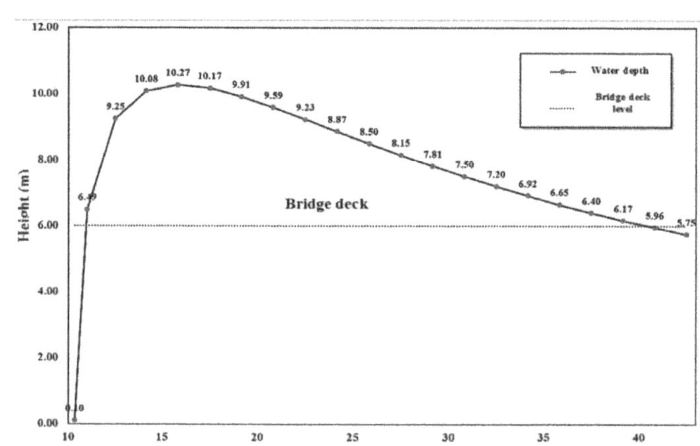

Figure 20. Water level variations at the B1 bridge.

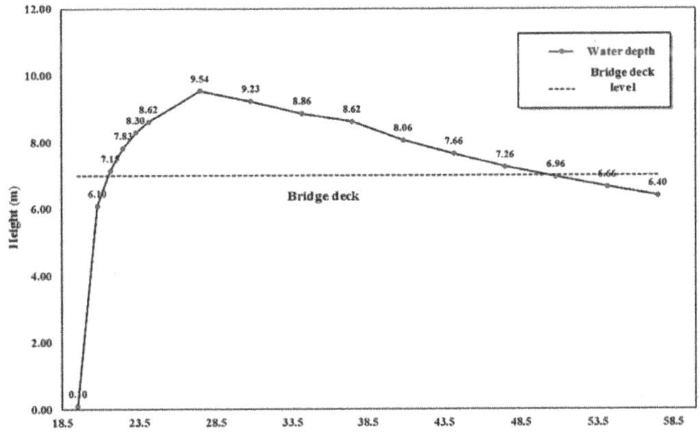

Figure 21. Water level variations at the B2 bridge.

7. Conclusions

Real-scale dam-break modeling using 3D numerical models has a high computational cost. It is very useful to use 2D models that can reduce the cost of calculations while being accurate. In this paper, a real-scale dam-break wave was simulated using the 2D finite volume Roe-TVD method. For this purpose, a numerical code was developed to solve the 2D depth average, shallow water equations on unstructured triangular cells considering turbulence terms and a dry bed front. In order to validate the code, at first, available experimental data were considered, the water level in the flume has been predicted and compared with the experimental data. Furthermore, a numerical tank was simulated to study the capability of the model in passing the flow over a barrier in the dry canal, conserving mass and to reach a steady flow case above the barrier level. Comparing the analytical and numerical solutions indicated that the conservation of mass is satisfied. After verifying the model, the real-scale dam break has been simulated, and the flow behavior from encountering the two bridges was analyzed along the pathway. The flood wave arrival time to the bridges, flooded area and the duration of flooding of the bridges were studied. The results of the dam break simulation showed that B1 and B2 bridges are at risk of flooding in the case of a dam break. Long waves affect the structures and can cause severe damage. Vehicle traffic should be banned at the moment of dam break. Furthermore, it is necessary to evacuate any vulnerable objects from these areas. The maximum required time for the evacuation of the two bridges is 12 and 21 min, respectively.

Author Contributions: Conceptualization, E.A.; methodology, E.A.; software, E.A.; validation, E.A., B.Đ. and S.D.; formal analysis, E.A., S.D. and B.Đ.; investigation, S.D. and E.A.; writing—original draft preparation, E.A.; visualization, E.A.; project administration, E.A. and B.Đ. All authors have read and agreed to the published version of the manuscript.

Funding: This research received no external funding.

Institutional Review Board Statement: Not applicable.

Informed Consent Statement: Not applicable.

Data Availability Statement: Data are available upon reasonable and unprofitable request.

Conflicts of Interest: The authors declare no conflict of interest.

References

1. Wang, Z.; Shen, H.T. Lagrangian Simulation of One-Dimensional Dam-Break Flow. *J. Hydraul. Eng.* **1999**, *125*, 1217–1220. [CrossRef]
2. Daneshfaraz, R.; Ghahramanzadeh, A.; Ghaderi, A.; Joudi, A.R.; Abraham, J. Investigation of the Effect of Edge Shape on Characteristics of Flow Under Vertical Gates. *J. Am. Water Work. Assoc.* **2016**, *108*, E425–E432. [CrossRef]

3. Zoppou, C.; Roberts, S. Numerical solution of the two-dimensional unsteady dam break. *Appl. Math. Model.* **2000**, *24*, 457–475. [CrossRef]
4. Bradford, S.F.; Sanders, B.F. Finite-volume model for shallow-water flooding of arbitrary topography. *Appl. Math. Model.* **2002**, *128*, 289–298.
5. Valiani, A.; Caleffi, V.; Zanni, A. Case Study: Malpasset Dam-Break Simulation using a Two-Dimensional Finite Volume Method. *J. Hydraul. Eng.* **2002**, *128*, 460–472. [CrossRef]
6. Ying, X.; Wang, S.S.Y.; Khan, A.A. Numerical Simulation of Flood Inundation due to Dam and Levee Breach. *World Water Environ. Resour. Congr.* **2003**, *2003*, 1–9. [CrossRef]
7. Sanders, B.F.; Schubert, J.E.; Gallegos, H.A. Integral formulation of shallow-water equations with anisotropic porosity for urban flood modeling. *J. Hydrol.* **2008**, *362*, 19–38. [CrossRef]
8. Ni, Y.; Cao, Z.; Borthwick, A.; Liu, Q. Approximate Solutions for Ideal Dam-Break Sediment-Laden Flows on Uniform Slopes. *Water Resour. Res.* **2018**, *54*, 2731–2748. [CrossRef]
9. Fent, I.; Zech, Y.; Soares-Frazão, S. Dam-break flow experiments over mobile bed: Velocity profile. *J. Hydraul. Res.* **2018**, *57*, 131–138. [CrossRef]
10. Park, K.M.; Yoon, H.S.; Kim, M.I. CFD-DEM based numerical simulation of liquid-gas-particle mixture flow in dam break. *Commun. Nonlinear Sci. Numer. Simul.* **2018**, *59*, 105–121. [CrossRef]
11. Issakhov, A.; Zhandaulet, Y.; Nogaeva, A. Numerical simulation of dam break flow for various forms of the obstacle by VOF method. *Int. J. Multiph. Flow* **2018**, *109*, 191–206. [CrossRef]
12. Yang, S.; Yang, W.; Qin, S.; Li, Q. Comparative study on calculation methods of dam-break wave. *J. Hydraul. Res.* **2018**, *57*, 702–714. [CrossRef]
13. Shadloo, M.; Oger, G.; Le Touzé, D. Smoothed particle hydrodynamics method for fluid flows, towards industrial applications: Motivations, current state, and challenges. *Comput. Fluids* **2016**, *136*, 11–34. [CrossRef]
14. Haltas, I.; Tayfur, G.; Elci, S. Two-dimensional numerical modeling of flood wave propagation in an urban area due to Ürkmez dam-break, İzmir, Turkey. *Nat. Hazards* **2016**, *81*, 2103–2119. [CrossRef]
15. Medeiros, S.C.; Hagen, S.C. Review of wetting and drying algorithms for numerical tidal flow models. *Int. J. Numer. Methods Fluids* **2012**, *71*, 473–487. [CrossRef]
16. Casulli, V.; Walters, R.A. An unstructured grid, three-dimensional model based on the shallow water equations. *Int. J. Numer. Methods Fluids* **2000**, *32*, 331–348. [CrossRef]
17. Hou, J.; Simons, F.; Mahgoub, M.; Hinkelmann, R. A robust well-balanced model on unstructured grids for shallow water flows with wetting and drying over complex topography. *Comput. Methods Appl. Mech. Eng.* **2013**, *257*, 126–149. [CrossRef]
18. Ji, Z.-G.; Morton, M.; Hamrick, J. Wetting and Drying Simulation of Estuarine Processes. *Estuar. Coast. Shelf Sci.* **2001**, *53*, 683–700. [CrossRef]
19. Hamrick, J.M.; Mills, W.B. Analysis of water temperatures in Conowingo Pond as influenced by the Peach Bottom atomic power plant thermal discharge. *Environ. Sci. Policy* **2000**, *3*, 197–209. [CrossRef]
20. Brufau, P.; Vázquez-Cendón, M.E.; García-Navarro, P. A numerical model for the flooding and drying of irregular domains. *Int. J. Numer. Methods Fluids* **2002**, *39*, 247–275. [CrossRef]
21. Brufau, P.; Garcia-Navarro, P. Two-dimensional dam break flow simulation. *Int. J. Numer. Methods Fluids* **2000**, *33*, 35–57. [CrossRef]
22. Cea, L.; Puertas, J.; Vázquez-Cendón, M.-E. Depth Averaged Modelling of Turbulent Shallow Water Flow with Wet-Dry Fronts. *Arch. Comput. Methods Eng.* **2007**, *14*, 303–341. [CrossRef]
23. Song, L.; Zhou, J.; Li, Q.; Yang, X.; Zhang, Y. An unstructured finite volume model for dam-break floods with wet/dry fronts over complex topography. *Int. J. Numer. Methods Fluids* **2010**, *67*, 960–980. [CrossRef]
24. Vichiantong, S.; Pongsanguansin, T.; Maleewong, M. Flood Simulation by a Well-Balanced Finite Volume Method in Tapi River Basin, Thailand, 2017. *Model. Simul. Eng.* **2019**, *2019*, 1–13. [CrossRef]
25. do Carmo, J.S.A.; Ferreira, J.A.; Pinto, L. On the accurate simulation of nearshore and dam break problems involving dispersive breaking waves. *Wave Motion* **2019**, *85*, 125–143. [CrossRef]
26. Do Carmo, J.A.; Ferreira, J.A.; Pinto, L.; Romanazzi, G. An improved Serre model: Efficient simulation and comparative evaluation. *Appl. Math. Model.* **2018**, *56*, 404–423. [CrossRef]
27. Gallegos, H.A.; Schubert, J.E.; Sanders, B.F. Two-dimensional, high-resolution modeling of urban dam-break flooding: A case study of Baldwin Hills, California. *Adv. Water Resour.* **2009**, *32*, 1323–1335. [CrossRef]
28. Spinewine, B.; Zech, Y. Small-scale laboratory dam-break waves on movable beds. *J. Hydraul. Res.* **2007**, *45*, 73–86. [CrossRef]
29. Di Cristo, C.; Evangelista, S.; Greco, M.; Iervolino, M.; Leopardi, A.; Vacca, A. Dam-break waves over an erodible embankment: Experiments and simulations. *J. Hydraul. Res.* **2018**, *56*, 196–210. [CrossRef]
30. Amicarelli, A.; Manenti, S.; Paggi, M. SPH Modelling of Dam-break Floods, with Damage Assessment to Electrical Substations. *Int. J. Comput. Fluid Dyn.* **2020**, 1–19. [CrossRef]
31. Dazzi, S.; Vacondio, R.; Mignosa, P. Internal boundary conditions for a GPU-accelerated 2D shallow water model: Implementation and applications. *Adv. Water Resour.* **2020**, *137*, 103525. [CrossRef]
32. Ratia, H.; Murillo, J.; García-Navarro, P. Numerical modelling of bridges in 2D shallow water flow simulations. *Int. J. Numer. Methods Fluids* **2014**, *75*, 250–272. [CrossRef]

33. USBR. *Prediction of Embankment Dam Breach Parameters*; Dam Safety Office: Denver, CO, USA, 1998.
34. Costa, J. *Floods from Dam Failures*; US Geological Survey: Reston, VA, USA, 1985.
35. Aliparast, M. Two-dimensional finite volume method for dam-break flow simulation. *Int. J. Sediment Res.* **2009**, *24*, 99–107. [CrossRef]
36. Torabi, M.A.; Shafieefar, M. An experimental investigation on the stability of foundation of composite vertical breakwaters. *J. Mar. Sci. Appl.* **2015**, *14*, 175–182. [CrossRef]
37. Shadloo, M.S. Numerical simulation of long wave runup for breaking and nonbreaking waves. *Int. J. Offshore Polar Eng.* **2015**, *25*, 1–7.
38. Bermúdez, A.; Dervieux, A.; Desideri, J.-A.; Vázquez, M. Upwind schemes for the two-dimensional shallow water equations with variable depth using unstructured meshes. *Comput. Methods Appl. Mech. Eng.* **1998**, *155*, 49–72. [CrossRef]
39. Alamatian, E.; Jaefarzadeh, M.R. Evaluation of turbulence models in simulation of oblique standing shock waves in supercritical channel flows. *Sharif* **2012**, *28*, 1–13. [CrossRef]
40. Frazao, S.S. Experiments of dam-break wave over a triangular bottom sill. *J. Hydraul. Res.* **2007**, *45*, 19–26. [CrossRef]
41. Ghani, A.A.; Zakaria, N.A.; Kiat, C.C.; Ariffin, J.; Hasan, Z.A.; Ghaffar, A.B. Revised equations for Manning's coefficient for Sand-Bed Rivers. *Int. J. River Basin Manag.* **2007**, *5*, 329–346. [CrossRef]

hydrology MDPI

Article

Modeling of the Geological Probability Procedure for the Prediction of High Flows in Small Streams, Case Study of Medvednica Mt., Croatia

Vedran Sudar [1], Tomislav Malvić [2], Tatjana Vujnović [1] and Josip Ivšinović [3,*]

[1] Croatian Meteorological and Hydrological Service, 10000 Zagreb, Croatia; vedran.sudar@cirus.dhz.hr (V.S.); tvujnovic@cirus.dhz.hr (T.V.)
[2] Faculty of Mining, Geology and Petroleum Engineering, University of Zagreb, 10000 Zagreb, Croatia; tomislav.malvic@rgn.unizg.hr
[3] Field Development, INA-Industry of Oil Plc., Av. V. Holjevca 10, 10000 Zagreb, Croatia
* Correspondence: josip.ivsinovic@ina.hr

Abstract: Floods are defined by maximum water levels or flow of high-water waves. Here, we defined the deterministic method for the calculation of the probability of a high discharge event, named as the Probability Of Success (POS). The POS method previously developed for petroleum subsurface systems has been modified for the surface hydrological system with the purpose of flood prediction. The case study of this research is the small basin of Kašina Stream on Medvednica Mt. (NW Croatia). The data are obtained upstream from the hydrological station Gornja Kašina. The POS model is defined by four categories. Each geological category is described with accompanied events and probabilities. Floods are defined by four categories: total precipitation, total water flow, basement, and maximal water capacity in soil. The categories total precipitation and basement were divided into two sub-categories each: quantity and duration; porosity and soil depth. Data are collected for a hydrometeorological event, namely an intensive convective storm on 24–25 July 2020, when Zagreb was locally hit by heavy urban floods. The presented probability method yielded a probability of 1.76% that such an event could happen to the station. However, the flooding was not recorded. A comparison of the real event and the predicted probability supported the adequacy and applicability of the method, showing it has high reliability. The presented probability model could be easily applied, with small modifications, to the entire area of Northern Croatia for the prediction of small basin flooding events.

Keywords: flood; geological categories; probability; Medvednica Mt.; Croatia

Citation: Sudar, V.; Malvić, T.; Vujnović, T.; Ivšinović, J. Modeling of the Geological Probability Procedure for the Prediction of High Flows in Small Streams, Case Study of Medvednica Mt., Croatia. *Hydrology* **2021**, *8*, 83. https://doi.org/10.3390/hydrology8020083

Academic Editors: Tommaso Caloiero, Carmelina Costanzo and Roberta Padulano

Received: 8 April 2021
Accepted: 12 May 2021
Published: 19 May 2021

1. Introduction

Occurrences of high waters defined by maximal water level or water wave flow caused by numerous factors that could initiate floods are described in [1,2]. High water studies through rainfall–runoff modeling are important for water resource management, water system design, real-time forecasting, flood design [3], and the construction of hydrotechnical objects [4]. Their frequencies/occurrences are used as input for hydrological forecasts, i.e., for protection and risk estimation of flooding. In recent years, Ref. [3] stated that there are 64 different hydrological models, though not all are suitable for every basin. For example, a continuous rainfall–runoff model is used operationally in the Croatian Meteorological and Hydrological Service, named the Sava Super Model [5], which also includes the research area of this paper. Other models used for determining the peak discharge are directed at specifying the characteristics of the Design Hydrograph [6]. However, the model designed in this paper is not similar to any of the mentioned models, although it uses Equation (3) of the Srebrenović unit hydrograph for peak discharge [1,2]. This emphasizes the need for an individual approach to almost every catchment area.

According to Žugaj [1], different factors influence runoff. They can be categorized into five major groups: geographical, geological, biological, climatological, and anthropological factors. In this paper, we confine the model to some geographical (size and slope of catchment area), geological (porosity of soils), and climatological (precipitation) factors.

The city of Zagreb is vulnerable to Sava River floods in its lower laying areas and to flash floods from the streams that flow towards the northern areas of the city from Medvednica Mt. [7]. A total of 19 retentions were constructed along Medvednica's streams to protect the city from the flashy flows of the 100-year recurrence discharges. The upper part of Kašina Stream is a rarity among Medvednica's streams in that it has no dam nor retention; thus, it can react naturally to precipitation. This is why it was selected for this research.

Here, we show how the basic POS methodology has been adapted for application in a hydrological flood probability model, conserving its simplicity. The presented probability model, with small corrections, could be applied to the entire area of this mountain.

2. The Study Area

The study area of the Kašina Basin is situated in the central northwestern part of Croatia, north of the city of Zagreb (see Figure 1) along the southeastern hills of Medvednica Mt. The basin covers 17.1 km^2, with altitudes of 195–625 m a.s.l. Its area covers the hilly part of the Kašina Basin, upstream from the "Gornja Kašina" stream gauge (586954.886 E, 5085893.794 N, 195.087 m a.s.l. (WGS84/UTM33N)). The drainage network consists of a few mountainous streams that emerge just below the main Medvednica Mt. ridge. The Kašina Stream starts after the confluence of the 3.7 km long Suhodol and 2.5-km long Šokot Stream. There are two additional small right tributaries—Vukov Dol, 3.2 km long, and Bjelak, 1.9 km long. The Kašina Stream by itself has a length of 3 km to the Gornja Kašina stream gauge.

A measuring instrumentation network that covers the studied watershed consists of two automatic meteorological stations in the vicinity—Zagreb Maksimir (123 m a.s.l.) in the eastern part of the city, RC Puntijarka at Medvednica Mt. (991 m a.s.l.)—and one stream gauge—Gornja Kašina (see Figure 1).

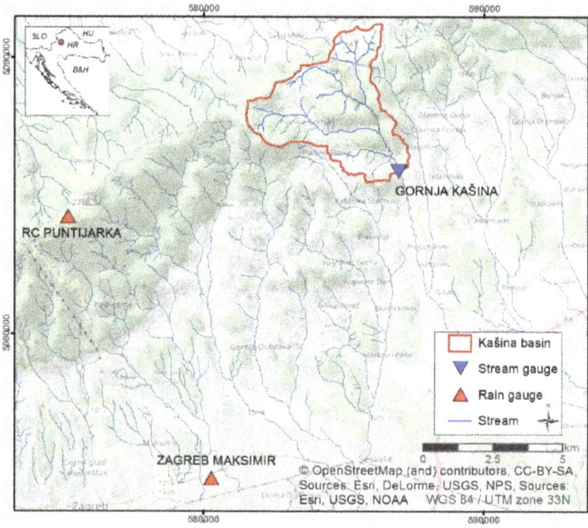

Figure 1. The location of the Kašina Basin margins, "Gornja Kašina" stream gauge, rain gauges and river network.

The climate of the study area is assigned to a warm temperate climate (Cfb in the Köppen–Geiger climate classification system) [8]. The long-term (1981–2015) climatological data for Zagreb Maksimir and Puntijarka meteorological stations are given in Table 1.

Table 1. Climatological data for Zagreb Maksimir and RC Puntijarka (1980–2015).

Meteorological Station	Annual Average Temperature (°C)	July Average Temperature (°C)	January Average Temperature (°C)	Annual Average Precipitation (mm)	Min Annual Precipitation (mm)	Max Annual Precipitation (mm)
Zagreb Maksimir	11.4	21.7	0.9	853.9	517	1317.8
RC Puntijarka	7	16.4	−1.9	1250.9	763.5	1638.3

RC Puntijarka has lower temperatures and higher precipitation than Zagreb Maksimir because of its significantly higher altitude. The warmest month was July and the coldest month was January at both meteorological stations. Most precipitation occurs during the warm period of the year (April–November) with two precipitation peak months—June and September. The year of 2014 gave the maximal annual precipitation of 1317.8 mm at Zagreb Maksimir but also the maximal monthly recorded precipitation in February, May and June. During the period of 1981–2015, the five "wet" years with precipitation above normal were 2014, 2010, 2013, 1998, 1996, while the years of 2011, 2003, 1990, 2000, 1997 were the five "dry" years, with precipitation below average. A similar pattern was recorded at RC Puntijarka where five "wet" years, with precipitation above normal were 2014, 2010, 1995, 2013, 2005, etc. and the five driest years were 2011, 2003, 2000, 2012 and 1990, with precipitation below average.

CORINE database for Land Cover (CLC 2018) GIS vector layer [9] was used to determine the land use data for the Kašina catchment. There were six CLC classes identified within the study area of which broad-leaved forest covers its majority with 61.8%, transitional woodland/shrub 12.4% and land principally occupied by agriculture, with significant areas of natural vegetation covering 9.5%.

Lithologically, there were [10–13] metamorphic diabase and gabbro (Paleozoic) transgressively overlaid with conglomerate (Miocene), including alternations of clay, marl, limestone and coal.

Slišković and Šarin [14] described permeable deposits in detail and listed impermeable ones. The mountain creek sediments are generally weakly permeable, and intrusive, highly metamorphic and clayey-marlitic rocks are impermeable. Miocene alternation of clastic and clastic-carbonate sediments [14,15] are weakly permeable without significant water reservoirs but, generally, the permeability of unconsolidated and weakly consolidated alluvium depends on granulometry.

Soil types in the analyzed basin could not be extracted in detail because large scale pedological maps do not exist. The most applicable were available in [16–18]. The following soils were recognized:

- rendzina on marl (flysch) or soft limestones;
- rendzina on dolomite and limestone;
- acid, brown soil over metamorphites and clastites;
- brown soil over dolomite.

Those soils are automorphic, where wetting happens solely by precipitation in free filtration through soil over a relatively short period [19]. Rendzina, as soil, is formed over weathered rocks in different climates, but always with high porosity, drainage and weak water retention capacity, over a depth of about 40 cm [19,20]. Acid soil is formed over acidic substrates in mountain ranges with humid or peri-humid climates. It could be with humus, loess, and/or exposed podzolic soil [19]. Brown soil is developed over the carbonates, in a forest environment, on different altitudes, and is relatively shallow [20], with loess and loam components [19].

3. Materials and Methods

The methodology of flooding risk estimation applied here is estimated adapting the general probability of success (POS), which is explained and used in numerous papers, e.g., [21–30]. The POS is a simple, flexible, robust, deterministical statistical method developed for the probability calculation of a selected statistical event, based on the multiplication of each method category selected to describe an observed event. The categories are statistical independent events (e.g., [23,27]). The methodology was developed primarily for hydrocarbon systems, but was later adapted in different estimations of numerous geological systems. The POS calculation is based on Equation (1):

$$POS = p_1 \times \ldots \times p_{n-1} \times p_n, \tag{1}$$

where:

POS—total probability of an observed geological event,

p—probability of a particular geological category,

1–n—number of geological categories.

Data are derived and a model had been examined for the flash/urban flood that happened in the city of Zagreb (Croatia), during the night of 24–25 July 2020, and compared with data collected in the same period on a hydrological station located upstream, on Medvednica Mt.

The model categories are selected according to geological and hydrological data available for the basin. The number and type are derived and simplified according to the scheme that describes total flow using components [2]. Consequently, the boundary conditions are determined from data on:

- precipitation in the basin and Medvednica Mt.;
- discharge at the section of the "Gornja Kašina" hydrological station;
- soil types and textures.

3.1. Precipitations in Basin and Medvednica Mt.

Data are taken from annual records collected on two automatic meteorological stations: (1) "Zagreb-Maksimir" and (2) "RC Puntijarka"; both for the period 1981–2015 (see Figure 2). Calculation of the annual mean showed a value of 1052.4 mm for the Kašina Stream basin.

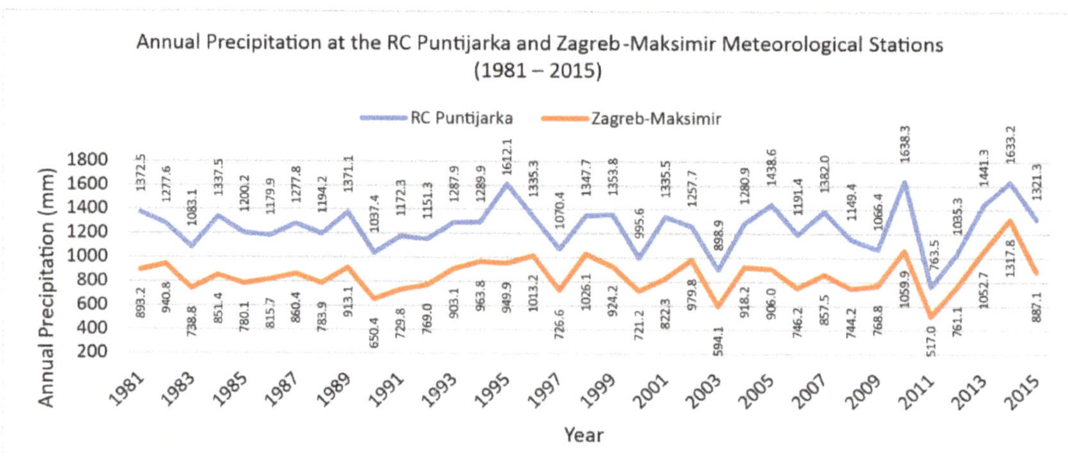

Figure 2. Annual mean precipitations on the "RC Puntijarka" and "Zagreb-Maksimir" station between 1981 and 2015.

3.2. Flow and Flow Section of the "Gornja Kašina" Hydrological Station

"Gornja Kašina" hydrological station has the discharge measured for the periods 1980–2015 and 2017–recent. The discharges are used for the calculation of the discharge curve as a function of water-level Q = f(H). We analyzed maximal annual discharge 1980–2015 (see Figure 3), with an observed weak positive trend of discharge. Based on annual maximums, three distribution curves were calculated and could be applied as return period (RP) sources for maximal discharges (see Figure 4). All three successfully approximated smaller discharges of lower RP, but we selected the Log Pearson type III curve as the most appropriate due to its good approximation of discharges with longer RP also. Consequently, the discharge of 10-year RP (Q_{10}) is 15.26 m^3/s, and the discharge of 100-year RP (Q_{100}) is 41.39 m^3/s.

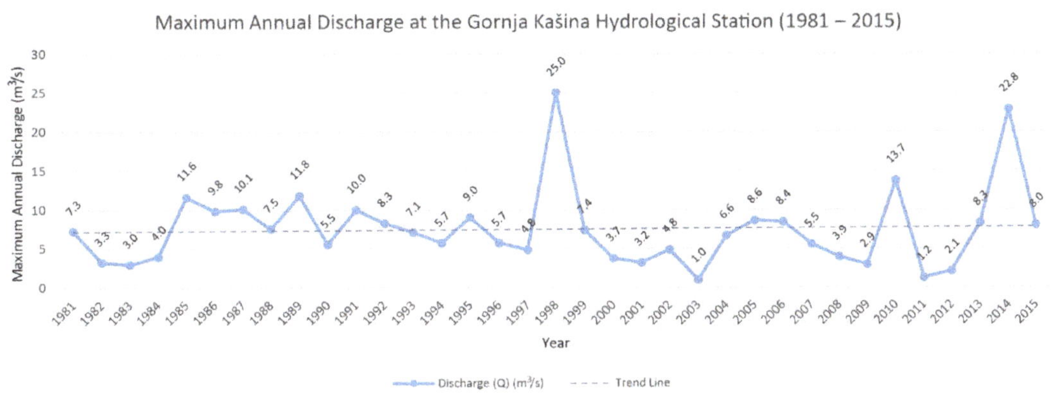

Figure 3. Maximal annual discharges at the "Gornja Kašina" stream gauge between 1981 and 2015.

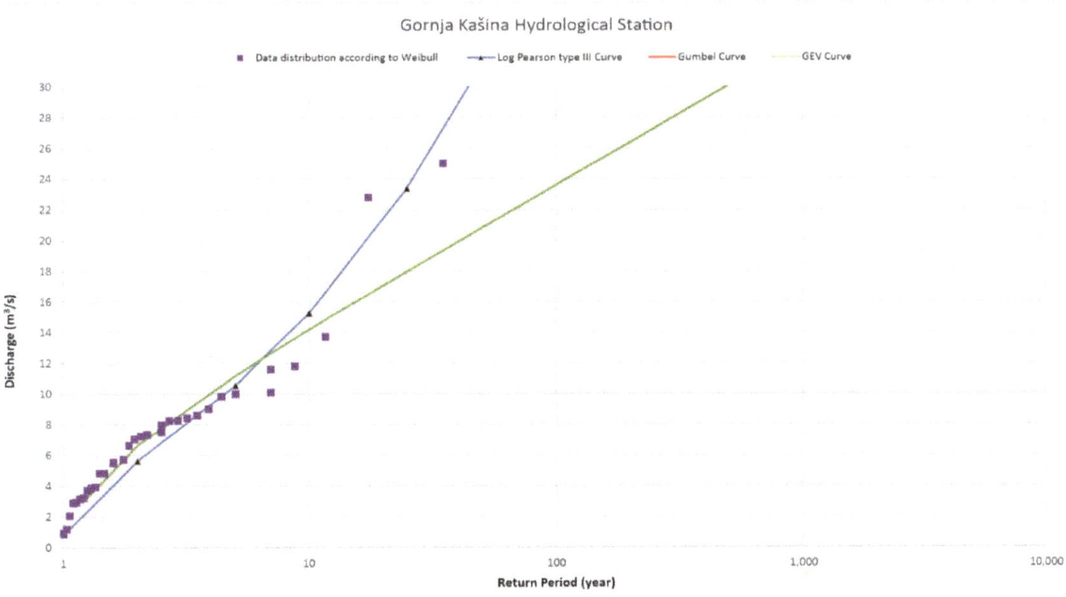

Figure 4. Frequencies of the maximal annual discharges on the "Gornja Kašina" hydrological station.

The vertical section of the "Gornja Kašina" stream gauge (see Figure 5) shows the marginal relative water-level just before start of overflowing—118 cm, with a corresponding bankfull discharge of 15.44 m^3/s. Figure 5 also shows marginal levels of high and medium water levels.

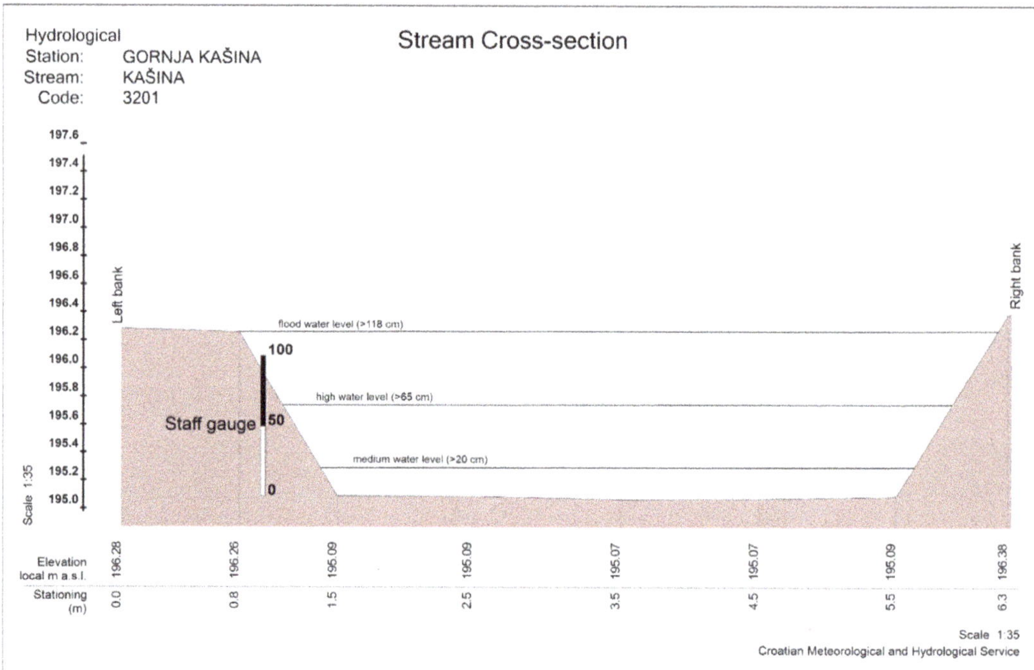

Figure 5. Cross-section of the "Gornja Kašina" hydrological station.

3.3. Soil Types and Texture Classes

As previously mentioned, the basin included four types of soils, dominantly automorphic. Systematic soil units were described in [16] including texture class and ecological depth (descriptive). Different, but also quantitative, depths are given by [31]. The applied depths in this paper are shown in Table 2.

Table 2. Comparison of ecological and applied soil depth in this work. Source: [31].

Ecological Soil Depth	Evaluation	Applied Description/Depth
<15 cm	Very shallow	Very shallow, <20 cm
15–30 cm	Shallow	Shallow, 25–50 cm
30–60 cm	Medium	Deep, 50–75 cm
50–120 cm	Deep	Very deep, >75 cm
>120 cm	Very deep	

3.4. Description of Categories

Malvić and Velić [23] determined 5 geological categories with accompanied probabilities, as 5 independent statistical events. Based on such an approach, here 4 hydrological categories are outlined that could describe flooding from components of total flow [2]. The categories shown in Table 3 "Total precipitation" and "Water Flow" are defined from 35-year long input data. The categories "Basement" and "Maximum water capacity in soil" are determined from the cited references.

Table 3. Geological–hydrological events presented in four categories with corresponding probabilities.

Basement		Maximal Water Capacity in Soil		Total Precipitation		Water Flow	
Porosity				**Quantity**			
Very low porosity, <30%	1.00	Low capacity, 21–42 mm/m	1.00	>35.0 mm	1.00	>15.44 m^3/s	1.00
Low porosity, 30–45%	0.75	Medium-low capacity, 42–83 mm/m	0.75	>26.25 mm	0.75	>10.31 m^3/s	0.75
Medium porosity, 45–60%	0.50	Medium-high capacity, 83–125 mm/m	0.50	>17.5 mm	0.50	>5.18 m^3/s	0.50
High porosity, >60%	0.25	High capacity, 125–167 mm/m	0.25	>8.75 mm	0.25	>0.04 m^3/s	0.25
Unknown porosity	0.05	Very high or unknown capacity, 167–208 mm/m	0.05	≤8.75 mm or unknown	0.05	≤0.04 m^3/s or unknown	0.05
Soil depth				Duration			
Very shallow, <25 cm	1.00			>3.11 h	1.00		
Shallow, 25–50 cm	0.75			>2.33 h	0.75		
Deep, 50–75 cm	0.50			>1.55 h	0.50		
Very deep, >75 cm	0.25			>0.78 h	0.25		
Unknown	0.05			≤0.78 h or unknown	0.05		

3.4.1. Category "Basement"

This category described one of the most important geological–pedological factors for flow/drain conditions. Here, it is considered that the basement is defined as a rocky base for water flow (see Figure 6). Moreover, this category are included two subcategories—"Porosity" and "Soil depth" as shown in Table 4.

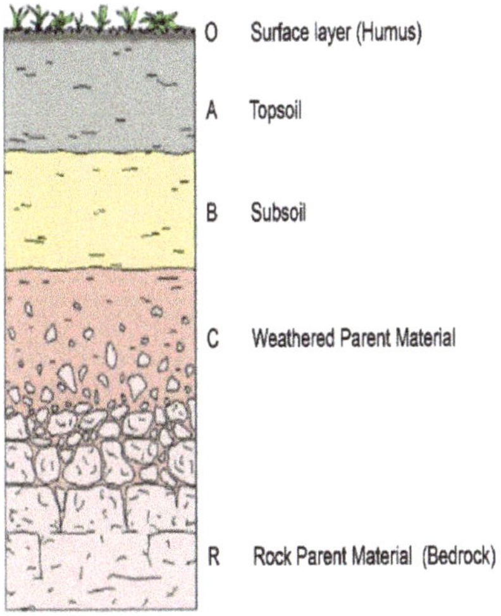

Figure 6. Soil horizons (layers), taken from [32].

The subcategory "Porosity" was defined according to [33] as soils evaluated in four types according to this variable. The event with $p = 1.0$ corresponds to the very low porosity soil basement (<30%). The $p = 0.75$ corresponds to the low value 30–45%, $p = 0.5$ to the medium value 45–60%, $p = 0.25$ to the high value of >60%. If porosity is unknown p is set at 0.05.

The subcategory "Soil depth" is defined according to [31], where soil depths are divided into four sections. Consequently, $p = 1.0$ corresponds to (very shallow) basement depths <25 cm, $p = 0.5$ to (shallow) 25–50 cm depths, $p = 0.5$ to (deep) 50–75 cm depths, $p = 0.25$ to (very deep) >75 cm depths, and $p = 0.05$ for events with no data.

Table 4. Geological–hydrological category "Basement" and corresponding subcategories with probability values.

Basement			
Porosity		**Soil Depth**	
Very low porosity, <30%	1.00	Very shallow, <25 cm	1.00
Low porosity, 30–45%	0.75	Shallow, 25–50 cm	0.75
Medium porosity, 45–60%	0.50	Deep, 50–75 cm	0.50
High porosity, >60%	0.25	Very deep, >75 cm	0.25
Unknown	0.05	Unknown	0.05

3.4.2. Category "Maximum Water Capacity in Soil"

The category "Maximum Water Capacity in Soil" is introduced with the purpose of representing the marginal quantity of precipitations that could be infiltrated (saturated) into soil before the subsurface and surface flows have been activated. After [34] infiltration, it depends on constants like soil texture and variables such as soil moisture. The second is expressed as the water quantity (i.e., height of the water column in mm) that exists in the soil in a 1 m deep column. If the soil is completely saturated with water, it has reached maximal water capacity. Saturation can change over the time and this is why it can hardly be predicted. Consequently, in this model, the saturation at the beginning was set on zero, i.e., the soil was considered as dry.

Ranges of [35] maximal water capacity in soil based on textures are given in Tables 5 and 6. The value $p = 1.0$ is valid for a capacity of 21–42 mm/m, when precipitations are sporadically infiltrated and mostly go (through the subsurface or over the surface) towards flow as shown in Table 5. The value $p = 0.75$ is appended to the interval 43–83 mm/m and $p = 0.5$ is appended to the interval 84–125 mm/m. p of 0.25 is appended to the interval 125–167 mm/m and if the capacity value is unknown, then $p = 0.05$.

Table 5. Geological–hydrological category "Maximal water capacity in soil" with probability values.

Maximal Water Capacity in Soil	
Low capacity, 21–42 mm/m	1.00
Medium-low capacity, 42–83 mm/m	0.75
Medium-high capacity, 83–125 mm/m	0.50
High capacity, 125–167 mm/m	0.25
Very high or unknown capacity, 167–208 mm/m	0.05

Table 6. Range of maximal water capacity in soil after texture soil classes.

Texture Class	Water Capacity in Soil (mm/m)
Coarse sand	21–63
Fine sand	63–83
Loamy sand	92–100
Sandy loam	104–117
Fine sandy loam	125–167
Silty loam	167–208
Silty-clayey loam	150–167
Sandy clay	125–142
Clay	100–125

3.4.3. Category "Total Precipitation"

This category is defined as part of total outflow/runoff created from effective precipitation. It is divided in two subcategories—"Quantity" and "Duration". All the presented equations are empirical.

"Quantity" is the total precipitation measured on the "Zagreb–Maksimir" and "RC Puntijarka" meteorological stations. The marginal value $p = 1$ is determined by the value of precipitation needed for a discharge of 15.44 m^3/s to occur. It is calculated by Equation (2) (Srebrenović's equations) taken from [1]:

$$P = P_e/\alpha \text{ (mm)}, \tag{2}$$

where:

P_e—effective precipitation in basin (mm);

α—flow coefficient.

Effective precipitation in the basin is calculated using Equation (3) [1]:

$$P_e = V/A \text{ (m)}, \tag{3}$$

where:

V—volume of water wave (m^3);

A—basin area (m^2).

The volume of water wave is defined according to Srebrenović Equation (4) [1]:

$$V = Q_m \tau \psi \text{ (m}^3\text{)}, \tag{4}$$

where:

Q_m—maximal flow of the water wave (m^3/s);

τ—period of surface water collecting, i.e., time of water wave creation (h);

ψ—function ratio between base and maximum flow φ.

The baseflow is not accounted for. Consequently, the variable ψ is set to 3744 [1]. The time of surface collecting is calculated after Equation (6) and, using the maximal flow (15.44 m^3/s), time of surface collecting (3.11 h) and ψ (3744.0) in Equation (4), the wave volume is 179,780.89 m^3. Applying that value in Equation (3), the effective precipitation quantity (P_e) is 0.0105 m (10.5 mm). The runoff coefficient (α) can be calculated using Srebrenović's Equation (5) [1]:

$$\alpha = 0.80[1 + 0.075(\log_{10}p - \beta)], \tag{5}$$

where:

p—back-period (years);

β—factor depending on permeability, afforestation, etc. (values ranging 1–3).

The factor β for low-permeability and low-vegetation soils is about 1, while for permeable and rich-vegetation it is 3. As the Kašina Stream basin is weakly populated, and located in forested parts of Medvednica Mt., the applied factor is $\beta = 2$. The RP of 15.44 m^3/s, according to the log Pearson III curve (see Figure 4), is 10.2 years. If the values of $\beta = 2$ and 10.2 years are input into Equation (5), the flow coefficient (α) is 0.74. After [36,37], such a value corresponds to flow from clayey soil and non-flat areas, but this is not like the researched terrain. Therefore, the value 0.30 is accepted, honoring the gentle surface dip and basement represented with medium textured soil (loam, silty-clay loam). Using such a runoff coefficient and effective precipitation (10.5 mm) as in Equation (2) the "Total precipitation" is 35 mm.

Consequently, the precipitation quantity of 35 mm, fallen during 3.11 h, was enough for a flow of 15.44 m^3/s, with a back-period of 10.2 years on the "Gornja Kašina" hydrological station. Therefore, the border value for $p = 1$ is 35 mm, and smaller probabilities are defined dividing the interval 0–35 into quartiles, i.e., $p = 0.75$ is valid for 26.25–34.99 mm, $p = 0.5$ 17.5–26.24, $p = 0.25$ 8.75–17.49, $p = 0.05$ for less than 8.75 or an unknown value.

The subcategory "Duration" is defined from the activity of total precipitation. The $p = 1$ is evaluated as necessary when the precipitation fallen inside the Kašina Stream basin flows through the vertical section of the "Gornja Kašina" hydrological station. According to [1], the precipitation duration for small basins is very similar to the time necessary for water collection and the creation of a water wave. The water collection time can be calculated from Picking's Equation (6) [1]:

$$T_c = 0.039\ (L/I_{max}{}^{0.25})^{0.76}\ (\text{min}), \tag{6}$$

where:

L—the longest path of water flow (m),

I_{max}—the basin dip that can be calculated after Equation (7).

$$I_{max} = \Delta H/L, \tag{7}$$

where:

ΔH—the elevation difference between the highest basin point and runoff section (m).

In the observed basin $\Delta H = 430$ m and $L = 26{,}087$ m, which are determined using GIS. According to Equation (7), the basin dip $I = 0.02$, i.e., 2%. According to Equation (6) time of concentration is 186.36 min (3.11 h). Such a value is a boundary for $p = 1$ (subcategory "Duration", 3.11 h) which is again divided into quartiles as shown in Table 7. Therefore, $p = 0.75$ is appended to the interval 3.10–2.33 h and $p = 0.5$ to the interval 2.32–1.55 h. p of 0.25 is appended to the interval 1.54–0.78 and $p = 0.05$ for less than 0.78 or unknown.

Table 7. Geological–hydrological category "Total Precipitation" and corresponding subcategories with probability values.

Total Precipitation			
Quantity		Duration	
>35.0 mm	1.00	>3.11 h	1.00
>26.25 mm	0.75	>2.33 h	0.75
>17.5 mm	0.50	>1.55 h	0.50
>8.75 mm	0.25	>0.78 h	0.25
≤8.75 mm or unknown	0.05	≤0.78 h or unknown	0.05

3.4.4. Category "Water Flow"

The hydrological category "Water flow" represents the total water quantity flowing through the section (see Figure 5) and just before the observed precipitation. The boundary condition for this category is defined from this section (see Figure 5) and the water-level (see Section 3.2). The flooding value, i.e., the value when outpouring from the bed starts is 118 cm, with a flow of 15.44 m^3/s. It is the lower boundary for $p = 1$. The minimal probability $p = 0.05$ is defined for less than 0.04 m^3/s and a stage of 4 cm. Other values are calculated as one-third of the interval 0.04–15.44 m^3/s shown in Table 8. Consequently, $p = 0.75$ is appended to the interval 15.44–10.31 m^3/s and $p = 0.5$ to 10.30–5.18 m^3/s. p of 0.25 is appended to the interval 5.17–0.04 m^3/s.

Table 8. Geological–hydrological category "Water flow" with probability values.

Water Flow	
>15.44 m^3/s	1.00
>10.31 m^3/s	0.75
>5.18 m^3/s	0.50
>0.04 m^3/s	0.25
≤0.04 m^3/s or unknown	0.05

4. Results

The Kašina Stream basin, upstream from the "Gornja Kašina" stream gauge, mostly is covered by rendzina on marl or soft limestone [16,17,19]. Since most precipitation reaches such soils, they were chosen as representative soils for the study area. According to [16], such soils have a silty–clayey to clayey loam texture.

Direct porosity data were not available. With the presumption that sandy soil contains more pores than clayey, i.e., more than rendzina on marl/flysch or soft limestone, the soil is considered as low-porosity with a range of 30–45%. According to Table 3, $p = 0.75$. Ecological soil depth [16] of the chosen soils is defined as medium, suggesting that in the interval 50–75 cm a probability value of $p = 0.50$ occurs. Multiplication of subcategories' p values results in a p value of the category "Basement" $p = 0.375$.

The texture class of soil rendzina on marl or soft limestone has a range of maximum capacity for water between 125 and 167 mm/m. The probability value of that event according to Table 3 is $p = 0.25$.

The quantity of total precipitation in the night of 24–25 July 2020 on the "Zagreb-Maksimir" station was 64.9 mm and 57.3 mm on the "RC Puntijarka" station. The mean value of 61.1 mm was used as input for the subcategory "Quantity". Duration, i.e., subcategory "Duration" was set to 3 h (as a mean of 2.67 h on the "Zagreb-Maksimir" and 3.33 h on the "RC Puntijarka"). The multiplication of the p values of subcategories according to Table 3 "Quantity" ($p = 1.0$) and "Duration" ($p = 0.75$) gave $p = 0.75$ as a probability value of the category "Total precipitation".

Just before the observed event on the "Gornja Kašina" hydrological station, the discharge was 0.15 m^3/s, which is the input value for the category "Water flow". According to Table 3, the p-value of this category is $p = 0.25$.

According to Equation (1), for the Kašina Stream basin, the probability value $p = 0.0176$. This result is interpreted as the probability of flooding during the event in the night of 24–25 July 2020 (1.76%). In reality, a flood did not happen, but rather the flow was slightly above the medium water level threshold (27 cm/1.19 m^3/s), which corresponds to the calculated probability. A flood could happen if the duration of precipitation or the amount of precipitation were higher.

5. Discussion

Each analytical method has a range of uncertainties, coming from different sources. It is valid for measurements as well as the calculation-based method. One simple, but robust, probability calculation method is presented here, built around hydrological categories. Such inputs are of mixed types, partially based on measurements, partially on the expert selection of data and data ranges. Therefore, two main uncertainties come from measurement and/or equipment errors, but also from the expert selection of the appropriate intervals described in the researched hydrological region. This second source is even more variable and sensitive, stressing the role of expert knowledge. The presented model is highly bounded for climate, relief, and population in the Medvednica Mt. area, Northern Croatia. Some other research groups could estimate the category events and ranges, probably in slightly different intervals, but the main estimation idea would be kept and comparable with the given results.

The basement porosity/soil depths, soil water capacity and total precipitation are clearly spatial and random variables. It means that they can be estimated as average or median from the total measured points, with no regard to their spatial location, but also firstly interpolated and later estimated from interpolated grid cells, where interpolated values are also considered as "hard data" as well as the original measurements. This means that uncertainties in sampling from interpolated maps also included uncertainties of interpolation methods. Consequently, they highly depend on the number of original "hard data" as the main factor in the selection of interpolation (e.g., inverse distance weighting, nearest neighborhood, Kriging, and even simulation). Such uncertainties must be estimated in each particular case, and a decision must be made regarding the use of only original measurements (like in this research) or interpolated values. A recommendation can be given that for highly dense hard datasets (n > 50 points); interpolation is the preferred option for obtaining values applied for a histogram, descriptive statistics, and cumulative distribution curves.

Regarding topography and climate, this model should be firstly confirmed several times in a similar climate and relief (in a broad sense, in Central Europe, excluding the Alps and Carpathians) and the success rate of such a calculation in the prediction of risky and extreme hydrological events should be determined. Only after such models prove their applicability in the entire Northern Croatian and the adjacent regions of Central Europe, could they be tested in other topographies and, especially, climate zones.

Other sources of uncertainties could not be in the spatial distribution of data or expert opinions about interval values, not even in the selected categories, but within the hydrological variables themselves. If, e.g., the runoff coefficient (Rc) and recession time (Tc) are observed as critical ones that control event runoff characteristics at a small catchment scale [38], the results (obtained in Lower Austria) showed that the Rc and their variability tend to be largest for the tile drainages and the main outlet, with larger values in January/February than in July/August. Tc variability was not oversized, but the largest at the main outlet and the smallest for the tile drainages. Therefore, the authors [38] explained changes in groundwater levels (due to shallow flows) as the main factors that influenced Rc and Tc, more important than soil moisture or precipitation. In our work, such additional subsurface flows were not considered for hydrological systems on Medvednica Mt., although they could exist, but they are not significant in the context of their contribution to the analyzed waterflow. Yet, similarly, in the system of the Gornja Kašina Stream, a critical variable event of extreme precipitation occurred, also in late July.

Moreover, in the same area (Lower Austria) another research study was performed to conclude the variability of the event runoff characteristics as a result of climate and hydro–geological conditions (for four "small" catchments with different hydro–geological settings mapped during field work [39]). A connection between the larger event runoff coefficients and flow peaks in catchments with high mean annual precipitation was observed. Interestingly, a weak poor relation between the previous precipitation and event runoff was proven, but the initial flow is the main factor defining the runoff coefficient and event peaks in all analyzed catchments. Those two examples pointed out the very local behavior of hydrological variables even in the catchments belonging to a very similar climate and covering comparable catchment areas.

6. Conclusions

In this article, definition and application of a model for flood prediction is presented. The model is developed for the "Gornja Kašina" hydrological station and its corresponding drainage area. It consists of geological–hydrological categories, compiled in a standard probability of success calculation procedure.

That approach was tested on the data collected during the most recent hydrometeorological extreme event on 24–25 July 2020 During that night, the city of Zagreb had localized urban floods in some of its parts, but the event did not register as a flood at the Gornja Kašina stream gauge. Such a scenario has been described with a probability calculation of $p = 1.76\%$ at the station. The obtained probability value proved the applicability of the methodology for the prediction of future possible flooding in the area of Medvednica Mt. The given methodology would also improve flood control mechanisms as to better define micro zones for flood defenses contrary to many hydrological models for broader areas. In the mentioned models, the catchment area of Kašina Stream would be just one cell on a flood risk map and unsuitable for individual observation. With this model, catchment risk assessment is independent and avoids interpolation algorithm errors.

The procedure, as its biggest advantage, can be easily adapted for similar catchments in Croatia, fed by data from other hydrological stations or meteorological prognoses. In such cases, defined categories, and their boundary values for "Basement" and "Maximal capacity" can be valid for any hydrological area. However, its application in other basins demands the modification of boundary values in categories "Precipitations" and "Water flow". The defined categories do not depend on climate zone, and therefore could be used in areas with different precipitations. The disadvantage of the given model is that it is not highly sensitive to all hydrological factors. Due to its resistance to various dynamic hydrological conditions, the model can be considered as robust.

Author Contributions: V.S. refined the idea for the study; collected, analyzed and interpreted data; carried out the literature and manual searches; wrote the paper. T.M. had the initial idea and later on refined the idea; provided input on the design and analyses; reviewed and provided substantive feedback on the paper and coordinated the study. J.I. provided input on the design and interpretation and reviewed and provided feedback on the paper. T.V. carried out literature searches; collected data and provided input on the analyses; reviewed and provided feedback on the paper. All authors have read and agreed to the published version of the manuscript.

Funding: This research received no external funding.

Acknowledgments: This research was partially carried out as part of the university project "Mathematical methods in geology V", funded in 2021 by the Faculty of Mining, Geology and Petroleum Engineering (Zagreb, HR).

Conflicts of Interest: The authors declare no conflict of interest.

References

1. Žugaj, R. *Hidrologija*; Faculty of Mining, Geology and Petroleum Engineering, University of Zagreb: Zagreb, Croatia, 2000; p. 407. (In Croatian)
2. Žugaj, R. *Hidrologija*, 2nd ed.; Faculty of Mining, Geology and Petroleum Engineering, University of Zagreb: Zagreb, Croatia, 2015; p. 538. (In Croatian)
3. Lavtar, K.; Bezak, N.; Šraj, M. Rainfall-Runoff modeling of the nested non-homogeneous Sava river sub-catchments in Slovenia. *Water* **2019**, *12*, 128. [CrossRef]
4. Srebrenović, D. *Problemi Velikih Voda*; Tehnička Knjiga: Zagreb, Croatia, 1970; p. 278. (In Croatian)
5. Sava, S.M. Available online: https://meteo.hr/infrastruktura.php?section=sustavi_upozorenja&param=sava_sm_upozorenja (accessed on 28 January 2021). (In Croatian)
6. Mlynski, D.; Walega, A.; Ksiazek, L.; Florek, J.; Petroselli, A. Possibility of using Selected rainfall-runoff models for determining the design hydrograph in mountainous catchments: A case study in Poland. *Water* **2020**, *12*, 1450. [CrossRef]
7. Croatian Disaster Risk Reduction Platform: Disaster Risk Assessment for the Republic of Croatia. 2019. Available online: https://civilna-zastita.gov.hr/UserDocsImages/DOKUMENTI_PREBACIVANJE/PLANSKI%20DOKUMENTI%20I%20UREDBE/Procjena%20rizika%20od%20katastrofa%20za%20RH.pdf (accessed on 26 April 2021).
8. Nimac, I.; Perčec Tadić, M. New 1981–2010 Climatological Normals for Croatia and Comparison to Previous 1961–1990 and 1971–2000 Normals. In Proceedings of the GeoMLA Conference, Belgrade, Serbia, 23–26 June 2016; Faculty of Civil Engineering, University of Belgrade: Belgrade, Serbia, 2016; pp. 79–85.
9. CORINE Database for Land Cover. Available online: https://land.copernicus.eu/pan-european/corine-land-cover/clc2018 (accessed on 26 April 2021).
10. Šikić, K.; Basch, O.; Šimunić, A. *(1978): Osnovna Geološka Karta SFRJ 1:100.000, List Zagreb L33–80.—Institut za Geološka Istraživanja, Zagreb (1972)*; Savezni Geološki Institute: Beograd, Serbia, 1977. (In Croatian)
11. Šimunić, A.; Pikija, M.; Hećimović, I. *(1983): Osnovna Geološka Karta SFRJ 1:100.000, List Varaždin L33–69.—Geološki Zavod, Zagreb, (1971–1978)*; Savezni Geološki Institute: Beograd, Serbia, 1982. (In Croatian)
12. Basch, O. *(1983): Osnovna Geološka Karta SFRJ 1:100.000, List Ivanić-Grad L33–81.—Geološki Zavod, Zagreb; OOUR za Geologiju i Paleontologiju (1969–1976)*; Savezni Geološki Institute: Beograd, Serbia, 1981. (In Croatian)

13. Šikić, K. *Geološki Vodič Medvednice;* Institut za Geološka Istraživanja and INA-Industrija Nafte d.d. Zagreb, Naftaplin: Zagreb, Croatia, 1995; p. 199. (In Croatian)
14. Slišković, I.; Šarin, A. *Osnovna Hidrogeološka Karta Republike Hrvatske 1:100.000, Tumač za List Ivanić Grad;* Institut za Geološka Istraživanja: Zagreb, Croatia, 1999; p. 47. (In Croatian)
15. Lovrić, F. Hydogeological Features of Water Springs in the Area of the Northeast Prigorje. Master's Thesis, Faculty of Geotechnical Engineering, University of Zagreb, Varaždin, Croatima, 19 September 2016. (In Croatian, with Abstract in English)
16. Husnjak, S. *Inventarizacija Poljoprivrednog Zemljišta Grada Zagreba i Preporuke za Poljoprivrednu Proizvodnju;* Faculty of Agriculture, University of Zagreb: Zagreb, Croatia, 2008; p. 137. (In Croatian)
17. Pedologija. Available online: http://pedologija.com.hr/iBaza/DPK-Hr_2019/index.html#5/45.948/16.265 (accessed on 24 January 2021). (In Croatian)
18. Tlo i Biljka. Available online: http://tlo-i-biljka.eu/iBaza/Pedo_HR/index.html (accessed on 24 January 2021). (In Croatian)
19. Klasifikacija Tala Hrvatske. Available online: http://os-akanizlica-pozega.skole.hr/upload/os-akanizlica-pozega/images/static3/3017/File/KLASIFIKACIJA%20TALA%20HRVATSKE.pdf (accessed on 24 January 2021). (In Croatian)
20. Osnove Pedološke Klasifikacije. Available online: http://www.fazos.unios.hr/upload/documents/Osnove%20Bilinogojstva%20002%20c%20osnove%20pedolo%C5%A1ke%20klasifikacije.pdf (accessed on 24 January 2021). (In Croatian)
21. Malvić, T. Oil-Geological Relations and Probability of Discovering New Hydrocarbon Reserves in the Bjelovar Sag. Ph.D. Thesis, Faculty of Mining, Geology and Petroleum Engineering, University of Zagreb, Zagreb, Croatia, 21 March 2003. (bilingual, In Croatian and English)
22. Malvić, T.; Rusan, I. Investment risk assessment of potential hydrocarbon discoveries in a mature basin; Case study from the Bjelovar sub-basin, Croatia. *Oil Gas Eur. Mag.* **2009**, *35*, 66–72.
23. Malvić, T.; Velić, J. Stochastically improved methodology for probability of success (POS) calculation in hydrocarbon systems. *RMZ Mater. Geoenviorenment* **2015**, *62*, 149–155.
24. Malvić, T.; Velić, J.; Režić, M. Geological probability calculation of new gas discoveries in wider area of Ivana and Ika Gas Fields, Northern Adriatic, Croatia. *Mater. Geoenvironment* **2016**, *63*, 127–138. [CrossRef]
25. Novak, K. Conditions Surface Transport and Geological Storage Modelling of Carbon-Dioxide into Northern Croatian Neogene Sandstone Reservoirs, Case Study Ivanić Field. Ph.D. Thesis, Faculty of Mining, Geology and Petroleum Engineering, University of Zagreb, Zagreb, Croatia, 13 April 2015. (In Croatian, with Abstract in English)
26. Ivšinović, J. Selection and Geomathematical Calculation of Variables for Sets with Less than 50 Data Regarding the Creation of an Improved Subsurface Model, Case Study from the Western Part of the Sava Depression. Ph.D. Thesis, Faculty of Mining, Geology and Petroleum Engineering, University of Zagreb, Zagreb, Croatia, 7 November 2019. (In Croatian, with Abstract in English)
27. Malvić, T.; Ivšinović, J.; Velić, J.; Sremac, J.; Barudžija, U. Increasing Efficiency of Field Water Re-Injection during Water-Flooding in Mature Hydrocarbon Reservoirs: A Case Study from the Sava Depression, Northern Croatia. *Sustainability* **2020**, *12*, 786. [CrossRef]
28. Malvić, T.; Pimenta Dinis, M.A.; Velić, J.; Sremac, J.; Ivšinović, J.; Bošnjak, M.; Barudžija, U.; Veinović, Ž.; Sousa, H.F.P. Geological Risk Calculation through Probability of Success (PoS), Applied to Radioactive Waste Disposal in Deep Wells: A Conceptual Study in the Pre-Neogene Basement in the Northern Croatia. *Processes* **2020**, *8*, 755. [CrossRef]
29. Ivšinović, J.; Malvić, T.; Velić, J.; Sremac, J. Geological Probability of Success (POS), case study in the Late Miocene structures of the western part of the Sava Depression, Croatia. *Arab. J. Geosci.* **2020**, *13*, 1–12. [CrossRef]
30. Ivšinović, J.; Pleteš, V. Risk and Financial Cost Management of Injection Wells in Mature Oil Fields. *J. Risk Financ. Manag.* **2021**, *14*, 184. [CrossRef]
31. Rai, R.K.; Singh, V.P.; Upadhyay, A. Soil Analysis. In *Planning and Evaluation of Irrigation Projects*, 1st ed.; Elsevier: Amsterdam, The Netherlands, 2017; pp. 505–523. [CrossRef]
32. Baletić, I. Sjetvena Grbica (Lepidium sativum L.) as a Potential Hyperacumulator of Copper. Master's Thesis, Faculty of Chemistry and Technology, University of Split, Split, Croatia, 12 July 2017. (In Croatian, with Abstract in English)
33. Gračanin, M. *Pedologija (Tloznanstvo), II dio, Fiziografija Tala;* Poljoprivredni Nakladni Zavod: Zagreb, Croatia, 1947; p. 234. (In Croatian)
34. Brouwer, C.; Goffeau, A.; Heibloem, M. *(1985): Irrigation Water Management: Training Manual No. 1—Introduction to Irrigation;* FAO: Rome, Italy, 1985.
35. Plant and Soil Sciences eLibrary. 2021. Available online: http://passel-test.unl.edu/beta/pages/informationmodule.php?idinformationmodule=1130447039&topicorder=10&maxto=10&minto=1 (accessed on 19 January 2021).
36. Mahmoud, S.H.; Alazba, A.A. Hydrological Response to Land Cover Changes and Human Activities in Arid Regions Using a Geographic Information System and Remote Sensing. *PLoS ONE* **2015**, *10*, e0125805. [CrossRef]
37. Tsutsumi, K.J.A.; Berndtsson, R. Surface and subsurface water balance estimation by the groundwater recharge model and a 3-D two-phase flow model/Estimation de bilan hydrologique de surface et de subsurface à l'aide de modèles de recharge de nappe et d'écoulement diphasique 3-D. *Hydrol. Sci. J.* **2004**, *49*, 205–226. [CrossRef]
38. Chen, X.; Parajka, J.; Széles, B.; Strauss, P.; Blöschl, G. Spatial and temporal variability of event runoff characteristics in a small agricultural catchment. *Hydrol. Sci. J.* **2020**, *65*, 2185–2195. [CrossRef] [PubMed]
39. Chen, X.; Parajka, J.; Széles, B.; Valent, P.; Viglione, A.; Blöschl, G. Impact of Climate and Geology on Event Runoff Characteristics at the Regional Scale. *Water* **2020**, *12*, 3457. [CrossRef]

hydrology MDPI

Article

Effects of River Discharge and Sediment Load on Sediment Plume Behaviors in a Coastal Region: The Yukon River, Alaska and the Bering Sea

Kazuhisa A. Chikita [1,*], Tomoyuki Wada [2], Isao Kudo [3], Sei-Ichi Saitoh [1] and Mitsuhiro Toratani [4]

1 Arctic Research Center, Hokkaido University, Sapporo 0010021, Japan; ssaitoh@arc.hokudai.ac.jp
2 Earth System Science, Inc., Tokyo 1600022, Japan; wada-tomoyuki@ess-jpn.co.jp
3 Faculty of Fisheries Sciences, Hokkaido University, Hakodate 0418611, Japan; ikudo@fish.hokudai.ac.jp
4 Department of Optical and Imaging Science and Technology, Tokai University, Hiratsuka 2591292, Japan; tora@keyaki.cc.u-tokai.ac.jp
* Correspondence: chikita@sci.hokudai.ac.jp; Tel.:+81-11-706-2764

Abstract: In the Bering Sea around and off the Yukon River delta, surface sediment plumes are markedly formed by glacier-melt and rainfall sediment runoffs of the Yukon River, Alaska, in June–September. The discharge and sediment load time series of the Yukon River were obtained at the lowest gauging station of US Geological Survey in June 2006–September 2010. Meanwhile, by coastal observations on boat, it was found out that the river plume plunges at a boundary between turbid plume water and clean marine water at the Yukon River sediment load of more than ca. 2500 kg/s. Grain size analysis with changing salinity (‰) for the river sediment indicated that the suspended sediment becomes coarse at 2 to 5‰ by flocculation. Hence, the plume's plunging probably occurred by the flocculation of the Yukon suspended sediment in the brackish zone upstream of the plunging boundary, where the differential settling from the flocculation is considered to have induced the turbid water intrusion into the bottom layer.

Keywords: Yukon River plume; river discharge; sediment load; flocculation; plunging; underflow

Citation: Chikita, K.A.; Wada, T.; Kudo, I.; Saitoh, S.-I.; Toratani, M. Effects of River Discharge and Sediment Load on Sediment Plume Behaviors in a Coastal Region: The Yukon River, Alaska and the Bering Sea. *Hydrology* **2021**, *8*, 45. https://doi.org/10.3390/hydrology8010045

Academic Editor: Carmelina Costanzo

Received: 17 February 2021
Accepted: 9 March 2021
Published: 12 March 2021

1. Introduction

One of actions of river discharge and sediment load into the ocean includes the sedimentation by energy dispersal in the estuarine area and through subsequent formation and advection of sediment plume in coastal and offshore regions. The sedimentary processes are often accompanied by the flocculation of suspended sediment, which affects nutrient and organic matters' cycles connected to the food chain or ecosystem in the ocean [1]. Behaviors of river-suspended sediment can also be controlled by how the river water mixes with ocean water under shearing, where the flocculation again plays the important role on slow or rapid deposition of nutrient and organisms [2]. Hetland and Hsu [3] proposed a conceptual model of sedimentation in the estuarine, near-field plume, and far-field plume associated with the flocculation. However, there are a few quantitative descriptions of the flocculation effect on sediment dispersion and deposition [2,4]. Meanwhile, dynamic behaviors of the whole river plumes, including buoyant jets or bulges of small scale, have been explored by field observations, satellite image analysis, and numerical simulations, for example, for Mackenzie River plumes [5], Yellow River sediment plumes [6,7], Pearl River plumes [8], Columbia River plumes [9], small-mouth Kelvin number plumes [10], and Amazon River plumes [11,12]. With respect to the ecosystem in coastal regions, there are some remote-sensing studies of river sediment plume dynamics from the spatio-temporal variations of suspended sediment concentration (SSC; mg/L) [13,14]. Dean et al. [15] pointed out that satellite image analyses can be utilized for studies of the Yukon River plume behaviors. By using the river discharge time series from 2002–2014, Pitarch et al. [16]

related satellite-derived SSC of the Tiber River plume to the magnitude of the river discharge. However, the correlation between the discharge and SSC is very low at $R = 0.5$, because there is a lack of continuous data of river sediment load such as those of the river discharge and investigations of sediment depositional processes offshore from the river mouth.

In this study, considering a response of suspended sediment to a salinity change, behaviors of a river sediment plume in a coastal region are investigated by using time series of daily mean river discharge and sediment load in 2006–2010, which are here exemplified by the Yukon River, Alaska, and the Bering Sea.

2. Study Area

Most of the Yukon River drainage basin (area, 8.55×10^5 km^2) belongs to the subarctic region south of the Arctic Circle (66°33′ N) and is occupied by 74.8% forest with discontinuous permafrost and 1.1% glacierized area in Alaska Range, Wrangell Mts., St. Elias Mts., etc. [17–20]. There are three USGS (U.S. Geological Survey) gauging stations along the main channel of the Yukon River. Site PLS is the lowest gauging station at ca. 170 km upstream of sites ALK and site EMK, base villages for the coastal observations (Figure 1). The Yukon River delta is radially formed with branching the Yukon River channel into some smaller river channels. Site EMK is located at ca. 18 km upstream of the delta front. The eastern shelf region of the Bering Sea (2.29×10^6 km^2 in area) is 50–70 m deep on average. The region at a distance of less than 170 km off the delta front is very shallow at less than 30 m in depth, which is due to sedimentation from sediment load of the Yukon River and the surrounding rivers. Behaviors of the Yukon sediment plume and the consequent dispersion of the plume-suspended sediment are controlled by the northern movement of the "Alaskan Coastal Water" [15,21], which is connected to the bottom distribution of clay minerals on the Alaskan-Chukchi margin [22].

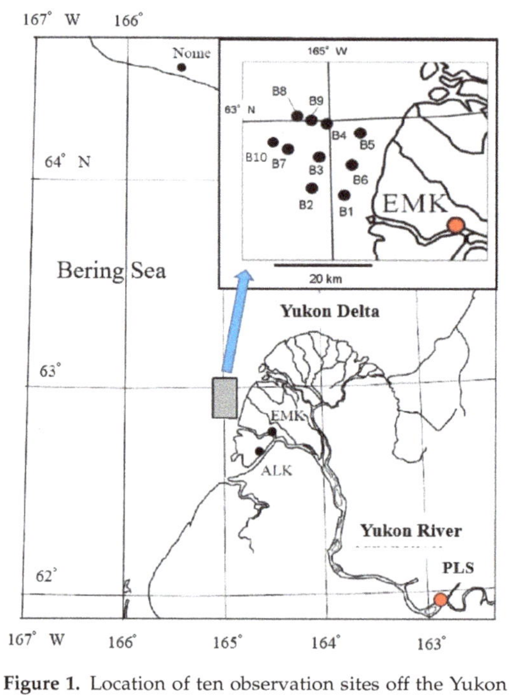

Figure 1. Location of ten observation sites off the Yukon delta front. The label "PLS" shows the location of the lowest USGS (U.S. Geological Survey) gauging station in the Yukon River.

With respect to the discharge and sediment load time series of the Yukon River, Chikita et al. [20] quantified the contribution of runoffs from the glacierized regions or the non-glacial regions to the total Yukon discharge and sediment load by the tank model, one of the runoff models. As a result, it was revealed that the contribution in the 1.1% glacierized area occupies 21.7–37.5%, and that most of the Yukon River suspended sediment originates from the glacial bedrock erosion in the glacierized regions and the resuspension in the river channels.

3. Methods

3.1. Field Observations

In order to obtain time series of suspended sediment load for the Yukon River, water turbidity was monitored at 1 h intervals at site PLS from June 2006–September 2010 by fixing a self-recording turbidimeter of infrared-ray back scattering type (model ATU3-8M, Alec Electronics, Inc.; range, 0–20,000 ppm; accuracy, ±2% of measured value) near the riverbank (Figure 1; Chikita et al. [19]). The turbidimeter had a wiper to keep the sensor window clean, which worked just before every hourly measurement. Thus, the dirt of the sensor window by organisms was prevented. Each of the hourly turbidity values was stored as an average of ten values measured instantaneously at 1 sec interval. The water turbidity (ppm) from the turbidimeter was converted into suspended sediment concentration (SSC: mg/L) by using the relationship between the turbidity, x, and the SSC, y, from the water sampled simultaneously at mid-channel ($y = 5.621x^{0.6762}$; $R^2 = 0.793$, $p < 0.01$) (Figure 2a). The regression curve is close to the one-to-one correspondence between turbidity and SSC. Water temperature was simultaneously measured every hour at site PLS by fixing a temperature logger, TidbiT v2 (Onset computer, Inc., Bourne, MA, USA; accuracy, ±0.2 °C), to the turbidimeter. In order to know the representativeness of SSC and water temperature by the one-point method, during their monitoring, fluctuations of SSC and water temperature across the river channel were assessed by lowering a TTD (turbidity-temperature-depth) profiler from a boat [23]. As a result, differences between recorded SSC and water temperature and cross-sectionally averaged SSC and water temperature were less than 50 mg/L and less than 0.2 °C, respectively, being independent of their magnitude. The small differences between the two SSCs and the two water temperatures are due to the slow settling of suspended sediment (more than 90% silt and clay) and the complete turbulence of riverflow, respectively. Daily mean discharge data at site PLS were downloaded from the National Water Information System on the USGS web site (URL: http://waterdata.usgs.gov/nwis/dv/?site_no=15565447&agency_cd=USGS&referred_module=sw (accessed on 11 February 2021)). The sediment load, L (kg/s), was calculated by multiplying SSC, C (g/L in this calculation), by discharge, Q (m^3/s). The electric conductivity at 25 °C (EC25) and salinity of the Yukon River water were also obtained at site PLS using a portable EC meter (Type CM-14P, TOA Corporation; accuracy, ±0.1 mS/m) for sampled water.

A relation between turbidity, x, and SSC, y, was obtained also for the sediment plume off the delta front. The regression curve was given by $y = 1.597x^{0.8119}$ ($R^2 = 0.932$, $p < 0.01$; Figure 2b), indicating that, in the region of brackish to oceanic waters, SSC is ca. 30% smaller than turbidity. This is probably due to an optical feature of marine water different from that of river water.

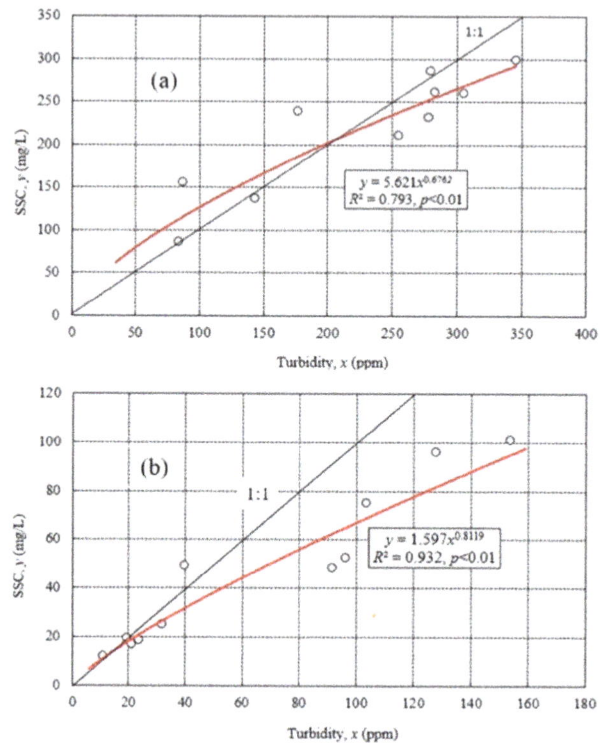

Figure 2. Relations between turbidity (ppm) and SSC (suspended sediment concentration; mg/L) at (**a**) site PLS and (**b**) the coastal region off the Yukon delta front.

At sites B1 to B10 in Figure 1, which were located at less than 30 km from the delta front near site EMK, profiler observations and water and sediment samplings were performed by boat on 6 September 2009 and 22 June 2010. A TCTD (turbidity-conductivity-temperature-depth) profiler (model ASTD687, JFE Advantech, Co., Ltd., Japan) was then utilized for the vertical measurements at 0.1 m pitch. The B5–B8 line and B6–B10 line correspond to two longitudinal lines offshore from the fan-shaped Yukon River delta.

With respect to the offshore extent of the Yukon sediment plume, relatively clear RGB (Red-Green-Blue color model) composite images of MODIS/aqua were downloaded from the NASA oceanic web site (URL: https://oceancolor.gsfc.nasa.gov (accessed on 19 and 24 February 2021)), which were selected on dates closest to those of the coastal observations.

3.2. *Laboratory Experiments*

During the turbidity record, river-suspended sediment and river-channel sediment were sampled at site PLS, and their grain size analyses were conducted by the photo-extinction method with a centrifuge for particles of grain size $d \leq 44$ μm (micrometers) and by the sieving method for $d > 44$ μm particles [19]. The channel sediment was a new one deposited in the glacier-melt and rainfall season of the year. With respect to the flocculation of river-suspended sediment in the coastal region, an effect of salinity on grain size of sediment was explored by suspending the river channel sediment in pure water of 0‰ and NaCl water of 2–35‰. In the grain size analyses, a sediment sample of $d \leq 44$ μm in a quartz cell was repeatedly used with increasing the salinity from 0 to 35‰. Also, in order to know the existence of cohesive sediment such as kaolinite, illite, etc., clay particles of $d \leq 4$ μm in river-suspended and plume-suspended sediments were mineralogically identified by the X-ray diffraction (XRD) method. In the XRD method, a peak at a certain 2θ value (θ: Bragg angle) indicates the existence of one to three minerals specified.

4. Results

4.1. Time Series of Discharge and Sediment Load

Figure 3 shows temporal variations of (a) daily mean water temperature and SSC, and (b) discharge and sediment load at site PLS June 2006–September 2009. In late May or early June, the first large peaks of SSC, discharge, and sediment load in the year were produced by snowmelt runoffs just after the breakup of river ice. Then, water temperature greatly increased at more than 0 °C. Thereafter, glacier-melt and rainfall runoffs resulted in some peaks of SSC and sediment load in July–September. When glacier-melt sediment runoffs prevailed in July–August, water temperature decreased by 1–2 °C after the peak in mid-July. In Figure 3b, it is seen that the sediment load at more than 3000 kg/s occurs for more 1 month as a whole. The water temperature was recorded at less than 0 °C for a while in the ice-covered season. This means that the temperature logger was then enclosed by river ice, due to the ice growth.

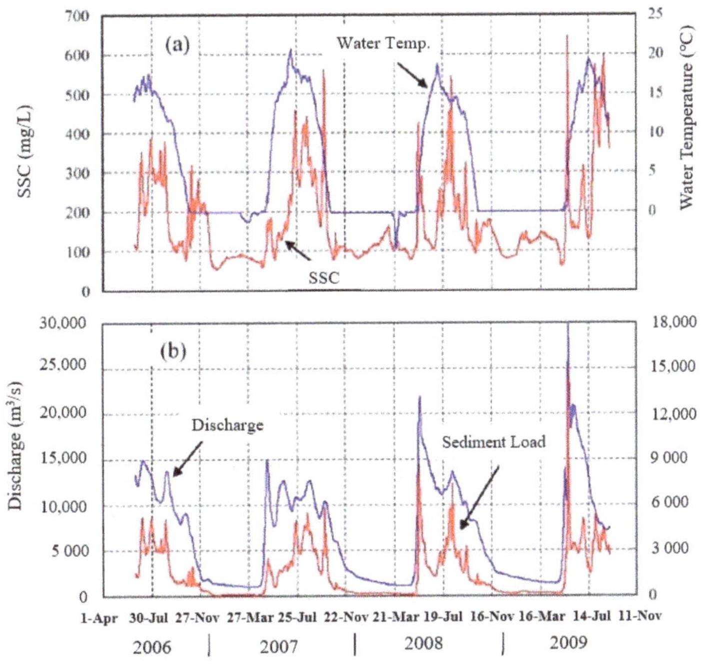

Figure 3. Temporal variations of (**a**) daily mean water temperature and SSC, and (**b**) daily mean discharge and sediment load at site PLS in June 2006–September 2009 (modified after Chikita et al. [19]).

4.2. Observational Results of Plume Behaviors

Figure 4 shows cross-sections of SSC, water temperature, salinity, and water density σ_T from the profiler observations on sites B5 to B8 line obtained on 6 September 2009 (Figure 1). Here, water density σ_T was calculated by $\sigma_T = \rho(S, T, 0)$—1000, where ρ(S, T, P) is water density (kg/m^3) and a function of salinity S (‰), water temperature T (°C), and pressure P (= 0 at 1 atm). Here, the lag time of 2 days (equal to the distance, 170 km, divided by rough riverflow speed, 1 m/s) was considered as an approximate time needed to flow downstream from site PLS to the coastal region. The correspondent, daily mean SSC, discharge, sediment load, and water temperature at site PLS were 391 mg/L, 7.39×10^3 m^3/s, 2.89×10^3 kg/s, and 11.7 °C on 4 September 2009, respectively.

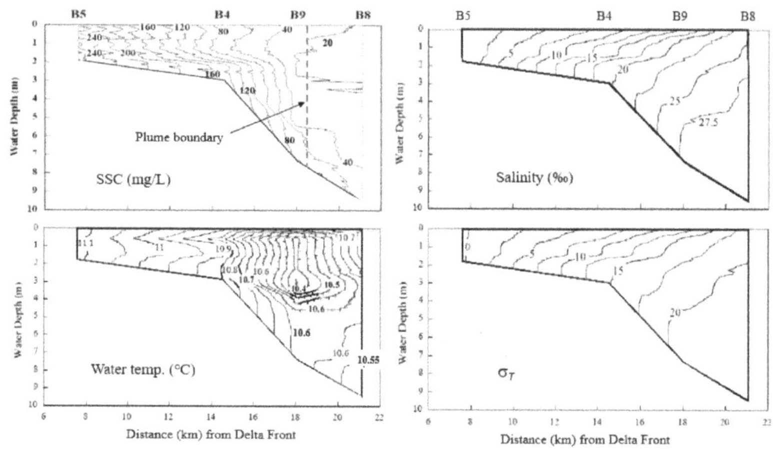

Figure 4. Longitudinal cross-sections of SSC, water temperature, salinity, and water density σ_T by the coastal observation on the B5–B8 line on 6 September 2009 (Figure 1). The dotted line in the SSC cross-section shows a boundary between turbid plume water and relatively clean ocean water observed in situ.

A boundary between the plume water and relatively clean marine water was clearly seen near site B9 as shown by the dotted vertical line in Figure 4. The SSC on the boundary, corresponding to a plunging point, was ca. 30 mg/L. Offshore from the boundary, a bottom layer at 40 mg/L or more SSC was observed probably as a downslope bottom current or underflow. This plunging seems to have been preceded on the gentle slope between sites B5 and B4, since the turbid water at SSC of 160–240 mg/L was accumulated near the bottom as a nepheloid layer. The water temperature distribution indicates that vertical mixing due to the entrainment occurred at the plunging point [24–26]. A boundary between the plume and marine water and the correspondent plunging were also observed between site B3 and site B7 (Figure 5). However, the resultant bottom current appears to have been advanced wholly in the lower layer of more than 27.5‰ salinity, since the SSC cross-section indicates a gradual SSC decrease offshore, accompanied by the vertical uniformity.

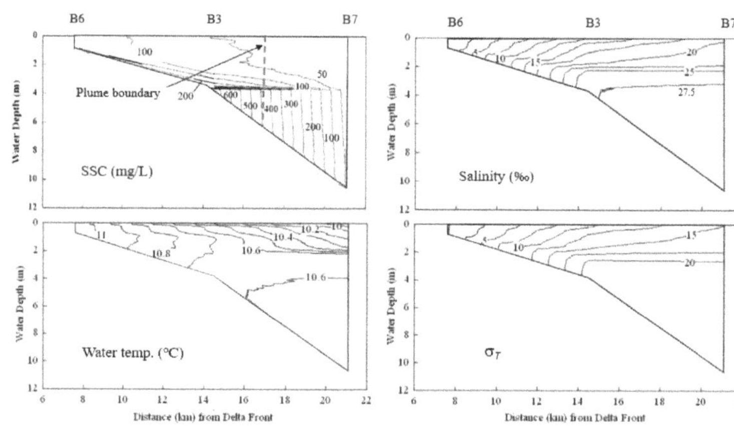

Figure 5. Longitudinal cross-sections of SSC, water temperature, salinity, and water density σ_T by the coastal observation on the B6–B7 line on 6 September 2009 (Figure 1). The dotted line in the SSC cross-section shows a boundary between turbid plume water and relatively clean ocean water observed in situ.

The traverse cross-section along the B1–B5 line (Figure 6) indicates that there is a zone with relatively high SSC of 140–250 mg/L between sites B6 and B5. This zone corresponds to the mixing zone upstream of the plunging points in Figures 4 and 5 [24], which is here accompanied by the vertical uniformity of SSC, water temperature, and salinity. Then, the SSC and water temperature decrease and the salinity increases gradually toward site B6 from site B5. This reflects the vertical mixing by relatively high river water flux at site B5. The longitudinal center of the sediment plume is, thus, located near site B5 rather than site B6 (Figure 1). If the high SSC zone on the B5–B6 line was produced by the resuspension of bottom sediment by tidal currents, wind-driven current, or littoral currents, then a high SSC zone should appear in the relatively shallow B6–B1 zone. Hence, the plunging is considered to have been raised through the mixing responding to the magnitude of the Yukon River sediment load.

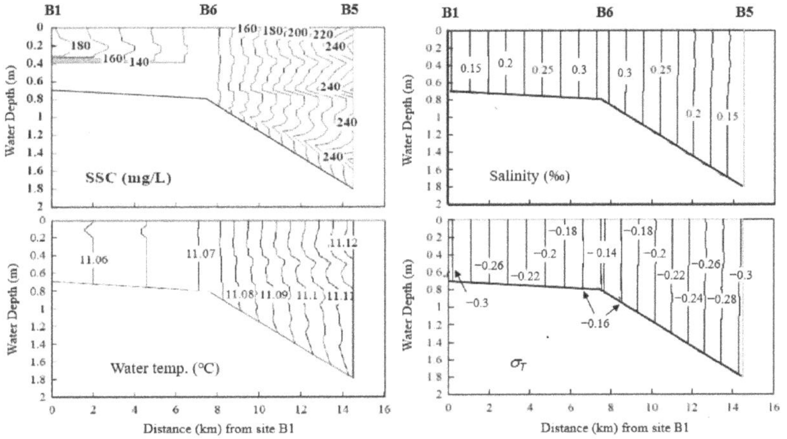

Figure 6. Traverse cross-sections of SSC, water temperature, salinity, and marine water density σ_T by the coastal observation on the B1–B5 line on 6 September 2009 (Figure 1).

Figure 7 shows longitudinal cross-sections of SSC, water temperature, salinity, and σ_T along the B6–B10 line by the coastal observation on 22 June 2010. The daily mean SSC discharge, sediment load, and water temperature at site PLS were 240 mg/L, 9.12 × 10^3 m^3/s, 2.19 × 10^3 kg/s, and 14.6 °C on 20 June 2010, respectively. The discharge was 19% larger than that on 4 September 2009, but sediment load was 24% smaller because SSC was 39% smaller than that on 4 September 2009. Then, a boundary between the plume and marine water was not clearly seen near site B3 or farther. Such a sediment plume's plunging as in Figures 4 and 5 is, thus, judged not to have occurred. In fact, the SSC in the cross-section was relatively very small at 10–60 mg/L. The SSC was then vertically uniform, being probably due to turbulent mixing by relatively strong river-induced current from the higher river discharge.

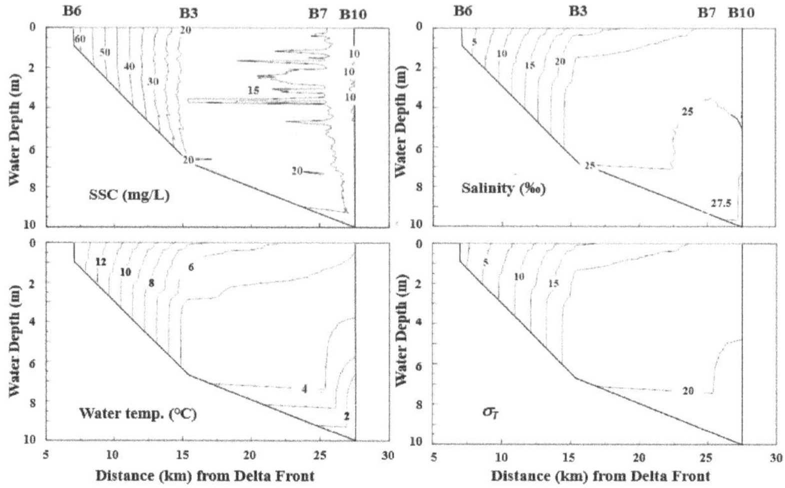

Figure 7. Longitudinal cross-sections of SSC, water temperature, salinity, and σ_T on the B6–B10 line by the coastal observation on 22 June 2010 (Figure 1).

4.3. Experimental Results for Yukon Sediment

Figure 8 shows a salinity effect on (**a**) cumulative grain size distributions of the Yukon channel sediment and (**b**) their mean size and standard deviation. When the salinity increased from 2 to 5‰, the suspended sediment became coarse at grain size d < 16.5 μm with increasing the mean size from 16.6 to 19.0 μm. This means that the differential settling from flocculation or coagulation occurred at more than 2‰ salinity for the Yukon suspended sediment of d < 16.5 μm [27]. At salinity of 10‰ or more, the grain size did not change greatly with mean size of 17.9 to 18.7 μm. The standard deviation decreased from 13.6 μm at 0‰ to 12.0 μm at 2‰, and then varied at a small range of 10.8 to 12.5 μm at 2–35‰. Hence, the addition of salinity to the suspension tends to increase the uniformity in grain size of suspended sediment, reflecting the flocculation. This experiment dealt with the behaviors of suspended sediment in still water. If a certain shearing by a flow was given, the flocculation could be more enhanced [28].

Figure 9 shows the mineralogy of Yukon River and plume suspended sediments at less than 4 μm in gain size by the XRD method. Such clay minerals as kaolinite, chlorite, smectite, and illite could easily produce the flocculation of suspended sediment in the brackish region offshore from the Yukon River mouth. The flocculation of clay particles induces the coarseness of the river-suspended sediment, which more rapidly accumulates the sediment near the bottom as an agent for the plunging. In fact, the plume-suspended sediment includes few clay minerals, suggesting their gravitational settling or deposition offshore from the delta front. Especially, kaolinite, smectite, and illite are more cohesive to easily flocculate.

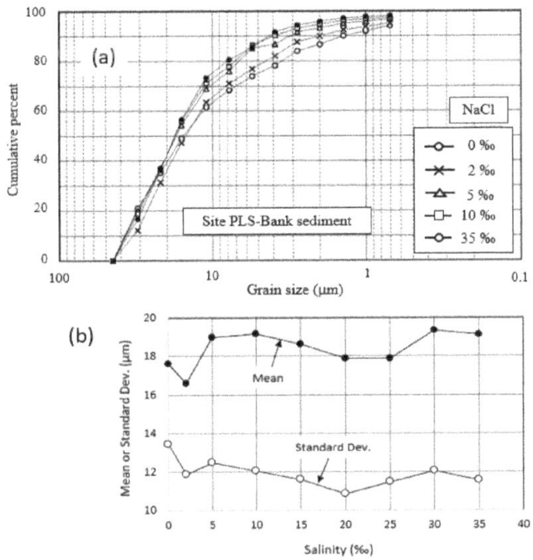

Figure 8. Changes in (**a**) cumulative grain size distributions and (**b**) correspondent mean size and standard deviation of the Yukon River bank sediment at grain size d ≤ 44 μm by increasing the salinity from 0 to 35 ‰.

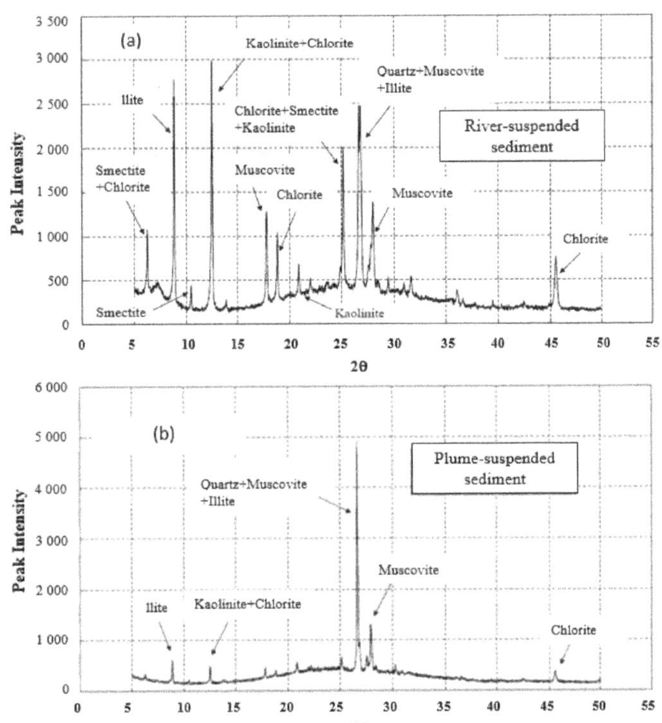

Figure 9. Mineralogy of suspended sediment at grain size d ≤ 4 μm by the X-ray diffraction method. The sediments were sampled at (**a**) site PLS and (**b**) site B9 in September 2009. θ: Bragg angle (°).

5. Discussion

River-induced plumes released from a river mouth are dynamically dispersed by littoral, tidal, and wind-driven currents, and thereby control the ecosystem and sedimentation in coastal and offshore regions. However, the observational results in Figures 4–7 indicate that, in the estuarine and near-field plume regions, the magnitude of river discharge and sediment load could control the level of mixing between river water and ocean water and the subsequent dynamic behaviors of river-induced plume. The Yukon River has only water density $\sigma_T = -0.133$ at SSC of 391 mg/L, water temperature of 11.7 °C, and EC25 of 27 mS/m (salinity, 0.127‰) at site PLS on 4 September 2009. Thus, the plunging into the lower ocean water as in Figures 4 and 5 cannot occur under pycnal condition between the river water and ocean water. Otherwise, the differential settling from the flocculation of suspended sediment probably produced the nepheloid layer near the bottom in the mixing zone, followed by the underflow. The flocculation shown in Figure 8 could be enhanced by the shearing in the mixing zone and the existence of cohesive clay minerals such as kaolinite, smectite, and illite (Figure 9).

The plunging and subsequent underflow in Figures 4 and 5 appear to have been produced on the gentle slope between sites B5 and B4 or sites B6 and B3. By the comparison between Figures 4 and 6, it is seen that the plunging may occur at the Yukon River sediment load of more than ca. 2500 kg/s at site PLS. The sediment load of more than 2500 kg/s can be recorded in July–September (Figure 3), when the suspended sediment contains more fine-grained particles supplied by the glacier-melt and rainfall runoffs [20].

A two-dimensional hydrodynamic condition of the underflow in Figure 4 is judged by using the densimetric Froude number F_d and Reynolds number R_e in the following:

$$F_d = U/(g'H_p)^{1/2}, \tag{1}$$

$$R_e = UH_p/\nu, \tag{2}$$

where U is the underflow's velocity (m/s), H_p is the water depth (m) at the plunging point, $g' = g\,(\rho-\rho_0)/\rho_0$ (g: acceleration due to gravity in m/s^2, ρ: underflow's density in kg/m^3, ρ_0: ambient water density in kg/m^3), and ν is the kinematic viscosity (m^2/s). From the σ_T distribution in Figure 4, ρ = 1025 kg/m^3, ρ_0 = 1020 kg/m^3, H_p= 7.5 m, and $\nu = 1.32\times10^{-6}$ m^2/s for salinity at 25‰ and water temperature at 10.6 °C gave F_d = 0.017–0.17 and $R_e = 5.7 \times 10^4$–5.7×10^5 in Equations (1) and (2). Then, U = 0.01–0.1 m/s.

Similarly, for the underflow downstream of the plume boundary in Figure 5, ρ = 1022 kg/m^3, ρ_0 = 1015 kg/m^3, H_p= 4.5 m, and $\nu = 1.33 \times 10^{-6}$ m^2/s for 27‰ salinity and 10.6 °C water temperature, which gave F_d =0.018–0.18 and $R_e = 3.4 \times 10^4$–3.4×10^5 at U = 0.01–0.1 m/s. Thus, the underflow may be subcritical and turbulent, indicating the low entrainment from the ambient water above. This indicates that, in spite of the small scale, the underflow could flow down in a relatively long travel distance.

When the river discharge increased and the sediment load decreased, the mixing zone was extended offshore as in Figure 7, where the plunging was not observed. As far as seeing the RGB satellite images of the Yukon sediment plume (Figure 10) on the dates closest to those of the coastal observations, the extent of the surface plume appears to be much larger on 4 September 2009.

The Yukon River surface plume could be wholly advected northwestward by the "Alaskan Coastal Water" [21] (Figure 10). In Figures 4 and 5, there was no turbid surface layer in the location of the plunging and subsequent underflow formation. Thus, the liftoff of the sediment plume into the surface layer is considered to have occurred in more northeastern regions, where the surface sediment plume is extended northward (Figure 10a).

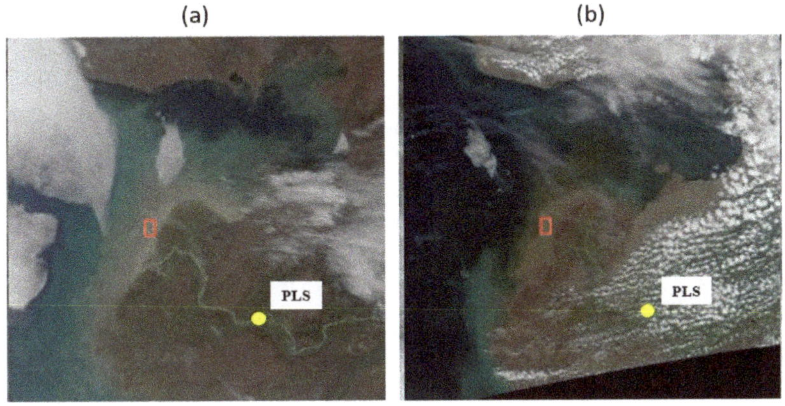

Figure 10. RGB (Red-Green-Blue color model) composite images of MODIS/Aqua on (**a**) 4 September 2009 and (**b**) 22 June 2010. The red rectangles show the area of the coastal observations on 6 September 2009 and 22 June 2010. The label "PLS" shows the location of the USGS gauging station at the Pilot Station village.

6. Conclusions

By the observations of Yukon River sediment plumes in the coastal region of the Bering Sea, it was found out that, at Yukon River sediment load of more than ca. 2500 kg/s, the plume plunges at a boundary of the plume and ocean water. The production of the bottom current or underflow from the plunging is probably due to the differential settling from the flocculation of river-suspended sediment. Both the suspended sediment concentration (SSC) and sediment load of the Yukon River were relatively high in the glacier-melt and rainfall runoffs of July–September. In such runoffs, suspended sediment becomes relatively fine, including clay minerals such as kaolinite, smectite, chlorite, and illite to easily make flocs in the brackish zone. Hence, temporal variations of glacier-melt and rainfall sediment runoffs of the Yukon River could change behaviors of the sediment plume in the coastal region. It was experimentally demonstrated that the flocculation of suspended sediment occurs at 2‰ or more salinity. This flocculation may produce the underflow in the deeper zone by accumulating the flocculated sediment on the onshore gentle slope. It should be needed to experimentally and theoretically clarify the mechanism of the underflow's production from the flocculation.

Author Contributions: K.A.C., T.W. and I.K. participated in all the filed surveys to set and manage field instruments; K.A.C. wrote the paper; S.-I.S. conceived and designed the Yukon River—Bering Sea research project; M.T. analyzed many images of MODIS/Aqua. All authors have read and agreed to the published version of the manuscript.

Funding: This study was financially supported by the Japan Aerospace Exploration Agency (JAXA) as a research in the joint project of IARC/JAXA.

Acknowledgments: We appreciate the official support of Emeritus S. Akasofu, L. Hinzman and Y. Kim, the International Arctic Research Center (IARC), the University of Alaska at Fairbanks (UAF) and the welcome data supply of the U.S. Geological Survey.

Conflicts of Interest: The authors declare no conflict of interest.

References

1. Milligan, T.G.; Hill, P.S.; Law, B.A. Flocculation and the loss of sediment from the Po River plume. *Cont. Shelf Res.* **2007**, *27*, 309–321. [CrossRef]
2. Strom, K.; Keyvani, A. Flocculation in a decaying shear field and its implications for mud removal in near-field river mouth discharges. *J. Geophys. Res. Oceans* **2016**, *121*, 2142–2162. [CrossRef]

3. Hetland, R.; Hsu, T. Freshwater and sediment dispersal in large river plumes. In *Biogeochemical Dynamics at Major River-Coastal Interfaces: Linkages with Global Change*; Bianchi, T., Allison, M., Cai, W., Eds.; Cambridge University Press: Cambridge, UK, 2013; pp. 55–85. [CrossRef]
4. Markussen, T.N.; Elberling, B.; Winter, C.; Andersen, T.J. Flocculated meltwater particles control Arctic land-sea fluxes of labile iron. *Sci. Rep.* **2016**, *6*, 24033. [CrossRef] [PubMed]
5. Milligan, R.P.; Perrie, W.; Solomon, S. Dynamics of the Mackenzie River plume on the inner Beaufort shelf during an open water period in summer. *Estuar. Coast. Shelf Sci.* **2010**, *89*, 214–220. [CrossRef]
6. Qiao, S.; Shi, X.; Zhu, A.; Liu, Y.; Bi, N.; Fang, X.; Yang, G. Distribution and transport of suspended sediments off the Yellow River (Huanghe) mouth and the nearby Bohai Sea. *Estuar. Coast. Shelf Sci.* **2010**, *86*, 337–344. [CrossRef]
7. Wang, H.; Bi, N.; Wang, Y.; Saito, Y.; Yang, Z. Tide-modulated hyperpycnal flows off the Huanghe (Yellow River) mouth, China. *Earth Surf. Process. Landf.* **2010**, *35*, 1315–1329. [CrossRef]
8. Pan, J.; Gu, Y.; Wang, D. Observations and numerical modeling of the Pearl River plume in summer season. *J. Geophys. Res. Oceans* **2014**, *119*, 2480–2500. [CrossRef]
9. Saldıas, G.; Shearman, R.K.; Barth, J.A.; Tufillaro, N. Optics of the offshore Columbia River plume from glider observations and satellite imagery. *J. Geophys. Res. Oceans* **2016**, *121*, 2367–2384. [CrossRef]
10. Cole, K.L.; Hetland, R.D. The effects of rotation and river discharge on net mixing in small-mouth Kelvin number plumes. *J. Phys. Oceanogr.* **2016**, *46*, 1421–1436. [CrossRef]
11. Varona, H.L.; Veleda, D.; Silva, M.; Cintra, M.; Araujo, M. Amazon River plume influence on Western Tropical Atlantic dynamic variability. *Dyn. Atmos. Oceans* **2019**, *85*, 1–15. [CrossRef]
12. Gouveia, N.A.; Gherardi, D.F.M.; Aragão, L.E.O.C. The role of the Amazon River plume on the intensification of the hydrological cycle. *Geophys. Res. Lett.* **2019**, *46*, 12221–12229. [CrossRef]
13. de Oliveira, E.N.; Knoppers, B.A.; Lorenzzetti, J.A.; Medeiros, P.R.P.; Carneiro, M.E.; de Souza, W.F.L. A satellite view of riverine turbidity plumes on the NE-E Brazilian coastal zone. *Braz. J. Oceanogr.* **2012**, *60*, 283–298. [CrossRef]
14. Restrepo, J.D.; Park, E.; Aquino, S.; Latrubesse, E.M. Coral reefs chronically exposed to river sediment plumes in the southwestern Caribbean: Rosario Islands, Colombia. *Sci. Total Environ.* **2016**, *553*, 316–329. [CrossRef]
15. Dean, K.G.; McRoy, C.P.; Ahlnäs, K.; Springer, A. The plume of the Yukon River in relation to the Oceanography of the Bering Sea. *Remote Sens. Environ.* **1989**, *28*, 75–84. [CrossRef]
16. Pitarch, J.; Falcini, F.; Nardin, W.; Brando, V.E.; Di Cicco, A.; Marullo, S. Linking flow-stream variability to grain size distribution of suspended sediment from a satellite-based analysis of the Tiber River plume (Tyrrhenian Sea). *Sci. Rep.* **2016**, *9*, 19729. [CrossRef]
17. Brabets, T.P.; Wang, B.; Meade, R.H. *Environmental and Hydrologic Overview of the Yukon River Basin, Alaska and Canada*; Water-Resources Investigations Report; U. S. Geological Survey: Anchorage, AK, USA, 2000. [CrossRef]
18. Striegl, R.G.; Dornblaser, M.M.; Aiken, G.R.; Wickland, K.P.; Raymond, P.A. Carbon export and cycling by the Yukon, Tanana, and Porcupine Rivers, Alaska, 2001–2005. *Water Resour. Res.* **2007**, *43*, W02411. [CrossRef]
19. Chikita, K.A.; Wada, T.; Kudo, I.; Kim, Y. The intra-annual variability of discharge, sediment load and chemical flux from the monitoring: The Yukon River, Alaska. *J. Water Resour. Prot.* **2012**, *4*, 173–179. [CrossRef]
20. Chikita, K.A.; Wada, T.; Kudo, I. Material-loading processes in the Yukon River basin, Alaska: Observations and modelling. *Low Temp. Sci.* **2016**, *74*, 43–54. [CrossRef]
21. Nelson, H.; Creager, J.S. Displacement of Yukon-derived sediment from Bering Sea to Chukchi Sea during Holocene time. *Geology* **1977**, *5*, 141–146. [CrossRef]
22. Ortiz, J.D.; Polyak, L.; Grebmeier, J.M.; Darby, D.; Eberl, D.D.; Naidu, S.; Nof, D. Provenance of Holocene sediment on the Chukchi-Alaskan margin based on combined diffuse spectral reflectance and quantitative X-ray diffraction analysis. *Glob. Planet. Chang.* **2009**, *68*, 73–84. [CrossRef]
23. Chikita, K.A.; Kemnitz, R.; Kumai, R. Characteristics of sediment discharge in the subarctic Yukon River, Alaska. *Catena* **2002**, *48*, 235–253. [CrossRef]
24. Chikita, K. A field study on turbidity current initiated from spring runoffs. *Water Resour. Res.* **1989**, *25*, 257–271. [CrossRef]
25. Chikita, K. Sedimentation by river-induced turbidity currents: Field measurements and interpretation. *Sedimentology* **1990**, *37*, 891–905. [CrossRef]
26. Knoblauch, H. *Overview of Density Flows and Turbidity Currents*; Water Resources Research Laboratory, U.S. Bureau of Reclamation: Denver, CO, USA, 1999; pp. 1–27.
27. Whitehouse, U.G.; James, L.M.; Debbrecht, J.D. Differential settling tendencies of clay minerals in saline waters. *Clays Clay Miner.* **1958**, *7*, 1–79. [CrossRef]
28. Lick, W.; Huang, H.; Jepsen, R. Flocculation of fine-grained sediments due to differential settling. *J. Geophys. Res. Oceans* **1993**, *98*, 10279–10288. [CrossRef]

hydrology

MDPI

Technical Note

WaterbalANce, a WebApp for Thornthwaite–Mather Water Balance Computation: Comparison of Applications in Two European Watersheds

Elisa Mammoliti [1], Davide Fronzi [2,*], Adriano Mancini [3], Daniela Valigi [1] and Alberto Tazioli [2]

1 Dipartimento di Fisica e Geologia, Università di Perugia, via Faina 4, 06123 Perugia, Italy; e.mammoliti90@gmail.com (E.M.); daniela.valigi@unipg.it (D.V.)
2 Department of Science and Engineering of Matter, Environment and Urban Planning (SIMAU), Marche Poly-Technic University, via Brecce Bianche 12, 60131 Ancona, Italy; a.tazioli@staff.univpm.it
3 Department of Information Engineering (DII), Marche Polytechnic University, via Brecce Bianche 1, 60131 Ancona, Italy; a.mancini@univpm.it
* Correspondence: d.fronzi@pm.univpm.it

Abstract: Nowadays, the balance between incoming precipitation and stream or spring discharge is a challenging aspect in many scientific disciplines related to water management. In this regard, although advances in the methodologies for water balance calculation concerning each component of the water cycle have been achieved, the Thornthwaite–Mather method remains one of the most used, especially for hydrogeological purposes. In fact, in contrast to physical-based models, which require many input parameters, the Thornthwaite–Mather method is a simple, empirical, data-driven procedure in which the error associated with its use is smaller than that associated with the measurement of input data. The disadvantage of this method is that elaboration times can be excessively long if a classical MS Excel file is used for a large amount of data. Although many authors have attempted to automatize the procedure using simple algorithms or graphical user interfaces, some bugs have been detected. For these reasons, we propose a WebApp for monthly water balance calculation, called WaterbalANce. WaterbalANce was written in Python and is driven by a serverless computing approach. Two respective European watersheds are selected and presented to demonstrate the application of this method.

Keywords: water balance; Thornthwaite–Mather method; Python; serverless computing approach; watershed hydrology

Citation: Mammoliti, E.; Fronzi, D.; Mancini, A.; Valigi, D.; Tazioli, A. WaterbalANce, a WebApp for Thornthwaite–Mather Water Balance Computation: Comparison of Applications in Two European Watersheds. *Hydrology* **2021**, *8*, 34. https://doi.org/10.3390/hydrology8010034

Received: 12 January 2021
Accepted: 16 February 2021
Published: 20 February 2021

1. Introduction

The hydrological balance is useful in many disciplines, from agriculture to hydrology and hydraulic engineering or more generally for water management purposes. Water balance refers to the balance between incoming water from precipitation and the outflow of water by evapotranspiration, groundwater recharge, and stream flow [1]. Even if several methods for monthly water balance calculation are available in the literature, the one introduced by Thornthwaite and Mather [2,3] is widely accepted. Although this method is empirical and outdated, it is used in several disciplines, especially in hydrogeology and for teaching purposes. As assessed by some authors [4,5], the Thornthwaite–Mather method performs well in humid regions where the precipitation and air temperature are the only input data. In accordance with other authors [6–10], this method underestimates monthly potential evapotranspiration (PET) under dry and arid climates, because the equation does not consider the saturation vapor deficit of the air. The issue was highlighted by the overestimation of monthly *PET* in the equatorial humid climate conditions of the Amazon [11]. Another important issue concerns its application to complex geological frameworks, where high permeability variations within the watershed may exist—for example, in carbonate and karstic aquifers, feeding streams, and springs or even highly fractured rocks [12,13]. With regard to the computation time for large datasets, many computer programs have been developed to accelerate the computation—for example, WTRBLN [1], EVAP [14], an MS Excel spreadsheet-based approach developed by [15], and the Thornthwaite monthly water-balance program of the United States Geological Survey (USGS) [16]. However, several aspects of these programs should be highlighted. WTRBLN was one of the first methods developed for water balance calculation; it was written in BASIC 3.0, and it is not compatible with modern computer systems. EVAP calculates only the *PET* and the monthly actual evapotranspiration (*AET*), starting with mean monthly temperature and monthly rainfall. The USGS program, on the other hand, is probably one of the most complete programs, as it considers various components of the hydrologic cycle (precipitation, evapotranspiration, and runoff) and calculates the potential evapotranspiration using the Hamon equation [17]. However, some data treatment is necessary in order to load input data and export results. In this paper, an alternative WebApp for monthly water balance calculation, based on the original Thornthwaite–Mather method, is presented. The developed solution relies on a serverless approach, exploiting a large set of cloud-based micro-services. This type of approach enables asynchronous processing (from request to result) using a queue manager that integrates and decouples distributed software components. To provide an example of its application, the basic water balance components were calculated on two small watersheds located in the Northern Apennines (Central Italy) and in Northwestern Slovenia. In similar areas in Europe, many authors [18,19] demonstrated that the error associated to the use of this method is smaller than the one due to the rainfall measurements. In addition, [20] demonstrated that, despite its poor data requirements, the Thornthwaite–Mather approach is a feasible and accurate approach to fill in the gaps of monthly flow series and to estimate monthly flows at ungauged catchments [21]. In the tested areas, the discharge calculated by WaterbalANce was compared to the measured values, and it was continually recorded at the closing section of the watershed by the Servizio Idrometeorologico—ARPA Emilia Romagna Region and the Slovenian Environmental Agency. Regarding the Italian catchment, named Reno at Pracchia, two different time series of temperature and precipitation were selected: the first covered January 1971 until December 1976 and the second one covered January 2008 until December 2018. For the Slovenian catchment, named Savica at Ukanc, the method was applied to a time series covering January 2016 to December 2019. WaterbalANce may be used as a research tool, for water assessment and classroom instruction, and it is freely available for non-commercial and academic purposes at https://thornwaterbalance.com/, accessed on 12 January 2021.

2. Materials and Methods

2.1. Outline of the Method

The WebApp is based on the Thornthwaite–Mather method, an empirical method used to estimate monthly water balance and originally presented by Thornthwaite and related itself [3,22–24]. It requires the following input parameters: the latitude of the study area (*LAT*, decimal degrees), the mean monthly temperature (*Tm*, degrees Celsius), and the monthly total precipitation (*P*, millimeters). By adding the soil moisture storage capacity (*SM*, millimeters), the rainfall snowfall temperature threshold (*SRT*, degree Celsius), and runoff factor (*beta*, percentage), it is possible to calculate the watershed runoff. Beginning with temperature and latitude, the potential evapotranspiration (*PET*, millimeters) is calculated as follows Equations (1)–(3):

$$PET = 16 \times k \times \left(\frac{10 \times Tm}{I}\right)^{a} \quad (1)$$

$$I = \sum_{n=1}^{12} i \text{ where } i = \left(\frac{Tm}{5}\right)^{1.514} \quad (2)$$

$$a = \left(675 \times 10^{-9} \times I^{3}\right) - \left(771 \times 10^{-7} \times I^{2}\right) + \left(1792 \times 10^{-5} \times I\right) + 0.49239 \quad (3)$$

where k is the latitude-dependent correction factor accounting the number of days in the month and the actual number of hours of insolation; I is the annual heat index; a is a coefficient strictly proportional to I. For the computation of the actual evapotranspiration (*AET*, millimeters), the soil moisture storage capacity must be selected in reason of the soil and crop type [1,3]. After that, according to the difference between monthly precipitation and *PET* (millimeters) the soil moisture content changes to satisfy the water request from the soil system. The water surplus is generated only after the soil moisture conditions are satisfied and the temperature is above a rainfall–snowfall threshold (generally temperature >-1 °C) [25]. The runoff factor (*beta*) is the percentage of the monthly water surplus implied in the generation of the basin runoff, and it is usually settled equal to the 50% of the monthly water surplus [24]. In fact, following the classical rational method, it depends on surface cover and it is a constant value, strictly related to the percentage of impermeable catchment area [26–28]. Other authors have demonstrated a seasonal variability of the *beta* parameter over different climatic regions [29]. Although the setting of the runoff coefficient and soil moisture initial conditions is extremely important for predicting catchment response, this work aims to merely present an online application, which speeds up a widely accepted routine, and it does not consider the calibration of such parameters. The issue related to the snow accumulation and its melt rate is avoided through a simple mass balance. In fact, the snowmelt is generally calculated through energy balance methods or physical-based approaches, which require the availability of several meteorological and hydrological input parameters, which are usually difficult to acquire [30]. The steps involved in the calculation are summarized in the workflow presented in Figure 1. Using this method, it is possible to calculate the components of the water cycle according to the Thornthwaite–Mather method and, in addition, the monthly snowmelt runoff (*SMRO*, millimeters) and the monthly total runoff (*TOT RO*, millimeters).

2.2. Exporting Algorithm to the Cloud

Today, the serverless computing approach is transforming attitudes towards computing. In particular, the serverless approach is widely used in common applications but is attracting the attention of scientists owing to its capability of allocating resources in real time according to the number of requests/users, as reported in [31,32]. Several service providers offer function-as-a-service (FaaS) solutions that could be used also for scientific computation. A set of functions could be easily exposed to the cloud owing to the availability of different runtime environments, such as Python, NodeJS, .NET core, and Java. In particular, the use of Python by scientists is increasing, and this is demonstrated by the wide availability of code and notebooks written in Python. The ability to convert programmed functions or methods into ready-to-use cloud services is leading to a seemingly serverless development and deployment experience for application software engineers. Without the necessity of allocating resources beforehand, the prototyping of new features and workflows is becoming faster and more convenient for application service providers. These advantages have served to boost the industry trend consequently called serverless computing. The more precise, almost overlapping term in accordance with everything-as a-service (XaaS) cloud computing taxonomies is function-as-a-service (FaaS).

The first implementation of the algorithm was performed in MATLAB. We decided to open up the application to external users, and the export of MATLAB code directly to the cloud is not straightforward. For this reason, we decided to port the code into the Python language using libraries such as numpy, matplotlib, and pandas. After the porting, we designed a cloud architecture based on the serverless approach to reduce the complexity as much as possible in order to maintain and scale the computing. The development of a WebApp simplifies the interaction with users, who are focused on the analysis of data without installing additional software modules. Figure 2 shows the adopted architecture that relies on cloud-based services. In particular, the user provides the required information through the front-end (see Figure 3): (a) *LAT*; (b) *SM*; (c) *beta*; (d) *P*; (e) *Tm*.

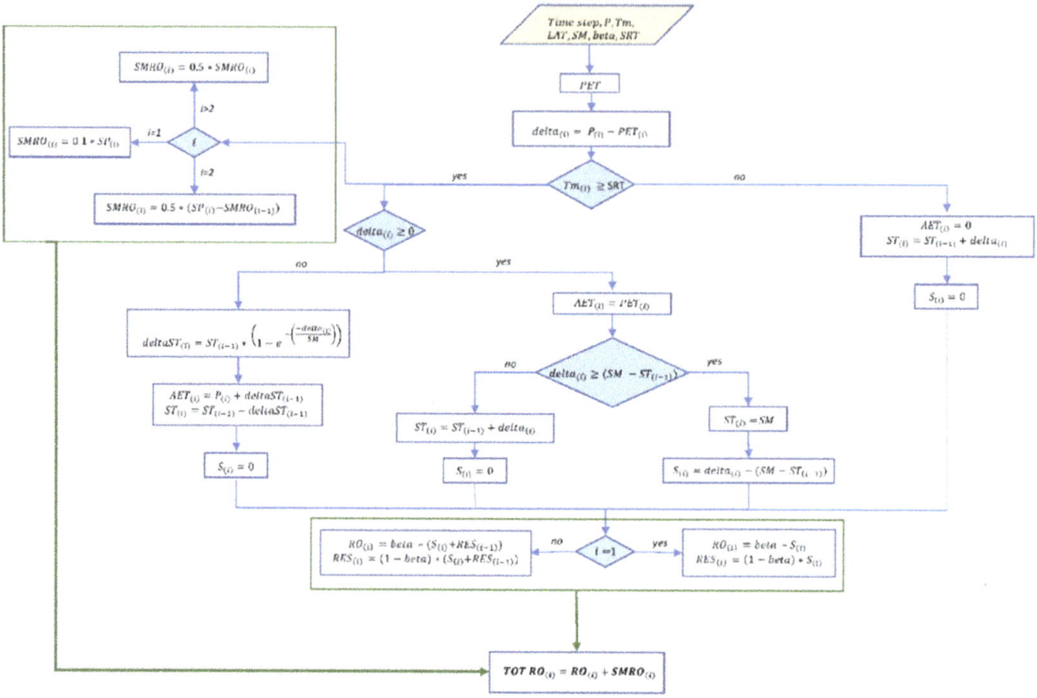

Figure 1. Workflow of the WaterbalANce code. *P* = monthly precipitation (mm); *Tm* = mean monthly temperature (°C); *LAT* = latitude (°); *SM* = soil moisture storage capacity value (mm); *beta* = dimensionless runoff coefficient (%); *SRT* = snowfall rainfall threshold (°C); *PET* = monthly potential evapotranspiration (mm); *delta* = *P-PET* (mm); *AET* = monthly actual evapotranspiration (mm); *ST* = monthly soil moisture (mm); *S* = monthly water surplus (mm); *RO* = monthly runoff (mm); $RES(i-1)$ = dynamic water stored in the basin in the previous month (mm); *RES* = dynamic water storage available for the next month (mm); *SMRO* = monthly snow melt runoff (mm); *TOT RO* = monthly total runoff (mm).

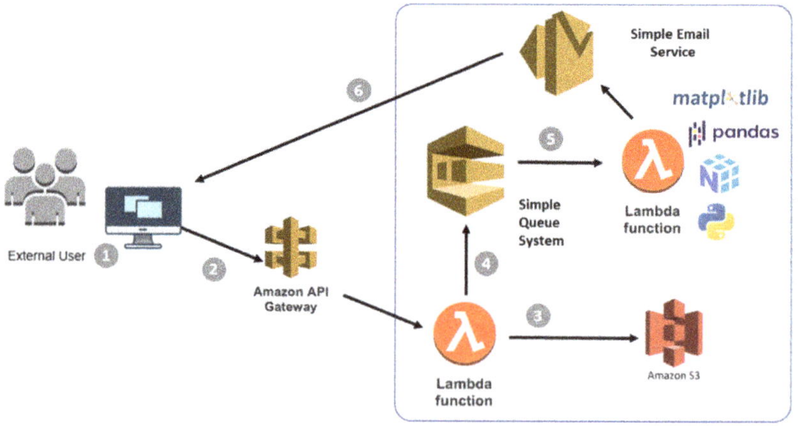

Figure 2. Graphical representation of the adopted architecture to manage the WaterbalANce application. The architecture is based on the interconnection of lambda functions and cloud services to perform the computation.

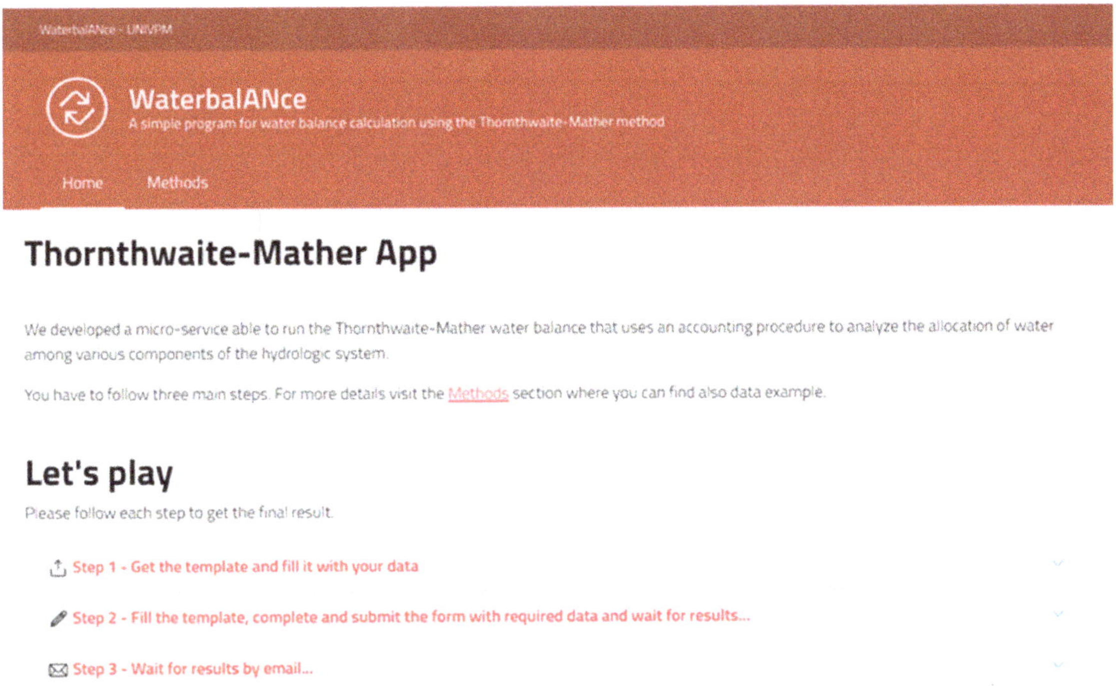

Figure 3. Developed front-end using a standard (adopted for Italian application) to interact with the end-user to specify the parameters and upload the data [https://italia.github.io/bootstrap-italia/ accessed on 12 January 2021].

Data are uploaded according to a template that we provide to the end-user. In this phase, the application interacts with API gateway that triggers the lambda function. The lambda function stores the file in a storage service (S3). After the file is stored, the algorithm is executed using a custom layer on the lambda function to include the required dependencies mentioned above. At the end of execution, the final report is stored in the cloud storage, and using a simple notification system service, a message is sent to the end-user, allowing them to retrieve the results in a format such as an Excel spreadsheet file and a graphical plot (Figure 4). In this way, it is possible to scale the processing of multiple requests. We decided to send the results by using an e-mail considering that the user wishes to start the processing in an asynchronous way, without actively waiting. We decided to use the lambda function as limits in terms of resources (e.g., memory) for our application were expected. An alternative approach in the case of large datasets could rely on an on-demand node that could be activated and managed by Kubernetes as a task, but for our scenario, this configuration was not applicable. We also released the source code under the GNU GPLv3 license as a public git-hub repository (public repository used to share the code developed in the context of this work can be accessed via the following URL: https://github.com/vrai-group/thornwaterbalance accessed on 12 January 2021) for users who wish to integrate the developed algorithm into their pipeline.

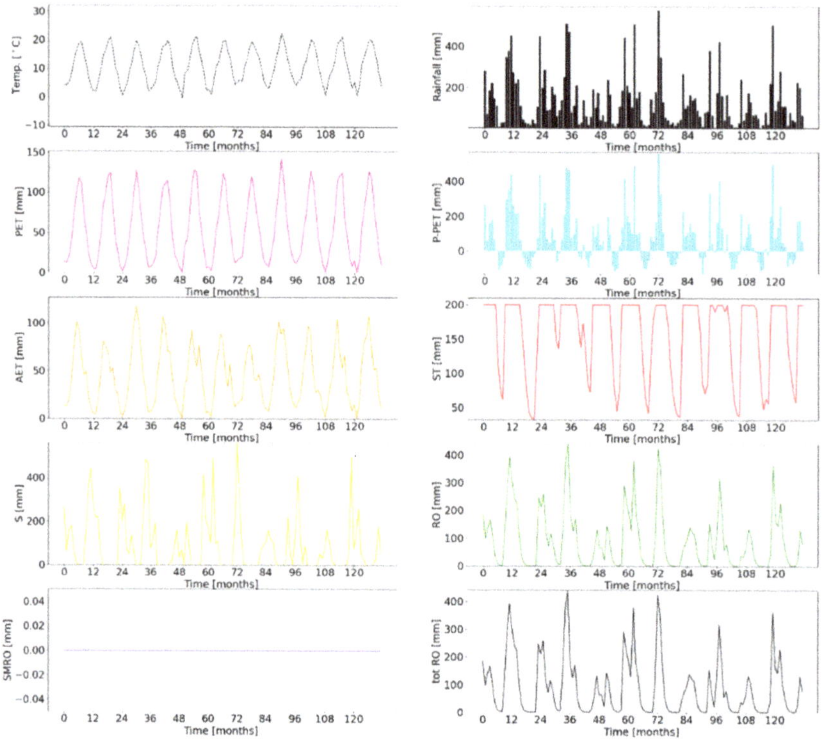

Figure 4. Example of the output plots. Months are plotted against the inputs (*P*, *Tm*) and output variables of the program.

2.3. The Practical Case of the Reno at Pracchia Watershed (Northern Apennines, Central Italy)

The first application of the model was performed on a watershed with a drainage are of roughly 39.8 km^2, located in the highest region of the Reno river (Northern Apennines, Italy), upstream of Pracchia village (610 m a.s.l.). Its maximum elevation is approximately 1640 m a.s.l., while the basin's mean elevation is around 890 m a.s.l. From a geological point of view, the catchment is characterized by the lithologies of the Tuscan Nappe and Cervarola Unit (Figure 5). The Tuscan Nappe Unit crops out extensively in the Northern Apennines and comprises a calcareous to shaly succession (Triassic–Oligocene) and, at the top the Macigno Fm. (Upper Oligocene–Lower Miocene), a thick arenaceous turbidite succession that is late Oligocene–Early Miocene in age. The Cervarola Unit (Lower–Middle Miocene) covers broad areas of the Northern Apennines and is mainly formed by a thick arenaceous turbidite succession. The permeability of these lithologies is mainly driven by widespread tectonic deformation, demonstrated by extensive fracturing [33]. Consequently, the hydrogeological framework consists of small springs, with a local groundwater recharge system emerging from the passage between the sandstone and marly and clay lithologies [34]. The area is characterized by uniform morphometric setting, and due to its small size and the scarce anthropization around the riverbed, it is well suited to this type of analysis. The watershed is equipped for total rainfall and temperature and river discharge measurements, and the devices are owned by Servizio Idrometeorologico—ARPA Emilia Romagna Region; the daily and the mean monthly temperature, precipitation, and river discharge have been monitored and published in the Annali Idrologici reports (Part I, Part II) since 1968. Two different, continuous discharge datasets for the Apennines watershed were selected: the first covered January 1971 to December 1976 (Series I) and the second covered January 2008 to December 2018 (Series II). A descriptive statistical analysis was conducted (Table 1).

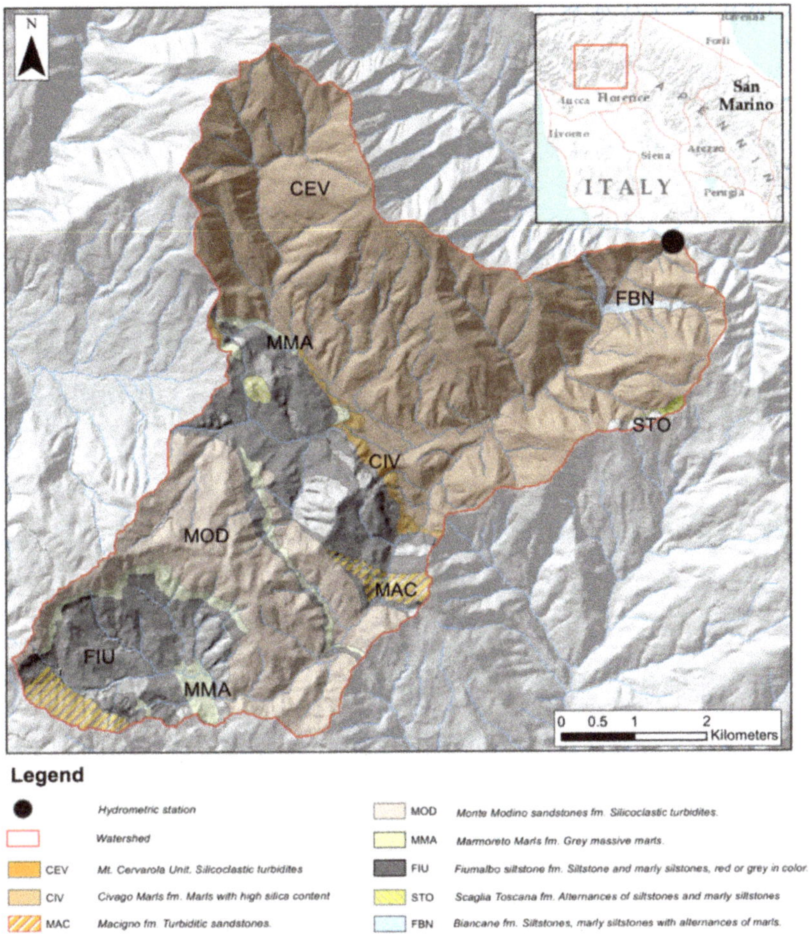

Figure 5. Geological sketch map of the Reno at Pracchia watershed.

Table 1. Descriptive statistics and comparison of modelled vs. measured discharge basic statistical parameters (Series I and Series II). Values refer to mean monthly calculation.

	Series I (January 1971–December 1976 Period)					Series II (January 2008–December 2018 Period)				
	Mean	**Min**	**Max**	**Std**	**Median**	**Mean**	**Min**	**Max**	**Std**	**Median**
Modelled discharge (m^3/s)	1.572	0.004	5.945	1.285	1.621	1.389	0.000	6.550	1.633	0.827
Measured discharge (m^3/s)	1.591	0.108	7.145	1.362	1.345	1.599	0.050	8.610	1.771	0.845
Tm (°C)	9.9	0.9	22.0	6.0	8.85	10.9	−0.5	22.3	6.2	10.7
P (mm)	155.4	27.9	496.4	94.5	133.7	136.4	4.6	576.8	127.8	96.85

To run the model and to provide an example to the WebApp stakeholders (presented in the "Methods" section of the WebApp), the *LAT*, *SM*, *SRT*, and *beta* parameters were set based on expert knowledge of the geographical, geological, and climatic features of the area [35]. Respectively, *LAT* = 43 degrees, the *SM* = 200 mm, *SRT* = −1 °C and *beta* = 70%. Starting with *TOT RO* (mm) obtained as output, the modelled discharge (m^3/s) was calculated by multiplying the latter with the drainage area (39.8 km^2).

2.4. The Practical Case of the Savica at Ukanc Watershed (Northwestern Slovenia)

The second application of the WebApp was performed on the highest region of the Savica river (Northwestern Slovenia, Figure 6).

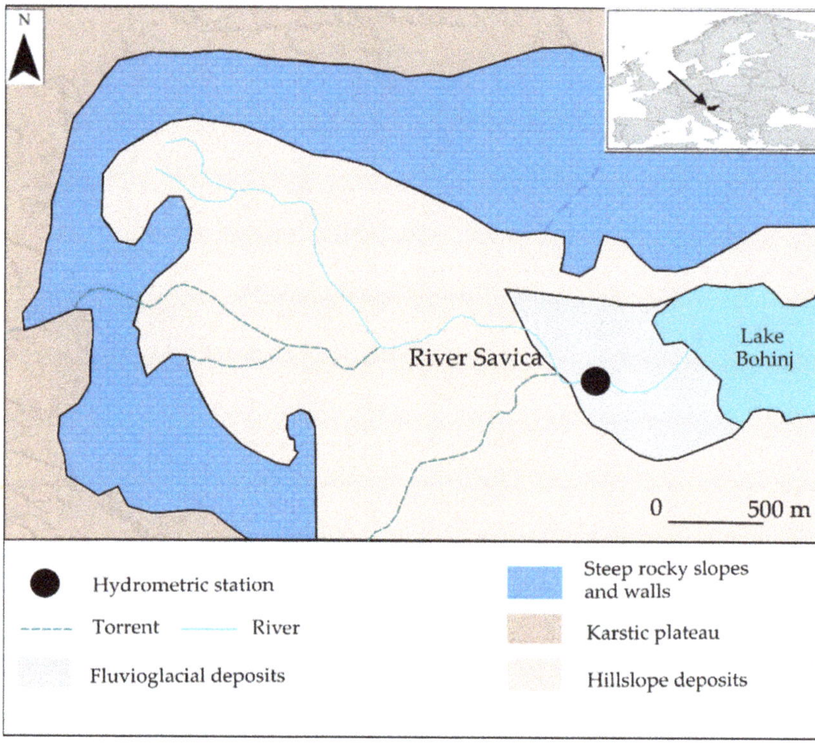

Figure 6. Geological sketch map of Savica at Ukanc watershed (modified from [36]).

The drainage area is roughly 67 km^2, and it extends from 520 to 2800 m a.s.l. [37]. The river provides the main recharge of Bohinj Lake, the largest natural lake in Slovenia, and it is one of the two main sources of the River Sava, which constitutes the main trans-boundary river basin in the West Balkans and is part of the Danube basin [36]. From a geological point of view, the catchment is predominantly Dachstein limestone of Upper Triassic age, subordinated by a small number of dolomite beds [38]. In accordance with many other authors [39–41], the main river recharge area is positioned on the high karstified mountainous plateau that extends into the rugged high mountain chain. Data on river discharge are available only at the gauging station, Savica Ukanc (Elev: 528.83 m a.s.l), which is positioned 720 m before the confluence of the river with Bohinj Lake. Rainfall measurements are provided by the Slovenian Environmental Agency, and the rainfall station gauge available in the watershed is Bohinjska Cesnjica. *LAT*, *SM*, *SRT*, and *beta* parameters were set, respectively, to: *LAT* = 46 degrees, the *SM* = 100 mm, *SRT* = −1 °C, and *beta* = 50%. Starting with *TOT RO* (mm) obtained as output, the modelled discharge (m^3/s) was calculated by multiplying the latter with the drainage area.

We emphasize that in both applications, *SM* and *beta* are not calibrated. Modelled discharge values were compared to the measured values by using Pearson's coefficient of determination (R^2), a standard means of measuring the error of a model in predicting quantitative data. The application of the model to the Reno at Pracchia watershed was also validated in a mean annual hydrologic year. This option can be set directly through the graphical user interface (Figure 3). Moreover, to identify if the discrepancy between measured and modelled discharge is linked to the peculiar meteoclimatic conditions, the difference between measured and modelled discharge data was plotted against the mean monthly precipitation (Figure 9).

3. Results and Discussion

3.1. Reno at Pracchia Watershed

For each time series, mean monthly modelled discharge values with an associated error of 5% were compared to the observed values (Figure 7).

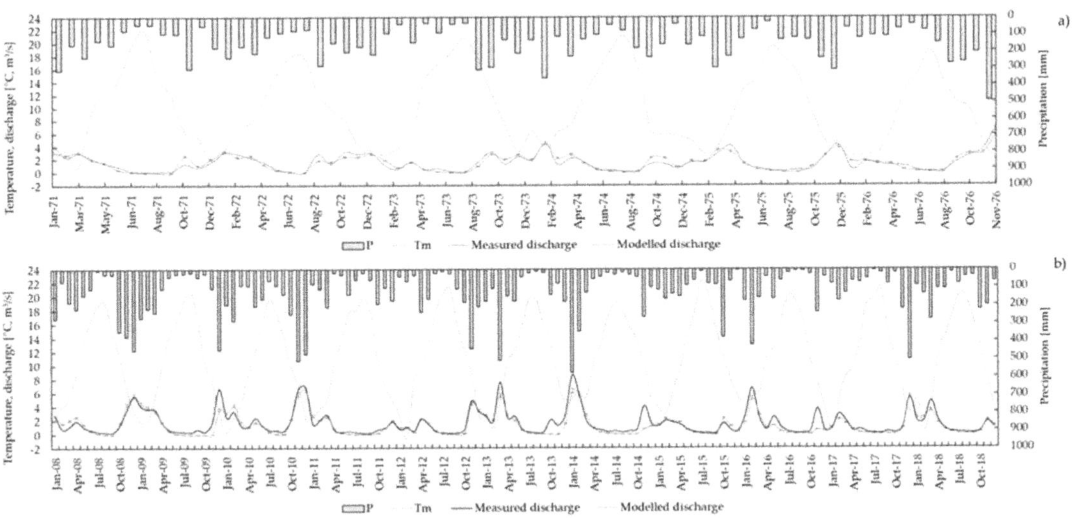

Figure 7. Comparison of mean monthly measured and modelled discharge, monthly precipitation, and mean monthly temperature data for time series I (**a**) and II (**b**).

In Figure 8a,b, mean monthly measured discharge data are plotted against the modelled data. The correlation is slightly stronger for Series I ($R^2 = 0.87$) with respect to Series II ($R^2 = 0.83$). This result is also evidenced by Figure 9a, for which the difference between measured and modelled discharge is close to zero for all the observed periods. Regarding Series II, the larger differences between measured and modelled data were collected during the rainiest months (e.g., November 2009, 2014, 2016, and December 2013). In all these periods, we observe a positive difference of around 2 cubic meters, always suggesting a higher measured discharge with respect to the modelled value (Figure 9b). This finding could be connected to extreme rainfall events that occurred after prolonged dry periods.

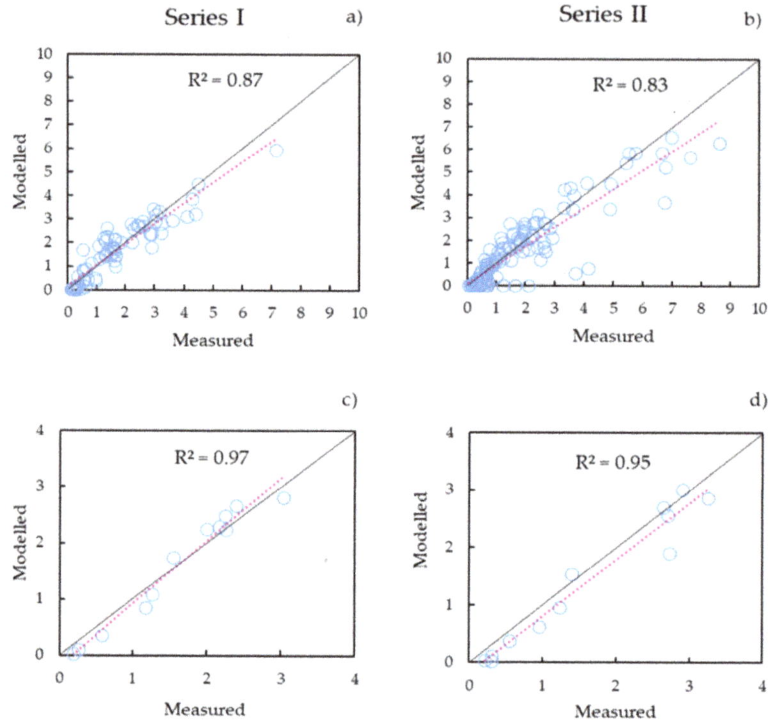

Figure 8. Measured vs. modelled discharge data (m^3/s) for each time series, Reno at Pracchia watershed. (**a**,**b**) all datasets for Series I and Series II, respectively; (**c**,**d**) mean annual hydrologic year for Series I and II. The dashed line represents the linear data interpolation, while the black solid line is the reference line (1:1).

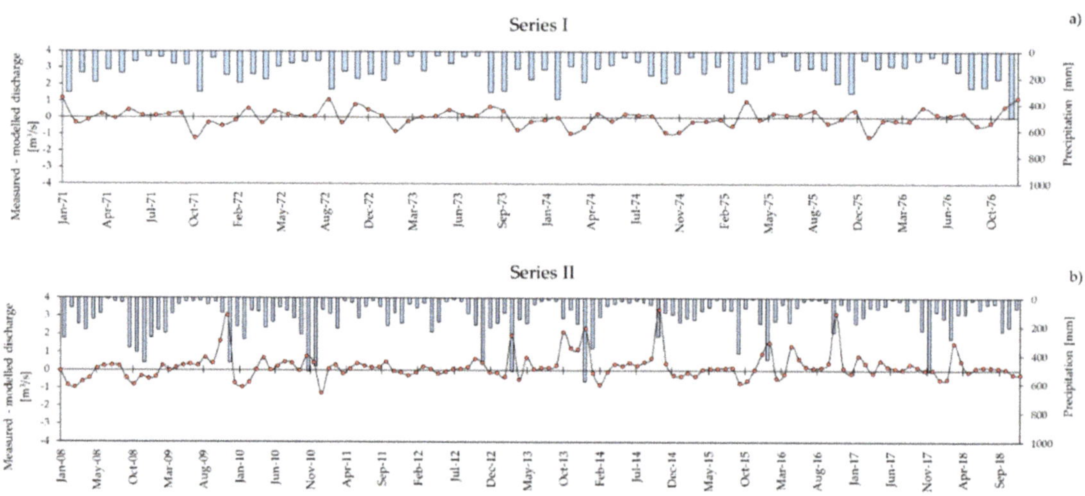

Figure 9. Measured minus modelled discharge and precipitation for time series I (**a**) and II (**b**).

Regarding the correlation of discharge data calculated in a mean hydrological year (Figure 8c,d), the Series I again demonstrates a slightly stronger correlation ($R^2 = 0.97$) with respect to Series II ($R^2 = 0.95$), both stronger than the monthly results.

The choice of *SM* and *beta* parameters, although not calibrated, have led to good simulation of the flow rates of the Reno at Pracchia watershed.

Table 1 shows the basic statistical results for the analyzed time series, with particular attention given to the model's results and the hydrological conditions during the monitoring period. As supported by the Pearson's coefficient of determination (Figure 7), the measured and modelled statistical parameters are extremely similar to one another. For instance, the mean discharge values for both the time series are quite similar (1.572 m^3/s vs. 1.591 m^3/s in Series I and 1.389 m^3/s vs. 1.599 m^3/s in Series II), also confirmed by the median values.

3.2. Savica at Ukanc Watershed

The results for the Savica at Ukanc watershed, presented in Figure 10, highlight the good performance of the Thornthwaite–Mather method during the winter and autumn seasons, while a discrepancy between modelled and measured discharge was observed between April and June.

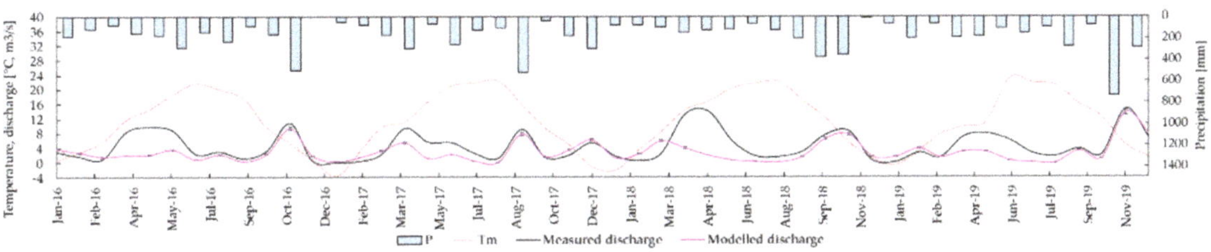

Figure 10. Comparison of mean monthly measured and modelled discharge, monthly precipitation, and mean monthly temperature data for Savica at Ukanc.

Descriptive statistics of the temperature and precipitation data used as input in the model, coupled with basic statistics of mean monthly modelled discharge vs. measured discharge, are shown in Table 2.

Table 2. Descriptive statistics of meteorological data and comparison of modelled vs. measured discharge basic statistical parameters. Values refer to mean monthly calculation.

	Mean	Min	Max	Std	Median
Modelled discharge (m^3/s)	2.957	0.023	13.683	2.814	2.155
Measured discharge (m^3/s)	4.544	0.098	14.610	3.886	2.876
Tm (°C)	11.2	−3.5	23.1	7.8	11.2
P (mm)	179.4	0.0	739.0	140.4	150

Considering the correlation between the mean monthly measured and modelled discharge data of the whole dataset from January 2016 to December 2019 (Figure 11a), Pearson's correlation coefficient is equal to 0.39. If the period between April and June for each year (red crosses in Figure 11b) is not considered, R^2 is equal to 0.83. This behavior can be explained by the underestimation of the precipitation that falls in solid form using the rain gauge available in the watershed. This discharge regime is classified in accordance with [42], as predominately dominated by snow melting; then, the rainfall's contribution to the water balance is subordinated.

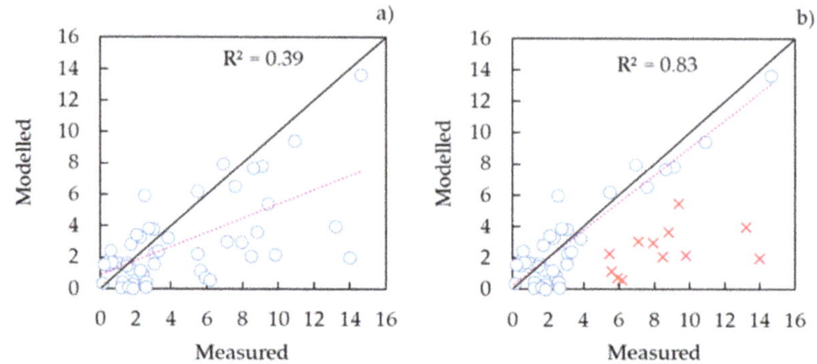

Figure 11. Mean monthly measured vs. modelled discharge data (m^3/s) for Savica at Ukanc watershed. (**a**) All datasets; (**b**) discarding the months between April and June for each year (red crosses). The dashed line represents the linear data interpolation, while the black solid line is the reference line (1:1).

Regarding the applicability of the Thornthwaite–Mather method, although some limitations to its use are reported in the literature. For instance, the method is usually applied where the geological site is characterized by the presence of soil cover [43], but in a few cases, it has been used in soil-less environments such as karst limestone aquifers [44,45].

In our study, the discharge data for the Reno at Pracchia watershed are quite well simulated (particularly in Series I) using the Thornthwaite–Mather method. In contrast, the Savica at Ukanc watershed, which is affected by the underestimation of snowfall and snow melting, is not a suitable area for which to simulate river discharge in the period between April and June.

4. Conclusions

The implementation of a WebApp for automatic water balance calculation allows the users to more efficiently carry out computation for large datasets. End-users will be provided with a template where data can be easily uploaded. The use of a serverless approach also represents a new means of processing data in case of scientific use. Data (input and output) are managed through several cloud services decoupling each component, also enabling automatic scalability. An Excel spreadsheet file and a graphical plot are sent to the users by e-mail, allowing the process to be run several times using different input parameters. Thanks to this automatic procedure for water balance calculation, it is possible to more quickly test the applicability of the Thornthwaite–Mather method in different geological and hydrogeological contexts and to discuss the important issue, which deserves to be studied in greater depth. In fact, in contrast to the classic MS Excel spreadsheet file, where time is proportional to the amount of data, the use of a WebApp permits the execution of the Thornthwaite–Mather method in a few minutes, regardless of the amount of input data. Moreover, no additional software modules need to be installed, and this simplifies the interaction with the final user, who can focus on the analysis of data. The availability of the source code as a public git-hub repository is useful for users who wish to integrate the developed algorithm into their pipeline. Future improvements of the WebApp may involve consideration of fraction of monthly precipitation that becomes snow, the function of the temperature, geographic location, elevation, and the aspect, as suggested by [46], in addition to testing the method in different contexts. For example, particular attention should be paid to the infiltration process in the soil or even in soil-less situations to improve the calibration of input parameters. The soil moisture storage capacity (*SM*) and the *beta* parameters, which influence the model discharge performance, should be improved by an automatic calibration procedure in order to minimize the associated mean absolute error between modelled and measured discharge. Once the calibration is performed on a basin with certain physical characteristics (e.g., soil and crop type, land use, morphometric characteristics, and lithology), the parameters obtained can aid in the simulation of the discharges of ungauged watersheds for water management.

Author Contributions: All authors made a significant contribution to the final version of the manuscript. Conceptualization, E.M. and D.F.; methodology, E.M., D.F., and D.V.; software, E.M. and A.M.; validation, E.M., D.F., D.V., and A.T.; writing—original draft preparation, E.M., D.F., and A.M.; writing—review and editing, all authors; supervision, A.T. and D.V.; project administration, A.T.

Funding: This research received no external funding.

Data Availability Statement: The data of Reno at Pracchia watershed presented in this study are available in the article. Publicly available datasets of Savica at Ukanc were analyzed in this study. This data can be found here: [https://nucleus.iaea.org/wiser, https://www.protezionecivile.fvg.it/it/rete-idrometeorologica] (accessed on 12 January 2021).

Acknowledgments: This work was supported by Progetto Strategico di Ateneo 2017, named "The network of the Botanical Gardens of Ancona", Dipartimento di Scienze Agrarie, Alimentari e Ambientali, Università Politecnica delle Marche, Italy.

Conflicts of Interest: The authors declare no conflict of interest.

References

1. Donker, N.H.W. WTRBLN: A computer program to calculate water balance. *Comput. Geosci.* **1987**, *13*, 95–122. [CrossRef]
2. Thornthwaite, C.W.; Mather, J.R. *The Water Balance*; Laboratory in Climatology, Johns Hopkins University: Baltimore, MD, USA, 1955; Volume 8, pp. 1–104.
3. Thornthwaite, C.W.; Mather, J.R. *Instructions and Tables for Computing Potential Evapotranspiration and the Water Balance*; Laboratory in Climatology, Johns Hopkins University: Baltimore, MD, USA, 1957; Volume 10, pp. 181–311.
4. Sepaskhah, A.R.; Razzaghi, F. Evaluation of the adjusted Thornthwaite and Hargreaves-Samani methods for estimation of daily evapotranspiration in a semi-arid region of Iran. *Arch. Agron. Soil Sci.* **2009**, *55*, 51–66. [CrossRef]
5. Bonacci, O.; Andrić, I. Karst spring catchment: An example from Dinaric karst. *Environ. Earth Sci.* **2015**, *74*, 6211–6223. [CrossRef]
6. Pelton, W.L.; King, K.M.; Tanner, C.B. An Evaluation of the Thornthwaite and Mean Temperature Methods for Determining Potential Evapotranspiration 1. *Agron. J.* **1960**, *52*, 387–395. [CrossRef]
7. Pruitt, W.O. Evapotranspiration: A guide to irrigation. *Calif. Turfgrass Cult.* **1964**, *14*, 27–32.
8. Doorenbos, J.; Pruitt, W.O. *Crop Water Requirements*; FAO Irrigation and Drainage Paper No. 24; FAO: Rome, Italy, 1977; pp. 34–37.
9. Hashemi, F.; Habibian, M.T. Limitations of temperature-based methods in estimating crop evapotranspiration in arid-zone agricultural development projects. *Agric. Meteorol.* **1979**, *20*, 237–247. [CrossRef]
10. Malek, E. Comparison of alternative methods for estimating ETp and evaluation of advection in the Bajgah area, Iran. *Agric. For. Meteorol.* **1987**, *39*, 185–192. [CrossRef]
11. Camargo, A.P.; Marin, F.R.; Sentelhas, P.C.; Picini, A.G. Adjust of the Thornthwaite's method to estimate the potential evapotranspiration for arid and superhumid climates, based on daily temperature amplitude. *Rev. Bras. Agrometeorol.* **1999**, *7*, 251–257.
12. Scozzafava, M.; Tallini, M. Net infiltration in the Gran Sasso Massif of central Italy using the Thornthwaite water budget and curve-number method. *Hydrogeol. J.* **2001**, *9*, 461–475. [CrossRef]
13. Nanni, T.; Rusi, S. Idrogeologia del massiccio carbonatico della montagna della Majella (Appennino centrale). *Bollettino-Società Geologica Italiana* **2003**, *122*, 173–202.
14. Sellinger, C.E. *Computer Program for Estimating Evapotranspiration Using the Thornthwaite Method*; NOAA Technical Memorandum ERL GLERL-101; United States Department of Commerce: Washington, DC, USA, 1996.
15. Armiraglio, S.; Cerabolini, B.; Gandellini, F.; Gandini, P.; Andreis, C. Calcolo informatizzato del bilancio idrico del suolo. *Nat. Brescia. Ann. Mus. Civ. Sci. Nat. Brescia* **2003**, *33*, 209–216.
16. McCabe, G.J.; Markstrom, S.L. *A Monthly Water-Balance Model Driven by a Graphical User Interface (Vol. 1088)*; US Geological Survey: Reston, VA, USA, 2007.
17. Hamon, W. Estimating potential evapotranspiration. *Trans. Am. Soc. Civ. Eng.* **1963**, *128*, 324–338. [CrossRef]
18. Di Matteo, L.; Dragoni, W. Climate change and water resources in limestone and mountain areas: The case of Firenzuola Lake (Umbria, Italy). In Proceedings of the 8th Conference on Limestone Hydrogeology, Neuchatel, Switzerland, 21–23 September 2006.
19. Di Matteo, L.; Dragoni, W.; Maccari, D.; Piacentini, S.M. Climate change, water supply and environmental problems of headwaters: The paradigmatic case of the Tiber, Savio and Marecchia rivers (Central Italy). *Sci. Total Environ.* **2017**, *598*, 733–748. [CrossRef] [PubMed]

20. Portela, M.M.; Santos, J.; de Carvalho Studart, T.M. Effect of the Evapotranspiration of Thornthwaite and of Penman-Monteith in the Estimation of Monthly Streamflows Based on a Monthly Water Balance Model. In *Current Practice in Fluvial Geomorphology-Dynamics and Diversity*; IntechOpen: Rijeka, Croatia, 2019.
21. Vandewiele, G.L.; Win, N.L. Monthly water balance models for 55 basins in 10 countries. *Hydrol. Sci. J.* **1998**, *43*, 687–699. [CrossRef]
22. Mather, J.R. *The Climatic Water Budget in Environmental Analysis*; Free Press: Wong Chuk Hang, Hong Kong, 1978.
23. Mather, J.R. Use of the climatic water budget to estimate streamflow. In *Use of the Climatic Water Budget in Selected Environmental Water Problems*; Elmer, N.J., Ed.; C.W. Thornthwaite Associates, Laboratory of Climatology: Centerton, NJ, USA, 1979; Volume 32, pp. 1–52.
24. McCabe, G.J.; Wolock, D.M. Independent effects of temperature and precipitation on modeled runoff in the conterminous United States. *Water Resour. Res.* **2011**, *47*. [CrossRef]
25. Pinna, M. *Climatologia*, XV ed.; UTET: Torino, Italy, 1977; p. 442.
26. Dooge, J.C.I. The rational method for estimating flood peaks. *Engineering* **1957**, *184*, 311–313, 374–377.
27. Schaake, J.C.; Geyer, J.C.; Knapp, J.W. Experimental examination of the rational method. *J. Hydraul. Div.* **1967**, *93*, 353–370. [CrossRef]
28. Boughton, W.C. Evaluating Partial Areas of Watershed runoff. *J. Irrig. Drain. Eng.* **1987**, *113*, 356–366. [CrossRef]
29. Longobardi, A.; Villani, P.; Grayson, R.B.; Western, A.W. On the relationship between runoff coefficient and catchment initial conditions. In Proceedings of the MODSIM, Townsville, Australia, 14–17 July 2003; pp. 867–872.
30. Ferguson, R.I. Snowmelt runoff models. *Prog. Phys. Geogr.* **1999**, *23*, 205–227. [CrossRef]
31. Spillner, J.; Mateos, C.; Monge, D.A. Faaster, better, cheaper: The prospect of serverless scientific computing and hpc. In Proceedings of the Latin American High-Performance Computing Conference, Buenos Aires, Argentina, 20–22 September 2017; Springer: Cham, Switzerland, 2017; pp. 154–168.
32. Pawlik, M.; Figiela, K.; Malawski, M. Performance considerations on execution of large-scale workflow applications on cloud functions. *arXiv* **2019**, arXiv:1909.03555.
33. Cervi, F.; Corsini, A.; Doveri, M.; Mussi, M.; Ronchetti, F.; Tazioli, A. Characterizing the recharge of fractured aquifers: A case study in a flysch rock mass of the northern Apennines (Italy). In *Engineering Geology for Society and Territory-Volume 3*; Springer: Cham, Switzerland, 2015; pp. 563–567.
34. Tazioli, A.; Cervi, F.; Doveri, M.; Mussi, M.; Deiana, M.; Ronchetti, F. Estimating the isotopic altitude gradient for hydrogeological studies in mountainous areas: Are the low-yield springs suitable? Insights from the northern Apennines of Italy. *Water* **2019**, *11*, 1764. [CrossRef]
35. Valigi, D. La Simulazione Delle Portate Mensili nei Bacini in Facies di Flysch dell'Appennino Centro-Settentrionale con Particolare Riguardo al F. Nestore. Ph.D. Thesis, Università degli Studi di Perugia, Perugia, Italy, 1995.
36. Brenčič, M.; Vreča, P. Hydrogeological and isotope mapping of the karstic River Savica in NW Slovenia. *Environ. Earth Sci.* **2016**, *75*, 651. [CrossRef]
37. Ogrinc, N.; Kanduč, T.; Stichler, W.; Vreča, P. Spatial and seasonal variations in δ18O and δD values in the River Sava in Slovenia. *J. Hydrol.* **2008**, *359*, 303–312. [CrossRef]
38. Buser, S. Basic geological map of SFRJ 1:100,000. In *Guidebook of Sheets Tolmin and Videm (Udine): L 33–63, 33–64*; Zvezni Geološki Zavod, Beograd: Ljubljana, Slovenian, 1986.
39. Trišič, N.; Bat, M.; Polajnar, J.; Pristov, J. Water balance investigations in the Bohinj region. In *Tracer Hydrology 97, Proceedings of the 7th International Symposium on Water Tracing, Portorož, Slovenia, 26–31 May 1997*; CRC Press: Boca Raton, FL, USA, 1997; pp. 26–31.
40. Urbanc, J.; Brancelj, A. Tracing experiment in lake Jezero v Ledvici, valley Triglavska jezera. *Geologija* **1999**, *42*, 207–214. [CrossRef]
41. Bat, M. Hidrološke raziskave ARSO na območju Bohinja. *Dela* **2007**, 165–181. [CrossRef]
42. Brenčič, M.; Vreča, P. Influence of snow thawing regime changes on the outflow from karstic aquifer. In Proceedings of the EGU General Assembly Conference Abstracts, Vienna, Austria, 7–12 April 2013.
43. Dourado-Neto, D.; Jong van Lier, Q.D.; Metselaar, K.; Reichardt, K.; Nielsen, D.R. General procedure to initialize the cyclic soil water balance by the Thornthwaite and Mather method. *Sci. Agric.* **2010**, *67*, 87–95. [CrossRef]
44. Dragoni, W.; Giontella, C.; Melillo, M.; Cambi, C.; Di Matteo, L.; Valigi, D. Possible response of two water systems in Central Italy to climatic changes. *Adv. Watershed Hydrol.* **2015**, 397–424.
45. Guardiola-Albert, C.; Martos-Rosillo, S.; Pardo-Igúzquiza, E.; Duran Valsero, J.J.; Pedrera, A.; Jiménez-Gavilán, P.; Linan Baena, C. Comparison of recharge estimation methods during a wet period in a karst aquifer. *Groundwater* **2015**, *53*, 885–895. [CrossRef]
46. Legates, D.R.; Bogart, T.A. Estimating the proportion of monthly precipitation that falls in solid form. *J. Hydrometeorol.* **2009**, *10*, 1299–1306. [CrossRef]

hydrology

MDPI

Article

Numerical Modeling of Venturi Flume

Mehdi Heyrani [1,*], Abdolmajid Mohammadian [1], Ioan Nistor [1] and Omerul Faruk Dursun [2]

[1] Department of Civil Engineering, University of Ottawa, 75 Laurier Ave E, Ottawa, ON K1N 6N5, Canada; amohamma@uottawa.ca (A.M.); inistor@uottawa.ca (I.N.)

[2] Department of Civil Engineering, Inonu University, 44280 Malatya, Turkey; faruk.dursun@inonu.edu.tr

* Correspondence: mheyr054@uottawa.ca

Abstract: In order to measure flow rate in open channels, including irrigation channels, hydraulic structures are used with a relatively high degree of reliance. Venturi flumes are among the most common and efficient type, and they can measure discharge using only the water level at a specific point within the converging section and an empirical discharge relationship. There have been a limited number of attempts to simulate a venturi flume using computational fluid dynamics (CFD) tools to improve the accuracy of the readings and empirical formula. In this study, simulations on different flumes were carried out using a total of seven different models, including the standard k–ε, RNG k–ε, realizable k–ε, k–ω, and k–ω SST models. Furthermore, large-eddy simulation (LES) and detached eddy simulation (DES) were performed. Comparison of the simulated results with physical test data shows that among the turbulence models, the k–ε model provides the most accurate results, followed by the dynamic k LES model when compared to the physical experimental data. The overall margin of error was around 2–3%, meaning that the simulation model can be reliably used to estimate the discharge in the channel. In different cross-sections within the flume, the k–ε model provides the lowest percentage of error, i.e., 1.93%. This shows that the water surface data are well calculated by the model, as the water surface profiles also follow the same vertical curvilinear path as the experimental data.

Keywords: venturi flume; CFD; OpenFOAM; RANS; turbulence model; numerical simulation; Parshall flume

Citation: Heyrani, M.; Mohammadian, A.; Nistor, I.; Dursun, O.F. Numerical Modeling of Venturi Flume. *Hydrology* **2021**, *8*, 27. https://doi.org/10.3390/hydrology8010027

Academic Editors: Carmelina Costanzo, Tommaso Caloiero and Roberta Padulano

Received: 21 December 2020
Accepted: 1 February 2021
Published: 4 February 2021

1. Introduction

The Parshall flume is a simple static measuring device with no moving parts that is used to determine the flow rate in an open channel where a constant recording of discharge is required. The initial idea by Ralph Parshall in designing the Parshall flume was to make it easier for water users, like farmers, who do not have access to sophisticated types of equipment, to be able to determine how much water is delivered to them with an acceptable level of accuracy [1]. Currently, Parshall flumes are mainly used in irrigation and sewer systems to measure the flowrate [2]. In general, they are designed to generate a critical flow within the throat section, which affects the water level along the converging section upstream, implementing an empirical relationship between water surface elevation and discharge results in finding the discharge value at a specific time from the water surface elevation.

Variations of Parshall flumes are typically restricted to the dimensions of the geometries proposed by Ralph Parshall in 1936 for limited size numbers, i.e., 16 sizes, which vary in the opening from 1″ to 144″. The arbitrary dimensions of the flumes used in various open channels in recent years have not been comprehensively studied. Since then, technology has advanced significantly, especially in the development of computational tools. Numerical simulations have introduced a new revolutionary chapter to the design of hydraulic structures, allowing engineers to extend their full potential by designing a variety of different hydraulic structures with new arrangements in their dimensions and shapes. They allow designers to optimize the specifications of their design to suit the need of the project's objective. Computers and modeling software offer possibilities to perform extensive calculations, and they allow new designs to be tested with relatively low cost. For instance, a proposed hydraulic structure can be simulated to study its behavior under working conditions. This provides engineers with extensive variations in geometries and dimensions when designing a structure. Using computer models makes it possible to implement Parshall flumes with complex shapes and dimensional limitations where needed.

The authors of [3] used a numerical model to develop an alternative rating equation to be implemented at low discharge for different flume sizes. In a study on a submerged Montana flume, to prove the reliability of the CFD programs' accuracy, the flow rate was calculated with FLOW- 3D. It is shown that the numerical results closely matched the experimental results. Furthermore, it is revealed that a free-flow equation for a Parshall flume would also be a good fit for a Montana flume [4]. To determine the accuracy of a field-scaled Parshall flume at a wastewater system in Minneapolis, Minnesota, [5] implemented the large-eddy simulation (LES) and level-set method to compute the turbulent flow under two-phase flow conditions. A three-dimensional finite difference code, SOLA-FLUMP, is presented by [6] to assess the various effects due to the sloped channel, the upstream velocity profile distortion, and the geometry of the flume.

A parametric study was performed in terms of stability and accuracy on Reynolds-average Navier–Stokes (RANS) and hybrid RANS/LES turbulence models and numerical schemes offered in openFOAM, an open-source software, by [7]. From the results, the second-order upwind scheme and limiters were found to be the most stable with the lowest computational cost, thereby providing the highest level of accuracy for RANS models.

It is important to determine the degree of reliability of the model that is going to be used for a specific simulation. One of the most dependable approaches is to evaluate the simulated results with actual experimental data. In this paper, different simulated datasets generated from seven dissimilar methods, i.e., standard k–ε, RNG k–ε, realizable k–ε, and k–ω SST; large-eddy simulation (LES); and detached eddy simulation (DES) are examined versus four physical datasets obtained from physical experiments on different flumes with different discharges.

A high-order partial differential equation, like the momentum transport equation, that includes nonlinear terms cannot be solved analytically to obtain a general solution. Numerical solutions require discretization techniques to reshape the continuous partial differential equation into a discrete equation. Computational fluid dynamics (CFD) using the finite volume method (FVM) [8] can be very useful for this purpose. With the rapid increase in the computational capability of computers, large-eddy simulation (LES) is increasingly embedded in the CFD models used by researchers and engineers to solve turbulent flows [9]. LES is a compromise between the efficiency of Reynolds-average Navier–Stokes (RANS) and the prohibitive computational cost of direct numerical simulation (DNS). Approaches of LES or the variant hybrid family like RANS/LES (DES) are progressively taking over the computationally expensive DNS approach to solve problems with compound geometries and flow properties [10].

The authors of [11] tried to introduce a correction coefficient to a 24″ Parshall flume where the positions of the staff gauges were mislocated and the condition of the flume entrance was set up differently from the one introduced by [1]. They used numerical modeling to implement the correction factor for other sizes of Parshall flumes. Based on this study, a part of the small deviations, the physical model, and numerical simulation were aligned with one another.

In order to decrease the head loss in a curved flume, three flumes were studied on the basis of critical flow by [12]. The study was conducted on laboratory experimental results versus numerical simulation data. The hydraulic parameters such as velocity of the free surface and the depth of water were analyzed and compared. A maximum error of

4.7% in the water depth was obtained. It was shown that a good consistency was achieved between numerical simulation and experimental data.

The objective of this paper is to solve the Navier–Stokes equations using OpenFOAM to study the behavior of various numerical models in simulating Parshall flumes similar in size and flowrate to the flumes used in physical experiments by Dursun (2016).

Based on the literature review at the beginning, there is no comprehensive numerical solution that has been conducted on the models simulating the hydraulic structures, i.e., Parshall flume. The reliability of RANS LES and DES turbulence models in simulations has been studied in this research by evaluating the performance of seven turbulence models against the experimental data of different Parshall flume structures. Subsequently, the consistency of the simulations was determined in various scenarios in relation to the experimental data. Based on what was discussed in the literature review, seven turbulence models, i.e., RAS models (including standard k–ε, RNG k–ε, realizable k–ε, and k–ω SST) and hybrid RANS/LES models (such as k–ω, SST-DES, and an LES models, namely, the Smagorinsky method and dynamic K LES method), respectively, were selected due to their wide usage and advantages compared to other turbulence models.

This paper is organized as follows: the methodology is explained in the next section, including governing equations, turbulence models (including RANS, LES, and DES models), numerical setup (including initial and boundary conditions), mesh analysis, and data; the results and discussion are then presented, in which the performance of numerical models is discussed. Some concluding remarks complete the study.

2. Methodology

By the increasing power of processors, computational fluid dynamics (CFD) has become the most convenient tool to simulate fluid motion. In CFD, the fluid's flow establishment follows physical parameters, including pressure, viscosity, velocity, and temperature. In order to simulate a physical case related to the fluid flow, the physical properties should be taken into account accurately.

CFD approaches are used in solving the fluid flow equations as well as fluid interaction with solid bodies. The Euler equation for inviscid fluid and the Navier–Stokes equation for viscous fluid can be derived in their integral arrangement with respect to the conservation of energy, mass, and momentum [13].

OpenFOAM is one of the open-source solvers for CFD that is widely used for simulations. It is a platform including numerous C++ libraries and applications that are able to solve numerically the continuum mechanics problems [14]. It uses a tensorial method that implements an object-oriented programing approach and employs the finite volume Method (FVM).

2.1. Governing Equations

A viscous incompressible fluid flow is governed by a general three-dimensional system of equations, called the Navier–Stokes system, that consists of momentum and continuity equations. The system is described as follows [15,16]:

$$\frac{\partial u}{\partial x}+\frac{\partial v}{\partial y}+\frac{\partial w}{\partial z}=0 \tag{1}$$

$$\frac{\partial u}{\partial t}+u\frac{\partial u}{\partial x}+v\frac{\partial u}{\partial y}+w\frac{\partial u}{\partial z}=-\frac{1}{\rho}\frac{\partial p}{\partial x}+\nu\nabla^2 u \tag{2}$$

$$\frac{\partial v}{\partial t}+u\frac{\partial v}{\partial x}+v\frac{\partial v}{\partial y}+w\frac{\partial v}{\partial z}=-\frac{1}{\rho}\frac{\partial p}{\partial y}+\nu\nabla^2 v \tag{3}$$

$$\frac{\partial w}{\partial t}+u\frac{\partial w}{\partial x}+v\frac{\partial w}{\partial y}+w\frac{\partial w}{\partial z}=-\frac{1}{\rho}\frac{\partial p}{\partial z}+\nu\nabla^2 w-g \tag{4}$$

where ρ denotes density; p represents total pressure; u, v, and w represent the velocity in three different directions, i.e., x, y, and z; t is used for time; and gravitational acceleration is denoted by g. ρ is obtained using the following equation:

$$\rho = \alpha\rho_1 + (1-\alpha)\rho_2 \tag{5}$$

Here, ρ_1 and ρ_2 represent the air and water densities, the two phases of the involved fluid. The α value varies from 1 to 0 depending on the location, where 1 denotes the presence of water and 0 shows the presence of air. Any number between these two values represent the interface.

Finally,

$$\nabla^2 = \frac{\partial^2}{\partial x^2} + \frac{\partial^2}{\partial y^2} + \frac{\partial^2}{\partial z^2} \tag{6}$$

2.1.1. Equation of the Free Surface

With respect to the zero pressure at the surface, the free surface was analyzed with the volume-of-fluid (VoF) method. The following equation is used by VoF:

$$\frac{\partial\alpha}{\partial t} + \frac{\partial(\alpha u)}{x} + \frac{\partial(\alpha v)}{y} + \frac{\partial(\alpha w)}{z} = 0 \tag{7}$$

As stated earlier in this paper, a powerful open-source CFD, i.e., OpenFOAM application, was used for the purpose of the numerical simulations.

The flow motion in a Parshall flume was simulated in this study with the help of seven different turbulence models: standard, realizable, and RNG k–ε models; k–ω SST models; and detached eddy simulation (DES) models, such as k–ω SST-DES; as well as LES methods, including the Smagorinsky LES model and dynamic K LES model. In the following, a brief description of each selected model is provided.

2.1.2. Reynolds-Average Navier–Stokes (RANS) Approach

The Reynolds-average Navier–Stokes Model is currently the most popular approach for the simulation of fluid flow. This approach essentially uses a viscosity term to approximate the turbulence equations. K is a term in these models that represents the fluctuations of the turbulence kinetic energy per unit mass.

Standard k–ε Model

This model requires two additional transport equations: one for turbulent kinetic energy (k) and another for energy dissipation (ε). Apart from its poor performance in large adverse pressure gradient cases, it is known as one of the most popular turbulence models [17]. This model comes from the Reynolds-averaged Navier–Stokes (RANS) category where modeling is applied to all properties of fluid motion.

The equation for turbulent kinetic energy k and dissipation ε is shown below:

$$\frac{\partial(\rho k)}{\partial t} + \frac{\partial(\rho k u_i)}{\partial x_i} = \frac{\partial}{\partial x_j}\left[\frac{\mu_t}{\sigma k}\frac{\partial k}{\partial x_j}\right] + 2\mu_t E_{ij}\cdot E_{ij} - \rho\varepsilon \tag{8}$$

$$\frac{\partial(\rho\varepsilon)}{\partial t} + \frac{\partial(\rho\varepsilon u_i)}{\partial x_i} = \frac{\partial}{\partial x_j}\left[\frac{\mu_t}{\sigma\varepsilon}\frac{\partial\varepsilon}{\partial x_j}\right] + C_{1\varepsilon}\frac{\varepsilon}{k}2\mu_t E_{ij}\cdot E_{ij} - C_{2\varepsilon}\rho\frac{\varepsilon^2}{k} \tag{9}$$

ε is the component that controls the turbulence scale where k represents the turbulence kinetic energy. The reader is referred to [15] for further details and values of the coefficients.

Realizable k–ε

The latest improved form of the three k-epsilon models is the realizable k-epsilon model [18,19]. There are two significant differences when this model is compared to the standard k-epsilon model. Firstly, formulation of the turbulence viscosity has been revised. Secondly, the dissipation rate transport equation is explained based on the equation of transport of the mean-square vorticity [19,20].

$$\frac{\partial}{\partial t}(\rho k) + \frac{\partial}{\partial x_j}(\rho k u_j) = \frac{\partial}{\partial x_j}\left[\left(\mu + \frac{\mu_t}{\sigma_k}\right)\frac{\partial k}{\partial x_j}\right] + G_k + G_b - \rho\varepsilon - Y_M + S_K \tag{10}$$

$$\frac{\partial}{\partial t}(\rho\varepsilon) + \frac{\partial}{\partial x_j}(\rho\varepsilon u_j) = \frac{\partial}{\partial x_j}\left[\left(\mu + \frac{\mu_t}{\sigma_\varepsilon}\right)\frac{\partial \varepsilon}{\partial x_j}\right] + \rho C_1 S\varepsilon - \rho C_2 \frac{\varepsilon^2}{k+\sqrt{v\varepsilon}} + C_{1\varepsilon}\frac{\varepsilon}{k}C_{3\varepsilon}G_b + S_\varepsilon \tag{11}$$

The reader is referred to [21] for further details and value of the coefficients.

2.1.3. LES Approach

Smagorinsky LES Model

This model was originally developed within the metrological community to simulate atmospheric air currents [22]. As a well-known subgrid-scale model according to [23], the Smagorinsky model estimates the shear as

$$v_t^{Smag} = (C_{Smag}\Delta)^2|\overline{S}| \tag{12}$$

$$\overline{S_{ij}} = \frac{1}{2}\left(\frac{\partial \overline{u_i}}{\partial x_i} + \frac{\partial \overline{u_j}}{\partial x_j}\right) \tag{13}$$

where

$$|\overline{S}| = \sqrt{2\overline{S_{ij}}\overline{S_{ij}}} \tag{14}$$

The reader is referred to [22] for further details and values of the coefficients.

One of the main disadvantages of this model is the lack of ability to predict the energy transfer from subgrid-scale structures to the greater resolved scales; thus, the model is totally dissipative. Another problem with the Smagorinsky model is that its coefficient has to be adjusted for every flow field. However, it is still one of the well-known models in the field of CFD. The Smagorinsky model is easy to use in numerical simulations. If the Smagorinsky coefficient is adjusted based on the local characteristics of the fluid motion, it can generate more accurate results [23].

The Dynamic One-Equation Model

The SGS stresses determine how successful an LES model can be. As a simple model, in the Smagorinsky model, the factor of the proportionality is a fixed value that has to be determined before running the model. In reality, the factor is a flow-dependent value and is not defined as a single universal constant. The weak point of the model comes from this section. There have been attempts to improve this model [24,25]. Moreover, this model is completely dissipative, and there is always a transformation of large-to-small scale for the energy.

On the other hand, dynamic models are the best choice to substitute the Smagorinsky model. In this model, the C value of the subgrid eddy viscosity is determined while the simulation is computed [26]. In recent decades, a one-equation dynamic model has been presented [27]. The equation of the dynamic model is presented below.

$$\frac{D}{Dt}(\rho k) = \nabla \cdot (\rho D_k \nabla_k) + \rho G - \frac{2}{3}\rho k \nabla \cdot u - \frac{C_e \rho k^{1.5}}{\Delta} + s_k \tag{15}$$

where the C_e and S_k coefficients are derived from local flow properties [21]. The reader is referred to [28] for further details.

2.1.4. DES Approach

k–ω SST-DES

While the RANS model is derived through Reynolds temporal averaging, the LES model is the result of spatial filtering. Between these two methods, the difference is the magnitude of the generated eddy viscosity. This is one of the main reasons for the development of the DES model, which has the ability to cover the weaknesses of LES when it comes to treating wall regions with very fine mesh [29].

The RANS approach is used in the near-wall region by the DES method where, at the same time, the LES model is applied to the rest of the region excluding the wall region. The region associated with the LES model is usually the core turbulent area, where large-scale turbulences play a major role. Within this area, the DES approach uses a LES subgrid-scale model, while for the near-wall region, it uses the RANS model [30].

A DES-improved form of the k–ω SST method is the k–ω SST DES approach [31,32]. Recently, in the aerodynamic field, DES has been widely implemented due to its computational speed and quality of results. It is proven to be less computationally expensive, and it generates better results than steady RANS [33,34].

The turbulence specific dissipation rate equation is given by

$$\frac{D}{Dt}(\rho\omega) = \nabla \cdot (\rho D_w \nabla\omega) + \rho\gamma\frac{G}{\nu} - \frac{2}{3}\rho\gamma\omega(\nabla \cdot u) - \rho\beta\omega^2 - \rho(F_1 - 1)CD_{k\omega} + S_\omega \quad (16)$$

and the turbulence kinetic energy is calculated as

$$\frac{D}{Dt}(\rho k) = \nabla \cdot (\rho D_k \nabla k) + \min(\rho G, (c_1\beta^*)\rho k\omega) - \frac{2}{3}\rho k(\nabla \cdot u) - \rho\frac{k^{1.5}}{\overline{d}} + s_k \quad (17)$$

the length scale, $\overline{d}$, is given by

$$\min\left(C_{DES}\Delta, \frac{\sqrt{k}}{\beta^*\omega}\right) \quad (18)$$

and the turbulence viscosity is obtained using

$$\nu_t = \alpha_1\frac{k}{\max(\alpha_1\omega, b_1F_{23}S)} \quad (19)$$

The reader is referred to [35] for further details and value of the coefficients.

2.2. *Numerical Setup*

2.2.1. Numerical Solution Details

The combination of the finite volume method with the VoF method was used in the model. To solve the governing equation of motion, the "interFoam" solver was implemented in OpenFOAM. The temporal term was discretized with the help of a Eulerian scheme, besides the Gauss linear method, which is used for the gradient term. For the Laplacian, the corrected Gauss linear method was applied. The divergence terms were discretized using a Gauss vanLeer plus Gauss linear scheme. The linear scheme was used to discretize the interpolation term.

The tolerance level was defined for each individual variable, while the desired convergence was expected to be achieved where an iterative solver was used in the processing. The Gauss–Seidel technique was applied with a level of accuracy of 10^{-5} for the fraction of liquid (α), 10^{-8} for pressure, and 10^{-8} for the velocity.

2.2.2. Initial Conditions

The flume had a fixed inlet flow velocity for each case, i.e., a discharge of 20 L/s or 30 L/s as the initial conditions. There was no initial acceleration or dissipation defined in the model. There was no flow across the walls, i.e., there was no liquid coming in or getting out through the wall boundaries at the defined locations. Initial water surface was set to be constant everywhere.

2.2.3. Boundary Conditions

There were different boundary conditions employed in this simulation as illustrated in Figure 1. Hydraulically smooth walls were considered in this study, and standard wall functions were employed. Water discharge was specified at the inlet. Zero gradient condition was considered at the outlet. The free surface was tracked based on the volume-of- fluid method based on a zero pressure condition at the interface of air and water. Figure 1 below shows the boundary segments from the side view and top view of the simulated Parshall flume.

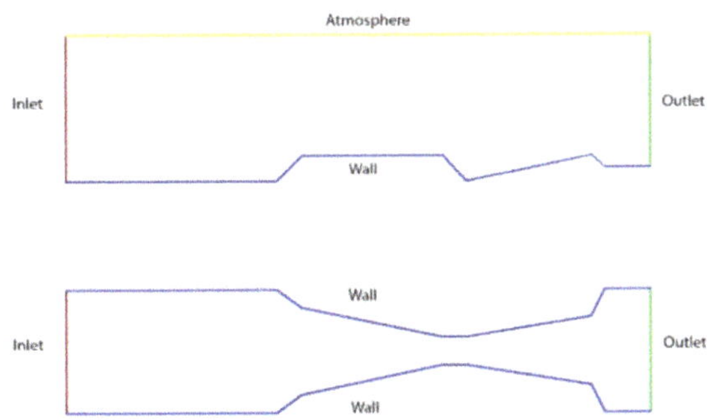

Figure 1. Side view and top view of the boundary condition of the modeled Parshall flume.

2.2.4. Mesh Analysis

In this study, in order to determine the optimum size of the mesh, mesh sensitivity analysis was conducted. The mesh grid used in this paper is a structured mesh. The purpose of this analysis is to find the finest mesh size for which the results will not be affected further.

Figure 2 shows the mesh sensitivity analysis performed on the Parshall flume. The maximum cell number was 263,700 cells for this structure. The optimal case in terms of computational cost and changes in results is that illustrated in Figure 2b, with a total of 74,496 cells. Greater reduction applied on the cell size did not significantly change the numerical results.

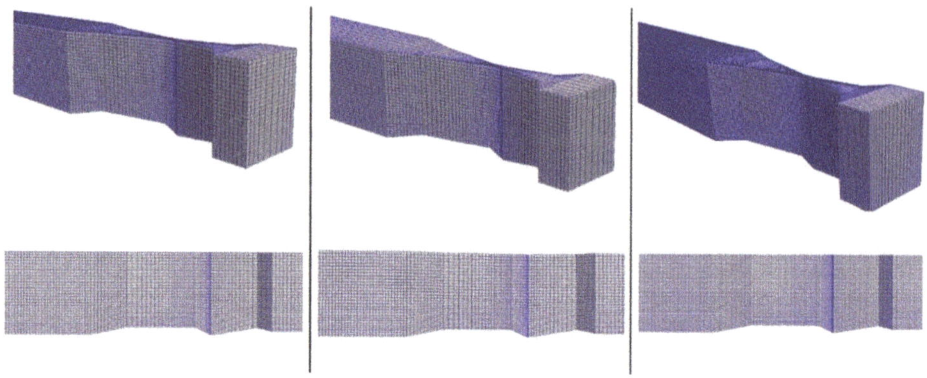

Figure 2. Mesh convergence analysis. Side view and 3D views of the Parshall flume mesh: (**a**) the mesh with a total of 52,200 cells; (**b**) the mesh with a total of 74,496 cells; (**c**) the mesh grid with a total of 263,700 cells.

In Figure 3, the results of the mesh sensitivity analysis are provided for the cross-sections 4–6. As the size of the original coarse mesh became finer, the simulated water levels approached the values of the experimental data. In this study, the initial number of cells was 52,200, and it was gradually increased to 74,496 and further to 263,700 cells. The results from the last two finer meshes show negligible difference; hence, the mesh containing 74,496 cells was selected and further used to implement the remaining turbulence models.

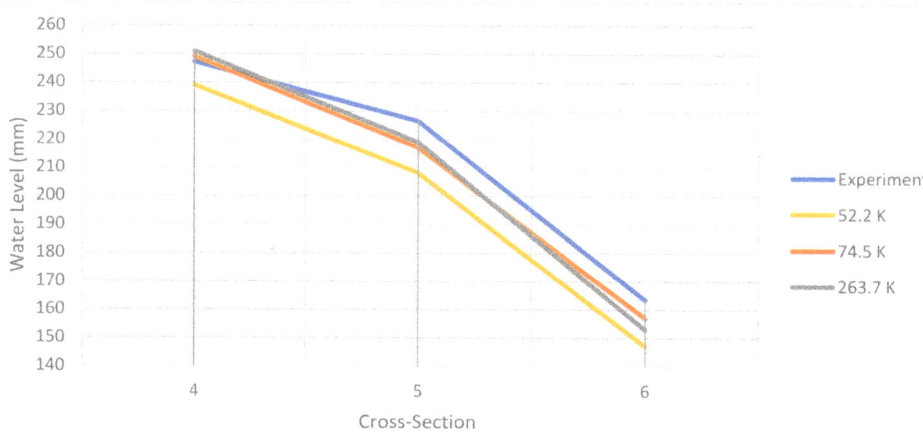

Figure 3. Mesh analysis graphs for the cross-section 4–6.

2.2.5. Data

The experimental tests were conducted in the hydraulic laboratory of Firat University, Elazig, Turkey (Dursun, 2016). The tests were performed in a rectangular channel with the dimensions of 0.4 by 5 by 0.6 m in width, length, and depth, respectively. The purpose of these experiments was to determine the changes in the quantity of the dissolved oxygen of the stream before and after a Parshall flume structure was introduced. The dimensions of the Parshall flumes that was used in this research were the same as the 3-inch (7.62 cm)

Parshall flume with a 45 degree wing wall. The discharge in the experiment was measured using an electromagnetic flow meter.

The model was run for 300 time-steps where, after 120 steps, water levels became steady; water surface fluctuations, for instance, were limited to +/−0.003 m. The water level is shown using a post-data analysis software called ParaView. At the point of interest, two vertical planes (perpendicular to each other) were introduced, where the intersection line between the two planes was set to pass the desired point. The Y-coordinate (water level) of the line was extracted using the calculator filter; for example, the water level at each desired time step is shown afterwards in the tabulated mode in a separate window in ParaView.

3. Results

The numerical simulation for the experimental case conducted by [2] is performed in this study. The Parshall flume used in this study is a 3-inch Parshall flume modified to meet the experimental criteria. OpenFOAM was used as an open source CFD tool to carry out the numerical simulation of the Parshall flumes.

Switching from a coarse size mesh grid to a finer grid size in the mesh sensitivity analysis led to some change seen in the numerical simulation model results. As the flow rate decreased from 30 to 10 L/s, the model tended to produce better results. This pattern is observed in the k–ε model as well as the other RANS models.

Water surface elevations in seven cross-sections were compared with the experimental results with three different flowrate values, i.e., 10, 20, and 30 L/s. Figure 4 shows water surface elevation for the flowrate of 10 L/s using seven various models. As shown in Figure 4, all models follow the same pattern as the experimental data. Figures 5 and 6 are introduced to show more clarified view of the water level data. Based on the results of various turbulence models used in simulations, the k–ε turbulence model provides the most accurate simulation compared to the other models used in this study.

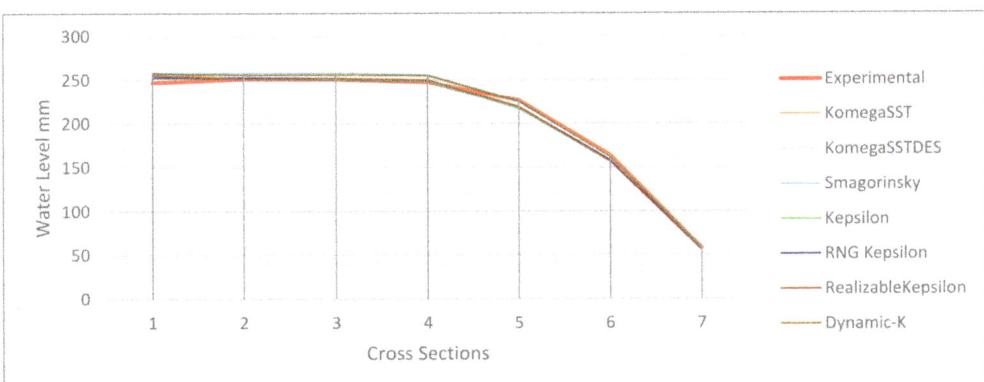

Figure 4. Comparison of experimental water level results (obtained with the average measuring error of 1.93–2.58%) with the numerical simulation results for the discharge of 10 L/s.

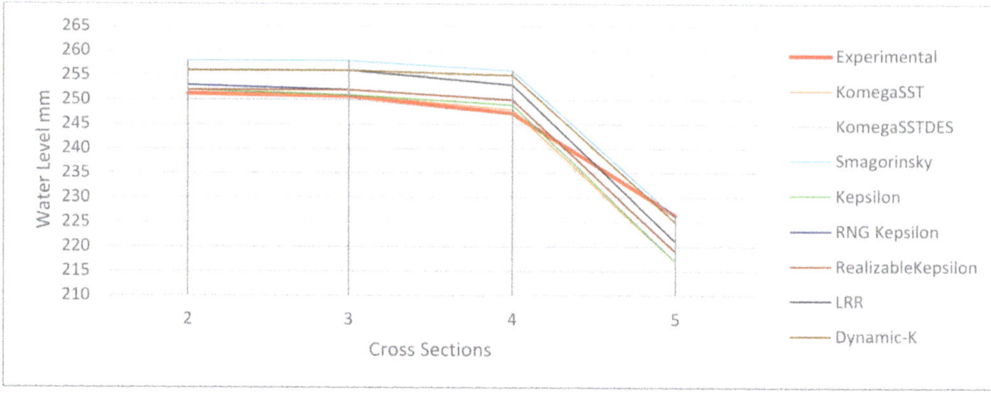

Figure 5. Comparison of experimental water level results (obtained with the average measuring error of 1.93–2.58%) with the numerical simulation results for the discharge of 10 L/s (for cross-sections 2–5).

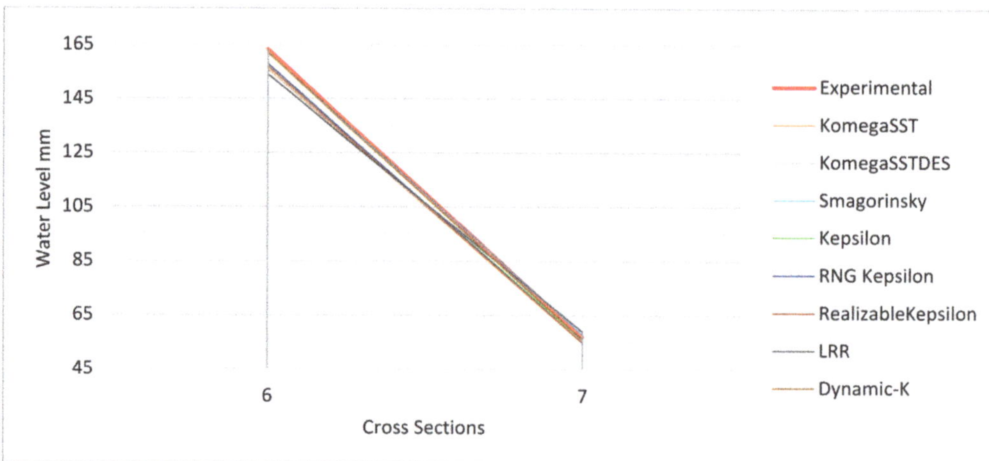

Figure 6. Comparison of experimental water level results (obtained with the average measuring error of 1.93–2.58%) with the numerical simulation results for the discharge of 10 L/s (for cross-sections 6 and 7).

The locations of the cross-sections in the Parshall flume are denoted in Figure 7 where at all critical locations, a cross-section is introduced. The reason for choosing the cross-section locations, as shown in Figure 7, is that in the laboratory experiment conducted by Dursun in 2016, the same locations were chosen; hence, water level data were also available for them.

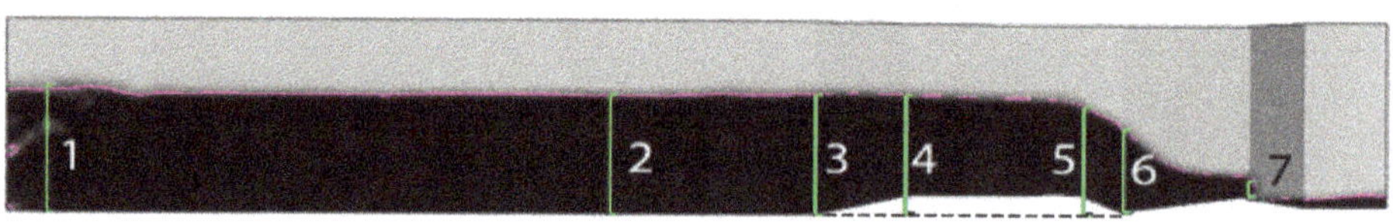

Figure 7. Location of cross-sections of Parshall flume.

In Figure 8, the fluid flow in the Parshall flume with a flowrate of 10 L/s is illustrated. As the stream lines indicate, the velocity of the fluid prior to the flume throat displays a maximum of 1 m/s, and by the time it reaches the narrow section with a declining floor, the speed rapidly increases by 50%. This is the section where the fluid experiences supercritical flow. It continues to increase by the end of the divergence section where maximum velocity is reached (red arrows), i.e., 2 to 2.5 m/s. Once the fluid reaches the inclined slope in the divergence section downstream, the flume forces it to develop a hydraulic jump.

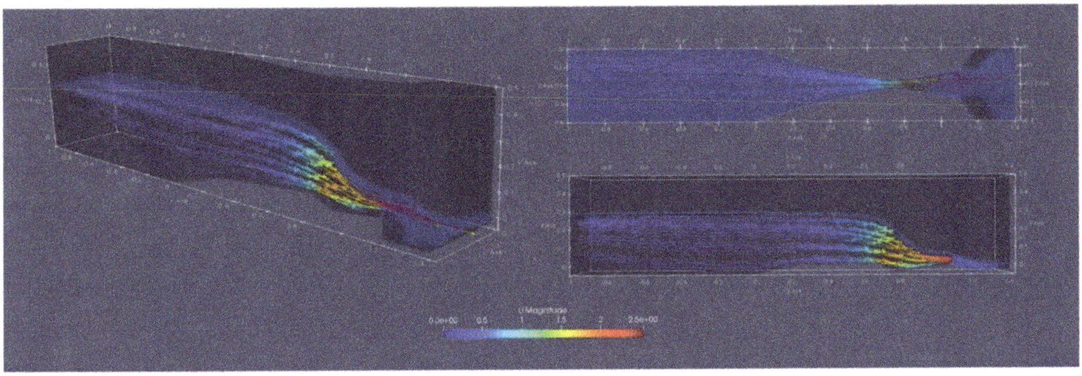

Figure 8. 3D, top, and side views of the Parshall flume.

The velocity profiles of all seven turbulence models at cross-section 5, where the numerical models exhibited the maximum velocity, are shown in Figure 9. The comparison between different velocity profile shows that in the seven different turbulence models, the values are almost identical, with a difference margin of 0.04 m/s at the maximum velocity points. The maximum recorded velocity in the flume for different models is within the range of 1.26–1.28 m/s in the different models. The "*y*" axis is the local coordinate of the flume throat defined in the OpenFOAM computational domain.

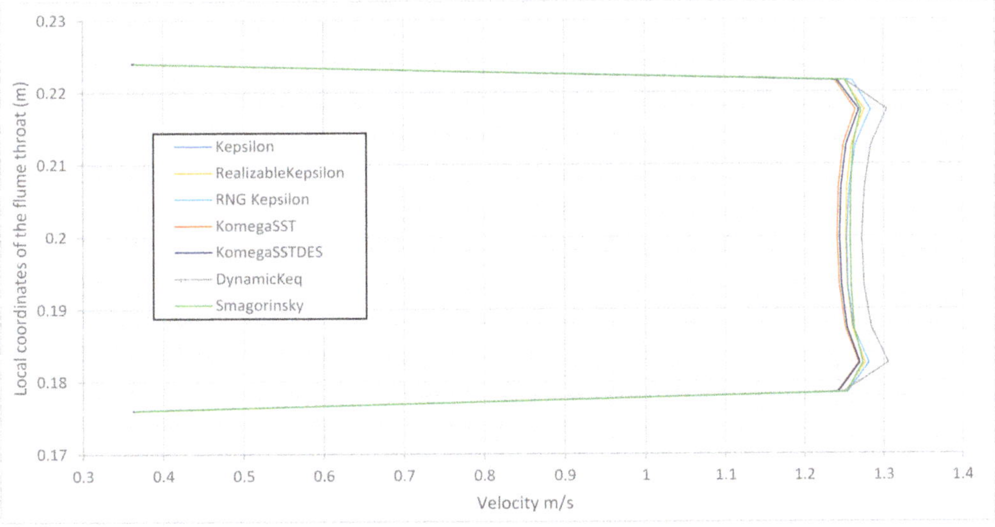

Figure 9. Comparison of the velocity profile at cross-section 5 and the results obtained using the investigated turbulence models.

As illustrated in Figure 10, the velocity is constant upstream of the midsection of the converging area. At the end of this section, the flow is forced to increase its velocity due to the narrow design of the throat combined with a sharply slopped bed. The velocity streamlines after the throat section of the flume show that the velocity magnitude is greater than that observed throughout the rest of the structure. Once the flow immediately exits the divergence section, it attains maximum velocity.

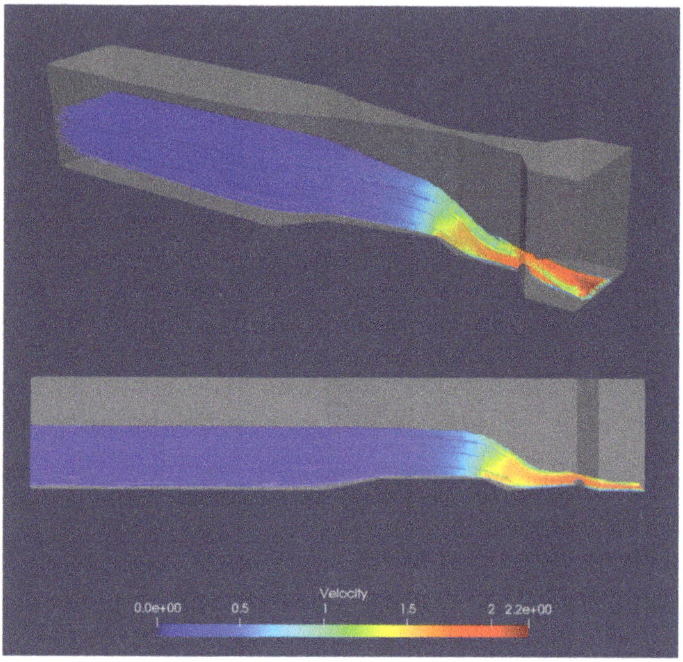

Figure 10. 3D and side view of the velocity field of the flume with a flowrate of 10 L/s.

The flume's pressure field is represented in Figure 11. In the section between cross-sections 5 and 6, the pressure is negative. As illustrated in this figure, due to the throat section, the pressure is built up in the upstream, while when passing cross-section 5 toward downstream, the value of the pressure drops rapidly.

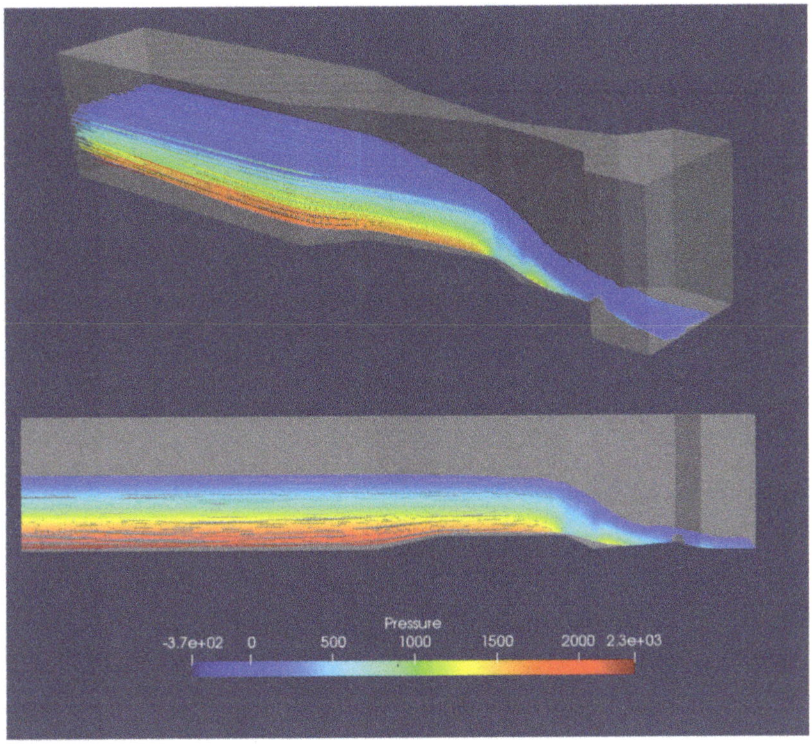

Figure 11. 3D and side view of the pressure field in the flume with a flowrate of 10 L/s.

4. Discussion

The percentage difference of water level values of the numerical models and the experimental data is specified in Table 1 below. The calculation is performed based on the following relationships [15]:

$$Error = \frac{|H_{exp} - H_{sim}|}{H_{exp}} \times 100 \quad (20)$$

$$RMSE = \sqrt{\frac{\sum (h_{exp} - h_{sim})^2}{n}} \quad (21)$$

$$R^2 = \frac{\sum(\hat{y} - \overline{y})^2}{\sum(y - \overline{y})^2} \quad (22)$$

$$CRMSE = \sqrt{\frac{\sum_{i=1}^{n}\left((X_i^{calc} - X_{mean}^{calc}) - (X_i^{expr} - X_{mean}^{expr})\right)^2}{n}} \quad (23)$$

Here, the root mean square error (RMSE), centered root mean square error (CRMSE), and correlation coefficient (R) are used in order to compare the simulation results with experimental data. The results from the above formulas are presented in Table 1, where different types of error analysis are applied in order to determine the reliability of the numerical simulations.

Table 1. The error percentage, root mean square error (RMSE), standard error, R^2 and centered root mean square error (CRMSE) of the simulated data vs. experimental data.

	Error Percentage							Avrg. Err %	RMSE	Stndr Err	R^2	CRMSE
Cross-Sections	1	2	3	4	5	6	7					
Kepsilon	2.30%	0.31%	0.11%	0.71%	4.16%	3.91%	2.03%	1.93%	0.49%	0.23%	0.996	0.47%
RNG Kepsilon	2.30%	0.70%	0.51%	1.12%	3.27%	3.30%	3.78%	2.14%	0.44%	0.18%	0.997	0.43%
Realizable Kepsilon	3.11%	0.31%	0.51%	1.12%	3.27%	3.91%	3.78%	2.29%	0.49%	0.00%	0.996	0.49%
Komega SST	4.73%	0.09%	0.11%	0.31%	4.16%	4.52%	3.78%	2.53%	0.64%	0.08%	0.993	0.63%
Komega SST DES	4.33%	0.49%	0.29%	0.10%	4.16%	4.52%	3.78%	2.52%	0.61%	0.14%	0.993	0.60%
Dynamic-K	3.92%	1.90%	2.10%	3.14%	0.62%	0.85%	2.03%	2.08%	0.55%	0.09%	0.998	0.43%
Smagorinsky	4.33%	2.69%	2.90%	3.54%	0.18%	0.85%	1.47%	2.28%	0.65%	0.08%	0.998	0.45%

Among the seven different turbulence models applied in this study, the Smagorinsky and dynamic K equation models provide the least average error values. This model provides the minimum error percentages not only in the average value of all of the cross-sections but also within the first three, i.e., 1, 2, and 3.

By calculating the root mean square error for all the turbulence models, the results show that two models, k–ε and dynamic K LES, provide the lowest values, 1.93% and 2.08%, respectively, while the k–ω SST model has a greater value compared to the others. However, for comparison of the R^2 values of this turbulence model, the Smagorinsky and dynamic K LES models have the closest value to 1, i.e., 0.998.

The ranges of the average error percentage for all of the turbulence models are within a narrow domain a minimum value of 1.93% and a maximum value is 2.53%. While most of the turbulence models used in this study provided reliable error percentages, the k–ω SST model was the only one that generated results with a high average error of 2.53% and RMSE of 0.64%, and the lowest R^2 was also recorded for this model.

Table 1 and Figure 12 tabulate the results of the different numerical simulations at various locations in the flume, i.e., at the seven cross-sections. The average values and the statistical analysis that were discussed earlier in this section are also included here.

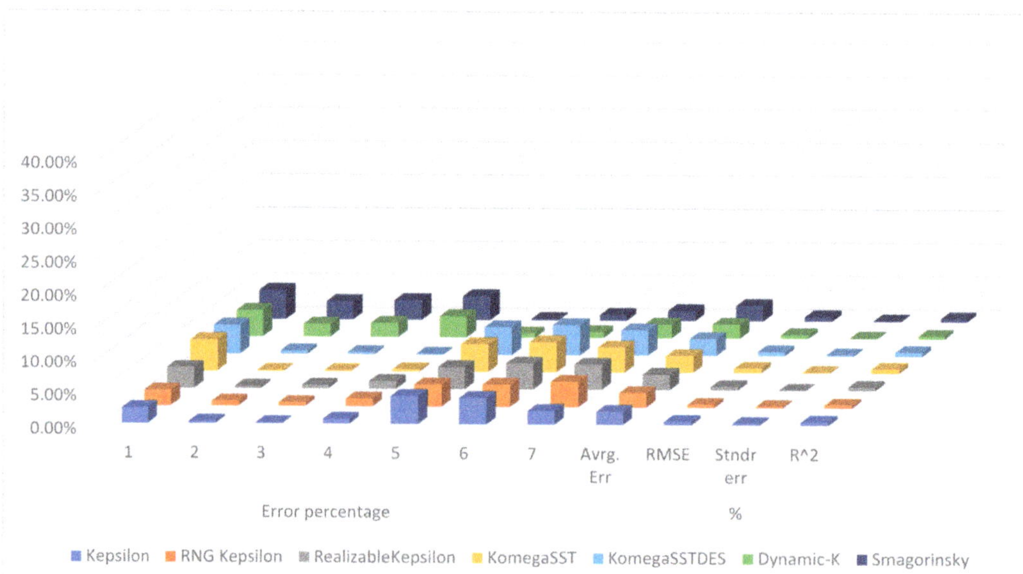

Figure 12. Errors calculated for the model results corresponding to the different turbulence models used.

At cross-section 1, the error percentages for all the models were larger than those calculated at the next cross-section (cross-section 2). Since the location of the first cross-section was chosen very close to the defined inlet of the flume, the unsteady water surface affected the results at this point.

4.1. RANS Models

The best average performance was obtained using the standard k–ε turbulence model. The other turbulence models that fall into this family showed less accuracy than the LES models. However, the difference percentages at Sections 2–4 are better than those obtained using the LES and DES models.

4.2. DES Model

The only DES model used in this study, k–ω SST DES , provided almost the same average result as those obtained using the k–ω SST , which has the greatest average error percentage among the models. The second highest average error percentage was observed for the DES family of turbulence models.

4.3. LES Model

Two LES models were used in this study. The water surface elevation results from the Smagorinsky model and dynamic K LES are similar to those of the RANS model family. The average error value, as mentioned earlier, is the lowest for dynamic K LES following the standard k–ε model.

The pressure and velocity fields for all the models follow same pattern, where the low velocity of the flow is converted to supercritical flow just after entering the throat section of the flume. The pressure field exhibits opposite behavior as the initial high-pressure flow enters the throat of the flume: the pressure drops and reaches a negative value in the throat over a short distance.

At the sections where the velocity is locally maximum, the pressure value drops dramatically. This is due the higher viscous losses occurring with the high velocity flow.

5. Conclusions

The main objective of this study was the numerical examination of fluid motion in a Parshall flume. Assessing the results of different turbulence models, such as standard k–ε, realizable k–ε, RNG k–ε, k–ω SST, k–ω SST DES, Smagorinsky, and the dynamic K LES. reveals that despite some poor performance of k–ω SST within two cross-sections, the same as the rest of the turbulence models except DES family models, this method estimates the water level accurately enough overall.

This study shows that the results from a turbulence model, in general, a CFD model like OpenFOAM, can provide reliable solutions to Parshall flume design problems. CFD models are able to simulate Parshall flumes to find the optimum design specifications for different scenarios like design changes to increase the efficiency of Parshall flumes. The results are provided with the lowest cost compared to experimental methods.

At the first location chosen to collect data (cross-section 1), the quality of simulated results was not adequate, as discussed in the previous section. To eliminate this issue, the authors recommend conducting further studies on this type of structure.

This study can be further expanded by proposing some slight changes in the design of the Parshall flume, for example, increasing the reliability in terms of the gauge reading of the stilling well at the converging section and also providing more accurate empirical relationships with respect to these changes.

It is recommended as a continuation of this study to conduct a detailed investigation into finding a proper correction factor for the numerical simulated models when different higher flow rates are used. In this study, the simulation data with a 10 L/s flowrate achieved the lowest error value compared to the experimental results, while those with larger flowrates, such as 20 or 30 L/s, where experimental data were available, demonstrated higher error values.

In order to conduct more accurate assessments of numerical model performance, it is recommended that future studies use more sources of experimental data to reduce the impact of experimental data error on the assessment of simulated results.

Author Contributions: Conceptualization, A.M. and I.N.; Data curation, M.H. and O.F.D.; Formal analysis, M.H.; Funding acquisition, A.M. and I.N.; Investigation, M.H.; Methodology, A.M. and I.N.; Project administration, A.M. and I.N.; Resources, M.H.; Software, M.H., A.M. and I.N.; Supervision, A.M. and I.N.; Validation, M.H.; Visualization, M.H.; Writing—original draft, M.H.; Writing—review & editing, A.M. and I.N. All authors have read and agreed to the published version of the manuscript.

Funding: This research received no external funding.

Conflicts of Interest: The authors declare no conflict of interest.

References

1. Parshall, R.L. *The Parshall Measuring Flume*; Agricultural Experiment Station—Colorado State University: Fort Collins, CO, USA, 1936.
2. Dursun, O.F. An experimental investigation of the aeration performance of Parshall flume and Venturi flumes. *KSCE J. Civ. Eng.* **2016**, *20*, 943–950. [CrossRef]
3. Wright, S.J.; Tullis, B.P.; Long, T.M. Recalibration of Parshall flumes at low discharges. *J. Irrig. Drain. Eng.* **1994**, *120*, 348–362. [CrossRef]
4. Willeitner, R.P.; Barfuss, S.L.; Johnson, M.C. Using numerical modeling to correct flow rates for submerged Montana flumes. *J. Irrig. Drain. Eng.* **2013**, *139*, 586–592. [CrossRef]
5. Khosronejad, A.; Herb, W.; Sotiropoulos, F.; Kang, S.; Yang, X. Assessment of Parshall flumes for discharge measurement of open-channel flows: A comparative numerical and field case study. *Measurement* **2020**, *167*, 108292. [CrossRef]
6. Davis, R.W.; Deutsch, S. A numerical-experimental study of Parshall flumes. *J. Hydraul. Res.* **1980**, *18*, 135–152. [CrossRef]
7. Robertson, E.D.; Chitta, V.; Walters, D.K.; Bhushan, S. *On the Vortex Breakdown Phenomenon in High Angle of Attack Flows over Delta Wing Geometries*; American Society of Mechanical Engineers Digital Collection: New York, NY, USA, 2015.
8. Tadayon, R. Modeling Curvilinear Flows in Hydraulic Structures. Ph.D. Thesis, Concordia University, Montreal, QC, Canada, 2009.
9. Adedoyin, A.A.; Walters, D.K.; Bhushan, S. Investigation of turbulence model and numerical scheme combinations for practical finite-volume large eddy simulations. *Eng. Appl. Comput. Fluid Mech.* **2015**, *9*, 324–342. [CrossRef]
10. Hanjalic, K. Will RANS Survive LES? A View of Perspectives. *J. Fluids Eng.* **2005**, *127*, 831–839. [CrossRef]
11. Savage, B.M.; Heiner, B.; Barfuss, S.L. Parshall flume discharge correction coefficients through modelling. *Proc. Inst. Civ. Eng. Water Manag.* **2014**, *167*, 279–287. [CrossRef]
12. Sun, B.; Zhu, S.; Yang, L.; Liu, Q.; Zhang, C.; Zhang, J. Experimental and numerical investigation of flow measurement mechanism and hydraulic performance on curved flume in rectangular channel. *Arab. J. Sci. Eng.* **2020**. [CrossRef]
13. Jasak, H. OpenFOAM: Open source CFD in research and industry. *Int. J. Nav. Archit. Ocean Eng.* **2009**, *1*, 89–94. [CrossRef]
14. Weller, H.G.; Tabor, G.; Jasak, H.; Fureby, C. A tensorial approach to computational continuum mechanics using object-oriented techniques. *Comput. Phys.* **1998**, *12*, 620. [CrossRef]
15. Imanian, H.; Mohammadian, A. Numerical simulation of flow over ogee crested spillways under high hydraulic head ratio. *Eng. Appl. Comput. Fluid Mech.* **2019**, *13*, 983–1000. [CrossRef]
16. Alfonsi, G. Reynolds-averaged navier–stokes equations for turbulence modeling. *Appl. Mech. Rev.* **2009**, *62*, 040802. [CrossRef]
17. Wilcox, D.C. *Turbulence Modeling for CFD*; DCW Industries: La Canada, CA, USA, 1998; Volume 2, ISBN 0-9636051-5-1.
18. Cable, M. An Evaluation of Turbulence Models for the Numerical Study of Forced and Natural Convective Flow in Atria. Master's Thesis, Queen's University, Kingston, ON, Canada, 2009; p. 148.
19. Shih, T.-H.; Liou, W.W.; Shabbir, A.; Yang, Z.; Zhu, J. A new k-ϵ eddy viscosity model for high reynolds number turbulent flows. *Comput. Fluids* **1995**, *24*, 227–238. [CrossRef]
20. Wang, W.; Wang, M. Application of KE model on the numerical simulation of a semi-confined slot turbulent impinging jet. In Proceedings of the 2011 Fourth International Joint Conference on Computational Sciences and Optimization, Yunnan, China, 15–19 April 2011; pp. 86–89.
21. Greenshields, C.J. *User Guide Version 6*; OpenFOAM Foundation Ltd.: London, UK, 2018.
22. Smagorinsky, J. General circulation experiments with the primitive equations: I. The Basic experiment. *Mon. Weather Rev.* **1963**, *91*, 99–164. [CrossRef]
23. Montazerin, N.; Akbari, G.; Mahmoodi, M. *Developments in Turbomachinery Flow: Forward Curved Centrifugal Fans*; Woodhead Publishing: Cambridge, UK, 2015.
24. Moin, P.; Kim, J. Numerical investigation of turbulent channel flow. *J. Fluid Mech.* **1982**, *118*, 341–377. [CrossRef]
25. Piomelli, U.; Zang, T.A. Large-eddy simulation of transitional channel flow. *Comput. Phys. Commun.* **1990**, *65*, 224–230. [CrossRef]

26. Sohankar, A.; Davidson, L.; Norberg, C. A dynamic one-equation subgrid model for simulation of flow around a square cylinder. In *Engineering Turbulence Modelling and Experiments 4*; Elsevier: Amsterdam, The Netherlands, 1999; pp. 227–236, ISBN 978-0-08-043328-8.
27. Davidson, L. *Large Eddy Simulations: A Note on Derivation of the Equations for the Subgrid Turbulent Kinetic Energies*; Technical Report 97/11; Department of Thermo and Fluid Dynamics, Chalmers: Gothenburg, Sweden, 1997.
28. Kim, W.-W.; Menon, S. A new dynamic one-equation subgrid-scale model for large eddy simulations. In *33rd Aerospace Sciences Meeting and Exhibit*; School of Aerospace Engineering, Georgia Institute of Technology: Atlanta, GA, USA, 1995; p. 356.
29. Brown, G.J.; Fletcher, D.F.; Leggoe, J.W.; Whyte, D.S. Application of hybrid RANS-LES models to the prediction of flow behaviour in an industrial crystalliser. *Appl. Math. Model* **2020**, *77*, 1797–1819. [CrossRef]
30. Baker, C.; Johnson, T.; Flynn, D.; Hemida, H.; Quinn, A.; Soper, D.; Sterling, M. Chapter 4—Computational techniques. In *Train Aerodynamics*; Baker, C., Johnson, T., Flynn, D., Hemida, H., Quinn, A., Soper, D., Sterling, M., Eds.; Butterworth-Heinemann: Oxford, UK, 2019; pp. 53–71, ISBN 978-0-12-813310-1.
31. Lindblad, D.; Jareteg, A.; Petit, O. *Implementation and Run-Time Mesh Refinement for the $k-\omega$ SST DES Turbulence Model When Applied to Airfoils*; Chalmers University of Technology: Gothenburg, Sweden, 2014.
32. Menter, F.R.; Kuntz, M.; Langtry, R. Ten years of industrial experience with the SST turbulence model. *Turbul. Heat Mass Transf.* **2003**, *4*, 625–632.
33. Li, X.; Zhao, J. Dam-break of mixtures consisting of non-Newtonian liquids and granular particles. *Powder Technol.* **2018**, *338*, 493–505. [CrossRef]
34. Morden, J.A.; Hemida, H.; Baker, C.J. Comparison of RANS and detached eddy simulation results to wind-tunnel data for the surface pressures upon a class 43 high-speed train. *J. Fluids Eng.* **2015**, *137*. [CrossRef]
35. Spalart, P.; Jou, W.-H.; Strelets, M.; Allmaras, S. *Comments on the Feasibility of LES for Wings, and on a Hybrid. RANS/LES Approach*; Greyden Press: Dayton, OH, USA, 1997.

hydrology

MDPI

Article

Assessment of Impacts of Climate Change on Tile Discharge and Nitrogen Yield Using the DRAINMOD Model

Golmar Golmohammadi [1,*], Ramesh P. Rudra [1], Gary W. Parkin [2], Priyantha B. Kulasekera [2], Merrin Macrae [3] and Pradeep K. Goel [4]

[1] School of Engineering, University of Guelph, Guelph, ON N1G 2W1, Canada; rrudra@uoguelph.ca
[2] School of Environmental Sciences, University of Guelph, Guelph, ON N1G2W1, Canada; gparkin@uoguelph.ca (G.W.P.); kulasekp@uoguelph.ca (P.B.K.)
[3] Department of Geography and Environmental Management, University of Waterloo, Waterloo, ON N2L 3G1, Canada; mmacrae@uwaterloo.ca
[4] Ontario Ministry of the Environment, Conservation and Parks, Etobicoke, ON M9P 3V6, Canada; Pradeep.goel@ontario.ca
* Correspondence: ggolmoha@uoguelph.ca; Tel.: +1-519-766-6786

Abstract: The detrimental impacts of agricultural subsurface tile flows and their associated pollutants on water quality is a major environmental issue in the Great Lakes region and many other places globally. A strong understanding of water quality indicators along with the contribution of tile-drained agriculture to water contamination is necessary to assess and reduce a significant source of non-point source pollution. In this study, DRAINMOD, a field-scale hydrology and water quality model, was applied to assess the impact of future climatic change on depth to water table, tile flow and associated nitrate loss from an 8.66 ha agricultural field near Londesborough, in Southwestern Ontario, Canada. The closest available climate data from a weather station approximately 10 km from the field site was used by the Ontario Ministry of Natural Resources and Forestry (MNRF) to generate future predictions of daily precipitation and maximum and minimum air temperatures required to create the weather files for DRAINMOD. Of the 28 models applied by MNRF, three models (CGCM3T47-Run5, GFDLCM2.0, and MIROC3.2hires) were selected based on the frequency of the models recommended for use in Ontario with SRA1B emission scenario. Results suggested that simulated tile flows and evapotranspiration (ET) in the 2071–2100 period are expected to increase by 7% and 14% compared to 1960–1990 period. Results also suggest that under future climates, significant increases in nitrate losses (about 50%) will occur along with the elevated tile flows. This work suggests that climate change will have a significant effect on field hydrology and water quality in tile-drained agricultural regions.

Keywords: field-scale hydrological models; DRAINMOD; tile drainage; nitrate leaching; climate change

Citation: Golmohammadi, G.; Rudra, R.P.; Parkin, G.W.; Kulasekera, P.B.; Macrae, M.; Goel, P.K. Assessment of Impacts of Climate Change on Tile Discharge and Nitrogen Yield Using the DRAINMOD Model. *Hydrology* **2021**, *8*, 1. https://dx.doi.org/10.3390/hydrology8010001

Received: 26 November 2020
Accepted: 21 December 2020
Published: 26 December 2020

Publisher's Note: MDPI stays neutral with regard to jurisdictional claims in published maps and institutional affiliations.

1. Introduction

Field agriculture is a significant land use in many areas of the world and has been documented as a contributor of non-point source pollution to surface waters. In a recent study in Uruguay, [1] researchers found a strong correlation between total phosphorus (P) and agriculture land use in a river. In central Asia, remote sensing was used to analyze effects of land use on surface water contamination [2]. Although the relationships were not straightforward, they did find that cropland was a significant contributor to surface water organic pollution. In another study in Europe, [3], analyses of nutrients and other components in river water found that the proportion of arable land in the watershed did affect water pollution levels. To further investigate agricultural land use as a non-point source of pollution, the hydrologic pathway of a pollutant from field to surface water body must be identified and quantified. This study focusses on quantifying tile flow and accompanying nitrate load to surface waters under future climate conditions of the Great Lakes region of North America.

Subsurface (tile) drainage is considered an effective and important practice in field agricultural water management to facilitate farm operations, especially in cold climatic regions. In Ontario, Canada, more than 50% of its arable land is under artificial drainage systems [4]. Tile drainage increases productivity of land by increasing crop yields, but also impacts the environment by degrading water quality. Indeed, it is widely recognized that nitrogen (N) and P leaching results in elevated nutrient concentrations in surface waters representing both environmental and economic (fertilizer loss) impacts. High levels of nitrate may even remain in tile drainage several years after nitrogen fertilizer reduction ([5,6]). Hence, to reduce contamination of surface waters, the impacts of changing cropping systems that may, for instance, occur under future climate regimes are essential. Although climate change is anticipated to play an important role in subsurface drainage, it is not possible to characterize the potential impacts of climate change on field hydrology and/or water quality in agricultural watersheds using field observations. However, modeling of these potential impacts does have potential to at least give acceptable trends in future tile flow and associated nitrate loss. For instance, [7] used the Root Zone Water Quality Model (RZWQM) to estimate the impact of agricultural management systems adaptations on gaseous and drainage nitrogen (N) losses in Iowa, USA. They found that the optimal N rate to minimize loss and maximize maize production was 120 kg N ha^{-1}. Using DRAINMOD, researchers [8] were able to suggest when to block tile outflow to keep the groundwater table at an acceptable depth in central-western Poland. In Illinois, [9], assessed impacts of changing atmospheric conditions using RZWQM along with Support Decision for Agrotechnology Transfer (DSSAT). They found that increasing rainwater nitrate concentration had a moderate impact on nitrate loss in drainage tiles.

In Canada mainly watershed-scale studies on impacts of climate change on water resources have been completed by [10–14]. In their study on a watershed in Quebec, [10], simulated future streamflows finding a slight decrease in annual runoff. In another study, [11], used a weather generator to recommend local water resource management adaptations for a watershed near the present study's site in Ontario. Using streamflow data and the LARS-WG weather generator, the Soil and Water Assessment Tool (SWAT) model was used by [12] to show that streamflow in a southern Ontario watershed could increase by 12% compared to base period 1961–1990. Also, in southern Ontario, [13] used model simulations to estimate the increase in winter streamflow in several large watersheds. A future assessment of groundwater nitrate concentrations in a sub-watershed of southern Ontario was conducted by [14]. They found that an agricultural BMP was very effective in reducing nitrate in groundwater under a crop rotation system.

In the Great Lakes region, where this study was conducted, there have been several recent studies on impacts of climate change on tile flow and nutrient losses at the watershed scale. The WEPP-WQ model was used by [15] to estimate future N and P losses in two small watersheds finding increases in losses of both nutrients. In another study [16], used DRAINMOD to simulate future tile flow in western Lake Erie basin. They found an average decrease of about 9% in subsurface drainage and recommended controlled drainage to retain more water in the soil profile as a BMP for crop production. In the same basin [17], using the SWAT model found an increase in subsurface drain flow but a decrease in P load from drains. On the other hand, relatively few field-scale studies, which use field data to first calibrate and then validate a tile-flow model to assess climate change impacts have been attempted in the region. Field-scale studies have the advantage over watershed-scale studies of a better-defined source area with associated soil and cropping properties for assessing impacts of future climate on tile flow and accompanying nutrient losses. Using DRAINMOD [18], examined impacts of different tillage practices on future tile flow and nitrate loss in eastern Ontario. They estimated greater future nitrate loss under no-tillage than conventional tillage. At the same field site as the work presented here, The SWAT model was applied to future climate data similar to the approach used in this study [19]. In addition, their results indicated a shift in seasonal water balance and an associated increase in annual P losses by year 2100. Main differences between this study and [19] are use of a different model and modeling approach and analysis of N loss in this study as opposed to P loss in the [19] study.

The DRAINMOD model was developed by [20] as a process-based distributed field-scale model. Many studies have assessed and applied DRAINMOD including, to name a few [21–24]. The DRAINMOD model has been extensively revised and updated over the years and is still undergoing improvements [25,26]. The main reasons why it was chosen for this study are three-fold: it includes impacts of winter-season processes on soil hydrology along with soil nitrogen dynamics and it has been previously tested and proven satisfactory in Ontario and other cold regions (e.g., [18,27–38]). For information on the structure and details on the applied processes in DRAINMOD see publications by [22,39].

Previous work at the same field site as used in this study includes analyses of P transport by [19,40–44]. As well, [45] examined the contribution of preferential flow to tile drainage.

Based on the preceding literature review, this study was undertaken to address the paucity of field-scale studies of how tile flow and nitrate loss may change under future climate regimes in the Great Lakes region. As climate changes there is the potential for shifts in agricultural practices, which may lead to increased nitrate leaching [46]. Hence more studies such as herein are needed to establish a baseline nitrate loading using current field crop practices. In a recent review by [47], it was noted that it is important to improve field-scale hydrology and water quality models under cold climates found in Canada and elsewhere in high-latitude regions. Studies that apply field data to calibrate and validate models, as done in this study, will lead to improvements in model accuracy in the future.

The objectives of this study were: (1) to calibrate and evaluate the latest version of DRAINMOD 6.1 using data from a field site in southern Ontario; and (2) to assess the impact of climate change on tile discharge and nitrogen yields at the same field site using the calibrated model. Using the CGCM3T47 with SRA1B emission scenario, we estimate the impact of future climate on tile discharge and its accompanying nitrate loading for the next century. This research may assist the farming community in developing adaptation strategies to minimize negative impacts of tile drainage systems on water resources.

2. Materials and Methods

2.1. Description of Study Area

The study site (LON) is in southern Ontario, Canada (UTM 472219 E and 4767583 N) near Londesborough. The study was conducted on an 8.66 ha agricultural field with both overland flow and tile drainage restricted to the study field. The site was under a reduced overland tillage (RT) and data collection spanned a range of years and therefore experienced a range of climatic conditions. Tile drainage systems in the field are systematically drained at 90 cm depth through 10-cm diameter laterals (13.5 m spacing) that connected to a larger main tile (20-cm diameter) that exits at the edge of the field. The contributing area to surface drainage within the field is 7.79 ha.

Long-term average annual precipitation measured at Blyth, Ontario, the closest Environment Canada weather station to LON site, is 1247 mm and long-term mean annual temperatures are 7.2 °C (https://climate.weather.gc.ca/historical_data/search_historic_data_e.html). Monthly temperatures vary seasonally across the year, with warm summers and cold winters. Daily maximum and minimum temperatures are variable throughout the year; however, daily mean air temperatures are generally below freezing between December and March. Due to the climate in Ontario, spring planted crops are seeded in early May, whereas fall seeded crops (winter wheat) are planted in September. Harvests are completed by early August (wheat), mid-September (soy) or November (corn).

The topography of the site is gently undulating with slopes ranging from 0.2 to 3.5% [48]. The field consists of soils from the Perth Clay Loam association [49]. These soils developed on clay loam glacial deposits with imperfect drainage. Soil samples collected at the site suggest that the average texture in the top 15 cm is silt (clay 0.9 ± 1.3%, silt 75.7 ± 2.1%, sand 23.4 ± 3.3%) [48]. Table 1 gives some basic soil properties for the LON site.

Table 1. Some soil Properties at Londesborough field site.

Soil Properties				
Depth (cm)	0–10	10–20	20–35	35–150
Soil hydraulic conductivity (cm/h)	0.26	0.07	0.2	0.26
Soil pH		7.7 ± 0.3		
Organic matter (%)		4.1 ± 0.7		

2.2. Model Formulation

The DRAINMOD model requires different inputs including air temperature and precipitation data, infiltration parameters, soil properties, crop information, and drainage system parameters. These data were used to calculate the relationships among the drained volume and depth to water table (WTD), and relationships between WTD and maximum steady upward flux. Infiltration was simulated using the Green–Ampt equation. Within the model, a soil moisture retention curve for the soil above the tile was used to calculate infiltration parameters as a function of WTD.

The model provides options to the user to use observed ET data or apply daily maximum and minimum temperatures to calculate ET using the Thornthwaite equation. In this study, Thornthwaite approach has been used. Details related to crop management and timing for the study site are given in Table 2.

Table 2. Management practices considered in the study.

Crop	Date	Activity Code
Soybeans	Mid-October 2010 (assume 15 October 2010)	Harvest
Winter Wheat	Mid-October 2010 (Considered 16 October 2010)	Plant
	April 11, 2011	Red Clover Air Seeded
	Late April 2011 (Considered 25 April 2011)	Fert App
	Late July 2011 (Considered 25 July 2011)	Harvest
Grain Corn	5 October 2011	Fert App
	5 October 2011	Spray
	11 November 2011	Tillage
	18 April 2012	Spray
	25 April 2012	Tillage
	26 April 2012	Plant
	26 April 2012	Fert App
	23 May 2012	Fert App
	9 June 2012	Spray
	21 October 2012 (estimate as no rain that day)	Harvest
Soybeans	9 November 2012	Tillage
	3 May 2013	Plant
	24 May 2016	Spray
	25 September 2013	Harvest

Table 2. *Cont.*

Crop	Date	Activity Code
Winter Wheat	27 September 2013	Plant
	19 April 2014	Plant CC
	26 April 2014	Fert App
	1 June 2014	Spray
	7–10 August 2014	Harvest
Grain Corn	9 October 2014	Fert App
	9 October 2014	Spray
	17 October 2014	Tillage
	19 April 2015	Spray
	27 April 2015	Tillage
	28 April 2015	Plant
	28 April 2015	Fert App
	May 15/15 (estimate)—check weather station	Spray
	25 May 2015	Fert App
	15 July 2015	Spray
	17 October 2015	Harvest
Soybeans	9 November 2015	Tillage
	3 May 2016	Plant
	24/05/2016 (estimate)—check weather station	Spray
		Harvest

Many of the input parameters for DRAINMOD are transferable between sites from nearby cold regions. Hence, parameters related to snow melt, ice content etc. were obtained from the literature, such as [29,30] and [18]. Crop-related constants including nitrogen uptake and transformation factors and organic matter dynamics were taken from [5,31,33,34,50,51].

The lower boundary at 1.2 m was assumed to be impermeable and at a constant soil temperature of 7 °C, which is approximately the long-term average air temperature [52].

The model requires initial conditions such as concentrations of NO_3-N and NH_4^+N, given in Table 3.

Table 3. Initial nitrogen concentrations in four soil depth ranges at the study site.

Initial NO_3-N concentration in soil (mg L^{-1})	
0–0.15 m	5.17
0.15–0.30 m	2.05
0.30–0.60 m	1.62
0.60–1.20 m	1.58
Initial NH_4^+-N concentration in soil (mgl^{-1})	
0–0.15 m	3.22
0.15–0.30 m	2.62
0.30–0.60 m	1.74
0.60–1.20 m	1.41
NH_4^+ sorption distribution coefficient (cm^3g^{-1}) K_d	
0–0.15 m	2.17
0.15–0.30 m	2.32
0.30–0.60 m	2.85
0.60–1.20 m	3.61

2.3. Model Calibration

To calibrate and assess the DRAINMOD model, measurements of tile discharge collected in the field from 2011–2016 were divided into two parts: data from the first two years (fall of 2011–summer of 2013), were used to calibrate DRAINMOD, and the remaining two years of the data (fall of 2013–early winter of 2016), were used to validate the model. The chosen calibration parameters were manually adjusted by minimizing differences between observed and simulated tile flow at first visually and then statistically. The most sensitive parameters were selected for calibration based on previous work using DRAINMOD [18,27,28,32,34–36,50]. As in other studies, hydrology was first used to calibrate DRAINMOD using K_{sat} of soil profile and restrictive bottom layer and maximum surface storage which controls water runoff (Table 4).

Table 4. Parameters Used in Calibration of DRAINMOD model.

Parameter			Value
Saturated hydraulic conductivity (4 layers, cm h^{-1})			Varies
Vertical hydraulic conductivity of restrictive layer (cm h^{-1})			0.0025
Maximum surface storage (mm)			1.6
Rooting depth (cm)	Month	Day	
	1	1	5.0
	5	5	5.0
	6	15	24.0
	7	15	36.0
	7	30	45.0
	8	31	45.0
	10	15	5.0
	12	31	5.0
Mineralization rate (d^{-1})			NT:0.00002 CT:0.00003
Maximum Nitrification rate (μgN$g^{-1}d^{-1}$)			NT:9.5 CT:7.7
Nitrification optimum temperature (°C)			20
Nitrification half saturation constant (μgN$g^{-1}d^{-1}$)			90
Maximum denitrification rate (μgN$g^{-1}d^{-1}$)			NT:4.0 CT:3.0
Nitrification empirical shape factor			0.5
Nitrification Optimum temperature (°C)			23.0
Nitrification half saturation constant (μgN$g^{-1}d^{-1}$)			30.0

Simulations were initially evaluated using visual comparisons between the observed and predicted values. These were subsequently assessed using various statistical parameters. The statistical parameters used were the coefficient of determination (R^2), the percent bias (PBIAS) [53] and the Nash-Sutcliffe efficiency (NSE; [54]). These parameters were calculated according to:

$$R^2 = \left[\frac{\sum_{i=1}^{i=n} (O_i - \overline{O})(P_i - \overline{P})}{\sqrt{\sum_{i=1}^{i=n} (O_i - \overline{O})^2} \sqrt{\sum_{i=1}^{i=n} (P_i - \overline{P})^2}} \right]^2 \quad (0 \leq R^2 \leq 1) \tag{1}$$

$$PBIAS = \frac{100 \sum_{i=1}^{i=n} O_i - P_i}{\sum_{i=1}^{i=n} O_i} \quad (2)$$

$$NSE = 1 - \left[\frac{\sum_{i=1}^{i=n} (O_i - P_i)^2}{\sum_{i=1}^{i=n} (O_i - \overline{O})^2} \right] \quad (3)$$

where, n is the total number of compared values, O_i is the ith observed value, $\overline{O}$ is the mean of observed values, P_i is the ith predicted value, $\overline{P}$ is the mean of predicted values.

2.4. Climate Change Simulations

Global Climate Models (GCM's) or Regional Climate Models (RCM's) are used to generate future climate data needed to assess the possible impacts of future climate changes on drainage and runoff in a watershed. These circulation models are based on physics and provide accurate predictions under different greenhouse gas emission criteria groups, defined by Intergovernmental Panel on Climate Change (IPCC) in their Special Report on Emission Scenarios (SRES) [55] The Ontario Ministry of Natural Resources and Forestry, Canada provided a facility through its web portal Aquamapper (http://climate/aquamapper.com/) to generate future climate data using up to 28 GCMs and RCMs under three gas emission scenarios of SRB1, SRA1B, and SRA2, and SRA2, used in future climate impact studies [56]. Emission scenario SRB1 describes a converging future world with constant population experiencing rapid structural changes towards an economy of service and information by introducing clean and resource-efficient technologies; whereas SRA1B describes a future scenario with rapid economic growth where population peaks at the middle of the century, and rapid introduction of new energy efficient technologies while development is being balanced across energy sources. A highly heterogeneous world with continuously increasing population with a fragmented and slower processing regional economic growth is described by emission scenario SRA2 [57,58].

For this study, we selected the emission scenario presented by SRA1B, which has been recommended for most of the 12 regions of Ontario considered in [23]. SRA1B is also considered to be one of the scenarios across Ontario by Ontario Climate Change Data Portal [59]. This emission scenario assumes the same level of socio-economic–technological growth throughout the region with a rapid economic growth, introduction of efficient new technologies, and has a balance among all energy sources. However, with variable grid sizes of hundreds of kilometers from model to model [60,61], the model predictions may lack the precision needed for smaller sites [62–64].

Since the high coarseness of the spatial (45 km) and temporal (daily) resolution of the GCMs to represent the physical processes of convection, land atmospheric interactions, and especially in predicting future rainfall intensity-duration-frequency characteristics, downscaling may be needed for obtaining a more accurate picture of the future climate scenario [61,64–66], which we have not attempted in this study.

For this study site near Londesborough, we selected Blyth (43°43′, 81°23′, Elevation = 350.5 m), the closest weather station presented in Aquamapper, to generate future predictions of daily precipitation, maximum and minimum air temperatures required to create the weather files for DRAINMOD. Out of the 28 models in Aquamapper, three models (CGCM3T47-Run5, GFDLCM2.0, and MIROC3.2 MedRes) were selected based on the frequency of the models recommended for use in Ontario with SRA1B emission scenario. The base data from 1971 to 2000 was used to generate the climate data for 2011–2040, 2041–2070, and 2071–2100.

DRAINMOD was run using the future climate data described above and weather data from 1960–1990 as a comparison. The DRAINMOD input parameters from the calibration exercise, also described above, were used in all simulations. By using the pre-existing soil properties and agricultural management system, the future simulations should yield differences between past and future tile flow and nitrate loading only due to climate change.

3. Results and Discussion

3.1. Model Calibration

Several trial-and-error runs were performed by adjusting K_{sat} of soil profile and restrictive bottom layer and maximum surface storage. The model results were the most sensitive, in order, as given in Table 4. Crop rooting depth was also varied to calibrate DRAINMOD (Table 4). Following the soil hydrology calibration, various nitrogen factors were determined using the same approach. Calibrated values are listed in Table 4, which are similar to those given in previous work (e.g., [21,33,35,51]).

3.2. Model Evaluation

3.2.1. Field-Scale Tile Discharge

Figures 1 and 2 show time series graphs of observed and simulated daily tile discharge and precipitation during the calibration and validation periods. Figure 1, showing the calibration period, indicates that measured tile-flow events correspond well with precipitation. DRAINMOD tile-flow peaks almost consistently match the timing of major precipitation events; however, it appears to underestimate more often than overestimate some peak flow values. Please note that DRAINMOD correctly shows no tile flow when precipitation occurs during dry soil conditions in the summer season. A similar pattern is shown in Figure 2 during the validation period. DRAINMOD does a very good job of when tile flow occurs.

The model performance during the calibration and validation stages at daily and monthly time steps are shown in Table 5.

Observed and simulated drain discharge were in close agreement at both daily and monthly time scales during the calibration period. The statistical values given in Table 5 for the monthly time interval are within the acceptable ranges for both NSE ($\geq$0.65) and PBIAS ($\leq \pm$15%), as suggested by [67]. The PBIAS values are very similar for monthly and daily time intervals.

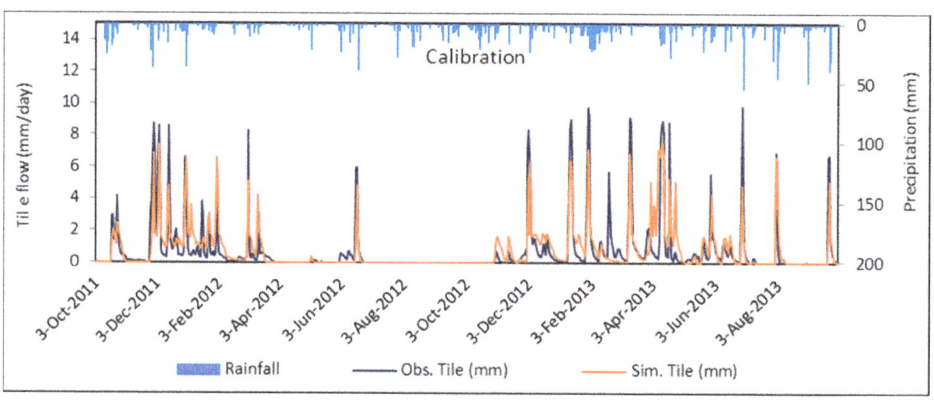

Figure 1. Measured tile discharge (blue line) and precipitation (blue bars) and simulated tile discharge (orange lines) during the model calibration.

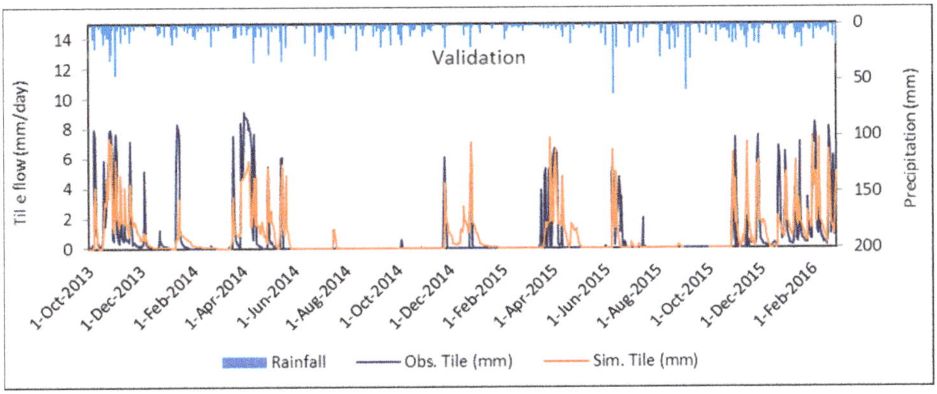

Figure 2. Measured tile discharge (blue line) and precipitation (blue bars) and simulated tile discharge (orange lines) during the model validation.

Table 5. Monthly and daily calibration and validation statistics.

Statistical Index	Monthly		Daily	
	Calibration	Validation	Calibration	Validation
R^2	0.78	0.74	0.65	0.54
NSE	0.76	0.70	0.62	0.62
PBIAS (%)	1.2	1.4	1.1	1.4

3.2.2. Field-Scale Nitrogen Yields

After DRAINMOD was calibrated and validated to successfully simulate tile discharge, a series of nitrogen simulations were done to calibrate the nitrogen component of the model. Ammonium yields were excluded from the model evaluation because both field measurements and model simulations exhibited very small $\mathbf{NH_4^+}$**-N** yields in tile drain effluent.

The calibration and validation results for daily and cumulative NO_3-N yields are shown in Figure 3. In general, observed and predicted NO_3-N yields were in reasonable agreement. During the calibration period, there were two major nitrate-loss events corresponding to significant tile-flow events. The June 2012 nitrate loss occurred during the first major precipitation event following fertilizer application in the spring. However, the major loss of nitrate in April 2013 was not associated with a particular fertilizer application but followed a winter season after corn was grown (Table 2). Figure 3 shows that there is a strong linear relationship between simulated and observed with the coefficient of determination only slightly lower during the validation period than calibration.

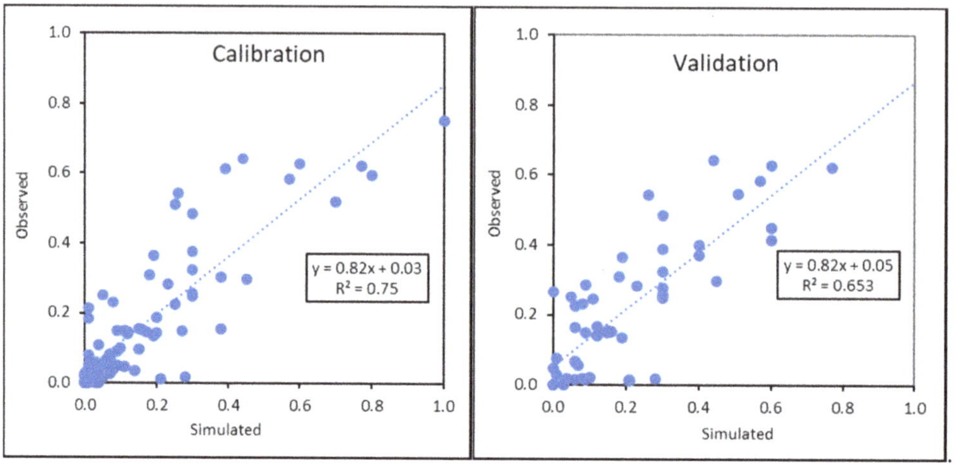

Figure 3. Observed and DRAINMOD-simulated daily NO_3-N (kg/ha) from tile outflow at Londesborough field site during calibration (**left**) and validation (**right**) periods.

3.3. Climate Change Analysis

3.3.1. Comparing Past and Future Climates

Figure 4a,d compares annual measured precipitation and DRAINMOD-estimated evapotranspiration for 1960–1990 to three different climate-model estimates of precipitation and DRAINMOD-estimated evapotranspiration for three future time periods. As well, Figure 4b,c compare past-measured and future-estimated daily average maximum and minimum air temperatures using the three different climate models. Please note that both precipitation and evapotranspiration increase in the future as determined in other studies. There are small differences between the three climate-model estimates of precipitation, but model CGCM3T47 does predict somewhat lower future temperature and corresponding evapotranspiration. Therefore, it has been selected as the scenario of future climate change assessment of tile flow and nitrate loss at the site. It represents the least temperature and evapotranspiration increase giving the most predicted future surplus water and, hence, presumably the worst-case scenario for increases in tile flow and corresponding nitrate loss.

Figure 5 shows past and future precipitation and temperature estimates from CGCM3T47 on annual and seasonal bases. Annual precipitation estimates increase with time with a difference of over 20% from the near- to far-future time periods. Precipitation is divided roughly evenly between seasons; however, summers become slightly drier and other seasons slightly wetter with time. These changes could have significant impacts on reducing crop production in summer and a greater potential for nitrogen loss through denitrification in non-cropping seasons. In terms of average daily temperatures, both minimum and maximum increase in every season almost without exception as time progresses. The summer season appears to show the greatest increase in temperatures from beginning to end of the modeled time period.

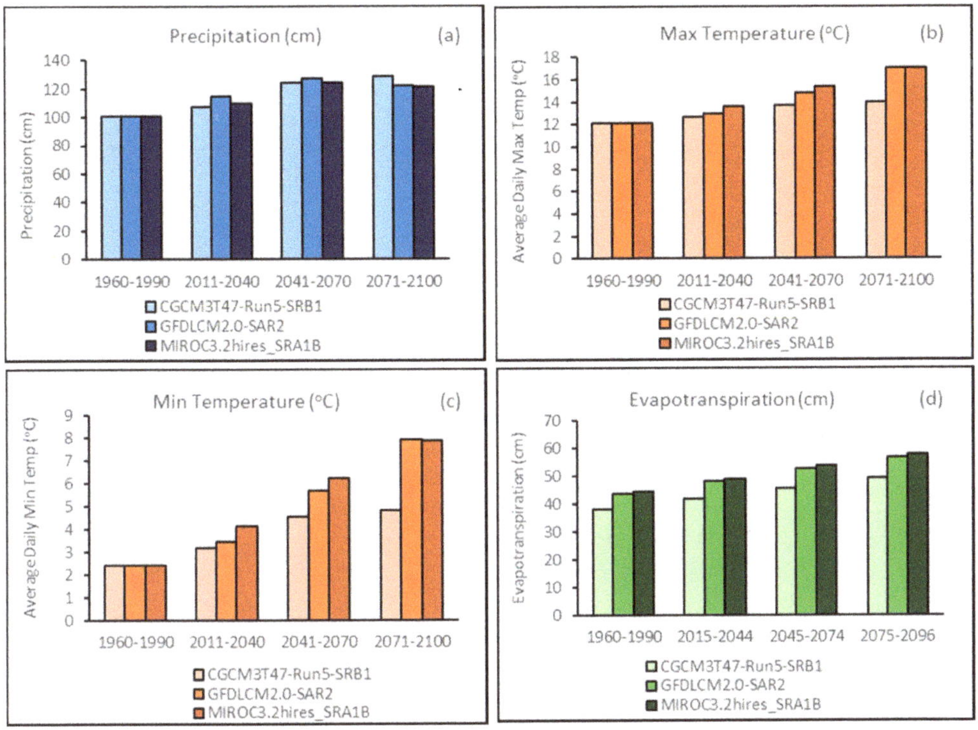

Figure 4. Comparing future climate scenarios ((**a**) precipitation; (**b**) maximum average daily temperature; (**c**) minimum average daily temperature; (**d**) evapotranspiration) of the three different climate models applied in this study. Please note that for comparison, measured average annual precipitation, average daily maximum and minimum temperatures for 1960–1990 were 100 cm, 12 °C and 2.3 °C, respectively.

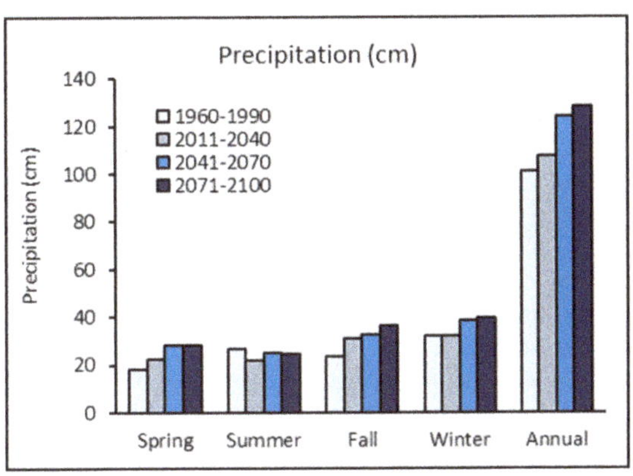

Figure 5. *Cont.*

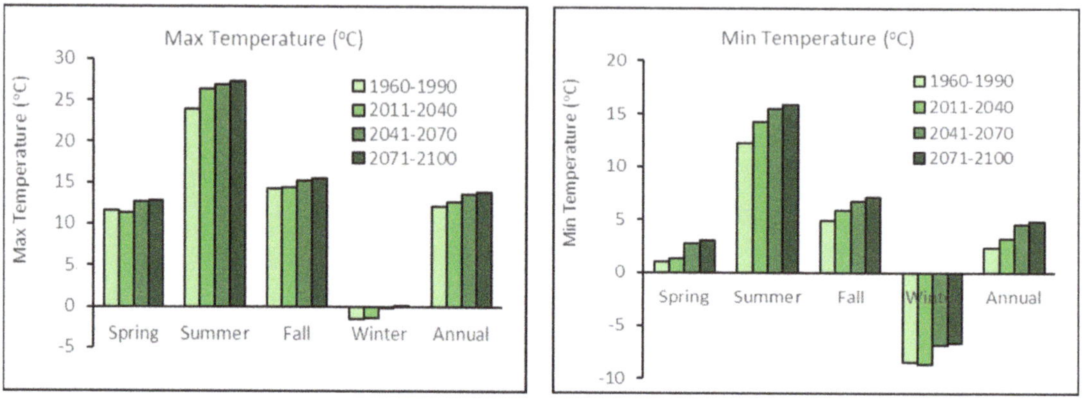

Figure 5. Historical (1960–1990) and future (2011–2100) predicted annual and seasonal mean precipitation (**upper figure**) and max/min daily average temperatures (**lower figures**) using CGCM3T47 model.

3.3.2. Effects of Climate Change on Water Balances

The DRAINMOD model was used to predict the impact of future climates on field-scale water balance (Figure 6) and the associated nitrate yields from the LON field site in Ontario.

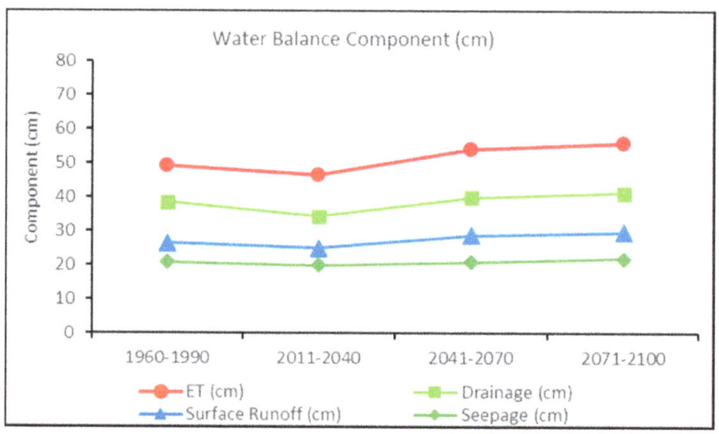

Figure 6. Simulated water balance components for 1960–1990, and 2011–2100 periods.

Evapotranspiration and tile flow first decreased and subsequently increased due to increased precipitation and temperatures under future climate conditions. The comparison of simulated results for the 2071–2100 period (late century) with 1960–1990 period (historical) show that tile flow and ET were found to increase by 7% and 14%.

3.3.3. Effects of Climate Change on Tile Discharge and Nitrogen Yields

Mean seasonal and annual tile discharge for the historical and future time periods is shown in Figure 7. As noted above, the DRAINMOD-estimated annual tile discharge increased slightly during the future period (vs. historical). The winter season appears to dominate this annual increase. The increase in tile flow during winter likely occurs because the estimated maximum daily temperature rises above 0 °C in the future (Figure 5). This would result in more snow melt and more precipitation occurring as rain. In support of this postulation, [13] predicted that streamflow will increase during winter season in the future in several southern Ontario watersheds. As well, [68] found that tile flow is expected to increase based on their study of four Lake Erie watersheds.

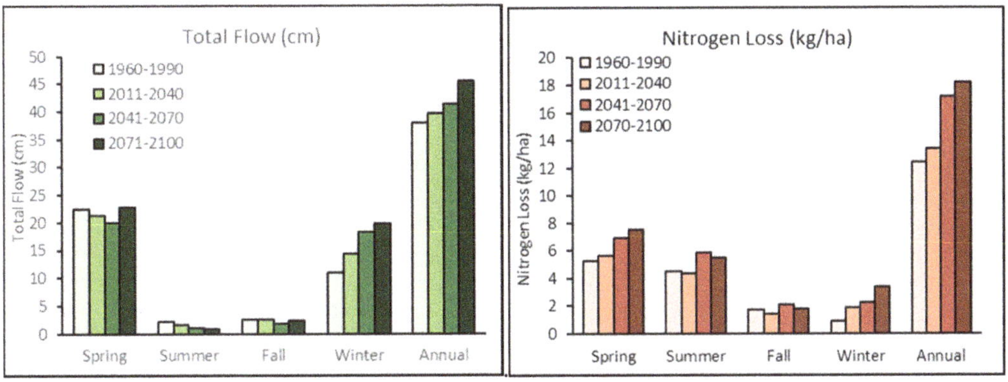

Figure 7. Historical (1960–1990) and future (2011–2100) simulated seasonal and annual average total tile flows (**left**) and NO_3-N loss (**right**).

Figure 7 also shows past and future nitrate loss as estimated by DRAINMOD. Increasing loss in winter season as time progresses matches well with tile-flow estimates. However, the steady increase in nitrate loss in spring does not correspond exactly with tile flow. Average annual estimated tile flow during spring season is variable with time. Hence, the concentration of nitrate must increase in tile water during spring to account for this steady increase in load. Other studies have revealed a range of results. For instance, [69] found very little change in future annual nitrogen loss from watersheds of various sizes in northeastern Indiana. In a study using SWAT model, [70], found that annual P loading would decrease in the future from a watershed in Lake Erie basin due to increased evapotranspiration and decreased snowfall. On the other hand, [14,18,19,68,71,72] all found that nutrient losses may increase in future in Ontario and Quebec watersheds.

4. Conclusions

This study has once again shown that DRAINMOD is an effective field-scale model for simulating tile flow in Ontario, Canada. The model calibrated and validated well in comparison to tile-flow measurements collected from a single field site. The calibrated model was used to predict future tile flow and nitrogen loss from the same study site. Tile flow is estimated to increase especially in winter in the future. This result is not uncommon when compared to other studies. However, nitrogen loss appears to be more complicated than tile flow. Although nitrate loss appears to increase consistently with increasing tile flow in the future, increases in nitrate loss through tiles does not correlate well with tile flow. Perhaps an increase in winter and spring season temperatures, especially minimum temperature in spring, is leading to enhanced nitrification and hence excessive nitrate loss during spring.

Author Contributions: Conceptualization, G.G., M.M., R.P.R., G.W.P., P.B.K., P.K.G.; Methodology, G.G., M.M., G.W.P., R.P.R., P.B.K., P.K.G.; Software, G.G., R.P.R., G.W.P., P.K.G.; Formal Analysis, G.G., M.M., G.W.P., R.P.R., P.B.K., P.K.G.; Data Curation, M.M.; Writing—Original Draft Preparation, G.G.; Writing—Review & Editing, P.K.G., G.W.P., P.B.K., M.M., G.G., R.P.R. Supervision, R.P.R., G.W.P.; Project Administration, G.W.P., M.M.; Funding Acquisition, G.W.P., M.M., R.P.R. All authors have read and agreed to the published version of the manuscript.

Funding: This research received no external funding.

Data Availability Statement: Not Applicable.

Acknowledgments: The funding received for this project from Ontario Ministry of Agriculture, Food and Rural Affairs (OMAFRA) is gratefully acknowledged.

Conflicts of Interest: The authors declare no conflict of interest.

References

1. Gorgoglione, A.; Gregorio, J.; Ríos, A.; Alonso, J.; Chreties, C.; Fossati, M. Influence of Land Use/Land Cover on Surface-Water Quality of Santa Lucía River, Uruguay. *Sustainability* **2020**, *12*, 4692. [CrossRef]
2. Wang, X.; Zhang, F. Effects of land use/cover on surface water pollution based on remote sensing and 3D-EEM fluorescence data in the Jinghe Oasis. *Sci. Rep.* **2018**, *8*, 1–13. [CrossRef]
3. Kändler, M.; Blechinger, K.; Seidler, C.; Pavlů, V.; Šanda, M.; Dostál, T.; Krása, J.; Vitvar, T.; Štich, M. Impact of land use on water quality in the upper Nisa catchment in the Czech Republic and in Germany. *Sci. Total Environ.* **2017**, *586*, 1316–1325. [CrossRef]
4. O'Neill, P. Modelling Tile Drains under Present and Future Climate Conditions. Master's Thesis, University of Waterloo, Waterloo, ON, Canada, 2008.
5. Patni, N.K.; Masse, L.; Jui, P.Y. Tile effluent quality and chemical losses under conventional and no tillage. Part1. Flow and Nitrate. *Trans. ASAE* **1996**, *39*, 1673–1679. [CrossRef]
6. Sharma, V.; Negi, S.C.; Rudra, R.P.; Yamg, S. Neural networks for predicting nitrate-nitrogen in drainage water. *J. Agric. Water Manag.* **2003**, *63*, 169–183. [CrossRef]
7. Wang, Z.; Qi, Z.; Xue, L.; Bukovsky, M. RZWQM2 simulated management practices to mitigate climate change impacts on nitrogen losses and corn production. *Environ. Model. Softw.* **2016**, *84*, 99–111. [CrossRef]
8. Sojka, M.; Kozłowski, M.; Kęsicka, B.; Wróżyński, R.; Stasik, R.; Napierała, M.; Jaskuła, J.; Liberacki, D. The Effect of Climate Change on Controlled Drainage Effectiveness in the Context of Groundwater Dynamics, Surface, and Drainage Outflows. Central-Western Poland Case Study. *Agronomy* **2020**, *10*, 625. [CrossRef]
9. Jeong, H.; Pittelkow, C.M.; Bhattarai, R. Simulated responses of tile-drained agricultural systems to recent changes in ambient atmospheric gradients. *Agric. Syst.* **2019**, *168*, 48–55. [CrossRef]
10. Quilbé, R.; Rousseau, A.; Moquet, J.; Trinh, N.; Dibike, Y.; Gachon, P.; Chaumont, D. Assessing the effect of climate change on river flow using general circulation models and hydrological modelling. Application to the Chaudiere River (Quebec, Canada). *J. Can. Water Resour.* **2008**, *33*, 73–93.
11. Prodanovic, P.; Simonovic, S.P. Impacts of changing climatic conditions in the Upper Thames River Basin. *Can. Water Resour. J.* **2007**, *32*, 265–284. [CrossRef]
12. Rahman, M.; Bolisetti, T.; Balachandar, R. Hydrologic modelling to assess the climate change impacts in a southern Ontario watershed. *Can. J. Civ. Eng.* **2012**, *39*, 91–103. [CrossRef]
13. Champagne, O.; Arain, M.A.; Leduc, M.; Coulibaly, P.; McKenzie, S. Future shift in winter streamflow modulated by the internal variability of climate in southern Ontario. *Hydrol. Earth Syst. Sci.* **2020**, *24*, 3077–3096. [CrossRef]
14. Saleem, S.; Levison, J.; Parker, B.; Martin, R.; Persaud, E. Impacts of climate change and different crop rotation scenarios on groundwater nitrate concentrations in a sandy aquifer. *Sustainability* **2020**, *12*, 1153. [CrossRef]
15. Wang, L.; Flanagan, D.C.; Wang, Z.; Cherkauer, K.A. Climate change impacts on nutrient losses of two watersheds in the Great Lakes region. *Water* **2018**, *10*, 442. [CrossRef]
16. Pease, L.A.; Fausey, N.R.; Martin, J.F.; Brown, L.C. Projected climate change effects on subsurface drainage and the performance of controlled drainage in the Western Lake Erie Basin. *J. Soil Water Conserv.* **2017**, *72*, 240–250. [CrossRef]
17. Mehan, S.; Aggarwal, R.; Gitau, M.W.; Flanagan, D.C.; Wallace, C.W.; Frankenberger, J.R. Assessment of hydrology and nutrient losses in a changing climate in a subsurface-drained watershed. *Sci. Total Environ.* **2019**, *688*, 1236–1251. [CrossRef]
18. Golmohammadi, G.; Rudra, R.P.; Madani, A.; Goel, P.K.; Mohammadi, K. Modeling the impacts of tillage practices on water table depth, drain outflow and nitrogen losses using DRAINMOD. *J. Comput. Electron. Agric.* **2016**, *124*, 73–83. [CrossRef]
19. Hanke, K. Impacts of Climate Change and Controlled Tile Drainage on Water Quality and Quantity in Southern Ontario, Canada. Master's Thesis, University of Waterloo, Waterloo, ON, Canada, 2018.
20. Skaggs, R.W. *A Water Management Model for Shallow Water Table Soils*; Report No. 134.; Water Resources Research Institute of the University of North Carolina: Chapel Hill, NC, USA, 1978.
21. Youssef, M.A.; Skaggs, R.W.; Chescheir, G.M.; Gilliam, J.W. Field evaluation of a model for predicting nitrogen losses from drained lands. *J. Environ. Qual.* **2006**, *35*, 2026–2042. [CrossRef]
22. Skaggs, R.W.; Youssef, M.A.; Chescheir, G.M. DRAINMOD: Model use, calibration, and validation. *Trans. ASABE* **2012**, *55*, 1509–1522. [CrossRef]
23. Brown, D.M.; Dadfar, H.; Fallow, D.J.; Gordon, R.J.; Lauzon, J.D.; Parkin, G.W. Temporal and Spatial Variability of Water Surplus in Ontario, Canada. *Isrn Soil Sci.* **2013**, *2013*, 362895. [CrossRef]
24. Malota, M.; Senzanje, A. Modelling mid-span water table depth and drainage discharge dynamics using DRAINMOD 6.1 in a sugarcane field in Pongola, South Africa. *J. Water Sa* **2015**, *41*, 325–334. [CrossRef]
25. Askar, M.H.; Youssef, M.A.; Chescheir, G.M.; Negm, L.M.; King, K.W.; Hesterberg, D.L.; Amoozegar, A.; Skaggs, R.W. DRAIN-MOD Simulation of macropore flow at subsurface drained agricultural fields: Model modification and field testing. *Agric. Water Manag.* **2020**, *242*, 106401. [CrossRef]

26. Adhikari, N.; Davidson, P.C.; Cooke, R.A.; Book, R.S. Drainmod-linked Interface for Evaluating Drainage System Response to Climate Scenarios. *Appl. Eng. Agric.* **2020**, *36*, 303–319. [CrossRef]
27. Singh, M.; Prasher, S.O.; Tan, C.; Tejawat, C. Evaluation of DRAINMOD for Southern Ontario conditions. *Can. Water Resour.* **1994**, *19*, 313. [CrossRef]
28. Shukla, M.B.; Prasher, S.O.; Madani, A.; Gupta, G.P. Field validation of DRAINMOD in Atlantic Canda. *J. Can. Agric. Eng.* **1994**, *36*, 205–213.
29. Luo, W.; Skaggs, R.W.; Chescheir, G.M. DRAINMOD modifications for cold conditions. *Trans. ASAE* **2000**, *43*, 1569–1582. [CrossRef]
30. Luo, W.; Skaggs, R.W.; Madani, A.; Cizikci, S.; Mavi, A. Predicting field hydrology in cold conditions with DRAINMOD. *Trans. ASAE* **2001**, *44*, 825–834.
31. Youssef, M.A.; Skaggs, R.W.; Reynolds, W.R.; Chescheir, G.M.; Gilliam, J.W. Field evaluation of DRAINMOD5.1 using six years of data from an artificially drained agricultural field in North Carolina. In Proceedings of the ASAE Annual International Meeting, Las Vegas, NV, USA, 27–30 July 2003.
32. Yang, C.; Prasher, S.; Wang, S.; Kim, S.; Tan, C.; Drury, C.; Patel, R. Simulation of nitrate-N movement in Southern Ontario, Canada with DRAINMOD-N. *Agric. Water Manag.* **2007**, *87*, 299–306. [CrossRef]
33. Salazar, O.; Wesstrom, I.; Youssef, M.A. Evaluation of DRAINMOD-N II model for predicting nitrogen losses in a loamy sand under cultivation in south-east Sweden. *J. Agric. Water Manag.* **2009**, *96*, 267–281. [CrossRef]
34. Dayyani, S.; Madramootoo, C.A.; Enright, P.; Simard, G.; Gullamudi, A.; Prasher, S.O.; Madani, A. Field evaluation of DRAINMOD 5.1 under a cold climate: Simulation of daily midspan water table depths and drain outflows. *J. Am. Water Resour. Assoc.* **2009**, *45*, 779–792. [CrossRef]
35. Dayyani, S.; Prasher, S.O.; Madramootoo, C.A.; Madani, A. Modeling water table depths, drain outflow, and nitrogen losses in cold climate using DRAINMOD 5.1. *Trans. ASABE* **2010**, *53*, 385–395. [CrossRef]
36. Golmohammadi, G.; Prasher, S.O.; Madani, A.; Rudra, R.P.; Youssef, M.A. SWATDRAIN, a new model to simulate the hydrology of agricultural lands, model development and evaluation. *J. Biosyst. Eng.* **2016**, *141*, 31–47. [CrossRef]
37. Golmohammadi, G.; Rudra, R.P.; Madani, A.; Goel, P.K.; Mohammadi, K. Modeling the effects of controlled drainage at a watershed scale using SWATDRAIN. *Arab. J. Geosci.* **2016**, *9*, 3–7. [CrossRef]
38. Saadat, S.; Frankenberger, J.; Bowling, L.; Ale, S. Evaluation of Surface Ponding and Runoff Generation in a Seasonally Frozen Drained Agricultural Field. *J. Hydrol.* **2020**, *588*, 124985. [CrossRef]
39. Skaggs, R.W. *DRAINMOD Reference Report. Methods for Design and Evaluation of Drainage Water Management Systems for Soils with High Water Tables*; USDA-SCS, South National Technical Report #329 Center; United States Department of Agriculture Soil Conservation Service: Washington, DC, USA, 1980.
40. Plach, J.M.; Macrae, M.L.; Ali, G.A.; Brunke, R.R.; English, M.C.; Ferguson, G.; Opolko, G. Supply and transport limitations on phosphorus losses from agricultural fields in the lower Great Lakes region, Canada. *J. Environ. Qual.* **2018**, *47*, 96–105. [CrossRef]
41. Plach, J.; Pluer, W.; Macrae, M.; Kompanizare, M.; McKague, K.; Carlow, R.; Brunke, R. Agricultural edge-of-field phosphorus losses in Ontario, Canada: Importance of the nongrowing season in cold regions. *J. Environ. Qual.* **2019**, *48*, 813–821. [CrossRef]
42. Lozier, T.M.; Macrae, M.L.; Brunke, R.; Van Eerd, L.L. Release of phosphorus from crop residue and cover crops over the non-growing season in a cool temperate region. *Agric. Water Manag.* **2017**, *189*, 39–51. [CrossRef]
43. Van Esbroeck, C.J.; Macrae, M.L.; Brunke, R.R.; McKague, K. Surface and subsurface phosphorus export from agricultural fields during peak flow events over the nongrowing season in regions with cool, temperate climates. *J. Soil Water Conserv.* **2017**, *72*, 65–76. [CrossRef]
44. Lozier, T. Potential and Observed Release of Phosphorus from Crop Residue and Cover Crops over the Non-Growing Season in a Cool Temperate Region. Master's Thesis, University of Waterloo, Waterloo, ON, Canada, 2016.
45. Pluer, W.T.; Macrae, M.; Buckley, A.; Reid, K. Contribution of preferential flow to tile drainage varies spatially and temporally. *Vadose Zone J.* **2020**, *19*, e20043. [CrossRef]
46. Eimers, M.C.; Liu, F.; Bontje, J. *Land Use, Land Cover, and Climate Change in Southern Ontario: Implications for Nutrient Delivery to the Lower Great Lakes*; Springer: Cham, Switzerland, 2020.
47. Rudra, R.P.; Mekonnen, B.A.; Shukla, R.; Shrestha, N.K.; Goel, P.K.; Daggupati, P.; Biswas, A. Currents Status, Challenges, and Future Directions in Identifying Critical Source Areas for Non-Point Source Pollution in Canadian Conditions. *Agriculture* **2020**, *10*, 468. [CrossRef]
48. Van Esbroeck, C.J. Edge of Field Phosphorus Export via Tile Drainage and Overland Flow from Reduced Tillage Systems in Ontario. Master's Thesis, University of Waterloo, Waterloo, ON, Canada, 2015.
49. Hoffman, D.W.; Richards, N.R. *Soil Survey of Perth County*; Experimental Farms Service; Canada Department of Agriculture and the Ontario Agricultural College: Guelph, ON, Canada, 1952; Volume 15.
50. Youssef, M.A.; Skaggs, R.W.; Chescheir, G.M.; Gilliam, J.W. The nitrogen simulation model, DRAINMOD-N II. *Trans. ASABE* **2005**, *48*, 611–626. [CrossRef]
51. Wang, S.; Prasher, S.O.; Patel, R.M.; Yang, C.C.; Kim, S.H.; Madani, A.; Macdonald, P.M.; Robertson, S.D. Fate and transport of nitrogen compounds in a cold region soil using DRAINMOD. *J. Comput. Electron. Agric.* **2006**, *53*, 113–121. [CrossRef]
52. Penrod, E.B.; Walton, W.W.; Terrel, D.V. A method to describe soil temperature variation. *J. Soil Mech. ASCE* **1958**, *1537*, 1–21.

53. Gupta, H.V.; Sorooshian, S.; Yapo, P.O. Status of automatic calibration for hydrologic models: Comparison with multilevel expert calibration. *J. Hydrol. Eng.* **1999**, *4*, 135–143. [CrossRef]
54. Nash, J.E.; Sutcliffe, J.V. River flow forecasting through conceptual models: Part 1. A discussion of principles. *J. Hydrol.* **1970**, *10*, 282–290. [CrossRef]
55. Nakićenović, N.; Swart, R. (Eds.) *Special Report on Emissions Scenarios. A Special Report of Working Group III of the Intergovernmental Panel on Climate Change*; Cambridge University Press: Cambridge, UK; New York, NY, USA, 2000; p. 599.
56. Das, S.; Millington, N.; Simonovic, S.P. Distribution choice for the assessment of design rainfall for the city of London (ON, Canada) under climate change. *Can. J. Civ. Eng.* **2013**, *40*, 121–129. [CrossRef]
57. Chan, A.; Chow, T. Energy and economic performance of green roof system under future climatic conditions in Hong Kong. *Energy Build.* **2013**, *64*, 182–198. [CrossRef]
58. Slamova, K.; Koehl, M. Measurement and GIS-based spatial modelling of copper corrosion in different environments in Europe. *Mater. Corros.* **2017**, *68*, 20–29. [CrossRef]
59. Wang, X.; Gordon, H. Ontario Climate Change Data Portal. 2013. Available online: http://www.ontarioccdp.ca (accessed on 12 September 2017).
60. Gebremeskel, S.; Liu, Y.B.; de Smedt, F.; Hoffmann, L.; Pfister, L. Analysing the effect of climate changes on streamflow using statistically downscaled GCM scenarios. *Int. J. River Basin Manag.* **2005**, *2*, 271–280. [CrossRef]
61. Coulibaly, P.; Shi, X. Identification of the Effect of Climate Change on Future Design Standards of Drainage Infrastructure in Ontario. Ministry of Transportation of Ontario, Highway Infrastructure Innovation Funding Program. 2005. Available online: http://www.cspi.ca/sites/default/files/download/Final_MTO_Report_June2005rv.pdf (accessed on 12 September 2017).
62. Chin, P.; Meyer, P. *Orangeville, Mono and Amaranth Water Quantity Risk Management and Climate Change Adaptation Assessment Piolet Study. Report Prepared for Toronto and Region Conservation Authority*; Matrix Solutions Inc.: Breslau, ON, Canada, 2014. Available online: https://ctcswp.ca/wp-content/uploads/2014/08/RPT_20140829_RMMEP.pdf (accessed on 11 September 2017).
63. Kassenaar, D.; Wexler, E.J. *Tier 3 Water Budget and Local Area Risk Assessment for the Region of York. Municipal Supplies—Risk Assessment Report. Prepared for "The Regional Municipality of York Transportation and Works Department"*; Earthfx Inc.: 3363 Young St., Toronto, ON, Canada, 2013. Available online: https://www.ctcswp.ca/wp-content/uploads/2017/05/RPT_201311_Earthfx_York_Tier3WBLocAreaRiskAssFNL.pdf (accessed on 11 September 2017).
64. Sunyer, M.A.; Madsen, H.; Ang, P.H. A comparison of different regional climate models and statistical downscaling methods for extreme rainfall estimation under climate change. *Atmos. Res.* **2012**, *103*, 119–128. [CrossRef]
65. Coulibaly, P.; Burn, D.H.; Switzman, H.; Henderson, J.; Fausto, E. A Comparison of Future IDF Curves for Southern Ontario: A Comparison of Future IDF Curves for Southern Ontario. 2016. Available online: https://climateconnections.ca/wp-content/uploads/2014/01/IDF-Comparison-Report-and-Addendum.pdf (accessed on 12 September 2017).
66. Chiotti, Q.; Tonto, F.; Perun, M. Ontario Regional Adaptation Collaborative Project- Weather and Water Information Gateway-Gaps Analysis Report. 2012. Available online: http://www.climateontario.ca/doc/ORAC_Products/WaterManagement/Gaps_Analysis_Report-Final_with_TOC.pdf (accessed on 12 September 2017).
67. Moriasi, D.N.; Arnold, J.G.; Van Liew, M.W.; Bingner, R.L.; Harmel, R.D.; Veith, T.L. Model evaluation guidelines for systematic quantification of accuracy in watershed simulations. *Trans. ASABE* **2007**, *50*, 885–900. [CrossRef]
68. Bosch, N.S.; Evans, M.A.; Scavia, D.; Allan, J.D. Interacting effects of climate change and agricultural BMPs on nutrient runoff entering Lake Erie. *J. Great Lakes Res.* **2014**, *40*, 581–589. [CrossRef]
69. Wallace, C.W.; Flanagan, D.C.; Engel, B.A. Quantifying the effects of conservation practice implementation on predicted runoff and chemical losses under climate change. *Agric. Water Manag.* **2017**, *186*, 51–65. [CrossRef]
70. Kalcic, M.M.; Muenich, R.L.; Basile, S.; Steiner, A.L.; Kirchhoff, C.; Scavia, D. Climate change and nutrient loading in the Western Lake Erie Basin: Warming can counteract a wetter future. *Environ. Sci. Technol.* **2019**, *53*, 7543–7550. [CrossRef] [PubMed]
71. Gombault, C.; Madramootoo, C.A.; Michaud, A.R.; Beaudin, I.; Sottile, M.F.; Chikhaoui, M.; Ngwa, F.F. Impacts of climate change on nutrient losses from the Pike River watershed of southern Québec. *Can. J. Soil Sci.* **2015**, *95*, 337–358. [CrossRef]
72. Dayyani, S.; Prasher, S.O.; Madani, A.; Madramootoo, C.A. Impact of climate change on the hydrology and nitrogen pollution in a tile-drained agricultural watershed in eastern Canada. *Trans. ASABE* **2012**, *55*, 398–401. [CrossRef]

hydrology MDPI

Article

An Assessment of the Recent Evolution of the Streamflow in a Near-Natural System: A Case Study in the Headwaters of the Mero Basin (Galicia, Spain)

M. L. Rodríguez-Blanco [1,*], M. M. Taboada-Castro [2] and M. T. Taboada-Castro [3]

1 History, Art and Geography Department, GEAAT Group, University of Vigo, Campus As Lagoas, 36310 Ourense, Spain
2 ETSIIAA, Area of Soil Science and Soil Chemistry, University of Valladolid, 34004 Palencia, Spain; mariamercedes.taboada@uva.es
3 Faculty of Sciences, Centre for Advanced Scientific Research (CICA), University of A Coruna, 15071 A Coruña, Spain; teresa.taboada@udc.es
* Correspondence: maria.luz.rodriguez.blanco@uvigo.es

Received: 30 September 2020; Accepted: 3 December 2020; Published: 8 December 2020

Abstract: Observational trend analysis is fundamental for documenting changes in river flows and placing extreme events in their longer-term historical context. Observations from near-natural catchments, i.e., with little or no alteration by humans, are of great importance in detecting and attributing streamflow trends. The purpose of this study is to analyze the annual and seasonal trends of stream discharge (mean, low and high flows) in a headwater catchment in NW Spain, i.e., in the wettest corner of the Iberian Peninsula. The results showed no significant decrease in the mean annual stream discharge. However, significantly lower summer and autumn mean stream discharge and an increase in low flow period were detected, in addition to lesser autumn high flow. The rainfall pattern followed an upward trend, but was not significant. This different pattern shown by rainfall and discharge indicates that is not sufficient to explain the observed trend in stream discharge. Air temperature, most notably by enhancing evapotranspiration, may explain the altered patterns of stream discharge.

Keywords: stream discharge; near-natural catchment; trend analysis; NW Spain

1. Introduction

Due to the incalculable value of water as an asset for all, both in the present and in the future, and the importance of streamflow as an indicator of long-term hydro-climatic changes, curbing and quantifying changes in streamflow series have become essential for water resource planning and management [1,2]. Streamflow integrates the influence of atmospheric variables over a catchment and, presumably, if consistent changes in rainfall and other climate variables (winds, atmosphere/heat fluxes) are observed, these should also be reflected in the catchment scale. However, this is not so obvious, because catchment characteristics (geology, slope, soils) impact rainfall-runoff transformation properties [3–5]. Additionally, human activities can greatly influence the river systems and make it difficult for trend detection and attribution [6,7]. Against this background, streamflow measures in natural or near-natural rivers become of prime importance in understanding hydrological processes in an area and identifying and attributing emerging trends [8–10]. Hydrological predictive modellings also require adequate and suitable local data to represent rainfall-runoff processes accurately. In fact, several studies have highlighted the importance of the hydrological modelling components in the rainfall-runoff modelling chain. Some of these studies investigated uncertainties linked to rainfall and the surface roughness component, and only a few focus on streamflow variations [5,11,12]. In this respect, Bermudez et al. [11] showed that a better representation of the hydrological processes occurring in the model domain and the availability of more accurate streamflow input data may reproduce a better response and improve model performance.

The term natural-regime stream is not well defined in the literature. Occasionally, it is simply described as rivers where little or no human intervention has taken place, thus rendering them highly unspoilt [13]. However, setting thresholds to differentiate natural-regime streams is not an easy task [14]. Various criteria relating to certain features describing the character of the riparian zone and river channel, the discharge regime and land use in the catchment were used to catalog human disturbance. For example, indicators of hydrologic alterations (IHA) are widely used to characterize human alterations of streamflow regimes [15], but they contain 33 individual parameters. The integrated connectivity status index (CSI) is also used to determine human interference in fluvial connectivity, defining the natural or free-flowing rivers as, i.e., systems largely unaffected by changes to fluvial connectivity, allowing unobstructed movement and exchange of water, energy, material and species within the river system and with surrounding landscapes [16]. As the impact is extensive, defining the threshold for distinguishing natural rivers is sometimes difficult. For example, Batalla et al. [17] define the degree of impoundment as an indicator of the extent to which reservoirs could change flow, while Nardi et al. [18] and Scheel et al. [19] used river–floodplain disconnectivity to evaluate man-made impacts. In the present study, the term "near-natural-regime stream" is used in the same sense as it has been used in previous research into this topic [8,20], meaning that it describes river flow regimes minimally affected by anthropogenic disturbances, such as reservoirs, dams, channelization, water extraction. Therefore, the term near-natural does not necessarily equate to totally undisturbed pristine conditions; but it might include minor disturbances, such as land-use and land management changes.

The request for a reference streamflow dataset for near-natural catchments has been largely recognized worldwide and supported by some international programs, like FRIEND [9]. In this context, a significant number of studies have been conducted over the last few years with the aim of detecting, measuring and evaluating streamflow trends in near-natural catchments [6–8,20–22]. Regarding trend studies in Europe, no uniform trend has been observed for streamflow [5,6]. Thus, whereas positive trends in annual, monthly and low flows were observed in near-natural catchments in Nordic countries [23], a generalized pattern of negative trends in annual and seasonal streamflows were reported for southern European rivers [5].

Most trend studies in the Iberian Peninsula focused on annual, seasonal, monthly and low flows. Lorenzo-Lacruz et al. [22] detected a generalized reduction in streamflow when studying a large database of Spanish mountain rivers not disrupted by major human interference. These results are in line with the Martínez-Fernández [20] study that found downward trends in annual and seasonal streamflows in an analysis of 74 near-natural rivers in Spain. Zabaleta et al. [24] also highlighted dominant-negative trends in various catchments in the Basque Country. Significant downward trends in annual streamflow were also identified in the headwaters of the Ribera Salada [21], Tagus [25] and Duero [26]. However, in the Águeda basin (north-central region of Portugal), Hawtree et al. [27] found no evidence of significant reductions in streamflow, despite wide afforestation that could, arguably, be explained by the presence of compensatory climate trends over the study period.

To summarize, most of the aforementioned trend studies in the streamflow series carried out in Spain focused on regions with a Mediterranean climate. Thus, these studies do not comprehensively analyze the hydro-climatic changes of streams in the humid region of Spain (NW Spain). There is general agreement that air temperatures have increased over the last century in NW Spain [28,29]. However, it is unclear how these changes may have affected streamflow. Evidently, the results obtained for the Mediterranean area are not directly applicable to more humid regions. Hence, it is a challenge to examine stream discharge in humid catchments.

In this paper, a streamflow trend analysis was carried out in a headwater catchment of NW Spain (i.e., under a humid climate) with the aim of contributing to filling a gap in research on a national scale. The analysis will be carried out at different time scales (annual, seasonal and monthly), paying special attention to the duration and severity of the low flow. The choice of the study area (a headwater catchment of the Mero basin) is based on: (i) not being affected by anthropogenic alterations such as reservoirs and dams, therefore qualifying it as an example of a near-natural regimen fluvial system, (ii) its characteristics, which respond to the archetype of the Galician rural environment: small population centers with highly dispersed single-family dwellings, and (iii) the strategic, environmental and social importance of its water resources. The Mero basin is important for the future socio-economic development of the city of A Coruña and its metropolitan area (NW Spain), because it is the main contributor to the Abegondo-Cecebre reservoir, the largest source of water supply for the area. This reservoir also provides high-quality habitats for a large number of species and was cataloged as a Special Area of Conservation of Natura 2000 in 2014.

2. Materials and Methods

2.1. Case Study and Data Description

This study was carried out at a headwater catchment (Corbeira) of the larger Mero basin (245 km^2), located near A Coruña in northwest Spain (Figure 1). The catchment is not affected by substantial human alterations directly modifying natural stream behavior, such as reservoirs and hydropower stations. It drains a catchment area of 16 km^2 with a river length of 10 km and mean slope of 19%. The gauging station lies at 65 m a.s.l., while the highest elevation stands at 470 m a.s.l. Soils and land use are typical of the region. Soils are formed on a uniform substrate consisting of materials belonging to the Series of Órdenes Complex [30]. They are generally deep, rich in organic matter, acid and can be classified as Umbrisols and Cambisols [31]. Their texture is silty loam. The catchment is predominantly rural, with a low population density (about 35 inhabitants km^2). The primary land uses include commercial forest (65%), mainly eucalyptus plantations, and agricultural land (30%). The climate is oceanic with a mean annual temperature of approximately 13 °C and mean annual rainfall of 1073.6 mm, ranging from 639.9 mm to 1788.2 mm (based on the period 1983/1984–2018/2019 for the nearest weather station of the Galician meteorological service about 8 km distant). Higher monthly rainfall values generally occur during the autumn-winter period (October–March), when 68% of the annual rainfall is recorded.

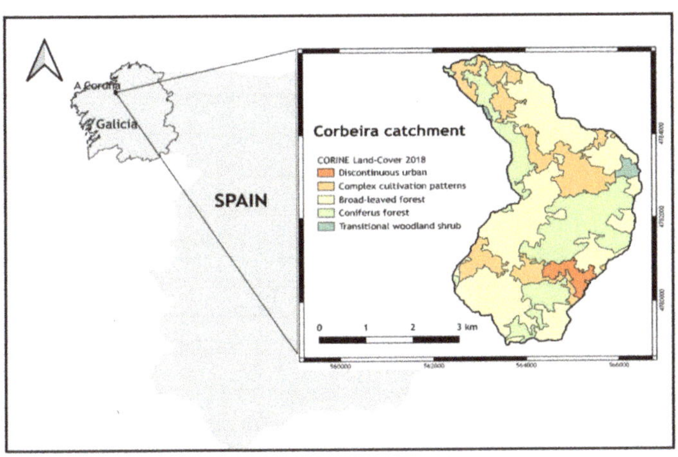

Figure 1. Geographic location and main types of land use within the headwater of the Corbeira catchment.

2.2. Instrumentation and Dataset

Monitoring the Corbeira headwater catchment started in 2004. Water level at the catchment outlet was measured using a water pressure sensor (registering data every minute and recording mean data every 10 min). The records were rendered as discharge through the level–discharge relationship, obtained by stream discharge measurements taken under different flow conditions.

Rainfall and temperature data series were used to compare with stream discharge, using climate data from the Mabegondo station, which is located near the study area and holds data from 1983.

2.3. Data Processing and Analysis

For this study, 15 successive years of hydro-meteorological data recorded from October 2004 to September 2019 were used. Taking into account the importance of low and high flows, in addition to the mean value, the following were extracted from the data series: the duration of low and high flows, i.e., number of days with discharge below 0.1 (Q10: mean daily discharge was exceeded 90% of the designated time) and 0.2 percentiles (Q20: mean daily discharge was exceeded 80% of the designated time) for low flows, and 0.8 (Q80: mean daily discharge was exceeded 20% of the designated time) and 0.9 (Q90: mean daily discharge was exceeded 10% of the designated time) for high flow [23] and its severity (discharge deficit), i.e., volume below Q10 and Q20 [32]. All parameters were calculated on a daily basis. Once these parameters were estimated, a trend analysis was performed with the non-parametric Mann–Kendall test and the Sen method, using the Makesens program developed by the Finnish Meteorological Office [33]. The presence of a monotonic increasing (positive) or decreasing (negative) trend was tested with the Mann–Kendall test and the slope of a linear trend (i.e., the magnitude) was estimated with the Sen method [34]. The Mann–Kendall test has been widely used in studies investigating trends in hydro-climatic series [35,36] due to it being more robust against outliers and being more powerful than other tests [37]. The analysis was applied on an annual, seasonal and monthly basis. For the seasonal trend, the year was divided in four seasons: autumn (October–December), winter (January–March), spring (April–June) and summer (July–September).

Changes in land use in the study period were also evaluated by analyzing differences in land use between 2000 and 2018, based on the Corine Land Cover for the years 2000 and 2018.

3. Results

3.1. Statistical Characteristics of the Climatic Conditions and Streamflow

Figure 2 presents the annual rainfall and temperature during the study period, while the mean annual stream discharge between 2004 and 2019 is shown in Figure 3. High interannual variability can be observed, especially for rainfall and stream discharge. The annual streamflow in the study catchment varied from 0.08 $m^3 s^{-1}$ (2016/2017) to 0.273 $m^3 s^{-1}$ (2009/2010), with a mean annual stream discharge of 0.181 $m^3 s^{-1}$, a standard deviation of 0.05 $m^3 s^{-1}$ and a variation coefficient of 29%, which denotes a marked inter-annual fluctuation. Regarding seasonal patterns, streamflow was observed to be concentrated in the winter season (0.306 $m^3 s^{-1}$), whereas summer exhibited the lowest stream discharge (0.058 $m^3 s^{-1}$). However, winter did not return the highest value for every year (Figure 3). For example, in 2004/2005 and 2007/2008 hydrological years, spring exhibited the highest streamflow due to the reduced rainfall during the rainy season (587.6 and 731.63 mm for 2004/05 and 2007/08 respectively v.731.6 mm of the mean). In regard to the monthly distribution of stream discharge, there is an outstanding intra-annual variability (Figure 4), with mean discharge varying from 0.037 $m^3 s^{-1}$ (September) to 0.345 $m^3 s^{-1}$ (February). The months of February, March and April presented the largest discharge, while the highest mean monthly rainfall was registered in November, October and December. Previous studies in this area, point to streamflow being largely dependent on water storage in the soils, which together with the distribution of rainfall and evapotranspiration, are key variables controlling the hydrological response of the catchment [38,39].

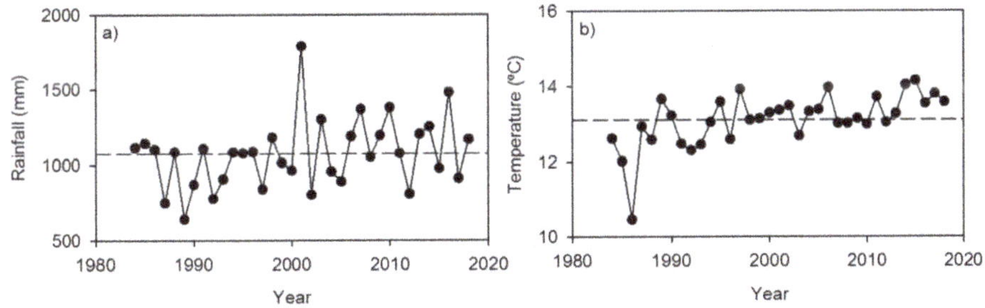

Figure 2. Annual series for (**a**) rainfall and (**b**) temperature. Horizontal dotted line indicates the long-term mean for each variable.

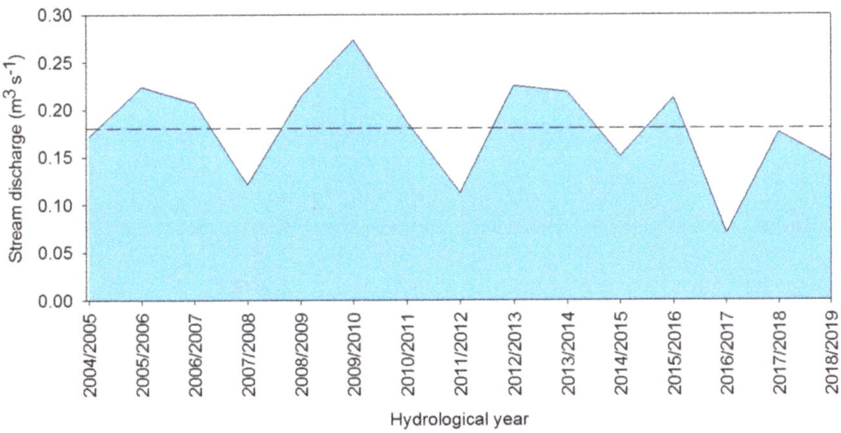

Figure 3. Variations in stream discharge in Corbeira catchment (2004/2005–2018/2019). Horizontal dotted line indicates the mean for the study period.

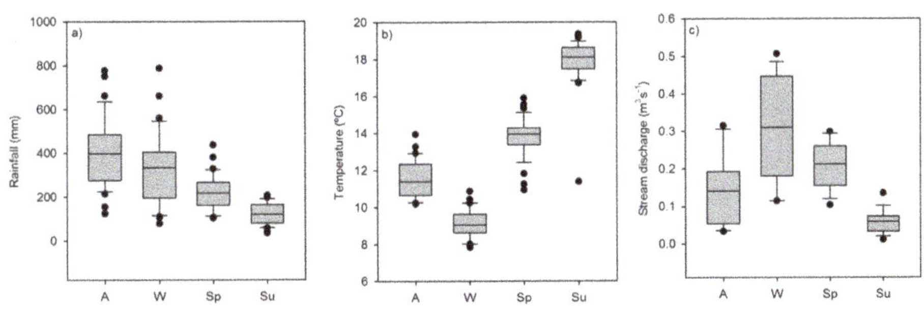

Figure 4. Boxplot of seasonal distribution of (**a**) rainfall, (**b**) temperature and (**c**) discharge in Corbeira catchment.

3.2. Land Use Changes

Figure 5 shows the main changes in land use in the Corbeira catchment based on Corine Land Cover. Forestry and agricultural land are the main land uses in the catchment. In the period between 2000 and 2018, a slight decrease was observed in the agricultural zone, whereas the forest area increased (about 3%). The findings from this study are in line with results from other studies indicating an expansion of forest due to the replacement of cultivated land by plantations, first with pine and from the 1960s, with eucalyptus [40,41]. In fact, the northwest of the Iberian Peninsula constitutes a representative area of the transformations of land-use systems in southern Europe [42], as it has undergone wide changes in the spatial distribution of land uses in recent decades, due to reduced traditional agricultural activity, rural abandonment and more plantations of fast-growing tree species.

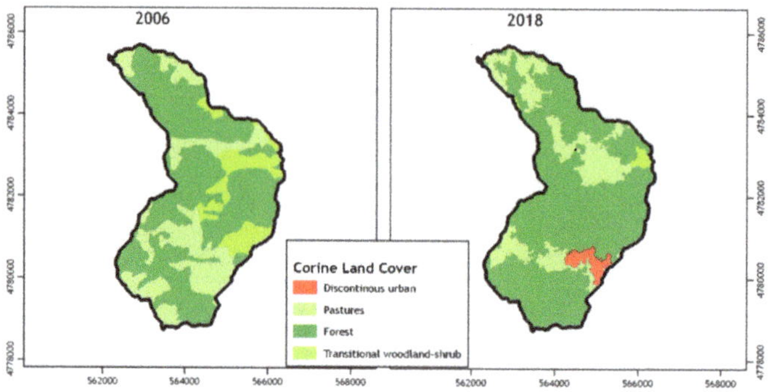

Figure 5. Land-use changes in the Corbeira catchment between 2000 and 2018.

3.3. *Trend Analysis of Climatic Variables and Stream Discharge*

The results for the Thiel–Sen/Mann–Kendall trend tests for rainfall and temperature are presented in Table 1, while the results for discharge are shown in Table 2. Upward trends were identified for rainfall and temperature data, with a very strong positive trend for temperature and annual results being significant at a 99% confidence level. This upward trend is closely related to the significant upward trend during summer. In contrast, the results for rainfall were not clear, and despite showing an upward trend signal, they did not prove to be significant (i.e., *p*-value > 0.1).

Table 1. Statistics (Z of the Mann–Kendall Test and Q of the Sen's Slope Estimator) for rainfall and temperature trend analysis.

	Rainfall			Temperature		
	Mann–Kendall Trend		Sen's Slope Estimate	Mann–Kendall Trend		Sen's Slope Estimate
	Test Z	**Significance**	**Q**	**Test Z**	**Significance**	**Q**
Annual	1.48		5.242	3.69	***	0.037
Autumn	0.61		1.989	1.35		0.027
Winter	1.43		3.688	1.79	+	0.027
Spring	−0.15		−0.198	3.81	***	0.055
Summer	−0.25		−0.208	2.60	**	0.036
October	0.69		0.873	1.87	+	−0.017
November	1.40		2.219	0.31		0.009
December	−1.08		−0.980	0.28		0.007
January	0.67		0.884	2.10	*	0.048
February	1.38		1.453	0.11		0.003
March	1.35		1.157	1.96	+	0.027
April	0.01		0.017	2.33	*	0.061
May	−0.04		−0.042	2.16	*	0.048
June	0.48		0.257	3.29	***	0.068
July	0.71		0.164	2.06	*	0.040
August	1.20		0.372	2.13	*	0.033
September	−1.27		−0.902	1.05		0.029

** test is significant at $p < 0.001$; * test is significant at $p < 0.05$; + test is significant at $p < 0.1$.

Table 2. Statistics (Z of the Mann–Kendall Test and Q of the Sen's Slope Estimator) for stream discharge trend analysis.

	Mann–Kendall Trend		Sen's Slope Estimate
	Test Z	**Significance**	**Q**
Annual	−1.09		−0.004
Autumn	−2.28	*	−0.011
Winter	0.20		0.002
Spring	−0.49		−0.003
Summer	−2.67	**	−0.003
October	−2.57	*	−0.008
November	−1.19		−0009
December	1.68	+	−0.020
January	0.00		−0.001
February	0.79		0.011
March	0.20		0.003
April	−0.49		−0.005
May	−1.09		−0.005
June	−1.29		−0.005
July	−1.39		−0.004
August	−2.28	*	−0.003
September	−2.67	**	−0.004

** test is significant at $p < 0.001$; * test is significant at $p < 0.05$; + test is significant at $p < 0.1$.

For the discharge data record, the results suggested a downward trend for annual discharge, although it was not found to be significant at a reliable confidence level (p-value > 0.1). However, when examining by season, a number of significant trends were found. Thus, a downward trend was detected in autumn, which was very clear in October, weaker and not significant in November and positive (upward) in December, although the test yielded results with a lower confidence level (p-value = 0.1). Summer also showed a downward trend due to those observed in September, August and July, although the trend in the latter month was weaker and not significant. Spring exhibited a downward trend, while results suggested an increase during winter. However, trends for these seasons were very weak and not statistically significant (Table 2).

3.4. *Trend in Low and High Flows*

Results of low and high flow trend analysis are displayed in Table 3. An upward trend for the low flow period was observed at different temporal scales, although some of these did not prove to be significant. If the seasonal scale is included, summer and autumn showed significant upward trends in the duration of the low flow period and the severity (volumetric deficit), due to the clear upward trend in the months of October, November, July, August and September. On the contrary, in winter and spring, there was no significant evidence at low flows.

Table 3. Statistics (Z of the Mann–Kendall Test and Q of the Sen's Slope Estimator) for low and high flow trend analysis.

		Mann–Kendall Trend		Sen's Slope Estimate
		Test Z	**Significance**	**Q**
Annual	Q10	2.79	**	6.875
	Q20	3.07	**	8.429
	Severity (Q10)	2.90	**	9.379
	Severity (Q20)	1.48	**	0.011
	Q80	−0.84		−1.000
	Q90	−0.10		−0.333

Table 3. *Cont.*

		Mann–Kendall Trend		Sen's Slope Estimate
		Test Z	Significance	Q
Autumn	Q10	2.59	**	1.00
	Q20	2.38	*	3.00
	Q80	−1.50		−1.444
	Q90	−1.89	+	−0.250
Winter	Q10	1.04		0.000
	Q20	0.17		0.000
	Q80	0.25		1.000
	Q90	0.40		0.222
Spring	Q20	1.04		0.000
	Q80	−1.19		−0.667
	Q90	−0.25		0.000
Summer	Q10	2.24	*	4.200
	Q20	2.31	*	4.200
October	Q10	2.88	**	0.875
	Q20	2.38	*	1.556
	Q80	−2.38	*	0.000
November	Q10	1.49		0.000
	Q20	1.80	+	0.000
	Q80	−1.88	+	−0.727
	Q90	−1.64		0.000
December	Q20	1.22		0.000
	Q80	−1.72	+	−0.889
	Q90	−1.69	+	0.000
January	Q20	0.17		0.000
	Q80	0.80		0.182
	Q90	0.68		0.000
February	Q80	0.75		0.400
	Q90	0.65		0.000
March	Q80	−0.05		0.000
April	Q80	−0.15		−0.125
	Q90	−0.10		0.000
May	Q80	−0.65		0.000
June	Q80	−1.00		0.000
July	Q10	1.14		0.000
	Q20	1.94	+	0.571
August	Q10	2.52	*	1.385
	Q20	2.40	*	1.333
September	Q10	2.68	**	2.231
	Q20	2.35	*	1.000

** test is significant at $p < 0.001$; * test is significant at $p < 0.05$; + test is significant at $p < 0.1$.

Concerning the high flow, a downward trend in the annual duration of the high flow period was observed, although it did not prove to be significant (Table 3). A negative (downward) trend was also observed in autumn and spring, but was only significant for Q80 in autumn, which can be justified by the downward trend observed during October.

4. Discussion

The results show that discharge in the Corbeira catchment does not show a marked annual fluctuation. However, the intra-annual variability is much larger, with the maximum variability occurring in autumn.

With respect to trend analysis of climate variables, upward trends were found for the temperature at the annual and seasonal scale, as has been already reported in previous studies in the Iberian Peninsula [9,43]. In contrast, no significant trends were detected for rainfall on any time scale, similar to the results of Rodrigo and Trigo [29] for the La Coruña station (located near the study area), which showed positive (non-significant) trends for annual and autumn rainfall and negative for spring and summer. Similarly, and in the context of long-term evaluation of rainfall in Galicia, Lago et al. [28] pointed out the absence of a unidirectional trend in annual rainfall. However, at a seasonal scale, these authors indicated a possible change in distribution throughout the year, with wetter autumns and less humid winters, mainly due to the decrease in rainfall in February and springs and summers of sub-dry or dry trends. These facts have a heavy impact because more than two-thirds of the water demand takes place in the period between April and September, in which scarcely one-third of the total annual rainfall is registered. Furthermore, it must be noted that in the autumn-winter semester, the period with the highest water contribution, it is possible to record a marked rainfall deficit. An example would be the winter 2004-05 (169.4 mm in winter 2004-05 vs. 333.4 mean rainfall in winter).

With hydrological variables, the annual discharge trend test revealed that there was no significant trend for stream discharge. However, significant downward trends were observed for autumn and summer. In the present study, the observed trends cannot be explained by changes in rainfall, since Z values, although negative, were close to zero, indicating no trends. The study of low and high flows suggested upward trends in the duration of low flow and severity (volumetric deficit) due to the increase in the number of days with low flow in summer and autumn (Table 3). In contrast, the duration of the high flow period seemed to decrease, particularly during autumn. The longer low flow duration and severity found in our study are in line with the results from other studies in temperate humid catchments [24,27] indicating an extension of the low flow period during autumn, which resulted in a longer period of time with low stream discharge, as well as a temporary increase in the volumetric deficit. Given the length of data, it is difficult to say whether the increase of evaporative demand during these seasons, induced by higher temperatures, has caused the apparent trend. Several studies in the Iberian Peninsula have already reported a reduction in streamflow due to the increase in air temperatures [8,9,20]. For example, Morán-Tejeda [43] observed an increase in the number of days with low flow and a reduction in the frequency and magnitude of high flows in the Duero basin. These authors argued that this behavior is a consequence of raising temperatures by enhancing evapotranspiration, and changes in the land-cover, as a result of re-growth of vegetation, whose effects are more evident during the growth period (spring and summer), as it is associated with the highest demand for water from the soil, and a greater capacity for rainfall interception by the canopy [43]. Other authors have also associated re-vegetation and land-cover expansion in headwaters as a primary cause of decreasing stream discharge in the absence of rainfall trends.

The observed discharge trends in the Corbeira catchment and hence the Mero basin may have important implications for water management in the study area. The Abegondo-Cecebre reservoir, which is mainly fed by water from the Mero basin, was built in the mid-1970s to guarantee the availability of water for the city of A Coruña and its metropolitan area during drought periods and to reduce negative effects from floods. The decrease in discharge, as well as increase in the low flow period, enhance future risk and vulnerability especially in summer, which is characterized by the low flow. In recent years, there has been a substantial increase in the population in the area (from 250,000 to 450,000) representing a large rise in water demand. Moreover, tourism is growing in the city of A Coruña and its surroundings, especially in summer when the population may even double, which makes it necessary to satisfy an additional water demand at what is undoubtedly the least rainy time of year. In some cases, this causes the Abegondo-Cecebre reservoir to be insufficient if there has not been enough rain to fill it in the previous months, as occurred in 2010, triggering water restrictions and threatening the ecological environment, society and economy. This highlights the fact that alternative strategies for water management may come to the fore in low flow seasons (the most critical period for water management in the area) in order to reconcile the heavier demand for water for human activities with maintaining functioning riparian ecosystems, as it also has to be borne in mind that the reservoir is a Special Area of Conservation.

5. Summary and Final Remarks

An analysis of trends in annual and seasonal stream discharge (mean, low and high flows) was conducted in the headwaters of the Mero river, a near-natural system representative of the climate and land use characteristics of the northwestern Iberian Peninsula catchments. For this, both the non-parametric Mann–Kendall test and the Sen methods were applied.

The study did not detect statistically significant trends in annual stream discharge. However, significant downward trends in mean discharge were observed for autumn and summer. In addition, a significant upward trend in the number of days with low flow was particularly evident in spring and summer. Additionally, a falling trend in the high flows was observed in autumn. On the contrary, rainfall showed a positive pattern, although it was not significant. The different behavior shown by rainfall and discharge may be explained by the reduction of water resources associated with the increase in temperature in the study area, although this must be interpreted with caution, given the absence of long-term measurements.

Author Contributions: Conceptualization, M.L.R.-B.; Formal Analysis, M.L.R.-B.; Investigation, M.L.R.-B., M.M. Taboada-Castro, M.T.T.-C.; Data Curation, M.L.R.-B.; Writing—Original Draft Preparation, M.L.R.-B.; Writing—Review & Editing, M.M.T.-C., M.T.T.-C.; Supervision, M.L.R.-B., M.M.T.-C., M.T.T.-C.; Funding Acquisition, M.L.R.-B., M.M.T.-C., M.T.T.-C. All authors have read and agreed to the published version of the manuscript.

Funding: This research is a contribution to the projects 10MDS103031 of the Xunta de Galicia and CGL2014-56907-R funded by the Spanish Ministry of Economy and Competitiveness (MINECO). M.L. Rodriguez-Blanco has been awarded a post-doctoral research contract (Juan de la Cierva Programme), which was funded by the MINECO. We thank the reviewers for their valuable suggestions that helped to improve the manuscript substantially.

Conflicts of Interest: The authors declare no conflict of interest.

References

1. Villarini, G.; Smith, J.A.; Serinaldi, F.; Ntelekos, A.A. Analyses of seasonal and annual maximum daily discharge records for central Europe. *J. Hydrol.* **2011**, *399*, 299–312. [CrossRef]
2. Svensson, C.; Kundzewicz, W.Z.; Maurer, T. Trend detection in river flow series: 2. Flood and low-flow index series. *Hydrol. Sci. J.* **2005**, *50*, 1–824. [CrossRef]
3. Wagener, T.; Wheater, H.; Gupta, H.V. *Rainfall-Runoff Modelling in Gauged and Ungauged Catchments*; World Scientific: Singapore, 2004.
4. Moradkhani, H.; Sorooshian, S. General Review of Rainfall-Runoff Modeling: Model Calibration, Data Assimilation, and Uncertainty Analysis. In *Water Science and Technology Library*; Springer: Berlin/Heidelberg, Germany, 2008; Volume 63, pp. 1–24.
5. De Paola, F.; Giugni, M.; Pugliese, F.; Annis, A.; Nardi, F. GEV Parameter Estimation and Stationary vs. Non-Stationary Analysis of Extreme Rainfall in African Test Cities. *Hydrology* **2018**, *5*, 28. [CrossRef]
6. Hannaford, J.; Marsh, T.J. High-flow and flood trends in a network of undisturbed catchments in the UK. *Int. J. Clim.* **2008**, *28*, 1325–1338. [CrossRef]
7. Bertola, M.; Viglione, A.; Lun, D.; Hall, J.; Blöschl, G. Flood trends in Europe: Are changes in small and big floods different? *Hydrol. Earth Syst. Sci.* **2020**, *24*, 1805–1822. [CrossRef]
8. Stahl, K.; Hisdal, H.; Hannaford, J.; Tallaksen, L.M.; Van Lanen, H.A.J.; Sauquet, E.; DeMuth, S.; Fendekova, M.; Jódar, J. Streamflow trends in Europe: Evidence from a dataset of near-natural catchments. *Hydrol. Earth Syst. Sci.* **2010**, *14*, 2367–2382. [CrossRef]
9. Morán-Tejeda, E.; Lorenzo-Lacruz, J.; López-Moreno, J.I.; Rahman, K.; Beniston, M. Streamflow timing of mountain rivers in Spain: Recent changes and future projections. *J. Hydrol.* **2014**, *517*, 1114–1127. [CrossRef]
10. Masseroni, D.; Camicis, S.; Cislaghi, A.; Vacchiano, G.; Massari, C.; Brocca, L. 65-year change of annual streamflow volumes across Europe with a focus on the Mediterranean basin. *Hydrol. Earth Syst. Sci.* **2020**, 1–16. [CrossRef]
11. Bermúdez, M.; Neal, J.; Bates, P.; Coxon, G.; Freer, J.E.; Cea, L.; Puertas, J. Quantifying local rainfall dynamics and uncertain boundary conditions into a nested regional-local flood modeling system. *Water Resour. Res.* **2017**, *53*, 2770–2785. [CrossRef]

12. Annis, A.; Nardi, F.; Volpi, E.; Fiori, A. Quantifying the relative impact of hydrological and hydraulic modelling parameterizations on uncertainty of inundation maps. *Hydrol. Sci. J.* **2020**, *65*, 507–523. [CrossRef]
13. LeRoy Poff, N.; Allan, J.D.; Bain, M.B.; Karr, J.R.; Prestegaard, K.L.; Richter, B.D.; Sparks, R.E.; Stromberg, J.C. The natural flow regime: A paradigm for river conservation and restoration. *BioScience* **1997**, *47*, 769–784. [CrossRef]
14. Chaves, M.L.; Costa, J.L.; Chainho, P.; Costa, M.J.; Prat, N. Selection and validation of reference sites in small river basins. *Hydrobiology* **2006**, *573*, 133–154. [CrossRef]
15. Richter, B.D.; Baumgartner, J.V.; Powell, J.; Braun, D.P. A Method for Assessing Hydrologic Alteration within Ecosystems. *Conserv. Biol.* **1996**, *10*, 1163–1174. [CrossRef]
16. Grill, G.; Lehner, B.; Thieme, M.; Geenen, B.; Tickner, D.; Antonelli, F.; Babu, S.; Borrelli, P.; Cheng, L.; Crochetiere, H.; et al. Mapping the world's free-flowing rivers. *Nature* **2019**, *569*, 215–221. [CrossRef] [PubMed]
17. Batalla, R.J.; Gómez, C.M.G.; Kondolf, G. Reservoir-induced hydrological changes in the Ebro River basin (NE Spain). *J. Hydrol.* **2004**, *290*, 117–136. [CrossRef]
18. Nardi, F.; Morrison, R.R.; Annis, A.; Grantham, T.E. Hydrologic scaling for hydrogeomorphic floodplain mapping: Insights into human-induced floodplain disconnectivity. *River Res. Appl.* **2018**, *34*, 675–685. [CrossRef]
19. Scheel, K.; Morrison, R.R.; Annis, A.; Nardi, F. Understanding the Large-Scale Influence of Levees on Floodplain Connectivity Using a Hydrogeomorphic Approach. *JAWRA J. Am. Water Resour. Assoc.* **2019**, *55*, 413–429. [CrossRef]
20. Martínez-Fernández, J.; Sánchez, N.; Herrero-Jiménez, C.M. Recent trends in rivers near-natural flow regime: The case of the river headwaters in Spain. *Prog. Phys. Geogr.* **2013**, *37*, 685–700. [CrossRef]
21. Buendia, C.; Bussi, G.; Tuset, J.; Vericat, D.; Sabater, S.; Palau, A.; Batalla, R.J. Effects of afforestation on runoff and sediment load in an upland Mediterranean catchment. *Sci. Total. Environ.* **2016**, *540*, 144–157. [CrossRef]
22. Lorenzo-Lacruz, J.; Serrano, S.M.V.; López-Moreno, J.I.; Morán-Tejeda, E.; Zabalza, J. Recent trends in Iberian streamflows (1945–2005). *J. Hydrol.* **2012**. [CrossRef]
23. Wilson, D.; Hisdal, H.; Lawrence, D. Has streamflow changed in the Nordic countries?—Recent trends and comparisons to hydrological projections. *J. Hydrol.* **2010**, *394*, 334–346. [CrossRef]
24. Zabaleta, A.; Morales, T.; Meaurio, M.; Gorria, C.; Antigüedad, I. Regional Hydrological Signs for Climate Change in Southern Europe (Basque Country). In Proceedings of the International Conference on Water, Climate and Environment (Balwois), Ohrid, Republic of Macedonia, 28 May–2 June 2012.
25. Lorenzo-Lacruz, J.; Serrano, S.M.V.; López-Moreno, J.I.; Beguería, S.; García-Ruiz, J.M.; Cuadrat, J.M. The impact of droughts and water management on various hydrological systems in the headwaters of the Tagus River (central Spain). *J. Hydrol.* **2010**, *386*, 13–26. [CrossRef]
26. Morán-Tejeda, E.; López-Moreno, J.I.; Barbancho, A.C.; Vicente-Serrano, S.M. River regimes and recent hydrological changes in the Duero basin (Spain). *J. Hydrol.* **2011**, *404*, 241–258. [CrossRef]
27. Hawtree, D.; Nunes, J.P.; Keizer, J.J.; Jacinto, R.; Santos, J.; Rial-Rivas, M.E.; Karine-Boulet, A.; Tavares-Wahren, F.; Feger, K.-H. Time series analysis of the long-term hydrologic impacts of afforestation in the Águeda watershed of north-central Portugal. *Hydrol. Earth Syst. Sci.* **2015**, *19*, 3033–3045. [CrossRef]
28. Lago, A.; Lage, A.; Cruz, R.; Pérez-Muñurizi, V. Estudo das varia- cións de temperatura e precipitación en Galicia nos últimos trinta anos no contexto do cambio climático global. *Rev. Real Acad. Galega Cienc.* **2006**, *25*, 27–58.
29. Rodrigo, F.S.; Trigo, R.M. Trends in daily rainfall in the Iberian Peninsula from 1951 to 2002. *Int. J. Clim.* **2007**, *27*, 513–529. [CrossRef]
30. Instituto Tecnologico Geominero de Espana (IGME). *Mapa GeoloÅLgico de Espana, 1:50,000*; Hoja 45: Betanzos, Spain, 1981.
31. IUSS Working Group WRB. *World Reference Base for Soil Resources 2014, Update 2015: International Soil Classification System for Naming Soils and Creating Legends for Soil Maps*; World Soil Resources Reports No. 106; FAO: Rome, Italy, 2015.
32. Ouarda, T.B.M.J.; Charron, C.; St-Hilaire, A. Statistical Models and the Estimation of Low Flows. *Can. Water Resour. J. Rev. Can. Ressour. Hydrol.* **2008**, *33*, 195–206. [CrossRef]

33. Salmi, T.; Määttä, A.; Anttila, P.; Ruoho-Airola, T.; Amnell, T. *Detecting Trends of Annual Values of Atmospheric Pollutants by the Mann-Kendall Test and Sen's Slope Estimates—The Excel Template Application MAKESEN*; Publication on Air Quality No 31; Air Quality Research; Finnish Meteorological Institute: Helsinki, Finland, 2002.
34. Gilbert, R.O. *Statistical Methods for Environmental Pollution Monitoring*; John Wiley & Sons: Hoboken, NJ, USA, 1987.
35. Fathian, F.; Dehghan, Z.; Bazrkar, M.H.; Eslamian, S. Trends in hydrological and climatic variables affected by four variations of the Mann-Kendall approach in Urmia Lake basin, Iran. *Hydrol. Sci. J.* **2016**, *61*, 1–13. [CrossRef]
36. Forootan, E. Analysis of trends of hydrologic and climatic variables. *Soil Water Res.* **2019**, *14*, 163–171. [CrossRef]
37. Kundzewicz, Z.W.; Graczyk, D.; Maurer, T.; Pińskwar, I.; Radziejewski, M.; Svensson, C.; Szwed, M. Trend detection in river flow series: 1. Annual maximum flow/Détection de tendance dans des séries de débit fluvial: 1. Débit maximum annuel. *Hydrol. Sci. J.* **2005**, *50*, 797–810. [CrossRef]
38. Palleiro, L.; Rodríguez-Blanco, M.; Taboada-Castro, M.M.; Taboada-Castro, M.T. Hydrological response of a humid agroforestry catchment at different time scales. *Hydrol. Process.* **2013**, *28*, 1677–1688. [CrossRef]
39. Rodríguez-Blanco, M.; Taboada-Castro, M. An overview of patterns and dynamics of suspended sediment transport in an agroforest headwater system in humid climate: Results from a long-term monitoring. *Sci. Total. Environ.* **2019**, *648*, 33–43. [CrossRef] [PubMed]
40. Jones, J.; Creed, I.F.; Hatcher, K.L.; Warren, R.J.; Adams, M.B.; Benson, M.H.; Boose, E.; Brown, W.A.; Campbell, J.L.; Covich, A.; et al. Ecosystem Processes and Human Influences Regulate Streamflow Response to Climate Change at Long-Term Ecological Research Sites. *Bioscience* **2012**, *62*, 390–404. [CrossRef]
41. Moreira, F.; Rego, F.C.; Ferreira, P.G. Temporal (1958–1995) pattern of change in a cultural landscape of northwestern Portugal: Implications for fire occurrence. *Landsc. Ecol.* **2001**, *16*, 557–567. [CrossRef]
42. Fuchs, R.; Herold, M.; Verburg, P.H.; Clevers, J.G.; Eberle, J. Gross changes in reconstructions of historic land cover/use for Europe between 1900 and 2010. *Glob. Chang. Biol.* **2014**, *21*, 299–313. [CrossRef] [PubMed]
43. Del Río, S.; Herrero, L.; Pinto-Gomes, C.; Penas, Á. Spatial analysis of mean temperature trends in Spain over the period 1961–2006. *Glob. Planet. Chang.* **2011**, *78*, 65–75. [CrossRef]

Publisher's Note: MDPI stays neutral with regard to jurisdictional claims in published maps and institutional affiliations.

MDPI
St. Alban-Anlage 66
4052 Basel
Switzerland
Tel. +41 61 683 77 34
Fax +41 61 302 89 18
www.mdpi.com

Hydrology Editorial Office
E-mail: hydrology@mdpi.com
www.mdpi.com/journal/hydrology

www.ingramcontent.com/pod-product-compliance
Lightning Source LLC
Chambersburg PA
CBHW041938260926
51850CB00018B/245
* 9 7 8 3 0 3 6 5 6 5 1 2 5 *